Viticulture and Winemaking under Climate Change

Viticulture and Winemaking under Climate Change

Special Issue Editor

Helder Fraga

MDPI • Basel • Beijing • Wuhan • Barcelona • Belgrade

MDPI

Special Issue Editor
Helder Fraga
Centre for the Research and Technology of Agro-Environmental and Biological Sciences
Portugal

Editorial Office
MDPI
St. Alban-Anlage 66
4052 Basel, Switzerland

This is a reprint of articles from the Special Issue published online in the open access journal *Agronomy* (ISSN 2073-4395) in 2019 (available at: https://www.mdpi.com/journal/agronomy/special_issues/ viticulture_winemaking_climate_change).

For citation purposes, cite each article independently as indicated on the article page online and as indicated below:

LastName, A.A.; LastName, B.B.; LastName, C.C. Article Title. *Journal Name* **Year**, *Article Number, Page Range.*

ISBN 978-3-03921-974-2 (Pbk)
ISBN 978-3-03921-975-9 (PDF)

Cover image courtesy of Helder Fraga.

Contents

About the Special Issue Editor

Helder Fraga (Dr.)   is a full Researcher at the Agronomy Department of the University of Trás-os-Montes e Alto Douro (UTAD, Portugal), as well as a member of the Centre for the Research and Technology of Agro-Environmental and Biological Sciences (CITAB). He has also been teaching several classes in this University, from courses of Agronomy and Oenology, where he has supervised several master and PhD students. He is also a partner at the Karlsruhe Institute of Technology (KIT, Germany). His current work focuses on the effects of climate and climate change on several relevant crops in Portugal, particularly on viticulture and olive trees. He has collaborated on over 15 research projects funded by both national and European funds. He is the author of over 40 JCR-indexed papers (h-index: 15), 1 book, 2 book chapters, and over 80 publications/proceedings in scientific meetings, and has been invited to speak at several scientific/technical seminars. He is part of the editorial team of the journal Agronomy and the journal IVES Technical Reviews.

agronomy

MDPI

Editorial

Viticulture and Winemaking under Climate Change

Helder Fraga

Centre for the Research and Technology of Agro-Environmental and Biological Sciences, CITAB, Universidade de Trás-os-Montes e Alto Douro, UTAD, 5000-801 Vila Real, Portugal; hfraga@utad.pt; Tel.: +351-259-350-000

Received: 12 November 2019; Accepted: 19 November 2019; Published: 21 November 2019

Abstract: The importance of viticulture and the winemaking socio-economic sector is acknowledged worldwide. The most renowned winemaking regions show very specific environmental characteristics, where climate usually plays a central role. Considering the strong influence of weather and climatic factors on grapevine yields and berry quality attributes, climate change may indeed significantly impact this crop. Recent-past trends already point to a pronounced increase in the growing season mean temperatures, as well as changes in the precipitation regimes, which has been influencing wine typicity across some of the most renowned winemaking regions worldwide. Moreover, several climate scenarios give evidence of enhanced stress conditions for grapevine growth until the end of the century. Although grapevines have a high resilience, the clear evidence for significant climate change in the upcoming decades urges adaptation and mitigation measures to be taken by the sector stakeholders. To provide hints on the abovementioned issues, we have edited a special issue entitled: "Viticulture and Winemaking under Climate Change". Contributions from different fields were considered, including crop and climate modeling, and potential adaptation measures against these threats. The current special issue allows the expansion of the scientific knowledge of these particular fields of research, also providing a path for future research.

Keywords: viticulture; winemaking; climatic influence; climate change; adaptation measures

---

1. Introduction

Viticulture and winemaking are largely recognized worldwide, having a strong socio-economic role for many countries. Globally, in 2018, wine production was 292×10^6 hl, which has remained relatively unchanged over the last decades [1]. Geographically, the winemaking regions are widespread, but are usually located in temperate climatic regions. Europe incorporates the largest vineyard area in the world (~40%), despite losing some of its dominance to Asia, USA, and some southern hemisphere areas (Argentina, Australia, Chile, South Africa). The world's top wine producing countries are France, Italy, and Spain, while it is worth noticing that China recorded the largest increases in production over the latest years.

Climate is an important forcing factor on grapevine (*Vitis vinifera* L.) physiological development [2], vegetative growth [3], phenology [4], production, and consequently on wine quality. Climatic factors also determine the geographical location of vineyards [5], and the variability in the weather parameters, such as air temperatures, precipitation, and solar radiation, leads to annual changes in productivity [6,7]. Weather extremes are also known to have detrimental impacts on grapevine productivity and quality, namely hail, late frost spells, and excessive rainfall [8].

Climate change is an anticipated challenge that winegrowers will have to deal with in the next decades. During the 20th century, significant changes in temperatures were found, including increases from 2 to 5 °C in Europe [9], which is home to world-renowned wine regions. Moreover, decreases in the precipitations over southern Europe [9] were also found. According to the latest report of the *International Panel on Climate Change* (IPCC), following different representative concentration pathways

(RCP), global temperature is expected to rise between 1 °C (RCP2.6—least severe scenario) and 5 °C (RCP8.5—most severe scenario), over the 21st century [10].

2. Climate Change Impacts on Viticulture

Given the projected modification to climatic conditions, it is expected that climate change will generally have a negative impact on grapevines and wine production. Grapevines will be strongly affected by the higher temperatures during the growing season. As temperatures are a major driver of the grapevine development stages [11], significant warming is expected to lead to earlier phenological events. The advance of the flowering stage may also have a strong impact on management practices. Moreover, a warming during the maturation period will most likely change wine quality attributes and typicity. Extreme heat during this period may abruptly reduce vine metabolism, affecting wine quality attributes. Higher sugar and lower acidity levels should be expected, potentially increasing the risk of wine spoilage [12], threatening wine production and quality. Furthermore, extreme heat and water stress, under future climates, may threaten final yields and productivity [13].

Given the mentioned climate change impacts on this crop, it becomes imperative to plan and implement suitable adaptation measures. Short-term adaptation measures imply changes in management practices, such as the application of irrigation, improving water use efficiency, or providing protection against sunburns. Long-term adaptation measures include more adequate varietal selection and vineyard geographical changes. Sector growers and stakeholders should become aware of this problem in order to timely plan and adopt these measures in order to ensure the future sustainability of this important crop.

3. The Special Issue

The current special issue collects contributions from several papers from colleagues worldwide, reporting how the effects of climate change can affect grapevines and how to deal with these changes on a regional level. The special issue contains reviews and original research articles devoted to the problem of climate change impacts on viticulture and winemaking. One article provides a review of the updated impacts of climate change on grapevines [14]. Some research articles are more focused on climatic factors, such as the possible impacts of decadal-scale cold waves over Europe on viticulture [15], and the grapevine response to natural hail events [16]. Other studies apply crop modeling to better understand the impact of climate change on grapevines [4,17,18]. Furthermore, this special issue contains articles devoted to improving water use efficiency [19] and the effect of a natural anti-transpirant on grapevines [20]. Additionally, another study makes use of reflectance indices to assess vine water status [21]. Other articles are devoted to new and innovative management practices that could prove beneficial under future climates, such as the semi-minimal pruned hedge [22], the application of kaolin clay [23], or even the introduction of unmanned aerial vehicles in vineyards to assess climate change impacts [24]. There are studies devoted to understanding the adaptation potential of some grapevine varieties under the context of climate change [25–27], while another article studies the effect of future enhanced CO_2 levels on specific vine pests. Overall, the current special issue incorporates several areas of research related to climate change impact on viticulture and winemaking, allowing the expansion of current scientific knowledge on this issue.

Funding: This work was funded by European Investment Funds (FEDER/COMPETE/POCI), POCI-01-0145-FEDER-006958, and by the Portuguese Foundation for Science and Technology (FCT), UID/AGR/04033/2013. Helder Fraga thanks the FCT for contract CEECIND/00447/2017.

Conflicts of Interest: The author declares no conflict of interest.

References

1. OIV. *2019 Statistical Report on World Vitiviniculture*; International Organisation of Vine and Wine: Paris, France, 2019.
2. Keller, M. *The Science of Grapevines: Anatomy and Physiology*; Elsevier, Inc.: Amsterdam, The Netherlands, 2010; p. 400.
3. Van Leeuwen, C.; Friant, P.; Choné, X.; Tregoat, O.; Koundouras, S.; Dubordieu, D. Influence of climate, soil, and cultivar on terroir. *Am. J. Enol. Vitic.* **2004**, *55*, 207–217.
4. Costa, R.; Fraga, H.; Fonseca, A.; de Cortazar-Atauri, I.G.; Val, M.C.; Carlos, C.; Reis, S.; Santos, J.A. Grapevine Phenology of cv. Touriga Franca and Touriga Nacional in the Douro Wine Region: Modelling and Climate Change Projections. *Agronomy* **2019**, *9*, 210. [CrossRef]
5. Fraga, H.; Pinto, J.G.; Santos, J.A. Climate change projections for chilling and heat forcing conditions in European vineyards and olive orchards: A multi-model assessment. *Clim. Chang.* **2019**, *152*, 179–193. [CrossRef]
6. Jones, G.V.; Davis, R.E. Climate influences on grapevine phenology, grape composition, and wine production and quality for Bordeaux, France. *Am. J. Enol. Vitic.* **2000**, *51*, 249–261.
7. Fraga, H.; Santos, J.A. Daily prediction of seasonal grapevine production in the Douro wine region based on favourable meteorological conditions. *Aust. J. Grape Wine Res.* **2017**, *23*, 296–304. [CrossRef]
8. Mosedale, J.R.; Wilson, R.J.; Maclean, I.M.D. Climate Change and Crop Exposure to Adverse Weather: Changes to Frost Risk and Grapevine Flowering Conditions. *PLoS ONE* **2015**, *10*, e0141218. [CrossRef]
9. Christensen, J.H.; Hewitson, B.; Busuioc, A.; Chen, A.; Gao, X.; Jones, R.; Kolli, R.K.; Kwon, W.T.; Laprise, R.; Magaña Rueda, V.; et al. Regional Climate Projections. In *Climate Change 2007: The Physical Science Basis. Contribution of Working Group I to the Fourth Assessment Report of the Intergovernmental Panel on Climate Change*; Solomon, S.D., Qin, M., Manning, Z., Chen, M., Marquis, K.B., Averyt, M.T., Miller, H.L., Eds.; Cambridge University Press: Cambridge, UK; New York, NY, USA, 2007; pp. 847–940.
10. IPCC. *Climate Change 2013: The Physical Science Basis. Summary for Policymakers. Working Group I Contribution to the IPCC Fifth Assessment Report*; The Intergovernmental Panel on Climate Change: Geneva, Switzerland, 2013.
11. Parker, A.; García de Cortázar-Atauri, I.; Chuine, I.; Barbeau, G.; Bois, B.; Boursiquot, J.-M.; Cahurel, J.-Y.; Claverie, M.; Dufourcq, T.; Gény, L.; et al. Classification of varieties for their timing of flowering and veraison using a modelling approach: A case study for the grapevine species *Vitis vinifera* L. *Agric. For. Meteorol.* **2013**, *180*, 249–264. [CrossRef]
12. Orduna, R.M. Climate change associated effects on grape and wine quality and production. *Food Res. Int.* **2010**, *43*, 1844–1855. [CrossRef]
13. Fraga, H.; García de Cortázar Atauri, I.; Santos, J.A. Viticultural irrigation demands under climate change scenarios in Portugal. *Agric. Water Manag.* **2018**, *196*, 66–74. [CrossRef]
14. Van Leeuwen, C.; Destrac-Irvine, A.; Dubernet, M.; Duchêne, E.; Gowdy, M.; Marguerit, E.; Pieri, P.; Parker, A.; de Rességuier, L.; Ollat, N. An Update on the Impact of Climate Change in Viticulture and Potential Adaptations. *Agronomy* **2019**, *9*, 514. [CrossRef]
15. Sgubin, G.; Swingedouw, D.; García de Cortázar-Atauri, I.; Ollat, N.; van Leeuwen, C. The Impact of Possible Decadal-Scale Cold Waves on Viticulture over Europe in a Context of Global Warming. *Agronomy* **2019**, *9*, 397. [CrossRef]
16. Petoumenou, G.D.; Biniari, K.; Xyrafis, E.; Mavronasios, D.; Daskalakis, I.; Palliotti, A. Effects of Natural Hail on the Growth, Physiological Characteristics, Yield, and Quality of *Vitis vinifera* L. cv. Thompson Seedless under Mediterranean Growing Conditions. *Agronomy* **2019**, *9*, 197. [CrossRef]
17. Andreoli, V.; Cassardo, C.; La Iacona, T.; Spanna, F. Description and Preliminary Simulations with the Italian Vineyard Integrated Numerical Model for Estimating Physiological Values (IVINE). *Agronomy* **2019**, *9*, 94. [CrossRef]
18. Schmidt, D.; Bahr, C.; Friedel, M.; Kahlen, K. Modelling Approach for Predicting the Impact of Changing Temperature Conditions on Grapevine Canopy Architectures. *Agronomy* **2019**, *9*, 426. [CrossRef]
19. Weiler, C.S.; Merkt, N.; Hartung, J.; Graeff-Honninger, S. Variability among Young Table Grape Cultivars in Response to Water Deficit and Water Use Efficiency. *Agronomy* **2019**, *9*, 135. [CrossRef]
20. Di Vaio, C.; Marallo, N.; Di Lorenzo, R.; Pisciotta, A. Anti-Transpirant Effects on Vine Physiology, Berry and Wine Composition of cv. Aglianico (*Vitis vinifera* L.) Grown in South Italy. *Agronomy* **2019**, *9*, 244. [CrossRef]

21. González-Flor, C.; Serrano, L.; Gorchs, G. Use of Reflectance Indices to Assess Vine Water Status under Mild to Moderate Water Deficits. *Agronomy* **2019**, *9*, 346. [CrossRef]

22. Molitor, D.; Schultz, M.; Mannes, R.; Pallez-Barthel, M.; Hoffmann, L.; Beyer, M. Semi-Minimal Pruned Hedge: A Potential Climate Change Adaptation Strategy in Viticulture. *Agronomy* **2019**, *9*, 173. [CrossRef]

23. Garrido, A.; Serôdio, J.; De Vos, R.; Conde, A.; Cunha, A. Influence of Foliar Kaolin Application and Irrigation on Photosynthetic Activity of Grape Berries. *Agronomy* **2019**, *9*, 685. [CrossRef]

24. Pádua, L.; Marques, P.; Adão, T.; Guimarães, N.; Sousa, A.; Peres, E.; Sousa, J.J. Vineyard Variability Analysis through UAV-Based Vigour Maps to Assess Climate Change Impacts. *Agronomy* **2019**, *9*, 581. [CrossRef]

25. Sancho-Galán, P.; Amores-Arrocha, A.; Palacios, V.; Jiménez-Cantizano, A. Genetical, Morphological and Physicochemical Characterization of the Autochthonous Cultivar 'Uva Rey' (*Vitis vinifera* L.). *Agronomy* **2019**, *9*, 563. [CrossRef]

26. Ahmed, S.; Ruffo Roberto, S.; Youssef, K.; Carlos Colombo, R.; Shahab, M.; José Chaves Junior, O.; Hideki Sumida, C.; Teodoro de Souza, R. Postharvest Preservation of the New Hybrid Seedless Grape, 'BRS Isis', Grown Under the Double-Cropping a Year System in a Subtropical Area. *Agronomy* **2019**, *9*, 603. [CrossRef]

27. Shahab, M.; Roberto, R.S.; Ahmed, S.; Colombo, C.R.; Silvestre, P.J.; Koyama, R.; de Souza, T.R. Anthocyanin Accumulation and Color Development of 'Benitaka' Table Grape Subjected to Exogenous Abscisic Acid Application at Different Timings of Ripening. *Agronomy* **2019**, *9*, 164. [CrossRef]

MDPI

Review

An Update on the Impact of Climate Change in Viticulture and Potential Adaptations

Cornelis van Leeuwen [1,*], Agnès Destrac-Irvine [1], Matthieu Dubernet [2], Eric Duchêne [3], Mark Gowdy [1], Elisa Marguerit [1], Philippe Pieri [1], Amber Parker [4], Laure de Rességuier [1] and Nathalie Ollat [1]

[1] EGFV, Bordeaux Sciences Agro, INRA, Univ. Bordeaux, ISVV, 210 Chemin de Leysotte, F-33882 Villenave d'Ornon, France
[2] Laboratoires Dubernet, ZA du Castellas, 35 Rue de la Combe du Meunier, F-11100 Montredon des Corbières, France
[3] UMR 1131 Santé de la Vigne et Qualité du Vin, INRA, Université de Strasbourg, F-68000 Colmar, France
[4] Department of Wine, Food and Molecular Biosciences, Faculty of Agriculture and Life Sciences, P.O. Box 85084, Lincoln University, Lincoln 7647, Christchurch, New Zealand
* Correspondence: vanleeuwen@agro-bordeaux.fr; Tel.: +33-557-575-911

Received: 1 August 2019; Accepted: 3 September 2019; Published: 5 September 2019

Abstract: Climate change will impose increasingly warm and dry conditions on vineyards. Wine quality and yield are strongly influenced by climatic conditions and depend on complex interactions between temperatures, water availability, plant material, and viticultural techniques. In established winegrowing regions, growers have optimized yield and quality by choosing plant material and viticultural techniques according to local climatic conditions, but as the climate changes, these will need to be adjusted. Adaptations to higher temperatures include changing plant material (e.g., rootstocks, cultivars and clones) and modifying viticultural techniques (e.g., changing trunk height, leaf area to fruit weight ratio, timing of pruning) such that harvest dates are maintained in the optimal period at the end of September or early October in the Northern Hemisphere. Vineyards can be made more resilient to drought by planting drought resistant plant material, modifying training systems (e.g., goblet bush vines, or trellised vineyards at wider row spacing), or selecting soils with greater soil water holding capacity. While most vineyards in Europe are currently dry-farmed, irrigation may also be an option to grow sustainable yields under increasingly dry conditions but consideration must be given to associated impacts on water resources and the environment.

Keywords: climate change; viticulture; adaptation; temperature; drought; plant material; rootstock; training system; phenology; modeling

1. Introduction

Like other agricultural crops, grape growing is impacted by environmental conditions, such as soil and climate [1]. The revenues from agricultural production are driven largely by yield, however, for wine grape growing the quality potential of the grapes is also important, as it can significantly affect the quality of the resulting wine and the prices consumers are willing to pay. In fact, wine prices can vary by a factor up to 1000 (e.g., from 1 to 1000 € per bottle), while yields generally vary by a factor of about 10 (e.g., from 3 to 30 tons/ha). Environmental conditions play an important role in determining not only yield, but also grape quality potential. In addition, depending on these conditions (and other factors like market access), profitability for growers in some regions can be driven by optimizing yields and reducing production costs, while in other regions it can be driven more by producing higher quality grapes for higher price wines.

The output of grape production in terms of yield and quality can be optimized through the choice of plant material, such as variety [2,3], clone [4,5], and rootstock [6], and through the choice of viticultural techniques, such training system [7], and vineyard floor management [8] (see also [1]). Production costs can be reduced largely through mechanization [9]. In established winegrowing regions, growers have historically adjusted their plant material selections and viticultural techniques through trial and error and research to achieve the best possible compromise between yield, quality, and production costs [3]. In each location environmental conditions are different, so there is no general recipe that can be applied everywhere. This explains why plant materials and viticultural techniques vary so much across winegrowing regions of the world.

High yields can be obtained when soil and climate provide for little or no limitation on photosynthesis, such as under moderately high temperature and non-limiting light, nitrogen and water conditions. However, if soil and climate induce a limitation on water and nitrogen, these can be augmented through irrigation and fertilization. Highest possible quality potential is generally achieved when environmental conditions are moderately limiting [10]. Ideal balance in grape composition at ripeness with regard to sugar/acid ratio, color, and aromas, is obtained when grape ripening occurs under moderate temperatures [3]. Excessive cool climatic conditions during ripening can result in green and acidic wines. High temperatures between véraison and harvest can result in unbalanced fruit composition, with sugar levels being too high, acidity too low, and an aromatic expression dominated by cooked fruit aromas [3,11], which result in wines lacking freshness and aromatic complexity.

Mild temperatures during grape ripening, which are favorable for better wine quality, are generally met late in the growing season, roughly between 10 September and 10 October in the Northern Hemisphere and in March or early April in the Southern Hemisphere. White wine production is optimized under cool ripening conditions, which are of particular importance in obtaining intense and complex aroma expression [12]. When varietal heat requirements match the critical temporal window to obtain ripeness, the best wine quality is obtained. For red wine production, water deficits at specific stages of grape development are favorable for wine quality, because they reduce berry size and increase phenolic compounds in grape skins [13–16]. Recently it has also been shown that vine water deficits positively influence aromatic expression in mature wines [17,18]. Moderate nitrogen uptake induces similar effects on grape composition, reducing berry size, and increasing skin phenolics [19]. For the quality of white wines, a limitation in vine water status is also desirable, although this limitation should be milder than for red wine production [20]. For white wine from thiol aroma driven varieties (e.g., Sauvignon blanc, Colombard, Sémillon, Riesling) vine nitrogen status should not be limiting [21].

Although soil and climate are both major environmental components in wine production, the latter is of greater importance for the development of yield components, vine phenology, and grape composition [19,22]. Until the end of the 20th century, soil and climate were considered stable in a given site, with the exception of year-to-year climatic variability. In the 1990s some European researchers became aware that the shifting climatic conditions due to climate change might possibly have a great impact on viticulture worldwide [23]. Progressively, over the first two decades of the 21st century, climate change has become a topic of increasing importance in the viticulture and enology research community. In 2011, 23 French research laboratories collaborated in the LACCAVE project to study the effect of climate change in viticulture and potential grower's adaptations [24]. Several peer reviewed scientific journals, including the Journal of Wine Economics [25], OENO One [24], and Agronomy (this issue, 2019) released special issues on this subject. Today, a substantial body of literature is available to assess the effects of climate change in viticulture and wine production, including effects on vine physiology, phenology, grape composition, and wine quality (among others see [2,26–28]). Several authors have also described potential impacts on pests and diseases [29,30].

Climate change will improve suitability in regions which are currently restricted by low summer temperatures (due to high latitude or elevation) and decrease suitability in warm and dry areas [28]. Several authors have produced suitability maps at global level [31], at the level of the European continent [32], or at the level of a country [33]. These studies, however, are most often conducted at low spatial resolution and underestimate fine-scale variability which may permit viticulture to remain viable under changing climatic conditions [28]. The impact of climate change on viticulture can also be studied by means of crop models which allow predicting the impact of changing temperatures, water availability, and ambient CO_2 levels on yield components and grape composition [34,35]. These predictions are complex, however, because all impacting factors interact. It has been shown that elevated CO_2 increases the optimum temperature for photosynthesis [36] and decreases transpiration [37]. Soil microbiology can also be modified under climate change and may indirectly impact drought resistance of crops [38]. Hence, to be accurate, these models need to be highly sophisticated. Beyond the study of traits related to adaptation, their responses to environmental variables could be studied as the phenotypic plasticity of these traits [39,40]. Given current climate change predictions, the selection of plant materials with an ability to adapt to environmental change will be of particular interest for perennial plants such as grapevine [28,41]. Such adaptative responses, (i.e., phenotypic plasticity), therefore need to be studied further to characterize the genetic variability available for selection [42]. Potential adaptations have been studied to help continued production of high quality wines with economically sustainable yields under changing climatic conditions, which is the main focus of this review

2. Temperature and Drought Effects of Climate Change

Temperature changes associated with climate change are not homogeneous around the globe. Temperatures are currently 1 °C higher on average compared to pre-industrial revolution [43], but the increase can be even higher in some regions. In Bordeaux for example, Average Growing Season Temperature (AvGST; [44]) has increased by approximately 2 °C over the past 70 years, with a remarkable jump between 1985 and 2006 (Figure 1a). Temperatures have become increasingly warmer during the period of grape ripening, as is shown by temperature summations >30 °C during 45 days before harvest (Figure 1b for Bordeaux). This can significantly affect the rate and timing of vine phenology and the final quality of the grapes. Additionally, as increased temperatures increase the evaporative demand driving both vine transpiration and soil evaporation, the soil water balance over the season will become increasingly negative (Figure 1d; [45]). In addition, while annual rainfall has not seen much change in long-term trends, there has been an increase in extreme wet and dry years (Figure 1c for Bordeaux). Taken together, increased temperatures resulting in higher reference evapotranspiration values (Figure 1d), and more frequent years with low rainfall have, and will continue to, induce more intense and frequent drought conditions for vineyards in Bordeaux and around the world.

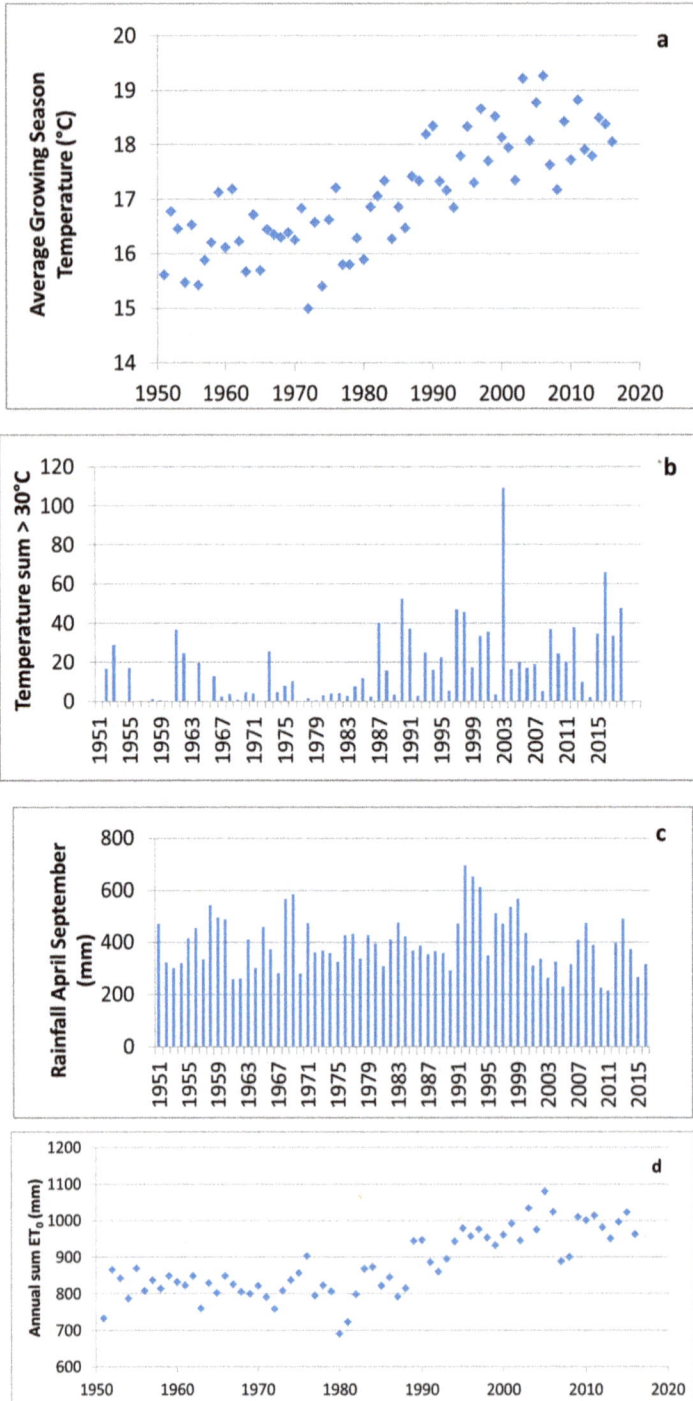

Figure 1. Climate data for Bordeaux (Bordeaux Mérignac weather station) from 1951 to 2018: (**a**) average growing season temperature, (**b**) temperature sum >30 °C during 45 days prior to harvest, (**c**) rainfall April–September, (**d**) annual sum of reference evopotranspiration (ET$_0$).

2.1. Temperature Effects

Temperature is the major driver of vine phenology [46]. Harvest dates have been used to reconstruct temperature series spanning several centuries [47]. Increased temperature as a consequence of climate change leads to advanced phenology [45,48]. In Alsace (France), over a 70-year timespan, budbreak has advanced by 10 days, flowering by 23 days, véraison by 39 days, and harvest by 25 days (Figure 2). Similar trends are observed in many winegrowing regions around the world [45]. Advanced budbreak may expose vines more frequently to spring frost, although this risk depends on the climatic situation of each specific winegrowing region [49–51]. Phenology varies widely among varieties [52,53], with varieties selected historically to perform best in a given winegrowing region based on their phenology [3]. With climate change, local varieties may move out of their ideal ripening window and, as a consequence, may be exposed to excessive temperatures during grape ripening [54]. Harvest in Alsace (France) for Riesling used to occur in the first two weeks of October. Today, in this region, harvests more frequently occur in the first week of September and sometimes even at the end of August (Figure 2). This evolution can be detrimental for the quality potential of the grapes, which are increasingly high in sugar content [48] and may eventually become less aromatic.

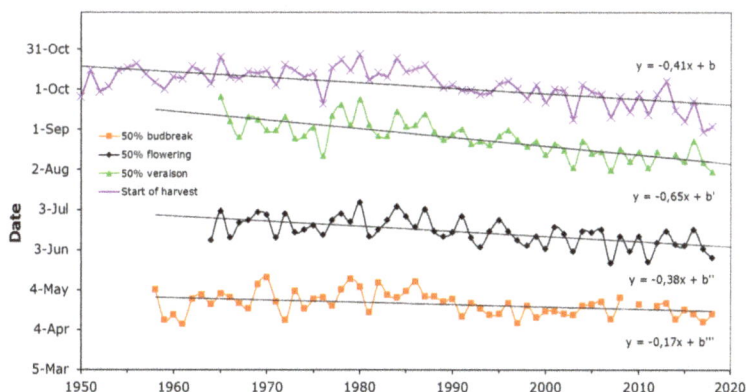

Figure 2. Long-term evolution of vine phenology for Riesling in Alsace. Data source: budbreak, flowering and veraison adapted from [48]; harvest dates from Conseil Interprofessionnel des Vins d'Alsace (CIVA).

In Bordeaux, major grapevine varieties include Sauvignon blanc, Merlot, Cabernet franc and Cabernet-Sauvignon. Harvest dates can be modelled by using the Grapevine Sugar Ripeness model (GSR) to predict sugar ripeness [55]. According to this model, 200 g/L of grape sugar is attained when a daily mean temperature summation reaches a value $F*$ (base temperature of 0 °C, start date day of the year 91, which is 1st of April in the Northern Hemisphere). $F*$ is variety specific, where a higher value indicates a later-ripening variety (Figure 3).

In the following example, the GSR model was used to predict the day when four major grapevine varieties grown in Bordeaux, i.e., Merlot, Cabernet-Sauvignon, Cabernet franc, and Sauvignon blanc, reach 200 g/L of sugar, with input temperature data from Bordeaux Mérignac weather station and $F*$ values retrieved from [55] (Figure 3). To predict harvest dates, five days were added for Sauvignon blanc, which is picked around 210 g/L of grape sugar (12.5% potential alcohol). For harvest dates of the three red varieties 15 days were added, because they are generally picked at 230 g/L of grape sugar (13.5% potential alcohol). When the model was run with average historical temperature data from 1951 to 1980, modelled ripeness was 22 September for Sauvignon blanc, 4 October for Cabernet franc, 7 October for Merlot, and 14 October for Cabernet-Sauvignon (Figure 4). These projections are perfectly in line with observed historical harvest dates from Bordeaux [45]. If the ideal window for grape ripeness is defined from 10 September to 10 October, when temperatures are not excessive but still high enough to achieve full ripeness, all varieties fall within this window except Cabernet-Sauvignon. This is consistent with the observation that during this period high-quality wines from Cabernet-Sauvignon could only be produced in early ripening locations on warm gravel soil. In the cooler parts of Bordeaux, wines from Cabernet-Sauvignon used to be green (high content in methoxypyrazines) and acidic. When the same projection is made with average climate data from 1981 to 2010, the following harvest dates were obtained: 7 September for Sauvignon blanc, 18 September for Merlot, 21 September for Cabernet franc, and 28 September for Cabernet-Sauvignon (Figure 4). At the turn of the millennium, Bordeaux has become suitable for growing high quality Cabernet-Sauvignon over most of the region, but has become marginally too warm for Sauvignon blanc. It is predicted that it will still be possible to grow high quality Sauvignon blanc in cooler locations of the region on North facing slopes or on cool soils. When 1 °C is added to the average 1981–2010 temperatures (which is close to temperature projections for around 2050), the Bordeaux climate is still perfectly suitable for producing high quality wines from Cabernet franc and Cabernet-Sauvignon (projected harvest 11 and 18 September respectively), but Merlot is moving out of the ideal ripening window (8 September) and the Bordeaux climate will be too warm to produce crisp and aromatic wines from Sauvignon blanc (29 August; Figure 4). Hence, among the traditional Bordeaux varieties, Sauvignon blanc and Merlot will be the first victims of climate change. During the past decade, Bordeaux wines containing a majority of Merlot, which is still the most widely planted variety in this region, are increasingly dominated by cooked fruit aromas and excessively high alcohol content [11].

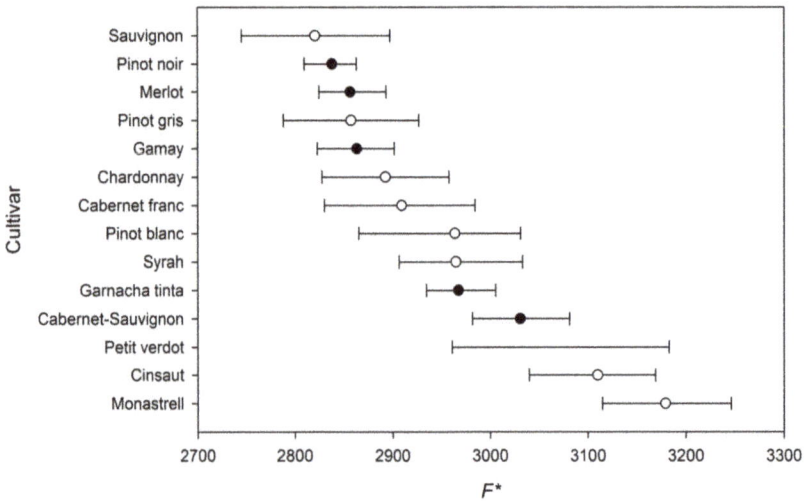

Figure 3. Temperature summation ($F*$) to reach 200 g/L of grape sugar according to Grapevine Sugar Ripeness (GSR) model for 15 major grapevine varieties. Horizontal bars represent 95% confidence intervals (CI) which were calculated using the optimization algorithm of Metropolis in PMP v5.4 and determined via the Fisher statistic ($p < 0.05$) as in [55]. Closed circles correspond to parameterizations where CI < 100, open circles correspond to CIs in the range 100–200 and no circle corresponds to CIs in the range of 201–350. (Cultivar synonyms: Monstrell = Mourvèdre, Sauvignon = Sauvignon blanc, Garnacha tinta = Grenache).

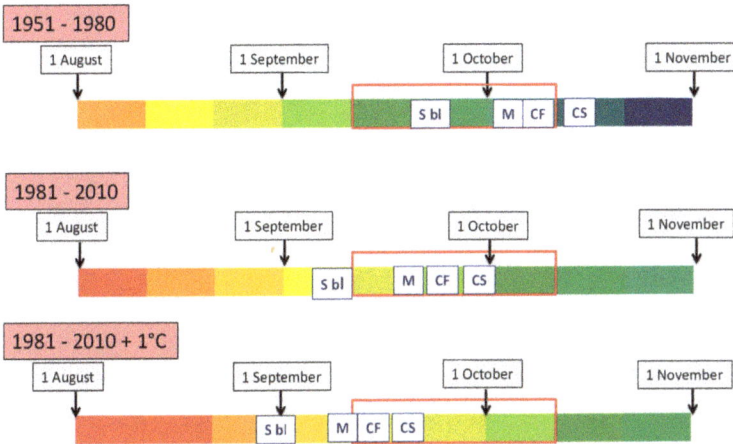

Figure 4. Modelled harvest dates for Sauvignon blanc (S bl), Merlot (M), Cabernet franc (CF), and Cabernet-Sauvignon (CS) in Bordeaux for the following periods: 1951–1980, 1981–2010, and 1981–2010 + 1 °C. Sugar ripeness is modelled with the grapevine Sugar Ripeness Model (GSR; [55]). Temperature data is from Bordeaux Mérignac weather station. Warm colors indicate higher temperatures and cold colors cooler temperatures.

In general, grape and wine compositions have dramatically changed over the past three decades worldwide. Mean data from Languedoc (France) shows that over a 35-year time span, alcohol in wine increased from 11% to 14%, pH from 3.50 to 3.75 and total acidity decreased from 6.0 to 4.5 g/L (Figure 5). Similar observations are made in many regions around the world [23,26,44,56].

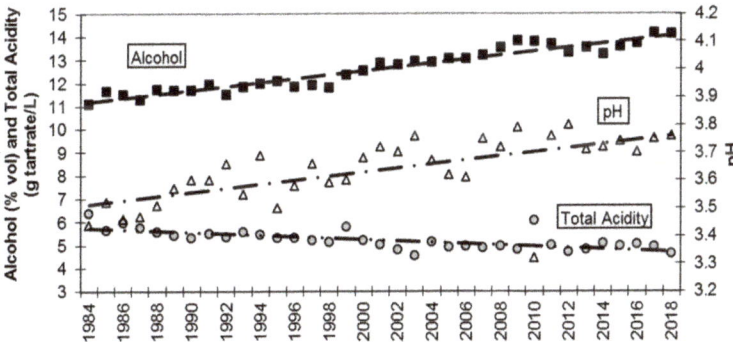

Figure 5. Evolution of red wine composition in the Languedoc region (France) from 1984 to 2018. Each data point is the average of several thousands of analyses of red wines just after alcoholic fermentation (data: Dubernet laboratory, F-11100 Montredon des Corbières).

2.2. Drought Effects

Climate change will also expose vines to increased drought, either because of reduced rainfall, or because of higher reference evapotranspiration due to elevated temperatures. This may lead to lower yields, because several yield parameters are impacted by water deficits, in particular berry size [14,15] and bud fertility [57]. On the other hand, water deficit has a positive effect on red wine quality because grape skin phenolics increase [14,15,58] and wines develop more complex aromas during bottle ageing [17,18]. So far, the best vintages in Bordeaux (where vines are not irrigated) are dry vintages [45]. The frequency of dry vintages has increased over the past three decades and this resulted in better vintage ratings in recent years. In white wine production only very mild water deficits are positive for wine quality, while more severe water deficits are detrimental [20]. For red wines, the general tendency under increased drought is lower yields and better quality (except situations of severe water stress); for white wine, not only yields can be negatively affected but quality can also be jeopardized.

In established wine growing regions, growers have optimized output in terms of quality and yield by choosing plant material, viticultural techniques, and wine making which are most adapted to their local environment. Now that the climate has become warmer and drier in most wine growing regions, this balance is threatened. Specific adaptations are needed to continue to produce optimum quality and yield in a changing environment.

3. Adaptations to Higher Temperatures

Higher temperatures advance grapevine phenology [46]. Hence, grapes ripen earlier in the season under warmer temperatures [49]. When grapes achieve full ripeness in the warmest part of the season (July–August in the Northern Hemisphere, January–February in the Southern Hemisphere) grape composition can be unbalanced (e.g., high sugar levels and low acidity), with red grapes containing less anthocyanins [59,60]. Wines from these grapes will lack freshness and aromatic complexity [12]. Hence adaptations to higher temperatures encompass all changes in plant material or modifications in viticultural techniques with the purpose of delaying ripeness [61].

3.1. Later Ripening Varieties

Grapevines have a wide phenotypic diversity regarding the timing of phenology [53]. In all traditional winegrowing regions in Europe, growers have planted varieties that ripen between 10 September and 10 October under local climatic conditions. This is the case for Riesling in the Rheingau (Germany), Chardonnay and Pinot noir in Burgundy (France), Merlot, Cabernet franc, and Cabernet-Sauvignon in Bordeaux (France), Grenache and Carignan in Languedoc (France), Tempranillo in La Rioja (Spain), Sangiovese in Tuscany (Italy), Nebbiolo in Barolo (Italy), Touriga nacional in Douro (Portugal), Agiorgitiko in Nemea (Greece), and Monastrell (Mourvèdre) in Alicante (Spain). Now that temperatures have increased, traditional varieties may move out of the ideal ripening window with detrimental effects on wine quality. In this context, potential adaptation to a changing climate is to plant later ripening varieties. The Ecophysiology et Génomique Fonctionelle de la Vigne research unit (EGFV) from the Institut des Sciences de la Vigne et du Vin (ISVV) near Bordeaux planted the VitAdapt vineyard experiment in 2009, where 52 varieties are planted with five replicates to study physiology, phenology, ripening dynamics, and wine quality (by small scale vinifications) to assess how these varieties behave differently in a warming climate [62]. The experimental set-up includes later-ripening varieties from warm locations, like Touriga nacional, Tinto Cao (Portugal, red varieties), and Assyrtiko (Greece, white variety; Figure 6). Data from this vineyard shows average véraison dates (2012–2018) span over 34 days, demonstrating the extent to which later ripening can be achieved by simply changing the variety (Figure 7).

In European wine appellations, the choice of varieties is regulated to allow only varieties that perform best in terms of quality and typicity under local climatic conditions. Under a changing climate, however, these regulations will need to be modified. Recently seven new varieties, including Touriga nacional, were accepted for planting in up to 5% of the area in Bordeaux winegrowing estates to allow testing with full-scale vinifications. This percentage may be increased if the experiments are conclusive. The choice of the varieties allowed for testing was based directly on results from the VitAdapt experiment. In New World winegrowing regions, grapevine varieties are not restricted by law, but surprisingly their diversity is even lower than in traditional Europeans winegrowing regions, due to the preeminence of well-known international varieties for marketing purposes. A wider range of grapevine varieties can be a useful tool for adaptation to climate change [63].

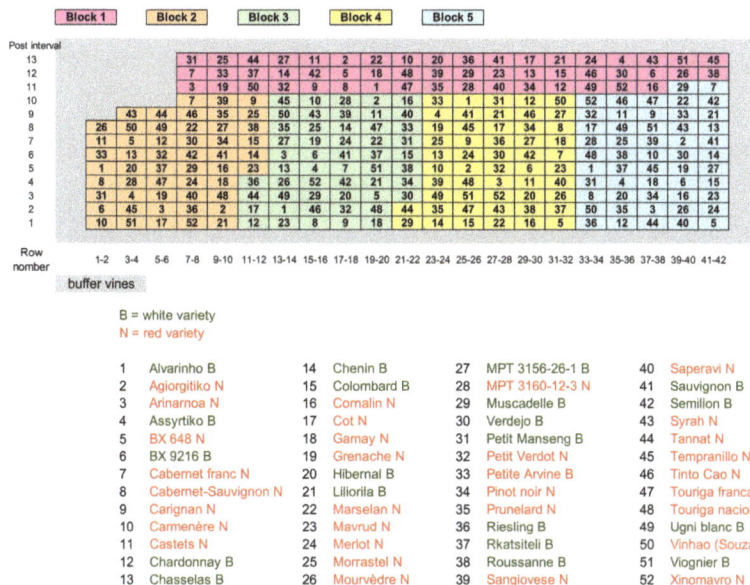

B = white variety
N = red variety

1	Alvarinho B	14	Chenin B	27	MPT 3156-26-1 B	40 Saperavi N
2	Agiorgitiko N	15	Colombard B	28	MPT 3160-12-3 N	41 Sauvignon B
3	Arinarnoa N	16	Cornalin N	29	Muscadelle B	42 Semillon B
4	Assyrtiko B	17	Cot N	30	Verdejo B	43 Syrah N
5	BX 648 N	18	Gamay N	31	Petit Manseng B	44 Tannat N
6	BX 9216 B	19	Grenache N	32	Petit Verdot N	45 Tempranillo N
7	Cabernet franc N	20	Hibernal B	33	Petite Arvine B	46 Tinto Cao N
8	Cabernet-Sauvignon N	21	Liliorila B	34	Pinot noir N	47 Touriga franca N
9	Carignan N	22	Marselan N	35	Prunelard N	48 Touriga nacional N
10	Carmenère N	23	Mavrud N	36	Riesling B	49 Ugni blanc B
11	Castets N	24	Merlot N	37	Rkatsiteli B	50 Vinhao (Souzao) N
12	Chardonnay B	25	Morrastel N	38	Roussanne B	51 Viognier B
13	Chasselas B	26	Mourvèdre N	39	Sangiovese N	52 Xinomavro N

Figure 6. Layout of the 52 varieties planted in the VitAdapt experiment, with five replicates per variety and 10 vines per replicate.

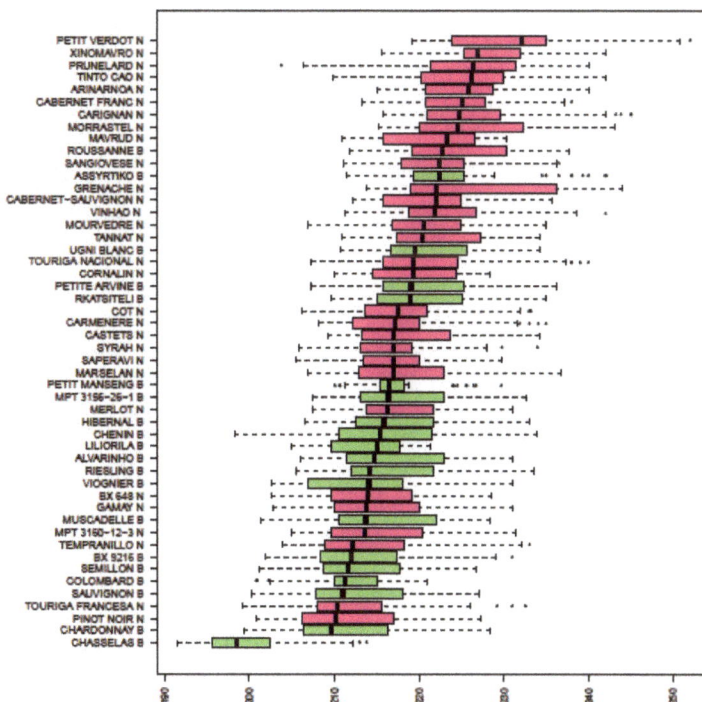

Figure 7. Boxplot of observed mid-véraison dates of varieties planted in the VitAdapt experiment (average day of the year from four replicates per variety over the period 2012–2018).

3.2. Later-Ripening Clones

Within a given variety a certain level of genetic variability exists, referred to as clonal variability. Historically, clones have been selected for traits such as high productivity, early ripening, and high sugar content in grapes. In the context of a changing climate it may be preferable to select new clones with the opposite characteristics. Sugar accumulation dynamics vary among clones, as shown from an example of a clonal selection trial on Cabernet franc [5] (Figure 8). At ripeness, differences in grape sugar concentration among clones can be over 17 g/L (1% potential alcohol). In the same clonal collection, differences in mid-véraison dates ranged from 6 to 9 days depending on the vintage (data not shown).

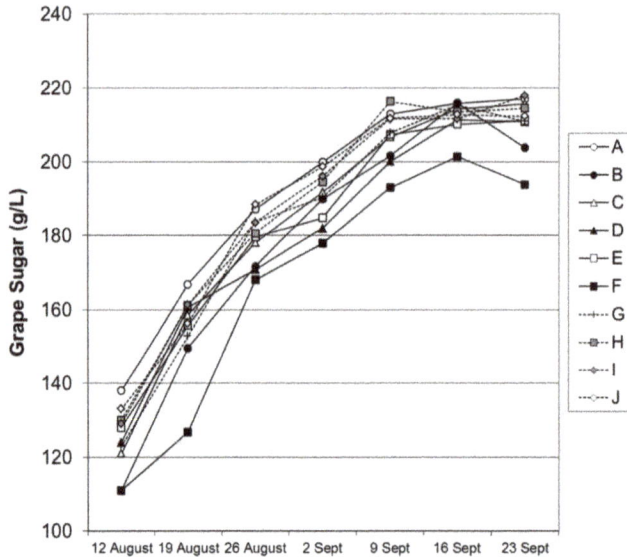

Figure 8. Sugar accumulation dynamics in 2013 from a private clonal selection program on Cabernet franc. A–J represent 10 different clones [5].

3.3. Later-Ripening Rootstocks

Rootstocks can influence the phenology of the grafted scion. Some rootstocks induce earlier phenology and ripening, while others induce a longer cycle [61,64]. Precise data on this effect is scarce in the scientific literature. In 2015 the GreffAdapt experiment was planted by the EGFV research unit from the ISVV. In this project, 55 rootstocks are phenotyped with five different scions in field condition. Each combination is planted with three replicates [65]. Over the coming years, this experimental vineyard will yield precise information regarding whether and how rootstocks may induce differences in grapevine phenology and timing of ripeness.

3.4. Increasing Trunk Height

Trunk height determines the distance from the soil to the grapes and can vary according to training systems from 30 cm to over 1 m. Maximum temperatures are higher close to the soil and the resulting vertical temperature gradient can be used to fine-tune the micro climate in the bunch zone through variations in trunk height. In Bordeaux, where the climate historically has been marginal for ripening Cabernet-Sauvignon, growers planted this variety on warm gravel soils and trained the vines with short trunks to have the bunches as close as possible to the soil. In warmer climatic conditions as caused by climate change, the temperatures may be too high close to the soil surface, in particular for early ripening varieties in the Bordeaux context like Sauvignon blanc and Merlot. An experiment was set up in the Saint-Emilion winegrowing region where temperature sensors were installed at 30, 60, 90, and 120 cm on vine posts with three replicates in four different vineyard blocks. The Winkler Index as measured in these canopies was 60 degree.days lower at 120 cm compared to 30 cm (Figure 9). Based on a 19 °C average temperature (which corresponds to 9 °C base of 10 °C) this difference may induce a delay of 7 days in grape ripening.

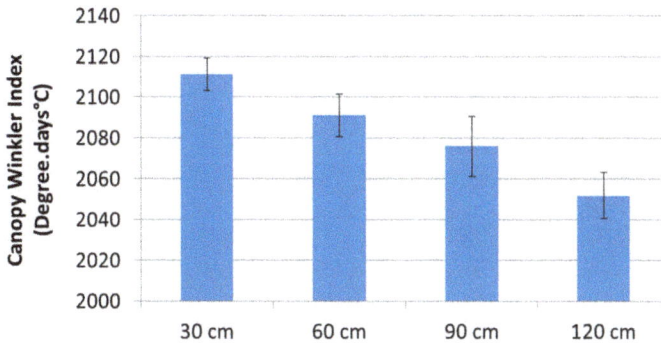

Figure 9. Variations in Canopy Winkler Index computed from temperature data acquired by sensors installed on vine posts at 30, 60, 90, and 120 cm in height.

3.5. Reducing Leaf Area to Fruit Weight Ratio

Leaf area to fruit weight ratio (LA:FW) is considered as an important parameter affecting the performance of a vineyard, both with regard to yield and grape composition [66]. A LA:FW of at least 1 m^2/kg is generally considered as necessary to ensure optimum ripening conditions and in particular sugar accumulation [67]. Lower LA:FW ratios can considerably delay véraison and sugar accumulation in grapes, with limited effect on total acidity [68,69]. Reduced LA:FW ratios, however, adversely affects anthocyanin accumulation in grapes, which makes this technique more easily applicable in white wine production than in red wine production. Studies in potted vines [70] and in field grown vines [71] found only a transient effect of leaf removal on vine physiology and small, or no effect of final grape composition. In these studies, however, leaf removal was less severe and LA:FW ratio was higher than 1 m^2/kg of fruit in all treatments. De Bei and co-authors [72] found an inconsistent effect of leaf removal on phenology and grape composition depending on the year and the grapevine variety, but LA:FW ratio was also above 1 m^2/kg of fruit in all treatments.

3.6. Late Pruning

When winter pruning is carried out late, budbreak is delayed by a few days [73]. However, differences tend to become smaller for subsequent phenological stages. Differences are more significant when pruning is carried out when the vines had 2–3 leaves, with no effect on yield or pruning weights the following season [74]. In this experiment, wine quality, as assessed by color intensity and sensory analysis, was improved by late pruning, probably because ripening occurred under lower temperature associated with delayed phenology [75]. Maturity is more substantially delayed when vines are pruned a second time, well after budburst [73,76,77]. This technique, however, is still experimental and long-term carry-over effects on vigor need to be studied.

3.7. Moving to Higher Altitudes

In mountainous areas, temperature decreases by 0.65 °C per 100 m of elevation. If other vineyard adaptations are not adequate, and if topography permits (Douro, Portugal; Mendoza, Argentina), moving vineyards to higher altitudes can be an effective adaptation to a warming climate. In Mendoza varieties are grown according to the altitude, where in very warm conditions at 800 m above sea level (a. s. l.) entry-level wines are produced from high-yielding vines. Finer wines are produced from Malbec and Cabernet-Sauvignon planted at 1100 m. a. s. l. and early ripening Chardonnay and Pinot noir planted at 1500 m. a. s. l. Moving vineyards to higher elevations, however, may have detrimental environmental effects associated with disruption to wildlife habitat and ecosystem services, which need to be considered [31].

3.8. Combination of Adaptations

The previously mentioned changes in plant material and viticultural techniques can be progressively implemented. Some of them do not require major changes in viticultural management (e.g., late pruning), while others may involve replanting vineyards with a potential change in wine typicity (e.g., change of varieties). To a certain extent, these techniques can be combined, but further research is needed to assess if the delaying effect by combining several techniques is additive. Overall, depending on the rate of climate warming, such adaptations should be effective for decades to come, except maybe for already very hot wine growing areas.

4. Adaptations to Increased Drought

Water deficits reduce yield but, except in situations of severe stress, it can have a positive effect by promoting red wine quality [10,58]. The production of high-quality white wines requires mild water deficits [78]. With increasing water deficits as a consequence of climate change, yields are negatively impacted, decreasing profitability of wine production. Hence, adaptations to drier growing conditions is becoming increasingly pertinent in viticulture worldwide. The vine is a highly drought resistant species. In the Mediterranean basin there are thousands of years of experience of growing vines in warm and dry conditions. In a context where water is an increasingly scarce resource it is important to take advantage of this expertise. Potential adaptations to increased drought include the use of drought resistant plant material, the implementation of specific training systems, locating vineyards where soils have greater soil water holding capacity, and possible use of irrigation.

4.1. Drought Resistant Rootstocks

Since phylloxera reached Europe in the second half of the 19th century, most vines in the world are grafted on rootstocks. Rootstocks vary considerably in their ability to resist drought. Several authors have addressed this issue [79] and recently a collation was made by Ollat et al. [6] (Table 1). Physiological mechanisms behind drought tolerance in rootstocks (as measured on the scion) were studied by Marguerit et al. [80]. This issue will be further investigated in field conditions in the GreffAdapt experiment in the EGFV research unit in Bordeaux [65]. The use of drought resistant rootstocks to sustain yields and avoid quality losses from excessive water stress is a powerful and environmentally friendly adaptation to increased drought, and once planted do not increase production costs.

Table 1. Drought tolerance among rootstocks (Adapted from Ollat et al. [6]).

Rootstocks	Usual Name	Phylloxera Resistance	Water Stress Adaptation
Riparia Gloire de Montpellier	Riparia Gloire	High to very High	Low
Grézot 1	G1	Low to Medium	Low
Foëx 34 École de Montpellier	34 EM	High	Low to Medium
Millardet et de Grasset 420 A	420 A	High	Very Low to Medium
Kober-Téléki 5 BB	5 BB	High	Low to Medium
Téléki 5 C	5 C	High	Low to Medium
Couderc 1616	1616 C	High	Low to Medium
Rupestris du Lot (St. George)	Rupestris	Medium to High	Low to Medium
Millardet et de Grasset 101-14	101-14 MGt	High	Very Low to Medium
Couderc 3309	3309 C	High	Very Low to High; mostly Low to Medium
Téléki-Fuhr Selection Oppenheim n°4	SO4	High	Very Low to High; mostly Low to Medium
Téléki 8 B	8 B	High	Low to Medium
Dog Ridge	Dog Ridge	High	Very Low to High
Schwarzmann	Schwarzmann	High to very High	Very Low to Medium
Couderc 1613	1613 C	Low to Medium	Low to Medium
Couderc 161-49	161-49 C	High	Low to Medium
Kober-Téléki 125 AA	125 AA	High	Medium
Millardet et de Grasset 41B	41 B	Medium to High	Very Low to High, mainly Medium
Castel 216-3	216-3 Cl	High	Medium
Fercal INRA Bordeaux	Fercal	Medium to High	Medium
Gravesac INRA Bordeaux	Gravesac	High to very High	Medium
Freedom	Freedom	Medium to High	Medium
Harmony	Harmony	Low to Medium	Medium to High
Foëx 333 École de Montpellier	333 EM	Medium to High	Low to High, mainly Medium to High
Richter 99	99 R	High	Medium to Very high
Börner	Börner	Very high	High
Castel 196-17	196-17 Cl	Low to Medium	Medium to High
Georgikon 28	Georgikon 28	High	High
Malègue 44-53	44-53 M	High	Medium to very High
Ramsey	Ramsey	High	Medium to very High
Paulsen 1103	1103 P	High	High to very High
Paulsen 1447	1447 P	High	High to very High
Richter 110	110 R	High	High to very High
Ruggeri 140	140 Ru	High	High to very High

4.2. Drought Resistant Varieties

Grapevine varieties are highly variable in their tolerance to drought [81]. This may be linked to the way different varieties regulate their water potential in response to increasing atmospheric demand and decreasing soil water content. Some varieties appear to control their water potential more closely (isohydric behavior) under drought conditions [82], although the characterization of this response has recently been challenged [83].

The way varieties modify their water use efficiency in response to drought is another useful indication of varietal drought tolerance. At the leaf level, water use efficiency is the amount of carbon assimilation (i.e., carbohydrates produced by photosynthesis) for a given amount of transpiration through the stomata (i.e., water loss). At the plant level, it is the yield of grapes and change in vine biomass compared to the amount of water consumed by the vine over the season [84]. Clonal differences in water use efficiency have been observed [85] and may be a useful tool for assessing the drought tolerance of different varieties. Analyzing the carbon isotope discrimination in grape berry juice sugars provides an integrative measure of the water use efficiency of a grapevine over the course of the berry ripening period [86] and comparison of changes in carbon isotope discrimination (i.e., water use efficiency) between wet versus dry years can help characterize the drought resistance of different varieties.

Most grapevine varieties originating from the Mediterranean basin (Grenache, Cinsault, Carignan) are considered drought tolerant, while varieties like Merlot, Tempranillo, or Sauvignon blanc are not. Some local varieties of Mediterranean islands, like Xinistery from Cyprus are reported to have a very high drought resistance and deserve experimentation outside this original region of production (Manganaris, personal communication). A study of the underlying physiological mechanisms of drought resistance is currently undertaken in the VitAdapt projects (EGFV research unit, ISVV Bordeaux; [87]. Planting drought resistant varieties in dry environments is a logical step in adapting to climate change, and therefore these varieties deserve increased attention.

4.3. Training Systems

Over centuries, wine growers in the Mediterranean basin have developed a training system which is particularly resistant to drought and high temperatures: the so-called Mediterranean goblet or bush vine. With this training system, it is possible to dry-farm vines in extremely dry environments, down to a mere 350 mm of rainfall/year [88,89]. Although goblet trained vines generally produce low yields, they are easy to cultivate at reduced production costs on a per hectare basis [9]. Hence, despite low yields, production costs expressed on a per kilogram basis are not necessarily high. They present the drawback, however, of being difficult to harvest by machine [90]. If harvesting goblet trained vines could be mechanized, this would further reduce production costs for this otherwise drought resistant training system.

An alternative solution to increasing drought resistance of a vineyard is to increase row spacing. Row spacing is traditionally high in regions where water deficit is not a major issue, like Bordeaux, Champagne, and Burgundy (France). Close row spacing optimizes sunlight interception, which allows producing high-quality wines at moderately high yields. When water is, or becomes, a limiting factor, close row spacing increases water use, because sunlight interception is providing the driving energy for transpiration. The effect of row spacing on water balance was recently modeled by van Leeuwen et al. [91] for three row spacings (2 m = 5000 vines/ha, 3 m = 3333 vines/ha, and 4 m = 2500 vines/ha) and three levels of total transpirable soil water (TTSW), a concept similar to soil water holding capacity [92]. The output of the water balance model is the fraction of transpirable soil water (FTSW), where the lower the FTSW, the greater the water deficit experienced by the vines. The output of the water balance modeling demonstrated that vine spacing had an important effect on water balance and water availability during grape ripening, except when TTSW was already low (Figure 10). It should be noted that increased vine spacing reduces both yield (and related revenue) and production cost, with profitability depending on the trade-off between these two effects. Modeling found production cost savings outweighing yield-related revenue loss when producing lower-value grapes, while the opposite is true for production of higher-value grapes [91].

4.4. Soil Water Holding Capacity

TTSW or soil water holding capacity has a major impact on vine water status. In the analysis described above and presented in Figure 10, average FTSW for the 30 days prior to modeled harvest is 0.43, 0.26, and 0.19 for TTSW of 300, 200, and 100 mm respectively. Note that vines do not face any water deficit when FTSW is between 1.00 and 0.40 and that water deficits are increasingly intense for FTSW between 0.40 and 0.00 [92]. TTSW depends on soil type (texture and content in coarse elements) as well as rooting depth. Under dry climates it makes sense to plant vineyards in soils with at least medium TTSW. Rooting depth can be promoted by through soil preparation, such as deep ripping [93].

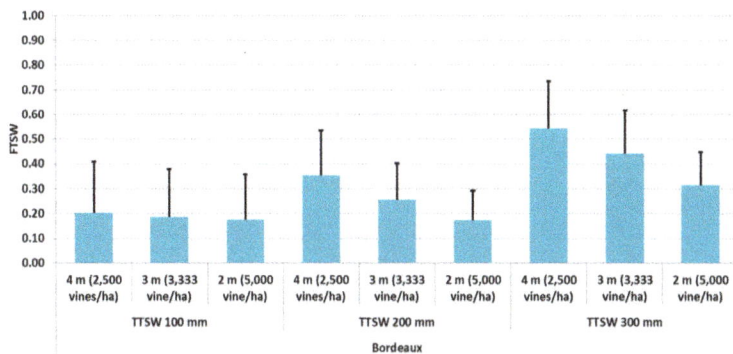

Figure 10. Modelled average fraction of transpirable soil water (FTSW) during 30 days prior to modelled harvest dates for three vines spacings (2, 3, and 4 m) and three levels of total transpirable soil water (100, 200, and 300 mm). Input weather data from 1981–2010, Bordeaux Mérignac weather station.

4.5. Irrigation

To avoid yield losses due to drought, irrigation is also an option when adequate water resources are available. Vineyard irrigation is not an historical technique in the Mediterranean basin, where the vast majority of vines are still dry-farmed. Although the acreage of irrigated vineyards is increasing, it is likely that there will never be enough water to irrigate the total area which is currently under vines. Hence, dry farming should be considered as a precious skill, of which the underlying mechanisms need to be better understood. Another drawback of irrigation is that in some situations (in particular when winters are dry), it can lead to increased soil salinity, which results in reduced long-term suitability of vineyard soils for cultivation.

When irrigation is chosen as a technique for vineyard management in dry climates, consideration must also be given to the potential negative impacts on regional surface and groundwater resources, including the effect on other potential users of water and the surrounding environment. If irrigation is implemented, techniques such as deficit irrigation should be used with precise vine water status monitoring (e.g., by measuring stem water potential) in order to limit, as much as possible, the amount of irrigation water applied. However, even with fine-tuned irrigation management, the blue water footprint of an irrigated vineyard is generally at least 100 times higher compared to a dry-farmed vineyard.

5. Conclusions

Due to climate change, vines are facing increasingly warm and dry growing conditions. The vine is, however, a plant of Mediterranean origin, which is well adapted to these conditions. However, higher temperatures shift phenology and the ripening period to a time in the season which is less favorable for the production of quality wine and increasingly dry conditions lead to yield reduction. In some situations, it promotes wine quality, in particular for the production of red table wines, while excessive water stress may jeopardize wine quality. Adaptations to climate change include modifications in plant material and viticultural techniques which delay phenology and grape ripening and increase drought tolerance. The use of late-ripening and drought resistant plant material (varieties, clones, and rootstocks) is an environmentally friendly and cost-effective tool for adaptation. The vast genetic diversity in vines for these traits constitutes a precious resource to continue to produce high-quality wines with sustainable yields in a changing climate.

Author Contributions: Writing, original draft preparation: C.v.L.; experimental results: A.D.I., M.D., E.D., M.G., E.M., P.P., A.P., L.d.R., N.O., C.v.L.; editing: M.G., N.O., A.P.; proofreading: A.D.I., E.M., A.P., M.G., L.d.R.; modeling water balance: P.P. climate database: L.d.R., C.v.L.

Funding: This study is implemented in the frame of the LACCAVE 2 project (towards integrated viticultural systems resilient to climate change). It has been carried out with financial support from the French National Research Agency (ANR) in the frame of the Investments for the future Programme, within the Cluster of Excellence COTE (ANR-10-LABX-45). Further financial support was obtained from the Conseil Interprofessionnel des Vines de Bordeaux (CIVB) and the regional government of Aquitaine (Conseil Régional d'Aquitaine).

Acknowledgments: We are grateful for support from the Unité Expérimentale de la Grande Ferrade (F-33883 Villenave d'Ornon) in managing the VitAdapt and GreffAdapt experiments.

Conflicts of Interest: The authors declare no conflict of interest.

References

1. Jackson, D.I.; Lombard, P.B. Environmental and Management Practices Affecting Grape Composition and Wine Quality—A Review. *Am. J. Enol. Vitic.* **1993**, *44*, 409–430.
2. Fraga, H.; Malheiro, A.-C.; Moutinho-Perreira, J.; Santos, J. An overview of climate change impacts on European viticulture. *Food Energy Secur.* **2012**, *1*, 94–110. [CrossRef]
3. Van Leeuwen, C.; Seguin, G. The concept of terroir in viticulture. *J. Wine Res.* **2006**, *17*, 1–10. [CrossRef]
4. Audeguin, L.; Boidron, R.; Bloy, P.; Grenan, S.; Leclair, P.; Boursiquot, J.-M. L'expérimentation des clones de vigne en France. Etat des lieux, méthodologie et perspectives. *Revue Française d'Oenologie* **2000**, *184*, 8–11.
5. Van Leeuwen, C.; Roby, J.-P.; Alonso-Villaverde, V.; Gindro, K. Impact of clonal variability in *Vitis vinifera* Cabernet franc on grape composition, wine quality, leaf blade stilbene content and downy mildew resistance. *J. Agric. Food Chem.* **2013**, *61*, 19–24. [CrossRef] [PubMed]
6. Ollat, N.; Peccoux, A.; Papura, D.; Esmenjaud, D.; Marguerit, E.; Tandonnet, J.-P.; Bordenave, L.; Cookson, S.; Barrieu, F.; Rossdeutsch, L.; et al. Rootstock as a component of adaptation to environment. In *Grapevine in a Changing Environment: A Molecular and Ecophysiological Perspective*; Geros, H., Chaves, M., Medrano, H., Delrot, S., Eds.; Wiley-Blackwell: Hoboken, NJ, USA, 2015.
7. Smart, R.E. Principles of Grapevine Canopy Microclimate Manipulation with Implications for Yield and Quality. A Review. *Am. J. Enol. Vitic.* **1985**, *36*, 230–239.
8. Wheeler, S.J.; Black, A.S.; Pickering, G.J. Vineyard floor management improves wine quality in highly vigorous *Vitis vinifera* 'Cabernet Sauvignon'in New Zealand. *N. Z. J. Crop. Hortic. Sci.* **2005**, *33*, 317–328. [CrossRef]
9. Roby, J.-P.; van Leeuwen, C.; Marguerit, E. *Références Vigne. Références Technico-Économiques de Systèmes de Conduite de la Vigne*, 2nd ed.; Synthèse, A., Ed.; Lavoisier: Paris, France, 2008.
10. Van Leeuwen, C.; Tregoat, O.; Chone, X.; Gaudillere, J.-P.; Pernet, D. Different environmental conditions, different results: The effect of controlled environmental stress on grape quality potential and the way to monitor it. In Proceedings of the 13th Australian Wine Industry Technical Conference, Adelaide, Australia, 29 July–2 August 2007.
11. Pons, A.; Allamy, L.; Schüttler, A.; Rauhut, D.; Thibon, C.; Darriet, P. What is the expected impact of climate change on wine aroma compounds and their precursors in grape? *OENO One* **2017**, *51*, 141–146. [CrossRef]
12. Drappier, J.; Thibon, C.; Rabot, A.; Gény, L. Relationship between wine composition and temperature: Impact on Bordeaux wine typicity in the context of global warming. *Crit. Rev. Food Sci. Nutr.* **2019**, *59*, 14–30. [CrossRef]
13. Matthews, M.; Anderson, M. Fruit ripening in *Vitis vinifera* L.: Responses to seasonal water deficits. *Am. J. Enol. Vitic.* **1988**, *39*, 313–320.
14. Ojeda, H.; Andary, C.; Kraeva, E.; Carbonneau, A.; Deloire, A. Influence of pre- and postveraison water deficit on synthesis and concentration of skin phenolic compounds during berry growth of Vitis vinifera cv. Syrah. *Am. J. Enol. Vitic.* **2002**, *53*, 261–267.
15. Van Leeuwen, C.; Trégoat, O.; Choné, X.; Bois, B.; Pernet, D.; Gaudillère, J.-P. Vine water status is a key factor in grape ripening and vintage quality for red Bordeaux wine. How can it be assessed for vineyard management purposes? *OENO One* **2009**, *43*, 121–134. [CrossRef]
16. Triolo, R.; Roby, J.-P.; Pisciotta, A.; Di Lorenzo, R.; van Leeuwen, C. Impact of vine water status on berry mass and berry tissue development of Cabernet franc (*Vitis vinifera* L.) assessed at berry level. *J. Sci. Food Agric.* **2019**, *99*. [CrossRef] [PubMed]
17. Picard, M.; van Leeuwen, C.; Guyon, F.; Gaillard, L.; de Revel, G.; Marchand, S. Vine water deficit impacts aging bouquet in fine red Bordeaux wine. *Front. Chem.* **2017**, *5*, 56. [CrossRef] [PubMed]

18. Le Menn, N.; van Leeuwen, C.; Picard, M.; Riquier, L.; de Revel, G.; Marchand, S. Effect of vine water and nitrogen status, as well as temperature, on some aroma compounds of aged red Bordeaux wines. *J. Agric. Food Chem.* **2019**. [CrossRef]

19. Van Leeuwen, C.; Roby, J.-P.; de Rességuier, L. Soil related terroir factors, a review. *OENO One* **2018**, *52*, 173–188. [CrossRef]

20. Peyrot des Gachons, C.; van Leeuwen, C.; Tominaga, T.; Soyer, J.-P.; Gaudillere, J.-P.; Dubourdieu, D. Influence of water and nitrogen deficit on fruit ripening and aroma potential of Vitis vinifera L. cv Sauvignon blanc in field conditions. *J. Sci. Food Agric.* **2005**, *85*, 73–85. [CrossRef]

21. Helwi, P.; Guillaumie, S.; Thibon, S.; Keime, C.; Habran, A.; Hilbert, G.; Gomes, E.; Darriet, P.; Delrot, S.; van Leeuwen, C. Vine nitrogen status and volatile thiols and their precursors from plot to transcriptome level. *BMC Plant Biol.* **2016**, *16*, 173. [CrossRef]

22. Van Leeuwen, C.; Friant, P.; Chone, X.; Tregoat, O.; Koundouras, S.; Dubourdieu, D. Influence of climate, soil and cultivar on terroir. *Am. J. Enol. Vitic.* **2004**, *55*, 207–217.

23. Schultz, H.R. Climate change and viticulture: A European perspective on climatology, carbon dioxide and UV-B effects. *Aust. J. Grape Wine Res.* **2000**, *6*, 2–12. [CrossRef]

24. Ollat, N.; van Leeuwen, C.; Garcia de Cortazar, I.; Touzard, J.-M. The challenging issue of climate change for sustainable grape and wine production. *OENO One* **2017**, *51*, 59–60. [CrossRef]

25. Storchmann, K. Introduction to the special issue devoted to wine and climate change. *J. Wine Econ.* **2016**, *11*, 1–4. [CrossRef]

26. Mira de Orduna, R. Climate change associated effects on wine quality and production. *Food Res. Int.* **2010**, *43*, 1844–1855. [CrossRef]

27. Xu, Y.; Castel, T.; Richard, Y.; Cuccia, C.; Bois, B. Burgundy regional climate change and its potential impact on grapevines. *Clim. Dyn.* **2012**, *39*, 1613–1626. [CrossRef]

28. Mosedale, J.R.; Abernethy, K.E.; Smart, R.E.; Wilson, R.J.; Maclean, I.M. Climate change impacts and adaptive strategies: Lessons from the grapevine. *Glob. Chang. Biol.* **2016**, *22*, 3814–3828. [CrossRef] [PubMed]

29. Caffarra, A.; Rinaldi, M.; Eccel, E.; Rossi, V.; Pertot, I. Modelling the impact of climate change on the interaction between grapevine and its pests and pathogens: European grapevine moth and powdery mildew. *Agric. Ecosyst. Environ.* **2012**, *148*, 89–101. [CrossRef]

30. Bois, B.; Zito, S.; Calonnec, A. Climate vs grapevine pests and diseases worldwide: The first results of a global survey. *OENO One* **2017**, *51*, 133–139. [CrossRef]

31. Hannah, L.; Roehrdanz, P.R.; Ikegami, M.; Shepard, A.V.; Shaw, M.R.; Tabor, G.; Zhi, L.; Marquet, P.A.; Hijmans, R.J. Climate change, wine, and conservation. *Proc. Natl. Acad. Sci. USA* **2013**, *110*, 6907–6912. [CrossRef]

32. Moriondo, M.; Jones, G.V.; Bois, B.; Dibari, C.; Ferrise, R.; Trombi, G.; Bindi, M. Projected shifts of wine regions in response to climate change. *Clim. Chang.* **2013**, *119*, 825–839. [CrossRef]

33. Fraga, H.; Malheiro, A.C.; Moutinho-Pereira, J.; Jones, G.V.; Alves, F.; Pinto, J.G.; Santos, J.A. Very high resolution bioclimatic zoning of Portuguese wine regions: Present and future scenarios. *Reg. Environ. Chang.* **2014**, *14*, 295–306. [CrossRef]

34. Poni, S.; Palliotti, A.; Bernizzoni, F. Calibration and evaluation of a STELLA software-based daily CO_2 balance model in *Vitis vinifera* L. *J. Am. Soc. Hortic. Sci.* **2006**, *131*, 273–283. [CrossRef]

35. Costa, R.; Fraga., H.; Malheiro, A.C.; Santos, J.A. Application of crop modelling to portuguese viticulture: Implementation and added-values for strategic planning. *Ciência e Técnica Vitivinícola* **2015**, *30*, 29–42. [CrossRef]

36. Schultz, H.R.; Stoll, M. Some critical issues in environmental physiology of grapevines: Future challenges and current limitations. *Aust. J. Grape Wine Res.* **2009**, *16*, 4–24. [CrossRef]

37. Ewert, F.; Rodriguez, D.; Jamieson, P.; Semenov, M.A.; Mitchell, R.A.C.; Goudriaan, J.; Weigel, H.J. Effects of elevated CO_2 and drought on wheat: Testing crop simulation models for different experimental and climatic conditions. *Agric. Ecosyst. Environ.* **2002**, *93*, 249–266. [CrossRef]

38. Rolli, E.; Marasco, R.; Vigani, G.; Ettoumi, B.; Mapelli, F.; Deangelis, M.-L.; Gandolfi, C.; Casati, E.; Previtali, F.; Gerbino, R.; et al. Improved plant resistance to drought is promoted by the root-associated microbiome as a water stress-dependent trait. *Environ. Microbiol.* **2015**, *17*, 316–331. [CrossRef] [PubMed]

39. Bradshaw, A.D. Evolutionary significance of phenotypic plasticity in plants. *Adv. Genet.* **1965**, *13*, 115–155.

40. Bradshaw, A.D. Unravelling phenotypic plasticity—Why should we bother? *New Phytol.* **2006**, *170*, 644–648. [CrossRef]

41. Nicotra, A.B.; Atkin, O.K.; Bonser, S.P.; Davidson, A.M.; Finnegan, E.J.; Mathesius, U.; Poot, P.; Purugganan, M.D.; Richards, C.L.; Valladares, F. Plant phenotypic plasticity in a changing climate. *Trends Plant Sci.* **2010**, *15*, 684–692. [CrossRef]

42. Chitwood, D.H.; Rundell., S.M.; Li, D.Y.; Woodford, Q.L.; Yu, T.T.; Lopez, J.R.; Greenblatt, D.; Kang, J.; Londo, J.P. Climate and developmental plasticity: Interannual variability in grapevine leaf morphology. *Plant Physiol.* **2016**, *170*, 1480. [CrossRef]

43. IPCC. *Climate Change Contribution of Working Groups I, II and III to the Fifth Assessment Report of the Intergovernmental Panel on Climate Change*; Pachauri, R., Meyer, L., Eds.; IPCC: Geneva, Switzerland, 2014.

44. Jones, G.; White, M.; Cooper, O.; Storchmann, K. Climate change and global wine quality. *Clim. Chang.* **2005**, *73*, 319–343. [CrossRef]

45. Van Leeuwen, C.; Darriet, P. The impact of climate change on viticulture and wine quality. *J. Wine Econ.* **2016**, *11*, 150–167. [CrossRef]

46. Parker, A.; Garcia de Cortazar, I.; van Leeuwen, C.; Chuine, I. General phenological model to characterise the timing of flowering and veraison of *Vitis vinifera* L. *Aust. J. Grape Wine Res.* **2011**, *17*, 206–216. [CrossRef]

47. Chuine, I.; Yiou, P.; Viovy, N.; Seguin, B.; Daux, V.; Leroy-Ladurie, E.L.R. Historical phenology: Grape ripening as a past climate indicator. *Nature* **2014**, *432*, 289–290. [CrossRef] [PubMed]

48. Duchêne, E.; Schneider, C. Grapevine and climatic change: A glance at the situation in Alsace. *Agron. Sustain. Dev.* **2005**, *25*, 93–99. [CrossRef]

49. Molitor, D.; Junk, J. Climate change is implicating a two-fold impact on air temperature increase in the ripening period under the conditions of the Luxembourgish grapegrowing region. *OENO One* **2019**, *5*. [CrossRef]

50. Mosedale, J.R.; Wilson, R.J.; Maclean, I.M.D. Climate change and crop exposure to adverse weather: Changes to frost risk and grapevine flowering conditions. *PLoS ONE* **2015**, *10*, e0141218. [CrossRef]

51. Sgubin, G.; Swingedouw, D.; Dayon, G.; Garcia de Cortazar, I.; Ollat, N.; Pagé, C.; van Leeuwen, C. The risk of tardive frost damage in French vineyards in a changing climate. *Agric. For. Meteorol.* **2018**, *250*, 226–242. [CrossRef]

52. McIntyre, G.N.; Lider, L.A.; Ferrari, N.L. The chronological classification of grapevine phenology. *Am. J. Enol. Vitic.* **1982**, *33*, 80–85.

53. Parker, A.; Garcia de Cortázar, I.; Chuine, I.; Barbeau, G.; Bois, B.; Boursiquot, J.-M.; Cahurel, J.-Y.; Claverie, M.; Dufourcq, T.; Gény, L.; et al. Classification of varieties for their timing of flowering and veraison using a modelling approach. A case study for the grapevine species *Vitis vinifera* L. *Agric. For. Meteorol.* **2013**, *180*, 249–264. [CrossRef]

54. Lereboullet, A.-L.; Beltrando, G.; Bardsley, D.K.; Rouvellac, E. The viticultural system and climate change: Coping with long-term trends in temperature and rainfall in Roussillon, France. *Reg. Environ. Chang.* **2014**, *14*, 1955–1966. [CrossRef]

55. Parker, A.; Garcia de Cortazar, I.; Gény, L.; Spring, J.-L.; Destrac, A.; Schultz, H.; Stoll, M.; Molitor, D.; Lacombe, T.; Graça, A.; et al. The temperature based Grapevine Sugar Ripeness (GSR) model for adapting a wide range of *Vitis vinfera* L. Cultivars in a Changing Climate. In Proceedings of the 21th International Giesco Meeting, Tessaloniki, Greece, 24–28 June 2019; Koundouras, S., Ed.; pp. 303–308.

56. Petrie, P.; Sadras, V. Advancement of grapevine maturity in Australia between 1993 and 2006: Putative causes, magnitude of trends and viticultural consequences. *Aust. J. Grape Wine Res.* **2008**, *14*, 33–45. [CrossRef]

57. Guilpart, N.; Metay, A.; Gary, C. Grapevine bud fertility and number of berries per bunch are determined by water and nitrogen stress around flowering in the previous year. *Eur. J. Agron.* **2014**, *54*, 9–20. [CrossRef]

58. Ollé, D.; Guiraud, J.L.; Souquet, J.M.; Terrier, N.; Ageorges, A.; Cheynier, V.; Verries, C. Effect of pre-and post-veraison water deficit on proanthocyanidin and anthocyanin accumulation during Shiraz berry development. *Aust. J. Grape Wine Res.* **2011**, *17*, 90–100. [CrossRef]

59. Spayd, S.; Tarara, J.; Mee, D.; Ferguson, J. Separation of sunlight and temperature effects on the composition of *Vitis vinifera* cv. Merlot berries. *Am. J. Enol. Vitic.* **2002**, *53*, 171–182.

60. Sadras, V.O.; Moran, M.A. Elevated temperature decouples anthocyanins and sugars in berries of Shiraz and Cabernet Franc. *Aust. J. Grape Wine Res.* **2012**, *18*, 115–122. [CrossRef]

61. Van Leeuwen, C.; Destrac, A. Modified grape composition under climate change conditions requires adaptations in the vineyard. *OENO One* **2017**, *51*, 147–154. [CrossRef]

62. Destrac-Irvine, A.; van Leeuwen, C. VitAdapt: An experimental program to study the behavior of a wide range of *Vitis vinifera* varieties in a Context of Climate Change in the Bordeaux Vineyards. In Proceedings of the Conference Climwine, Sustainable Grape and Wine Production in the Context of Climate change, Bordeaux, France, 11–13 April 2016; Ollat, N., Ed.; pp. 165–171.

63. Wolkovich, E.M.; de Cortázar-Atauri, I.G.; Morales-Castilla, I.; Nicholas, K.A.; Lacombe, T. From Pinot to Xinomavro in the world's future wine-growing regions. *Nat. Clim. Chang.* **2018**, *8*, 29. [CrossRef]

64. Bordenave, L.; Tandonnet, J.P.; Decroocq, S.; Marguerit, E.; Cookson, S.J.; Esmenjaud, D.; Ollat, N. Wild vitis as a germplasm resource for rootstocks. In Proceedings of the Exploitation of Autochtonous and More Used Vines Varieties—Oenoviti International Network Meeting, Geisenheim, Germany, 3 November 2014.

65. Marguerit, E.; Lagalle, L.; Lafargue, M.; Tandonnet, J.-P.; Goutouly, J.-P.; Beccavin, I.; Roques, M.; Audeguin, L.; Ollat, N. GreffAdapt: A relevant experimental vineyard to speed up the selection of grapevine rootstocks. In Proceedings of the 21th International Giesco meeting, Tessaloniki, Greece, 24–28 June 2019; Koundouras, S., Ed.; pp. 204–208.

66. Poni, S.; Bernizzoni, F.; Civardi, S.; Libelli, N. Effects of pre-bloom leaf removal on growth of berry tissues and must composition in two red *Vitis vinifera* L. cultivars. *Aust. J. Grape Wine Res.* **2009**, *15*, 185–193. [CrossRef]

67. Kliewer, W.; Dokoozlian, N. Leaf area/crop weight ratios of grapevines: Influence on fruit composition and wine quality. *Am. J. Enol. Vitic.* **2005**, *56*, 170–181.

68. Parker, A.; Hofmann, R.; van Leeuwen, C.; McLachlan, A.; Trought, M. Leaf area to fruit mass ratio determines the time of veraison in Sauvignon blanc and Pinot noir grapevines. *Aust. J. Grape Wine Res.* **2014**, *20*, 422–431. [CrossRef]

69. Parker, A.; Hofmann, R.; van Leeuwen, C.; McLachlan, A.; Trought, M. Manipulating the leaf area to fruit mass ratio alters the synchrony of soluble solids accumulation and titratable acidity of grapevines: Implications for modelling fruit development. *Aust. J. Grape Wine Res.* **2015**, *21*, 266–276. [CrossRef]

70. Poni, S.; Gatti, M.; Bernizzoni, F.; Civardi, S.; Bobeica, N.; Magnanini, E.; Palliotti, A. Late leaf removal aimed at delaying ripening in cv. Sangiovese: Physiological assessment and vine performance. *Aust. J. Grape Wine Res.* **2013**, *19*, 378–387. [CrossRef]

71. Palliotti, A.; Panara, F.; Silvestroni, O.; Lanari, V.; Sabbatini, P.; Howell, G.S.; Gatti, M.; Poni, S. Influence of mechanical postveraison leaf removal apical to the cluster zone on delay of fruit ripening in S angiovese (*Vitis vinifera* L.) grapevines. *Aust. J. Grape Wine Res.* **2013**, *19*, 369–377.

72. De Bei, R.; Wang, X.; Papagiannis, L.; Cocco, M.; O'Brien, P.; Zito, M.; Jingyun Ouyang, J.; Fuentes, S.; Gilliham, M.; Tyerman, S.; et al. Does postveraison leaf removal delay ripening in Semillon and Shiraz in a hot Australian climate? *Am. J. Enol. Vitic.* in press. [CrossRef]

73. Friend, A.; Trought, M. Delayed winter spur-pruning in New Zealand can alter yield components of Merlot grapevines. *Aust. J. Grape Wine Res.* **2007**, *13*, 157–164. [CrossRef]

74. Moran, M.; Petrie, P.; Sadras, V. Effects of late pruning and elevated temperature on phenology, yield components, and berry traits in Shiraz. *Am. J. Enol. Vitic.* **2019**, *70*, 9–18. [CrossRef]

75. Moran, M.A.; Bastian, S.E.; Petrie, P.R.; Sadras, V.O. Late pruning impacts on chemical and sensory attributes of Shiraz wine. *Aust. J. Grape Wine Res.* **2018**, *24*, 469–477. [CrossRef]

76. Martínez-Moreno, A.; Sanz, F.; Yeves, A.; Gil-Muñoz, R.; Martínez, V.; Intrigliolo, D.; Buesa, I. Forcing bud growth by double-pruning as a technique to improve grape composition of Vitis vinifera L. cv. Tempranillo in a semi-arid Mediterranean climate. *Sci. Hortic.* **2019**, *256*, 108614.

77. Petrie, P.R.; Brooke, S.J.; Moran, M.A.; Sadras, V.O. Pruning after budburst to delay and spread grape maturity. *Aust. J. Grape Wine Res.* **2017**, *23*, 378–389. [CrossRef]

78. Koundouras, S.; Marinos, V.; Gkoulioti, A.; Kotseridis, Y.; van Leeuwen, C. Influence of vineyard location and vine water status on fruit maturation of non-irrigated cv Agiorgitiko (*Vitis vinifera* L.). Effects on wine phenolic and aroma components. *J. Agric. Food Chem.* **2006**, *54*, 5077–5086. [CrossRef]

79. Carbonneau, A. The early selection of grapevine rootstocks for resistanceto drought conditions. *Am. J. Enol. Vitic.* **1985**, *36*, 195–198.

80. Marguerit, E.; Brendel, O.; Lebon, E.; Decroocq, S.; van Leeuwen, C.; Ollat, N. Rootstock control of scion transpiration and its acclimation to water deficit are controlled by different genes. *New Phytol.* **2012**, *194*, 416–429. [CrossRef] [PubMed]

81. Chaves, M.; Santos, T.; Souza, C.; Ortuño, M.; Rodrigues, M.; Lopes, C.; Maroco, J.; Pereira, J. Deficit irrigation in grapevine improves water-use efficiency while controlling vigour and production quality. *Ann. Appl. Biol.* **2007**, *150*, 237–252. [CrossRef]

82. Schultz, H.R. Differences in hydraulic architecture account for near-isohydric and anisohydric behaviour of two field grown *Vitis vinifera* L. cultivars during drought. *Plant Cell Environ.* **2003**, *26*, 1393–1405. [CrossRef]

83. Charrier, G.; Delzon, S.; Domec, J.-C.; Zhang, L.; Delmas, C.; Merlin, I.; Corso, D.; Ojeda, H.; Ollat, N.; Prieto, J.; et al. Drought will not leave you glass empty: Low risk of hydraulic failure revealed by long term drought observations in world's top wine regions. *Sci. Adv.* **2018**, *4*, 1. [CrossRef]

84. Tomás, M.; Medrano, H.; Pou, A.; Escalona, J.M.; Martorell, S.; Ribas-Carbó, M.; Flexas, J. Water-use efficiency in grapevine cultivars grown under controlled conditions: Effects of water stress at the leaf and whole-plant level. *Aust. J. Grape Wine Res.* **2012**, *18*, 164–172. [CrossRef]

85. Tortosa, I.; Escalona, J.; Bota, J.; Tomas, M.; Hernandez, E.; Escudero, E.; Medrano, H. Exploring the genetic variability in water use efficiency: Evaluation of inter and intra cultivar genetic diversity in grapevines. *Plant Sci.* **2016**, *251*, 35–43. [CrossRef]

86. Bchir, A.; Escalona, J.M.; Gallé, A.; Hernández-Montes, E.; Tortosa, I.; Braham, M.; Medrano, H. Carbon isotope discrimination ($\delta^{13}C$) as an indicator of vine water status and water use efficiency (WUE): Looking for the most representative sample and sampling time. *Agric. Water Manag.* **2016**, *167*, 11–20. [CrossRef]

87. Gowdy, M.; Destrac, A.; Marguerit, E.; Gambetta, G.; van Leeuwen, C. Carbon isotope discrimination berry juice sugars: Changes in response to soil water deficits across a range of *Vitis vinifera* cultivars. In Proceedings of the 21th International Giesco Meeting, Tessaloniki, Greece, 24–28 June 2019; Koudouras, S., Ed.; pp. 813–814.

88. Deloire, A. A few thoughts on grapevine training systems. *Wineland Mag.* **2012**, *274*, 82–86.

89. Santesteban, L.G.; Miranda, C.; Urrestarazu, J.; Loidi, M.; Royo, J.B. Severe trimming and enhanced competition of laterals as a tool to delay ripeining in Tempranillo vineyards under semiarid conditions. *OENO One* **2017**, *51*, 191–203. [CrossRef]

90. Champagnol, F. *Eléments de Physiologie de la Vigne et de Viticulture Génrale*; Dehan, Ed.; Saint-Gely-du-Fesc: Montpellier, France, 1984.

91. Van Leeuwen, C.; Pieri, P.; Gowdy, M.; Ollat, N.; Roby, C. Reduced density is an environmental friendly and cost effective solution to increase resilience to drought in vineyards in a context of climate change. *OENO One* **2019**, *53*, 129–146. [CrossRef]

92. Lebon, E.; Dumas, V.; Pieri, P.; Schultz, H. Modelling the seasonal dynamics of the soil water balance of vineyards. *Funct. Plant Biol.* **2003**, *30*, 699–710. [CrossRef]

93. Van Zyl, J.; Hoffman, E. Root development and the performance of grapevines in response to natural as well as man-made soil impediments. In Proceedings of the 21th International Giesco Meeting, Tessaloniki, Greece, 24–28 June 2019; Koudouras, S., Ed.; pp. 122–144.

agronomy

MDPI

Article

The Impact of Possible Decadal-Scale Cold Waves on Viticulture over Europe in a Context of Global Warming

Giovanni Sgubin [1,*], Didier Swingedouw [1], Iñaki García de Cortázar-Atauri [2], Nathalie Ollat [3] and Cornelis van Leeuwen [3]

[1] Environnements et Paléoenvironnements Océaniques et Continentaux (EPOC), University Bordeaux, 33615 Pessac, France
[2] Institut National de la Recherche Agronomique (INRA), US1116-Agroclim, F-84914 Avignon, France
[3] Ecophysiologie et Génomique Fonctionnelle de la Vigne (EGFV), Bordeaux Sciences Agro, University Bordeaux, INRA, ISVV, 33882 Villenave d'Ornon, France
*   Correspondence: giovanni.sgubin@u-bordeaux.fr; Tel.: +33-540-006-193

Received: 12 June 2019; Accepted: 16 July 2019; Published: 18 July 2019

Abstract: A comprehensive analysis of all the possible impacts of future climate change is crucial for strategic plans of adaptation for viticulture. Assessments of future climate are generally based on the ensemble mean of state-of-the-art climate model projections, which prefigures a gradual warming over Europe for the 21st century. However, a few models project single or multiple O(10) year temperature drops over the North Atlantic due to a collapsing subpolar gyre (SPG) oceanic convection. The occurrence of these decadal-scale "cold waves" may have strong repercussions over the continent, yet their actual impact is ruled out in a multi-model ensemble mean analysis. Here, we investigate these potential implications for viticulture over Europe by coupling dynamical downscaled EUR-CORDEX temperature projections for the representative concentration pathways (RCP)4.5 scenario from seven different climate models—including CSIRO-Mk3-6-0 exhibiting a SPG convection collapse—with three different phenological models simulating the main developmental stages of the grapevine. The 21st century temperature increase projected by all the models leads to an anticipation of all the developmental stages of the grapevine, shifting the optimal region for a given grapevine variety northward, and making climatic conditions suitable for high-quality wine production in some European regions that are currently not. However, in the CSIRO-Mk3-6-0 model, this long-term warming trend is suddenly interrupted by decadal-scale cold waves, abruptly pushing the suitability pattern back to conditions that are very similar to the present. These findings are crucial for winemakers in the evaluation of proper strategies to face climate change, and, overall, provide additional information for long-term plans of adaptation, which, so far, are mainly oriented towards the possibility of continuous warming conditions.

Keywords: climate change; *Vitis vinifera* L.; general circulation model; EURO-CORDEX; phenological model

---

1. Introduction

The production of high-quality wine represents a valuable cultural and economic patrimony for many local communities all over Europe, notably in France, Italy, and Spain, which together account for about half of the world production [1]. The reputation of currently recognized winegrowing regions mainly results from a complex combination of favorable climatic conditions [2]. Along with particular local soil compositions, typical grape varieties, and the expertise in vineyard management maturated and handed down over centuries, specific climatic conditions define the concept of *terroir* [3,4]. Premium wine production is, in this context, acknowledged by specific certifications in Europe and preserved by regional regulations, like, inter alia, the French AOC (*Appellation d'Origine Controlee*), the Italian DOCG (*Denominazione di Origine Controllata e Garantita*), and the Spanish DO (*Denominación de Origen*). The maintenance and the expansion of the European wine-making heritage is, however, a delicate matter in the context of global warming [5], as the equilibrium between the different climatic conditions may be altered in the future and therefore the *terroir* characteristics.

Temperature is the predominant driver of grapevine (*Vitis vinifera*) growing [6], as it primarily regulates the main phenological phases of the plant, i.e., bud break, flowering, veraison, and maturity, thus characterizing yield and quality parameters. Due to the ongoing climate change, earlier phenological events have been registered in the last decades over most of the traditional vineyards of Europe, e.g., in Bordeaux and Rhone Valley [7,8], northeast Spain [9], northeast Italy [10,11], and Piedmont [12].

By modulating the length of each phenological phase, temperature also plays a central role in determining the fruit composition [13] by outlining the ratio between sugar content and acidity [14], whose equilibrium is essential for high-quality wines [15]. Temperatures that are too high would produce precocious development of the fruit, resulting in wines with high alcohol and low organic acid contents [16]. On the contrary, conditions that are too cool would prevent the complete maturity of the fruits, yielding berries with high acidity, low sugar, and unripe flavors [7,17]. This is one of the reasons why climatic conditions primarily determine the potential for premium wine production in a given region [18,19]. Indeed, in order to accomplish a balanced development of the fruit, maturity in the northern hemisphere should occur between approximatively 10 September and the 10 October [2], thus implying the mean temperature during the growing season needs to be bounded within a narrow range. In [18], temperature limits between approximatively 12 °C and 22 °C were proposed to define suitable growing areas for *Vitis vinifera*, while for individual grapevine varieties, this range is much narrower down to 2 °C, e.g., for Pinot noir.

These temperature thresholds define the suitable climatic conditions for high-quality wine potential, thus identifying specific grapevine varieties for each particular winegrowing region and characterizing the geography of premium wine production. Early ripening grapevines varieties like Chardonnay, Pinot noir, and Riesling are typically cultivated in the northernmost vineyards of Europe, e.g., Germany and U.K., as well as in continental regions, e.g., Champagne, Alsace, and Burgundy (France), and mountains regions, e.g., Trentino Alto Adige (Italy). For their characteristics, these varieties are those classically selected for new plantations in the so-called "cool-climate wine" areas. Average ripening varieties are currently cultivated in the Atlantic sector of Europe, e.g., Merlot in Bordeaux (France), Tempranillo in Rioja (Spain), and Touriga in Douro (Portugal), and in hilly areas of the Mediterranean sector, e.g., Syrah in Rhone Valley (France), Sangiovese in Tuscany (Italy). These varieties potentially risk over-ripening under too-warm climate conditions. Late ripening varieties are currently cultivated in the Mediterranean region, e.g., Grenache in Languedoc (France), Sardinia (Italy), Arangon, and Navarra (Spain), and in the warmest regions of the Atlantic sector, e.g., Cabernet Sauvignon in Bordeaux (France). These varieties are expected to expand to northern regions under climate change.

The general warmer conditions registered all over the Europe in the last decades have already promoted new viticulture areas to emerge beyond 50° N. For example, the vineyard coverage of England and Wales has more than doubled since 2004 according to recent estimations [20]. A similar trend is observed in Denmark [21]. Moreover, warmer conditions, so far, appear to be generally beneficial for many traditional vineyards, since the optimum climate for their typical varieties has been approached. For example, Merlot and Cabernet Sauvignon in Bordeaux have tended to produce larger berry weights and higher sugar to total acid ratios, which corresponded to an increase of vintage rating [22]. This was likely due to earlier veraison dates, which enabled wine-makers to have a larger margin of time to establish when the optimal fruit composition was reached, with the possibility to pick fruits at greater levels of ripeness [7]. However, further warming over these traditional regions may push climatic conditions beyond the optimum for their typical varieties, which will likely force wineries to adapt and eventually switch to more appropriate varieties for warmer climates [23,24].

For the future, the main temperature over Europe is projected to continue to increase due to anthropogenic global warming [25]. Such an assessment is mainly based on the results of the different climate projections included in the fifth coupled model intercomparison project (CMIP5) [26], for different future emission scenarios, i.e., the representative concentration pathways (RCPs) [27,28]. Depending on the region, future warmer conditions may represent either an advantageous opportunity or a threat [29] by moving away from or by approaching the optimal climatic conditions for a given grapevine variety. This has the potential to overturn the geography of wine production by the end of the 21st century as suggested by many studies, e.g., [30–33], which prefigure a loss of suitability over the major present-day wine-producing areas and the establishment of new vineyards at higher latitudes or altitudes, however the extent of these changes is under debate [33,34]. These assessments have mainly been carried out by taking into account an ensemble of several climate model projections, e.g., the CMIP5. Yet, each model differs from the others due to different model parameterizations and numerical methods, defining a broad spectrum of possible climate projections. Their distribution states the inter-model uncertainty, while their ensemble mean is considered as the most reliable result. Indeed, comparisons between historical simulations and observational data demonstrated that multi-model mean generally outperforms most of, if not all, individual models [35,36]. This is likely because systematic biases intrinsically affecting individual models are, at least partly, cancelled by the averaging procedure [37]. This procedure, however, also cancels the internal variability out and all the large climatic oscillations reproduced by any individual model. Furthermore, an un-weighted multi-model mean tends to indiscriminately under-rate the probability of events that are physically plausible but scarcely reproduced by models due to their biases. For this reason, new methods are being developed to characterize the model response in relation to some emergent constraints [38], and to weight models according to their reliability for the simulation of a given phenomenon [39]. This approach eventually restricts the broad range of possible climate change scenarios and allows a better characterization of the uncertainty by dividing the models in different clusters depending on their response and on their reliability. Moreover, clustering enables the extraction of one or more model projections from the different subsets that can serve as case studies to analyze specific potential climate change scenarios and their impacts.

A similar approach has been adopted to analyze the North Atlantic temperature projections in the 40 CMIP5 models, which are characterized by a large uncertainty [40]. Sgubin et al. (2017) [41] found a strict link between the simulated temperature and the dynamical response of the subpolar gyre (SPG) oceanic convection, a key process for the heat exchange between the deep ocean and the atmosphere. Depending on the fate of the SPG convection in the projections, indeed, they identified three main cluster of models. Two models showed a large-scale Atlantic meridional overturning circulation (AMOC) disruption, provoking a gradual but strong temperature decrease all over the northern hemisphere, with peaks up to 4 °C in 50 years over Europe. Seven models exhibited an abrupt local collapse of the oceanic vertical convection in the SPG region, with temperature evolution characterized by a long-time increasing trend suddenly interrupted by single or multiple rapid drops, up to 3 °C over 10 years. The rest of the models, i.e., 31, did not show any abrupt change in the SPG convection, and were characterized by a continuous warming trend over the North Atlantic. Sgubin et al. (2017) [41] also argued that an assessment based only on an unweighted multi-model mean underestimates the occurrence of a SPG convection collapse, since the likelihood for such an event is enhanced if the model's reliability is accounted for. When considering only the most realistic models in simulating the present-day SPG ocean stratification, which has been shown to be an emergent constraint, the chance of an abrupt cooling event is almost as likely as a continuous warming trend, while the chance of a complete AMOC collapse is negligible. These findings highlight the necessity of specific impact analyses accounting for a scenario characterized by a SPG convection collapse. This is notably important for impact analyses over Europe, whose temperature changes are strictly connected to those in the North Atlantic Ocean [42]. Sudden temperature drops over the North Atlantic have actually already been reported around 1970 [43], yet an analysis on their impacts on grapevine production in Europe is missing.

Under these premises, the aim of the present study is to investigate the implications of potential large temperature variations over Europe on viticulture practices at regional scale. For this purpose, we analyze different downscaled projections provided by the EURO-CORDEX exercise [44,45], and we mainly focus on the CSIRO-Mk6-3-0 model, which belongs to that cluster of CMIP5 models exhibiting a SPG convection collapse during the 21st century [41]. We present results for the RCP4.5 scenario, whose level of global warming is the closest to the 2 °C limit, a threshold often proposed as a potentially safe upper bound on global warming. Our choice, however, was also dictated by the limited number of downscaled projections simulating a SPG convection collapse within the EURO-CORDEX database. After a dynamical downscaling, the projected temperature data are used to force a hierarchy of phenological models simulating the main developmental stages of the grapevine. Their future evolution defines the climatic suitability for premium wine production. Current and new potential suitable winemaking areas are evaluated under the climate scenario prefiguring a SPG convection collapse and compared with the results shown by the ensemble mean of CMIP5 models. This comparison clearly marks the different impacts on viticulture coming out from different clusters of models, which should be carefully accounted for adaptation management.

2. Methods and Material

The methodology on which the present work is based can essentially be summarized in four main points, which also contain information about the material adopted:

- Simulation of coarse-resolution future climate by means of 7 CMIP5 general circulation models (GCM) under the RCP4.5 scenario.
- Dynamical downscaling over Europe according to the RCA4-SMHI model of 7 coarse-resolution GCMs.
- Coupling of the downscaled air temperature projections with 3 phenological models for the main developmental stages of the grapevine.
- Definition of climatic suitability for premium wine production based on estimated maturity dates.

2.1. General Circulation Models: the CMIP5 Simulations

Climatic projections are based on simulations of all the GCM participating to the CMIP5 project [26], which provides a standard protocol of daily data from the end of the pre-industrial era (historical simulations) to 2100 (future projections). The historical simulations run from 1850 to 2006, and the external boundary conditions consist of a prescribed radiative forcing representing all the known aerosol and greenhouse gases concentrations in the atmosphere estimated from observational data. The initial conditions are those obtained from the O (1000)-year control simulations based on stationary climatological forcing. The future projections start in 2006 and are forced by a common pattern of greenhouse gas concentration trajectories until 2100 describing different possible emission scenarios, i.e., the RCP scenarios [27,28]. Here, results from the RCP4.5. scenario [46] are analyzed, which prefigures a stabilization of radiative forcing at 4.5 W m^{-2} by the end of the century. However, GCMs run at coarse spatial resolution, i.e., O (100) km, thus describing only large-scale processes and limiting impact analyses to global and continental scales.

2.2. Dynamical Downscaling with a Regional Circulation Model

For assessments at the regional scale, higher-resolution climate projections are required. For this scope, we use the EURO-CORDEX data (http://www.euro-cordex.net) [44,45], which provides CMIP5 climate projections at finer spatial grid, i.e., O(10) km, over Europe. These data are obtained by means of dynamical downscaling, a method consisting of running a regional circulation model

(RCM) starting from the GCM outputs over a limited area of the globe. The EURO-CORDEX data initiative offers an unprecedented number of simulations centered over Europe, thus constituting the benchmark dataset for future climate impact assessments. The whole dataset derives from 10 RCMs and 14 CMIP5 GCMs for the different RCP scenarios (updated on 2018) and is available at horizontal resolutions 0.44° (~50 km, EUR-44) and 0.11° (~12 km, EUR-11). Here, the outputs from the EUR-44 Rossby Centre regional atmospheric climate model (RCA4) [47,48] nested inside 7 different GCMs for the RCP4.5 scenario, i.e., CanESM2, CNRM-CM5, CSIRO-Mk3-6-0, GFDL-ESM2M, HadGEM2-ES, IPSL-CM5A-MR, MPI-ESM-LR, are analyzed. Furthermore, model outputs have been adjusted in order to ensure a statistical conformity between observational data and historical simulation. Our bias correction consists of aligning both mean and standard deviation of the model daily outputs to those calculated from WATCH observational data [49]. Such an adjustment has been carried out separately for each single month.

2.3. Phenological Models

Downscaled temperature projections over Europe have been successively used to force 3 different phenological models simulating the day of the year of occurrence for the main developmental stages of the grapevine. We carried out simulations for 4 different grapevine varieties, representative of different heat requirements for ripening [50,51], i.e., Chardonnay for early ripening variety, Syrah for middle ripening variety, and Cabernet Sauvignon and Grenache for late ripening varieties.

The phenological models used here assume that each developmental stage is exclusively induced by a sequence of certain temperature conditions. According to this approach, the day of occurrence of a given phenological stage t_p coincides with the fulfilment of a critical temperature forcing F^* formalized in terms of the cumulative daily forcing units F_u after a certain starting day t_0:

$$t_p : \sum_{t_0}^{t_p} F_u = F^* \tag{1}$$

Depending on the different formulations of the function F_u and on the different assumptions for t_0, three different phenological models have been here adopted: (i) A linear non-sequential model, (ii) a linear sequential model, and (iii) a non-linear sequential model.

The linear non-sequential model is a thermal time model [52], also known as a growing degree days (GDD) model [53], based on the cumulative heat forcing. In such a formulation, the forcing unit F_u is a linear growing function of the daily mean temperature T, when this latter is greater than a base temperature T_b:

$$F_u = \text{GDD}_{T_b} = \begin{cases} 0 \text{ if } T \leq T_b \\ T - T_b \text{ if } T > T_b \end{cases} \tag{2}$$

Moreover, the thermal summation is calculated from a constant starting time t_0, meaning that each developmental stage is independent of the previous one. The budburst is based on a GDD model with a base temperature $T_b = 10$, and a fixed starting time $t_0 = 1$ January. Its formulation, parameterization, and validation have been provided in [54] by using a collection of 616 budburst measurements for 10 different grapevine varieties. The flowering and veraison have been calculated according to the grapevine flowering veraison model (GFV) [55], which is also based on a GDD model (Equation (2)). The daily sum of the forcing unit starts at $t_0 = 1$ March, i.e., the 60th day of year (DOY), and T_b has been set at 0 °C. Its calibration and validation are based on a database corresponding to 81 varieties, 2278 flowering observations, and 2088 veraison observations, spanning from 1960 to 2007 and from 123 different locations over Europe. The maturity day has been instead assumed as occurring k days after the simulated day of veraison, where the constant k has been calculated as the average veraison-to-maturity period from more than 500 historical observations for the different grapevine varieties.

The linear sequential model is also based on a linear relation between daily temperature and forcing (chilling) unit, but the starting time of the sum of each phenological phase (Equation (1)) is not fixed a priori but depends on the previous phenological stages. The budburst model is based on the BRIN model [54], which includes dormancy and post-dormancy sub-models, thus allowing the simulation of the dormancy break from which the summation of F_u starts. The dormancy break t_{db} sub-model is based on the accumulation of chilling unit C_u until a critical value F^* is reached:

$$t_{db} : \sum_{t_0}^{t_{db}} C_u = C^* \tag{3}$$

with C_u formalized according with the Q_{10} Bidabe's formula [56]:

$$F_u = Q_{10}^{\frac{-T_{max}}{10}} + Q_{10}^{\frac{-T_{min}}{10}} \tag{4}$$

where T_{min} and T_{max} are, respectively, the minimal and the maximal daily temperatures, Q_{10} is an a-dimensional constant set at 2.17. The post-dormancy calculation follows the method of Richardson [57], which is based on the growing degree hours (GDH) cumulated over a day, so that the forcing unit F_u in Equation (2) is here approximated as:

$$F_u = \sum_{h=1}^{24} GDH_{T_b} \approx \begin{cases} \frac{T_{max}+T_{min}}{2} - T_b \text{ if } \frac{T_{max}+T_{min}}{2} \leq T_B \\ T_B - T_b \text{ if } \frac{T_{max}+T_{min}}{2} > T_B \end{cases} \tag{5}$$

where T_B is the upper base temperature, here set at 25 °C, beyond which development rate becomes constant [58], while the (lower) base temperature T_b is 5 °C. The parameterization and validation of the BRIN model are based on a database corresponding to 10 grapevine varieties and 616 budburst observations [54]. The flowering and the veraison are based on a GDD_{10} (Equation (2)), whose summation start when budburst is accomplished. The day of maturity is calculated according to a GDD_{10} model and a starting time coinciding with the budburst occurrence. The parameterization of the different critic temperature accumulation F^* and their validation have been provided in [59].

The non-linear sequential model is based on a curvilinear response to the temperature for the calculation of flowering, veraison, and maturity. As for the linear sequential model, the budburst model is also based on the BRIN model [54], which represents the only linear component of this model. The following phenological phases are instead based on a non-linear formulation of the forcing unit F_u, which is determined by three cardinal temperatures, i.e., a base temperature T_b, a limit temperature T_{lim}, and an optimal temperature T_{opt} [60]:

$$F_u = \begin{cases} \frac{2(T-T_b)^\alpha (T_{opt}-T_b)^\alpha - (T-T_b)^{2\alpha}}{(T_{opt}-T_b)^{2\alpha}} & \text{if } T_b \leq T \leq T_B \\ 0 \text{ if } T < T_b \text{ or } T > T_B \end{cases} \tag{6}$$

where

$$\alpha = \frac{\ln 2}{\ln\left(\frac{T_B-T_b}{T_{opt}-T_b}\right)} \tag{7}$$

Its curvilinear structure allows it to consider the effects of high temperatures on development slowdown [61]. Cardinal temperatures T_b and T_{lim} have been fixed, respectively, to 0 °C and 40 °C, while optimal temperature T_{opt} and the critical forcing F^* are obtained from [62]. The values of all the parameters for the different phenological models are summarized in Table 1.

Table 1. Values of parameters and calibration of the different phenological models for the different varieties.

		Linear Non-Sequential			Linear Sequential				Curvilinear Sequential				
		t_0	T_b (°C)	F^*	t_0	T_b (°C)	T_B (°C)	F^*	t_0	T_b (°C)	T_B (°C)	t_{opt} (°C)	F^*
Budburst	Chardonnay	DOY = 1	5	220.1	t_{DB}	5	30	6577	t_{DB}	5	30	/	6577
	Syrah			265.3				7819				/	7819
	Cabernet S.			318.6				/				/	9169
	Grenache			321.3				9174				/	/
Flowering	Chardonnay	DOY = 60	0	1217	t_{BUD}	10	40	253.9	t_{BUD}	0	40	30.3	18.8
	Syrah			1279				313.3				32.0	12.5
	Cabernet S.			1299				/				30.2	20.3
	Grenache			1277				327.7				/	/
Veraison	Chardonnay	DOY = 60	0	2547	t_{BUD}	10	40	951	t_{FLO}	0	40	24.3	56.2
	Syrah			2601				1012				27.0	52.8
	Cabernet S.			2689				/				24.3	63.0
	Grenache			2761				1148				/	/
Maturity	Chardonnay	DOY = t_{VER}	/	K = 41	t_{BUD}	10	40	1675	t_{VER}	0	40	24.3	46.0
	Syrah			K = 46				1685				27.0	43.2
	Cabernet S.			K = 52				/				24.3	51.5
	Grenache			K = 51				1926				/	/

2.4. Definition of Climatic Suitability for the Different Grapevine Varieties

We introduce the concept of climatic suitability for premium wine production by means of the definition of an optimal temporal window for the maturity day. Here, we assume this to range between 10 September and 20 October, similarly to the time interval proposed in [2], in which, however, the upper limit was fixed to 10 October. According to this assumption, hence, the climatic conditions are favorable for the production of high-quality wine if the maturity day falls within this specific period of the year. The definition of climatic suitability intrinsically states the stability of the traditional vineyards under climate change as well the opportunity for new regions to become appropriate for high-quality production. However, it is important to stress that our definition of suitability only accounts for the thermal conditions for ripening, yet other parameters can be also important.

3. Results

3.1. Uncertainty in Climate Projections and Model Clustering

As shown in the multi-model analysis in [41], different behaviors of the oceanic circulation in the North Atlantic SPG led to divergent temperature projections over that region, which defined three main distinct clusters of models. Their characterization implies three different temperature trends over the SPG as well as the occurrence or not of an abrupt cooling. In order to evaluate if these different temperature behaviors over the SPG also propagates in the surrounding regions and penetrate over the continents, Figure 1 shows the maximum 10-year temperature drop throughout the 21st century over Europe, against the 100-year temperature trend for each available CMIP5 projection. Such a 2D diagram groups model projections according to their inter-decadal variability and their long-term temperature change. The distribution of the single models and their ensemble means and spread are shown for the 37 non-downscaled RCP4.5 projections (Figure 1a) and for the 7 projections downscaled with the SMHI-RCA4 regional model (Figure 1b).

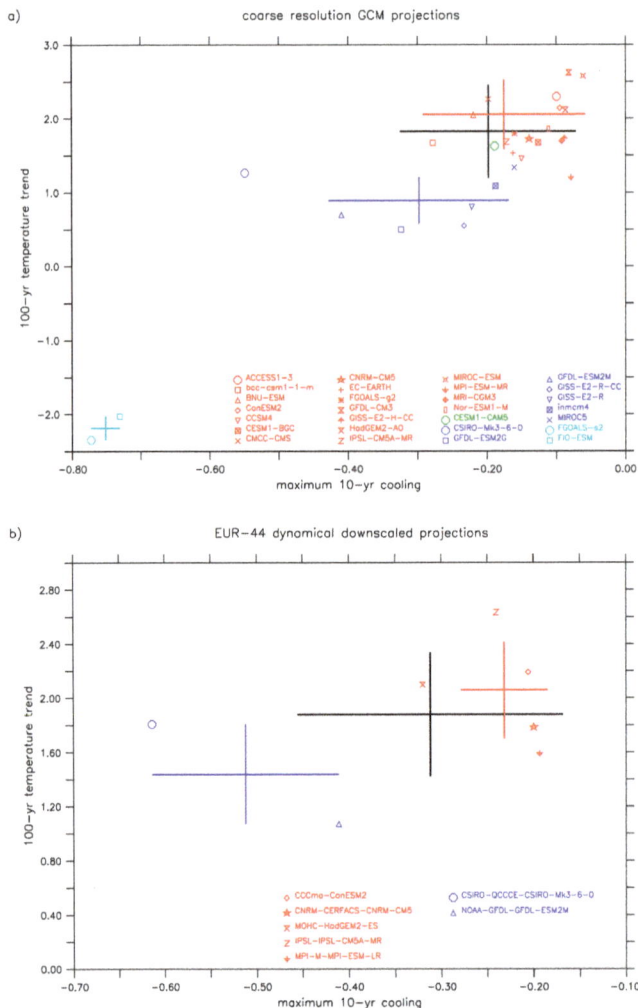

Figure 1. Scatterplot of the simulated 100-year temperature trend (in °C) versus the maximum 10-year cooling event (in °C) over western Europe for (**a**) the 37 coarse-resolution fifth coupled model intercomparison project (CMIP5) projections and (**b**) the 7 dynamical downscaled projections for the representative concentration pathways (RCP)4.5 scenario. Different colors follow the model clustering proposed in Sgubin et al., 2017, which groups projections not showing any abrupt change in the subpolar gyre (SPG) (red), projections producing a SPG convection collapse (blue), and projections simulating an Atlantic meridional overturning circulation (AMOC) collapse within 2100 (cyan). The green point in the upper panel corresponds to CESM1-CAM5, which is a model showing a SPG convection collapse, but only for RCP8.5. Crossing lines individuate the mean and the standard deviation of each subset of models, with the black lines corresponding to the ensemble of all the models. For multiple models developed at the same institute, we displayed just one point for a matter of readability in the diagram, e.g., among the different models developed by the Institut Pierre Simon Laplace (IPSL), we just displayed results of IPSL-CM5A-MR. However, all the projections have been considered in the calculation of the different clusters' ensemble mean and spread.

When all the 37 GCMs (7 downscaled projections) are considered, the ensemble mean temperature trend over Europe is 1.83 ± 0.63 (1.88 ± 0.46) °C/century while the maximum 10-year cooling is, on average, −0.20 ± 0.13 (−0.31 ± 0.14) °C/decade. The distribution of the single-model results on this diagram evidences that the clustering proposed in [41] for the characterization of the sub-polar North Atlantic temperature response also subsists when Europe is analyzed. Indeed, three main distinct subsets of models can clearly still be identified for the temperature response over the continent. In Figure 1a, models simulating an AMOC disruption (cyan) are characterized by a 100-year cooling trend (−2.2 ± 0.16 °C/century) and by a maximum 10-year cooling of −0.75 ± 0.02 °C/decade. The models projecting a SPG convection collapse (blue) are all characterized by a smaller than average warming trend, i.e., 0.89 ± 0.31 °C/century, and/or by a stronger than average 10-year cooling events, i.e., −0.30 ± 0.13 °C/decade. The rest of the models (red) exhibits higher temperature trends over Europe, i.e., 1.89 ± 0.37 °C/century as well as slighter or almost null cooling episodes, −0.13 ± 0.06 °C/decade. These models, being the majority, strongly influence the ensemble mean of all the models. A similar pattern is qualitatively valid when considering the responses of the seven downscaled projections (Figure 1b), which, however, just include two models showing a SPG convection collapse and five models not showing any abrupt change in the North Atlantic. The latter shows a mean temperature trend 2.06 ± 0.36 °C/century and a maximum 10-year cooling of −0.23 ± 0.05 °C/decade, while the former is characterized by a subdued warming trend, i.e., 1.43 ± 0.37 °C/century) and by larger 10-year temperature oscillations, i.e., −0.51 ± 0.10 °C/decade.

3.2. The Spatial and Temporal Features of the Cold Waves over Europe

Since the number of models showing an abrupt decadal-scale cooling is much lower than the models not showing any abrupt cooling, an assessment based on the ensemble mean of all the CMIP5 models covers, to some extent, the possibility of a SPG convection collapse and its associated temperature oscillations affecting the European climate. However, the likelihood of such an event has been actually assessed to be higher than what the unweighted CMIP5 multi-model ensemble mean shows [41]. In order to take into account such a possibility, we therefore differentiate the impact analysis by separating the results of one of the projections reproducing a SPG convection collapse from the results evidenced by the multi-model mean classical procedure. In Figure 2, the temperature evolutions simulated by the CSIRO-Mk3-6-0 model for different European regions are displayed and compared with the ensemble mean trend of the 37 projections. The response to the RCP4.5 emission scenario prefigured for the 21st century is characterized by a long-term warming trend all over the Europe, in line with all the models here analyzed (Figure 2). In addition, the CSIRO-Mk3-6-0 projection is also characterized by a strong inter-decadal variability, with multiple cooling events that interrupt, for a certain period, the long-term warming trend. It is possible, indeed, to identify three main decadal cold waves along the 21st century, which make this model the one featuring the largest multi-decadal variability over Europe among the downscaled projections (Figure 1b). It is worth emphasizing that these simulated cold waves over Europe occur in concomitance with an abrupt reduction of the oceanic convective activity in the SPG, which prevents the local heat exchange from the deep ocean to the surface normally occurring in winter, and cause temperature to drop locally, despite the global warming signal (see Figure S1 in Supplementary Materials). Although the identification of the driver of the cold wave is not the aim of this study, the fact that the three decadal cooling events over Europe simulated by the CSIRO-Mk3-6-0 model coincide with abrupt reductions of the convection activity in the SPG reinforces the hypothesis of a strict connection between ocean circulation changes in the North Atlantic and rapid climate oscillations over Europe.

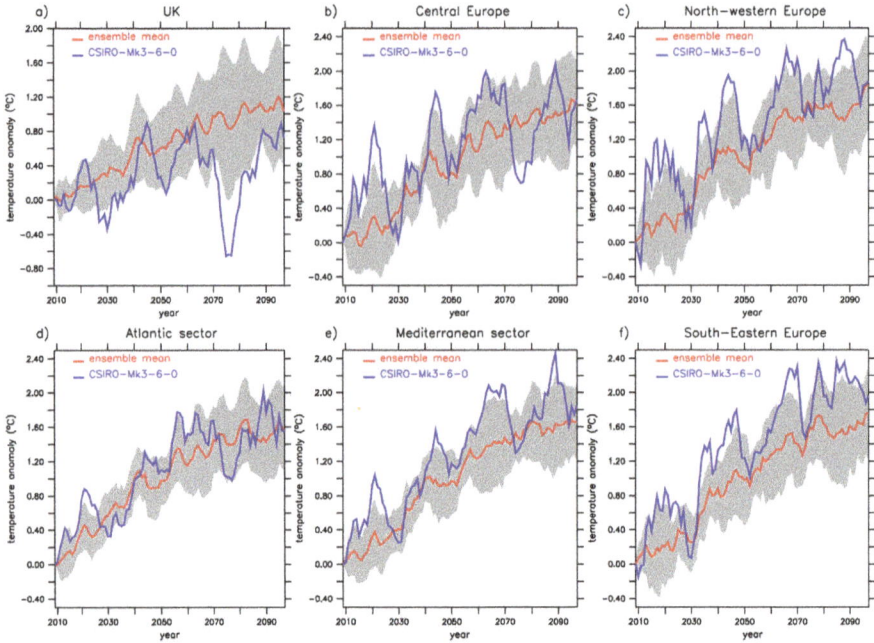

Figure 2. Evolution of the 2 m air temperature for the 21st century RCP4.5 projections over different regions of Europe: (**a**) The U.K., (**b**) Central Europe, (**c**) North-eastern Europe, (**d**) Atlantic sector, (**e**) Mediterranean sector, and (**f**) South-eastern Europe. Blue lines indicate the temperature evolution simulated with the CSIRO-Mk3-6-0 model. Red lines indicate the seven downscaled models ensemble mean temperature evolution, which is embedded in the grey portion indicating the inter-model spread, i.e., the standard deviation.

Despite a long-term warming trend in line with the other models, the temperature evolution simulated by the CSIRO-Mk3-6-0 model largely deviates from the continuous and gradual warming pattern characterizing the CMIP5 ensemble-mean. The cooling event simulated around 2075 even exceeds, by two times, the standard deviation of ensemble mean temperature over the U.K. and continental part of Europe, thus representing an outstanding case of study for the analysis of the impacts of large temperature oscillations over Europe. We, therefore, focus on this specific cooling event simulated by the CSIRO-Mk3-6-0 model, with the intent of characterizing its possible effects on climate over Europe, which can be overlooked by a multi-model ensemble procedure. In Figure 3, we show the anomaly of temperature over Europe during the 10-year cooling event (2069–2078) with respect to the previous 10 years (2059–2068). The pattern of such a long-lasting simulated "cold blob" appears to form in winter over the northern North Atlantic and to propagate towards Europe, losing its intensity in south-east direction, towards the Mediterranean Sea and the Black Sea, where it extinguishes. At the annual time-scale, the temperature drop mainly involves the U.K. and the continental region of Europe (Figure 3a), including areas indicated as the most suitable for new grapevine plantations. The intensity of the 10-year cooling is larger during the winter (Figure 3b), while it weakens with the following seasons (Figure 3c–e). In winter, the cold blob that formed over the North Atlantic Ocean mainly runs over the U.K. and most of the central regions of Europe, from France to Poland, where winter temperature in the decade 2069–2078 are, on average, 2 °C colder than in the previous decade. In spring, most of the Europe is touched by the cold wave, whose core is centered more south-eastern with respect to winter, notably involving the regions surrounding the Alps and the Balkans. The effect of the cold blob starts to fade in summer, while in autumn it appears to vanish over the continent, while still affecting the temperature of the North Atlantic Ocean.

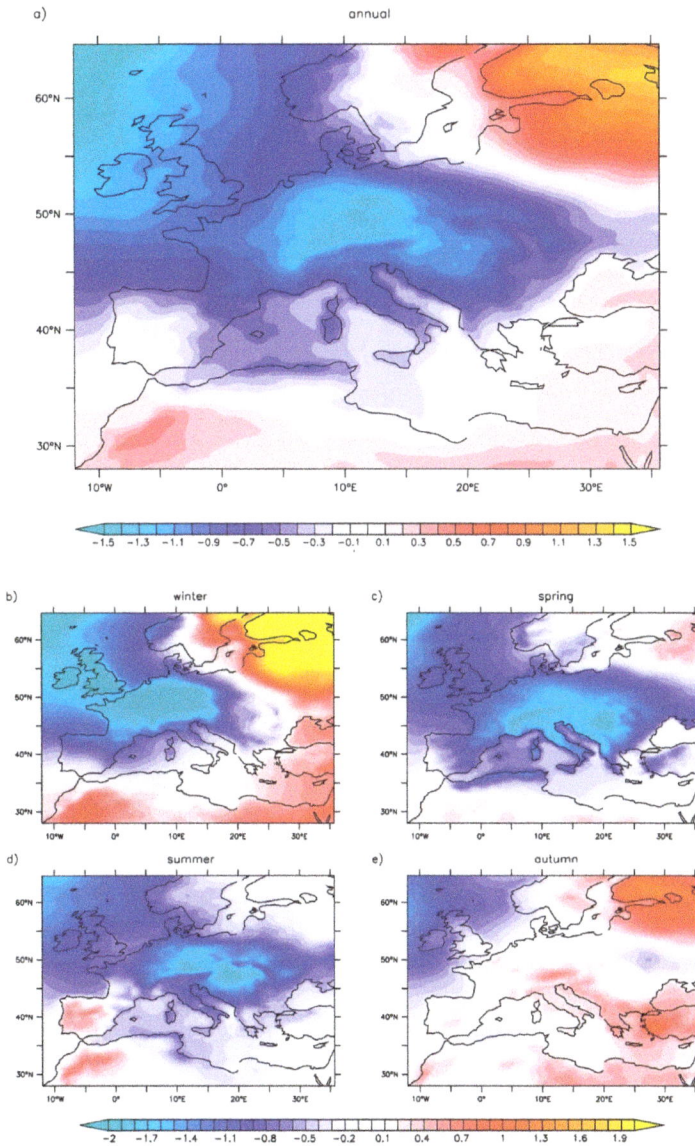

Figure 3. Pattern of the mean air temperature anomaly (in °C) between the decade 2069–2078 and the decade 2059–2068 as simulated by the downscaled CSIRO-Mk3-6-0 model for different periods: (**a**) Annual, (**b**) winter, (**c**) spring, (**d**) summer, and (**e**) autumn.

3.3. The Effect of the Rapid Cooling on Phenology of the Grapevine

The distinct seasonal responses imply that the different phenological phases of the grapevine are not affected in the same way during the occurrence of the cold wave. In Figure 4, we display the anomaly between the decades 2069–2078 and 2059–2068 of the simulated occurrence of the main phenological stages for Chardonnay (an early variety; for the other grapevine varieties, see Figures S2–S4). Such an anomaly is carried out by using the ensemble of the three phenological models used here, while details related to each of the single phenological models are illustrated in Figures S5–S7 of Supplementary Material. All the growing phases of the grapevine appear to occur later (positive anomaly) in the decade 2069–2078 all over Europe, as a consequence of the large cooling with respect to the previous decade.

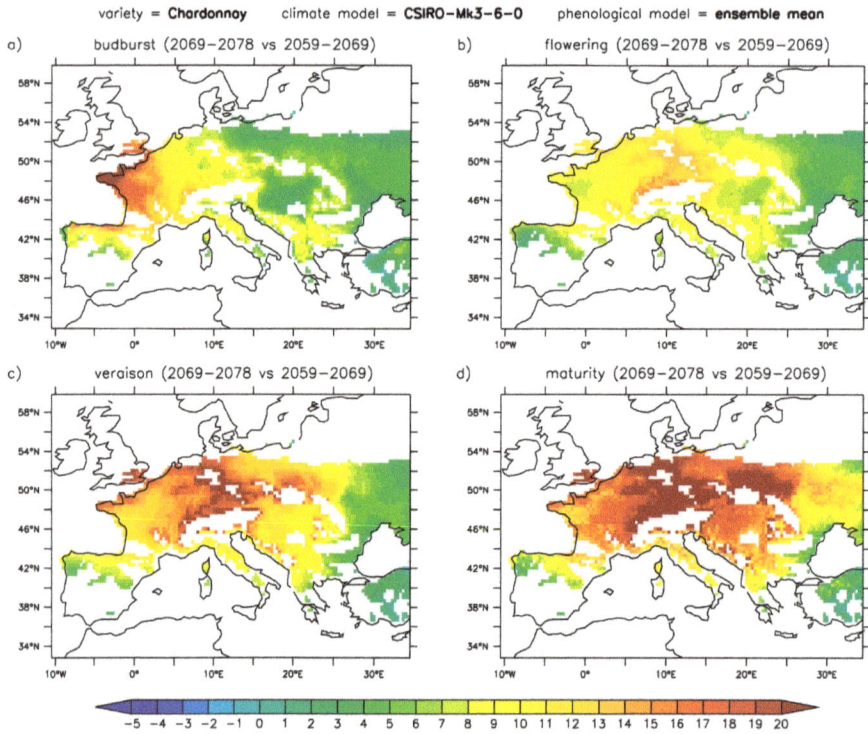

Figure 4. Pattern of the anomaly (in days) of the occurrence of the phenological stages, i.e., (**a**) budburst, (**b**) flowering, (**c**) veraison, and (**d**) maturity, for the Chardonnay variety between the decade 2069–2078 and the decade 2059–2068 as simulated by the downscaled CSIRO-Mk3-6-0 model. Results are based on the ensemble mean of the three phenological models here adopted.

The average delay of the budburst is about 10 days over Europe, with peaks of delay mainly concentrated over the Atlantic sector of France and the UK, where the anomaly is around 20 days (Figure 4a). In general, late budburst notably involves the western part of Europe, while in the eastern part of Europe, it is limited to a few days. The mean delay of the flowering is also about 10 days, but much more uniform in space, with peaks between 10 and 15 days, mainly located in the central part of Europe (Figure 4b). The cumulated lags led to large veraison anomalies, up to 15–20 days over most of the central part of Europe. This eventually led to maturity dates strongly delayed in the period 2069–2078 compared to the previous decade, i.e., on average by 15 days over Europe, with peaks of almost one month over the central part of Europe. This represents a significant delay, notably if we consider that such anomalies take place in less than 10 years. These abrupt changes may have high repercussions on the production and quality of wine.

3.4. The Climatic Suitability for Premium Wine Production During the Cold-Wave Events

A main threat for winemakers in the context of climate change concerns the conservation of *terroir* characteristics, and if the future climatic conditions will still be favorable for the production of high-quality wine. As illustrative examples, in Figure 5 we show the evolution of the main phenological phases for typical varieties in four different renowned wine production regions, according with the downscaled CSIRO-Mk3-6-0 projection.

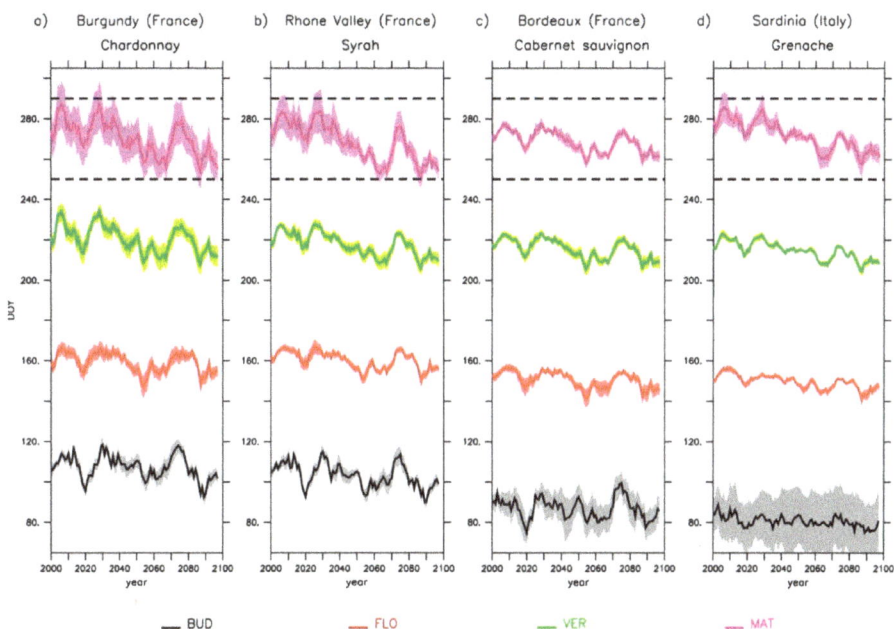

Figure 5. Evolution of day main phenological phases (in day of year (DOY)) of typical grapevine varieties in four selected renowned wine-making regions of Europe: (**a**) Chardonnay in Burgundy, (**b**) Syrah in Rhone Valley, (**c**) Cabernet Sauvignon in Bordeaux, and (**d**) Grenache in Sardinia, as simulated by the downscaled CSIRO-Mk3-6-0 model. Results are based on the ensemble mean of the three phenological models here adopted. Black lines indicate the budburst, red lines the flowering, green lines the veraison, and violet lines the maturity evolution, while their shaded intervals represent the two-sigma spread between the three phenological models. Dashed lines indicate the limits for maturity day within which climate conditions are suitable for premium wine production according to the definition in Methods and Materials.

At the centennial time-scale, the results evidence a widespread trend to an anticipation of all the phenological stages in all the four selected checkpoints grapevine varieties. This is the direct effect of the general warmer conditions throughout the 21st century. The trend towards earlier stages appears slight or even null for budburst, while it is more pronounced for successive stages, in particular for maturity. Superimposed on this trend, each phenological stages is characterized by a significant inter-decadal variability, mainly associated with the multiple temperature drops over Europe evidenced by the CSIRO-Mk3-6-0 model. This implies large oscillations in the characteristics of the fruit composition and therefore in the vintage rating, likely affecting the economical resilience of wine businesses. Nevertheless, by using the definition of climatic suitability for high-quality production we introduced in the section Methods and Materials, Figure 5 also shows that the maturity, despite its long-term trend towards precocity and its strong inter-decadal variability, always falls in the optimal range for premium wine production in the four illustrative examples here analyzed. We can, therefore, claim that these typical varieties in these traditional sites appear to be resilient according with the CSIRO-Mk3-6-0 model. The same feature is, to some extent, also valid when the ensemble mean is considered (see Figure S8). The simulated maturity day always falls within the range of optimal ripening. However, it approaches its lower limit, so that further warming during the 21st century due to a more severe emission scenario likely enshrines the necessity of varietal shifts.

Climate change may promote the settlement of new regions for high-quality wine production and force winemakers in the current vineyards to adapt to warmer conditions by replacing traditional varieties with later ripening varieties. In Figure 6, the same analysis as in Figure 5 was carried out but for potential new emerging wine regions (upper panels) and for traditional regions where we assumed hypothetical replantation with later ripening varieties (lower panels). The general pattern of the evolution of all the phenological phases is qualitatively similar to those shown in Figure 5. However, at beginning of the century, none of the maturity associated with the grapevine varieties selected for these regions falls within the optimal temporal window for high-quality wine production. This well-reflects the actual present-day suitable areas for the different grapevine varieties, whose northern limit generally does not exceed the 50° N parallel. As the temperature increase during the 21st century and the grapevine development becomes faster, maturity dates of the grapevine varieties selected for these six regions starts to fall within the suitable range for premium production. Therefore, climate change appears to be beneficial for plantations in cool climate regions as well as being compatible with variety replacements in the traditional regions. Nevertheless, during the abrupt cold event simulated in the decade 2069–2078, the optimal climatic conditions in these regions are not satisfied anymore, as they would produce too-late maturity dates. Depending on the specific region, the loss of climatic suitability appears to last from a few years to approximately 15 years, thus questioning the economic viability of those adaptation strategies presupposing northward varietal shifts [63], which are irreversible adaptations in the short-term. These results differ from those carried out by the multi-model ensemble mean (see Figure S9). Indeed, warming trend causes maturity dates to persistently fall within the range of suitability after a certain period, although at the beginning of the century, none of the selected regions are characterized by optimal climatic conditions for premium wine production. Overall, we can, thus, claim that while the long-term warming signal may represent an opportunity for new vineyards areas and may be compatible with plantations of later ripening varieties in traditional regions, the climate decadal variability represents a serious risk that could compromise the quality of wine production for a relatively long period and therefore the economic investments implied in these adaptation measures.

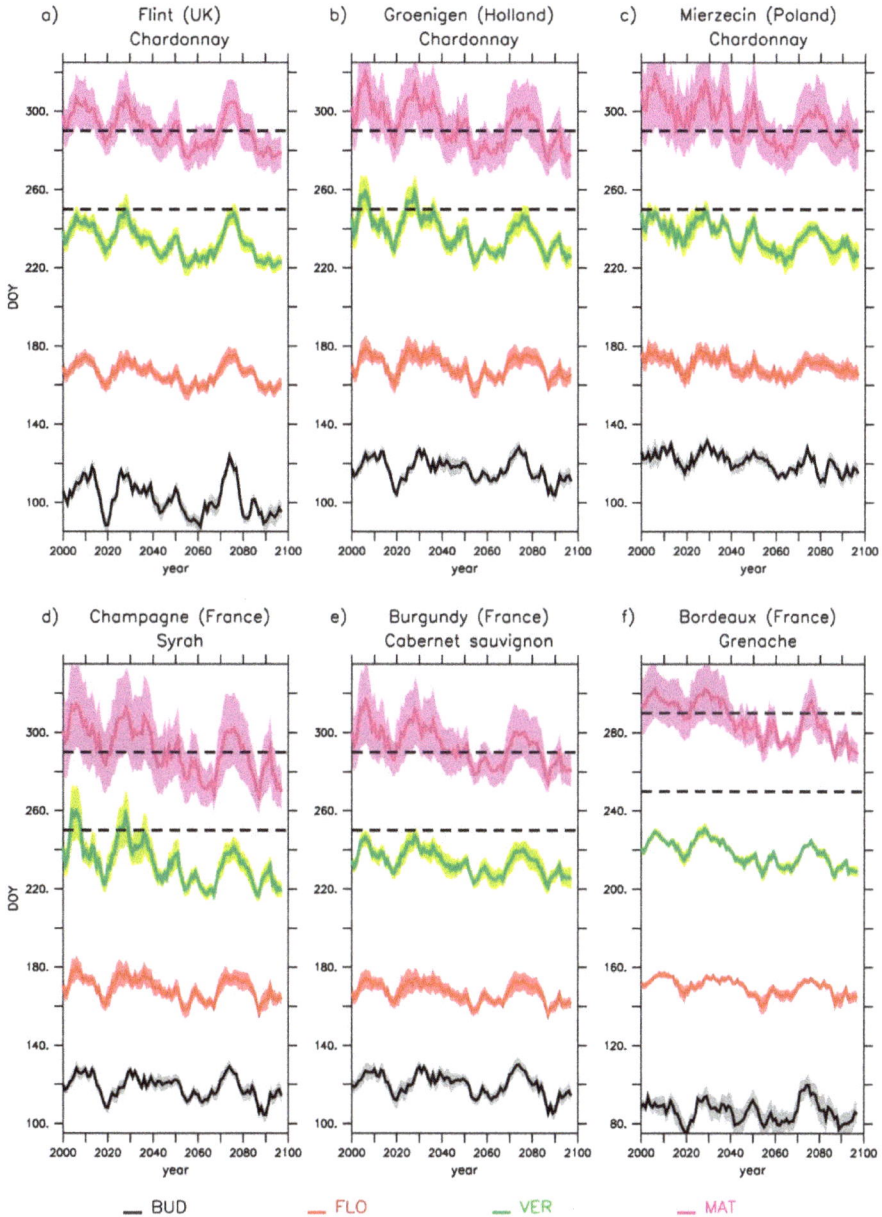

Figure 6. As Figure 5 but for different varieties and locations: (**a**) Chardonnay in UK, (**b**) Chardonnay in Holland, (**c**) Chardonnay in Poland, (**d**) Syrah in Champagne, (**e**) Cabernet Sauvignon in Burgundy, and (**f**) Grenache in Bordeaux.

Focusing on Chardonnay (see Figures S10–S12 for the other varieties), we summarize in Figure 7 the changes in climatic suitability for premium production over Europe during the 2069–2078 cold wave with respect to the previous decade. In this period, most of the supposed "new cool wine regions" [64,65] appears to lose their climatic suitability, from U.K. to Poland, as a consequence of the rapid cooling. In contrast, the south-west of France, part of the Balkans, and some hilly regions in Spain and Italy would benefit from the temperature decrease by recovering their climatic suitability lost throughout the 21st century. This pattern is nearly specular to the changes in climatic suitability between the decade 2059–2068 (just before the occurrence of the cold event), and the present-day decade, i.e., 2010–2019: The effect of 50-year climate changes produces a shift of climatic suitability towards the north (Figure S13). This means that the occurrence of cold waves has the potential to abruptly cancel, at least for a decade, all the beneficial effects of warmer conditions accumulated since the beginning of the century, which are projected by the totality of the state-of-the-art climate models for the RCP4.5 scenario.

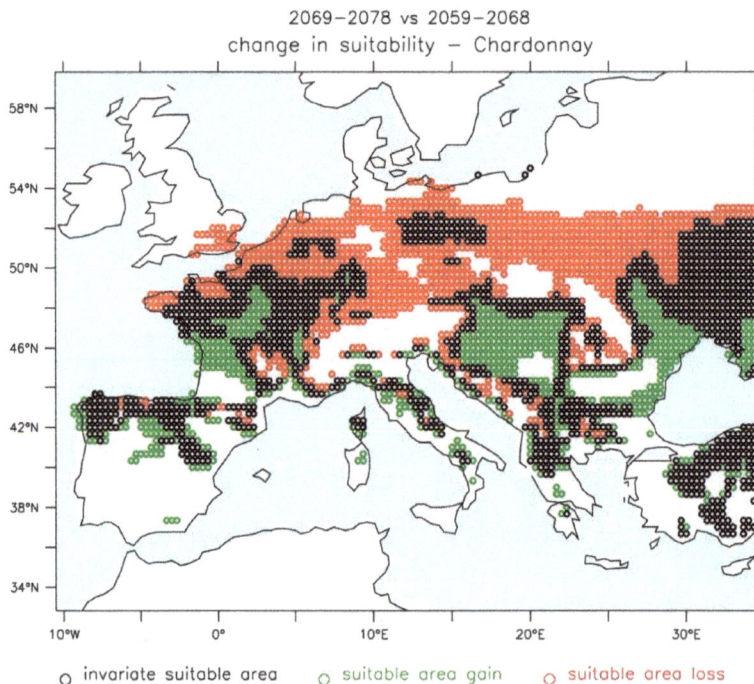

Figure 7. Anomaly in climatic suitability for premium wine production between the decade 2069–2078 and the decade 2059–2068 as simulated by the downscaled CSIRO-Mk3-6-0 model. Climatic suitability is based on the definition we introduced in Methods and Material. Over a decade, a location is assumed to be suitable when the climatic conditions are satisfied for at least 70% of the years. Results are based on the ensemble mean of the three different phenological models. Black circles indicate locations for which climatic conditions are suitable for both the 2059–2068 and the 2069–2078 periods. Green (red) circles indicate sites where climatic conditions are suitable for the decade 2069–2078 (2059–2078) but not for the decade 2059–2078 (2069–2078).

4. Discussion

The endorsement of measures of adaptation to climate change presupposes a deep investigation of both the beneficial and the deleterious potential return of all the proposed actions. This implies considering all the wide range of plausible future climatic changes at regional scale, in order to optimize the economic investments and to minimize the risks related to the occurrence of overlooked events. In this context, the assessment of a very likely long-term gradual temperature increase along the 21st century, which is based on the ensemble mean of the state-of-the-art climate projections, is orienting decision-makers and stakeholders to rethink the grapevine cultivation zoning, prefiguring a varietal shift at higher latitudes and/or at higher altitudes. The latter are long-term adaptation measures that entail considerable investments and a non-immediate benefit. However, in the assessment of a gradual long-term warming, the effects of decadal variability are missing. In Europe, rapid temperature variations may occur as a side effect of the climate change, producing decadal-scale large local cooling events in a context of global warming [41], as shown, for instance, by the CSIRO-Mk3-6-0 model. The present study arises from the need of considering the possibility of these cold waves in an impact analysis for viticulture. We highlight that the occurrence of sudden cooling events over Europe has implications for grapevine growing that largely differ from those associated with a gradual warming. This finding may promote the application of our methodology to other relevant crops for Europe, e.g., wheat and maize.

Taking into account, at the same time, the possibility of a long-term continuous warming and the possibility of rapid cold events poses further challenges in the planning of proper measures of adaptation for wine-producing sectors in Europe, and in the analysis of their economic impact and sustainability. Indeed, beyond the effective costs that the expected increase of temperature will presuppose for preserving the wine production over Europe, our results suggest a further economic evaluation of the risks associated with an amplified decadal variability. This suggests the research of producing systems that are able to face, at the same time, the warming trend and the possibility of unprecedented large and rapid changes over Europe. Such an approach implies an accurate analysis of the sustainability of radical choices, thus promoting a more rational strategy of adaptation, e.g., a diversified variety relocation, which takes into account the risk of sudden temperature changes for each specific region. Also, this approach can partly resize the pursuit of new wine regions, notably in those regions appearing particularly sensitive to large inter-decadal variability. For example, according to our results, hypothetical new vineyard settlements would be more suitable in the south-eastern part of Europe rather than in its north-western part, since the latter would be primarily impacted by the occurrence of the cold waves originated in the North Atlantic. These considerations become notably relevant if one considers the large economic effort presupposing the potential settlement of new vineyards in cool climate regions.

Under the hypothesis of a stabilization of the greenhouse gas emissions within 2100 and a level of global warming limited to approximatively 2 °C, our results also suggest that varietal changes in traditional wine-producing regions do not appear strictly necessary. Indeed, over most of these regions, climate conditions appear to remain always suitable for typical grapevine varieties, both in the case of a gradual warming as evidenced by the model ensemble mean, and in the case of rapid decadal cold waves throughout the 21st century as shown by the CSIRO-Mk3-6-0 model. However, for more severe emission scenarios like RCP8.5, a northward varietal shift may appear as the most proper adaptation measure. Hence, the potential benefits of such a strategy appear particularly conditional on the capacity of mitigation of climate change. Moreover, the actual feasibility of varietal replacements is strictly dependent on regional *terroir* regulations, whose conventions would be arduous to change due to the cultural legacy at the base of local wine identities.

It is worth stressing that our definition of suitability for a given grapevine variety is exclusively based on thermal conditions for optimal maturity. It considers the cumulative temperature forcing during the growing season, meaning that the risk of extreme meteorological events like spring frosts [66], heat waves, and droughts, which have detrimental effects on suitability, are not taken into account. The definition of an indicator that also includes these factors is the subject of further studies.

Independently of the change in climatic suitability, the occurrence of abrupt temperature oscillations may imply rapid changes on the wine composition and organoleptic features, likely implying negative repercussions on the wine market [67]. In this regard, progresses in oenological and viticultural practices may subdue the negative effects of the short-term large climate variations by adjusting the year-to-year wine composition according with the needs [68]. In parallel with assessments of the long-term climate change, an optimal management of the vineyards is therefore conditional on the capacity of anticipating with accuracy the decadal climate variability and therefore the occurrence of potentially forthcoming rapid climate changes. The recent advances in climate decadal predictions are promising in this sense, as a skilled forecast system can effectively support operational adaptation measures at the short-term time scale. For example, predicting the mean growing seasons for the following few years may promote strategic procedures of canopy management, aimed at optimizing leaf-area-to-fruit-weight ratio [69], and therefore the grape development and the wine composition. Also, an assessment of the risks associated with possible extreme decadal-scale events may prompt producers to deploy remedial infrastructures, e.g., wind machines, sprinkling water machines, gas-powered heaters, anti-hail nets, or to stipulate specific insurances, thus minimizing the possible economic damage of hail and frost events. The reliability of climate decadal predictions and their effective applicability in the context of adaptation strategies for European vineyards management will be the subject of future studies.

5. Conclusions

In this study, we analyzed, in detail, the impact of potential rapid temperature drops on viticulture over Europe, which have been previously demonstrated to be plausible events that can superimpose on the long-term warming trend along the 21st century [41]. We focused on the decadal-scale cold waves over Europe projected by the CSIRO-Mk3-6-0 model for the RCP4.5 scenario, and we compared their impact on grapevine growing with that resulting from the ensemble mean of seven climate projections producing a progressive warming signal. Our results evidenced that the occurrence of the cold waves yields significant changes in all the developmental stages of the grapevine, which would be able to overturn the long-term warming effect, at least for time steps of approximately a decade. During these cold events simulated in the future, climate conditions became rather similar to the present-day conditions in a very short period of time (a few years), thus rapidly cancelling out the previous warming that was gradually taking place since the beginning of the century. By defining the climatic suitability for premium wine production as those conditions satisfying the temperature requirements for the grapevine ripening to fall within a specific period of the year, we reported a potential loss of suitability during the occurrence of cold wave events over most of the central-western part of Europe. The same regions were those that became previously suitable, due to the simulated 21st century gradual warming. Our findings therefore disclosed the possibility that long-term adaptation measures like varietal northward shifts may not be the most appropriate in those regions potentially strongly hit by cold wave events. The most sensitive region includes the U.K. and different countries of central Europe (e.g., Holland, Germany, Poland), which have been often identified as new "cool-climate" viticulture areas [64,65]. Overall, the outcomes of this study integrate the debate on the impacts of climate change on viticulture in Europe, which so far, was mainly based on the paradigm that temperature will continue to gradually rise in the future.

Supplementary Materials: The following are available online at http://www.mdpi.com/2073-4395/9/7/397/s1, Figure S1: Relation between convection activity and temperature in the SPG, Figure S2: Anomalies (2069–2078 vs 2059–2068) of the occurrence of the phenological stages as simulated by CSIRO-Mk3-6-0 for Syrah according with the ensemble mean of the phenological models, Figure S3: Anomalies (2069–2078 vs 2059–2068) of the occurrence of the phenological stages as simulated by CSIRO-Mk3-6-0 for Cabernet sauvignon according with the ensemble mean of the phenological models, Figure S4: Anomalies (2069–2078 vs 2059–2068) of the occurrence of the phenological stages as simulated by CSIRO-Mk3-6-0 model for Grenache according with the ensemble mean of the phenological models, Figure S5: Anomalies (2069–2078 vs 2059–2068) of the occurrence of the phenological stages as simulated by CSIRO-Mk3-6-0 for Chardonnay according with the linear non-sequential model, Figure S6: Anomalies (2069–2078 vs 2059–2068) of the occurrence of the phenological stages as simulated by CSIRO-Mk3-6-0 model for Chardonnay according with the linear sequential phenological model, Figure S7: Anomalies (2069–2078 vs 2059–2068) of the occurrence of the phenological stages as simulated by CSIRO-Mk3-6-0 model for Chardonnay according with the curvilinear sequential phenological model, Figure S8: Evolution of the main phenological stages in four traditional winemaking regions based on the ensemble mean of climate projections, Figure S9: Evolution of the main phenological stages in six regions potentially involved in northward varietal shift based on the ensemble mean of climate projections, Figure S10: Anomaly (2069–2078 vs 2059–2068) in climatic suitability for premium production of Syrah, Figure S11: Anomaly (2069–2078 vs 2059–2068) in climatic suitability for premium production of Cabernet sauvignon, Figure S12: Anomaly (2069–2078 vs 2059–2068) in climatic suitability for premium production of Grenache, Figure S13: Anomaly (2069–2068 vs 2010–2019) in climatic suitability for premium production of Chardonnay.

Author Contributions: Experimental design, G.S. and D.S.; methodology, G.S., D.S., I.G.d.C.-A., N.O. and C.v.L.; formal analysis, G.S. and D.S.; investigation, G.S.; writing-original draft preparation, G.S.; writing-review and editing, D.S., I.G.d.C.-A.; N.O. and C.v.L.

Funding: This research was funded by the French National Research Agency (ANR), grant number ANR-10-LABX-45.

Acknowledgments: We thank the French National Research Agency (ANR) for the financial support in the frame of the Investments for the future Programme within the Cluster of Excellence (Labex COTE). We are grateful to three anonymous reviewers for their valuable and constructive comments that improved the manuscript.

Conflicts of Interest: The authors declare no conflict of interest.

References

1. OIV. *Statistical Report on World Vitiviniculture*; OIV: Paris, France, 2018.
2. van Leeuwen, C.; Seguin, G. The concept of terroir in viticulture. *J. Wine Res.* **2006**, *17*, 1–10. [CrossRef]
3. Gladstones, J. *Wine, Terroir and Climate Change*; Wakefield Press: Kent Town, Australia, 2011; p. 279.
4. van Leeuwen, C.; Friant, P.; Chone, X.; Tregoat, O.; Koundouras, S.; Dubourdieu, D. Influence of climate, soil, and cultivar on terroir. *Am. J. Enol. Vitic.* **2004**, *55*, 207–221.
5. Mozell, R.N.; Thach, L. The impact of climate change on the global wine industry: Challenges and solutions. *Wine Econ. Policy* **2014**, *3*, 81–89. [CrossRef]
6. Jones, G.V. Phenology: An Integrative Environmental Science. In *Winegrape Phenology*, 2nd ed.; Schwartz, M.D., Ed.; Springer: Berlin, Germany, 2013; p. 610.
7. van Leeuwen, C.; Darriet, P. The impact of climate change on viticulture and wine quality. *J. Wine Econ.* **2017**, *11*, 150–167. [CrossRef]
8. Garcia de Cortazar-Atauri, I.; Duchêne, E.; Destrac, A.; Barbeau, G.; de Resseguier, L.; Lacombe, T.; Parker, A.K.; Saurin, N.; van Leeuwen, C. Grapevine phenology in France: From past observations to future evolutions in the context of climate change. *OENO One* **2017**, *51*, 115. [CrossRef]
9. Ramos, M.C.; Jones, G.V.; Martinez-Casanovas, J.A. Structure and trends in climate parameters affecting winegrape production in northeast Spain. *Clim. Res.* **2008**, *38*, 1–15. [CrossRef]
10. Caffarra, A.; Eccel, E. Projecting the impacts of climate change on the phenology of grapevine in a mountain area. *Aust. J. Grape Wine Res.* **2011**, *17*, 52–61. [CrossRef]
11. Tomasi, D.; Jones, G.V.; Giust, M.; Lovat, L.; Gaiotti, F. Grapevine Phenology and climate change: Relationships and trends in the Veneto Region of Italy for 1964–2009. *Am. J. Enol. Vitic.* **2011**, *62*, 329–339. [CrossRef]
12. Andreoli, V.; Cassardo, C.; La Iacona, T.; Spanna, F. Description and preliminary simulations with the Italian Vineyard Integrated Numerical Model for estimating physiological values (IVINE). *Agronomy* **2019**, *9*, 94. [CrossRef]
13. Drappier, J.; Thibon, C.; Rabot, A.; Geny-Denis, L. Relationship between wine composition and temperature: Impact on Bordeaux wine typicity in the context of global warming: Review. *Crit. Rev. Food Sci. Nutr.* **2017**. [CrossRef]
14. Coombe, B.G. Influence of temperature on composition and quality of grapes. *Acta Hort.* **1987**, *206*, 23–36. [CrossRef]
15. Mira de Orduna, R. Climate change associated effects on wine quality and production. *Food Res. Int.* **2010**, *43*, 1844–1855. [CrossRef]
16. Sweetman, C.; Sadras, V.O.; Hancock, R.D.; Soole, K.L.; Ford, C.M. Metabolic effects of elevated temperature on organic acid degradation in ripening *Vitis vinifera* fruit. *J. Exp. Bot.* **2014**, *65*, 5975–5988. [CrossRef]
17. Dai, Z.W.; Ollat, N.; Gomès, E.; Decroocq, S.; Tandonnet, J.-P.; Bordenave, L.; Pieri, P.; Hilbert, G.; Kappel, C.; van Leeuwen, C.; et al. Ecophysiological, genetic, and molecular causes of variation in grape berry weight and composition: A review. *Am. J. Enol. Vitic.* **2011**, *62*, 413–425. [CrossRef]
18. Jones, G.V. Climate and terroir: Impacts of climate variability and change on wine. *Geosci. Can. Reprint Ser.* **2006**, *9*, 1–14.
19. Schultz, H.R.; Jones, G. Climate induced historic and future changes in viticulture. *J. Wine Res.* **2010**, *21*, 137–145. [CrossRef]

20. Nesbitt, A.; Kemp, B.; Steele, C.; Lovett, A.; Dorling, S. Impact of recent climate change and weather variability on the viability of UK viticulture—Combining weather and climate records with producers' perspectives. *Aust. J. Grape Wine Res.* **2016**, *22*, 324–335. [CrossRef]

21. Bentzen, J.; Smith, V. *Wine Production in Denmark Do the Characteristics of the Vineyards Affect the Chances for Awards?* Working Papers; University of Aarhus, Aarhus School of Business, Department of Economics: Aarhus, Denmark, 2009.

22. Jones, G.V.; Davis, R.E. Climate Influences on Grapevine Phenology, Grape Composition, and Wine Production and Quality for Bordeaux, France. *Am. J. Enol. Vitic.* **2000**, *51*, 249–261.

23. Duchêne, É.; Huard, F.; Dumas, V.; Schneider, C.; Merdinoglu, D. The challenge of adapting grapevine varieties to climate change. *Clim. Res.* **2010**, *41*, 193–204. [CrossRef]

24. Wolkovich, E.M.; García de Cortázar-Atauri, I.; Morales-Castilla, I.; Nicholas, K.A.; Lacombe, T. From Pinot to Xinomavro in the world's future wine-growing regions. *Nat. Clim. Chang.* **2018**, *8*, 29–37. [CrossRef]

25. Stocker, T.F.; Qin, D.; Plattner, G.-K.; Tignor, M.M.B.; Allen, S.K.; Boschung, J.; Nauels, A.; Xia, Y.; Bex, V.; Midgley, P.M. Technical Summary. In *Climate Change 2013: The Physical Science Basis. Contribution of Working Group I to the Fifth Assessment Report of the Intergovernmental Panel on Climate Change*; Cambridge University Press: Cambridge, UK; New York, NY, USA, 2013.

26. Taylor, K.E.; Stouffer, R.; Meehl, G.A. An overview of CMIP5 and the Experiment Design. *Bull. Am. Meteorol. Soc.* **2011**, *93*, 485–498. [CrossRef]

27. Moss, R.H.; Edmonds, J.A.; Hibbard, K.A.; Manning, M.R.; Rose, S.K.; van Vuuren, D.P.; Carter, T.R.; Emori, S.; Kainuma, M.; Kram, T.; et al. The next generation of scenarios for climate change research and assessment. *Nature* **2010**, *463*, 747–756. [CrossRef] [PubMed]

28. Meinshausen, M.; Smith, S.J.; Calvin, K.; Daniel, J.S.; Kainuma, M.L.T.; Lamarque, J.-F.; Matsumoto, K.; Montzka, S.A.; Raper, S.C.B.; Riahi, K.; et al. The RCP greenhouse gas concentrations and their extensions from 1765 to 2300. *Clim. Chang.* **2011**, *109*, 213–241. [CrossRef]

29. Ollat, N.; van Leeuwen, C.; Garcia de Cortazar, I.; Touzard, J.M. The challenging issue of climate change for sustainable grape and wine production. *OENO One* **2017**, *51*, 59–60. [CrossRef]

30. Malheiro, A.C.; Santos, J.A.; Fraga, H.; Pinto, J.G. Climate change scenarios applied to viticultural zoning in Europe. *Clim. Res.* **2010**, *43*, 163–177. [CrossRef]

31. Moriondo, M.; Jones, G.V.; Bois, B.; Di Bari, C.; Ferrise, R.; Trombi, G.; Bindi, M. Projected shifts of wine regions in response to climate change. *Clim. Chang.* **2013**, *119*, 825–839. [CrossRef]

32. Fraga, H.; Garcia de Cortazar-Atauri, I.; Malheiro, A.C.; Santos, J.A. Modelling climate change impacts on viticultural yield, phenology and stress conditions in Europe. *Glob. Chang. Biol.* **2016**, *22*, 3774–3788. [CrossRef]

33. Hannah, L.; Roehrdanz, P.R.; Ikegami, M.; Shepard, A.V.; Shaw, M.R.; Tabor, G.; Zhi, L.; Marquet, P.A.; Hijmans, R.J. Climate change, wine, and conservation. *Proc. Natl. Acad. Sci. USA* **2013**, *110*, 6907–6912. [CrossRef]

34. van Leeuwen, C.; Schultz, H.R.; Garcia de Cortazar-Atauri, I.; Duchêne, E.; Ollat, N.; Pieri, P.; Bois, B.; Goutouly, J.P.; Quénol, H.; Touzard, J.M.; et al. Why climate change will not dramatically decrease viticultural suitability in main wine-producing areas by 2050. *Proc. Natl. Acad. Sci. USA* **2013**, *110*, E3051–E3052. [CrossRef]

35. Gleckler, P.J.; Taylor, K.E.; Doutriaux, C. Performance metrics for climate models. *J. Geophys. Res.* **2008**, *113*. [CrossRef]

36. Reichler, T.; Kim, J. How well do coupled models simulate today's climate? *Bull. Am. Meteorol. Soc.* **2008**, *89*, 303–311. [CrossRef]

37. Tebaldi, C.; Knutti, R. The use of the multi-model ensemble in probabilistic climate projections. *Philos. Trans. R. Soc. A* **2007**, *365*, 2053–2075. [CrossRef]

38. Flato, G.; Marotzke, J.; Abiodun, B.; Braconnot, P.; Chou, S.C.; Collins, W.; Cox, P.; Driouech, F.; Emori, S.; Eyring, V.; et al. Evaluation of Climate Models. In *Climate Change 2013: The Physical Science Basis. Contribution of Working Group I to the Fifth Assessment Report of the Intergovernmental Panel on Climate Change*; Cambridge University Press: Cambridge, UK; New York, NY, USA, 2013.

39. Gillet, N.P. Weighting climate model projections using observational constraints. *Philos. Trans. R. Soc. A* **2015**, *373*. [CrossRef]

40. Mauritzen, C.; Zivkovic, T.; Veldore, V. On the relationship between climate sensitivity and modelling uncertainty. *Tellus A* **2017**, *69*. [CrossRef]
41. Sgubin, G.; Swingedouw, D.; Drijfhout, S.; Mary, Y.; Bennabi, A. Abrupt cooling over the North Atlantic in modern climate models. *Nat. Commun.* **2017**, *8*. [CrossRef]
42. Moffa-Sánchez, P.; Hall, I.R. North Atlantic variability and its links to European climate over the last 3000 years. *Nat. Commun.* **2017**, *8*, 1726. [CrossRef]
43. Thompson, D.W.J.; Wallace, J.M.; Kennedy, J.J.; Jones, P.D. An abrupt drop in Northern Hemisphere sea surface temperature around 1970. *Nature* **2010**, *467*, 444–447. [CrossRef]
44. Jacob, D.; Petersen, J.; Eggert, B.; Alias, A.; Christensen, O.B.; Bouwer, L.M.; Braun, A.; Colette, A.; Déqué, M.; Georgievski, G.; et al. EURO-CORDEX: New high-resolution climate change projections for European impact research. *Reg. Environ. Chang.* **2014**, *14*, 563–578. [CrossRef]
45. Kotlarski, S.; Keuler, K.; Christensen, O.B.; Colette, A.; Déqué, M.; Gobiet, A.; Goergen, K.; Jacob, D.; Lüthi, D.; van Meijgaard, E.; et al. Regional climate modeling on European scales: A joint standard evaluation of the EURO-CORDEX RCM ensemble. *Geosci. Model Dev.* **2014**, *7*, 1297–1333. [CrossRef]
46. Thomson, A.M.; Calvin, K.V.; Smith, S.J.; Kyle, G.P.; Volke, A.; Patel, P.; Delgado-Arias, S.; Bond-Lamberty, B.; Wise, M.A.; Clarke, L.E.; et al. RCP4.5: A pathway for stabilization of radiative forcing by 2100. *Clim. Chang.* **2011**, *109*, 77. [CrossRef]
47. Samuelsson, P.; Jones, C.G.; Willen, U.; Ullerstig, A.; Gollvik, S.; Hansson, U.; Jansson, C.; Kjellström, E.; Nikulin, G.; Wyser, K. The Rossby Centre regional climate model RCA3: Model description and performance. *Tellus A* **2011**, *63*, 4–23. [CrossRef]
48. Samuelsson, P.; Gollvik, S.; Jansson, C.; Kupiainen, M.; Kourzeneva, E.; van de Berg, W.J. *The Surface Processes of the Rossby Centre Regional Atmospheric Climate Model (RCA4)*; Report in Meteorology; SMHI: Norrköping, Sweden, 2014; Volume 157.
49. Weedon, G.P.; Balsamo, G.; Bellouin, N.; Gomes, S.; Best, M.J.; Viterbo, P. The WFDEI meteorological forcing data set: WATCH Forcing Data methodology applied to ERA-Interim reanalysis data. *Water Resour. Res.* **2014**, *50*, 7505–7514. [CrossRef]
50. van Leeuwen, C.; Garnier, C.; Agut, C.; Baculat, B.; Barbeau, G.; Besnard, E.; Bois, B.; Boursiquot, J.M.; Chuine, I.; Dessup, T.; et al. Heat Requirements for Grapewine Varieties Is Essential Information to Adapt Plant Material in a Changing Climate. In Proceedings of the 7th International Terroir Congress, Nyon, Switzerland, 19 May 2008; Volume 1, pp. 222–227.
51. Parker, A.; Garcia de Cortázar-Atauri, I.; Chuine, I.; Barbeau, G.; Bois, B.; Boursiquot, J.M.; Cahurel, J.Y.; Claverie, M.; Dufourcq, T.; Gény, L.; et al. Classification of varieties for their timing of flowering and veraison using a modelling approach: A case study for the grapevine species *Vitis vinifera* L. *Agric. For. Meteorol.* **2013**, *180*, 249–264. [CrossRef]
52. Robertson, G.W. A biometeorological time scale for a cereal crop involving day and night temperatures and photoperiod. *Int. J. Biometeorol.* **1968**, *12*, 191–223. [CrossRef]
53. Wang, J.Y. A critique of the heat unit approach to plant response studies. *Ecology* **1960**, *41*, 785–789. [CrossRef]
54. Garcia de Cortazar-Atauri, I.; Brisson, N.; Gaudillere, J.P. Performance of several models for predicting budburst date of grapevine (*Vitis vinifera* L.). *Int. J. Biometeorol.* **2009**, *53*, 317–326. [CrossRef]
55. Parker, A.; Garcia de Cortazar-Atauri, I.; van Leeuwen, C.; Chuine, I. General phenological model to characterize the timing of flowering and veraison of *Vitis vinifera* L. *Aust. J. Grape Wine Res.* **2011**, *17*, 206–216. [CrossRef]
56. Bidabe, B. Contrôle de l'époque de floraison du pommier par une nouvelle conception de l'action de températures. *CR Acad. Agric. France* **1965**, *49*, 934–945.
57. Richardson, E.A.; Seeley, S.D.; Walker, D.R. A model for estimating the completion of rest for Redhaven and Elberta peach trees. *HortScience* **1974**, *9*, 331–332.
58. Moncur, M.W.; Rattigan, K.; Mackenzie, D.H.; Mc Intyre, G.N. Base temperatures for budburst and leaf appareance of grapevines. *Am. J. Enol. Vitic.* **1989**, *40*, 21–26.
59. Garcia de Cortazar-Atauri, I. Adaptation du Modèle STICS à la Vigne (*Vitis vinifera* L.): Utilisation Dans le Cadre D'une Étude D'impact du Changement Climatique à L'échelle de la France. Ph.D. Thesis, École National Supérior Agronomique, Montpellier, France, 2006.
60. Wang, E.L.; Engel, T. Simulation of phenological development of wheat crops. *Agric. Syst.* **1998**, *58*, 1–24. [CrossRef]

61. Greer, D.H.; Weedon, M.M. The impact of high temperatures on *Vitis vinifera* cv. Semillon grapevine performance and berry ripening. *Front. Plant Sci.* **2013**, *4*, 491. [CrossRef]

62. Garcia de Cortazar-Atauri, I.; Chuine, I.; Donatelli, M.; Parker, A.K.; van Leeuwen, C. A curvilinear process-based phenological model to study impacts of climatic change on grapevine (*Vitis vinifera* L.). *Proc. Agron.* **2010**, *11*, 907–908.

63. Ollat, N.; Touzard, J.M.; van Leeuwen, C. Climate change impacts and adaptations: New challenges for the wine industry. *J. Wine Econ.* **2016**, *11*, 139–149. [CrossRef]

64. Dunn, M.; Rounsevell, M.D.A.; Boberg, F.; Clarke, E.; Christensen, J.; Madsen, M.S. The future potential for wine production in Scotland under high-end climate change. *Reg. Environ. Chang.* **2017**, *19*, 723–732. [CrossRef]

65. Maciejczak, M.; Mikiciuk, J. Climate change impact on viticulture in Poland. *Int. J. Clim. Chang. Strateg. Manag.* **2019**, *11*, 254–264. [CrossRef]

66. Sgubin, G.; Swingedouw, D.; Dayon, G.; García de Cortázar-Atauri, I.; Ollat, N.; Pagé, C.; Van Leeuwen, C. The risk of tardive frost damage in French vineyards in a changing climate. *Agric. For. Meteorol.* **2018**, *250–251*, 226–242. [CrossRef]

67. Fuentes Espinoza, A. Vin, Réchauffement Climatique et Stratégies des Entreprises: Comment Anticiper la Réaction des Consommateurs? Ph.D. Thesis, Economies et finances, Université de Bordeaux, Bordeaux, France, 2016.

68. Tilloy, V.; Cadière, A.; Ehsani, M.; Dequin, S. Reducing alcohol levels in wines through rational and evolutionary engineering of Saccharomyces cerevisiae. *Int. J. Food Microbiol.* **2015**, *213*, 49–58. [CrossRef]

69. Kliewer, W.M.; Dokoozlian, N.K. Leaf area/crop weight ratios of grapevines: Influence on fruit composition and wine quality. *Am. J. Enol. Vitic.* **2005**, *56*, 170–181.

MDPI

Article

Effects of Natural Hail on the Growth, Physiological Characteristics, Yield, and Quality of *Vitis vinifera* L. cv. Thompson Seedless under Mediterranean Growing Conditions

Despoina G. Petoumenou [1,*], Katerina Biniari [2], Efstratios Xyrafis [1], Dimitrios Mavronasios [2], Ioannis Daskalakis [2] and Alberto Palliotti [3]

[1] Department of Agriculture Crop Production and Rural Environment, Laboratory of Viticulture, University of Thessaly, 38446 Volos, Greece; xyrafis@agr.uth.gr
[2] Department of Crop Science, Laboratory of Viticulture, Agricultural University of Athens, Iera Odos, 75, 11855 Athens, Greece; kbiniari@aua.gr (K.B.); dimmavronasios@gmail.com (D.M.); john-daskalakis@hotmail.com (I.D.)
[3] Dipartimento di Scienze Agrarie e Ambientali, Università di Perugia, Borgo XX Giugno 74, 06128 Perugia, Italy; alberto.palliotti@unipg.it
* Correspondence: petoumenou@uth.gr; Tel.: +30-2421-093-180

Received: 25 March 2019; Accepted: 15 April 2019; Published: 17 April 2019

Abstract: Hailstorms are typically localized events, and very little is known about their effect on crops. The objective of this study was to examine the physiological and vine performance responses to natural hail, registered four weeks after full bloom, of field-grown Thompson seedless (*Vitis vinifera* L.) grapevines, one of the most important table grape varieties cultivated in Greece and especially in the Corinthian region in northeastern Peloponnese. Leaf gas exchange, vegetative growth, vine balance indices, cane wood reserves, yield components, and fruit chemical composition were recorded from hail-damaged vines and compared with control vines. Visibly, the extent of the hailstorm damage was great enough to injure or remove leaves as well as cause partial stem bruising and partial injury or total cracking of berries. Our results indicated that natural hail did not affect leaf photosynthesis, berry weight, total acidity, and cane wood reserves but significantly reduced the total leaf area, yield, and the total phenolics of berries at harvest. At the same time, hail-damaged vines increased the leaf area of lateral canes and presented a higher total soluble solid (TSS) accumulation, while no effect on the next year's fertility was registered. The present work is the first attempt to enhance our understanding of the vegetative yield, berry quality, and physiological responses of grapevines to natural hail, which is an extreme and complex natural phenomenon that is likely to increase due to climate change.

Keywords: leaf area; table grapes; photosynthesis; berry composition; phenolics; natural hail

---

1. Introduction

The Thompson seedless table grape cultivar (*Vitis vinifera* L.), introduced in Greece in 1838 [1], is by far one of the most important crops in the region of Corinthia, in northeastern Peloponnese (Greece). Owing to the optimal characteristics of the soil and environmental conditions from the climate of the Gulf of Corinth, this cultivar rapidly reached high-level quality standards. From there, Thompson seedless cultivation expanded to several other regions of Greece. Thus, it is now the most cultivated seedless cultivar for fresh consumption and second only to Black Corinth for raisin production in Greece [2].

High-impact weather events in Greece, including hail, are well documented [3,4] and result in significant subsequent socioeconomic impacts [4,5]. Nevertheless, very little is known about the effects of hailstorms on crops. Indeed, numerous studies have been conducted on the effects of simulated hail damage on different crops [6–9], but the effects of natural hail have scarcely been investigated [10,11].

There are significant variations in terms of crop damage due to different factors concerning both hail and crops. Therefore, the amount of damage can vary in terms of hailstone size and intensity, the kinetic energy of the hailstones [12], and the presence of wind accompanying hailfall. In addition, some crop conditions can influence the extent of the damage (i.e., the growth stage and the elasticity of the vegetation).

The cultural practices required to obtain the optimum yield and quality of the Thompson seedless cultivar have already been achieved by several studies on training system, defoliation [13], cluster thinning, girdling, the application of gibberellins, or a combination of all of these [14–16]. However, modern table grape growers are faced with front vineyard management problems that are amplified by climate change effects, such as heat waves [17], drought [18], frost [19], flooding [20], wind [21], and hailstorms. In particular, hailstorms can be very detrimental to crops and lead to the complete loss of the harvest. In Greece, during the period 1999–2011, damage costs from hail accounted for 26.2% of total insured crop losses [5]. In order to reduce these losses, anti-hail nets have been used as a protective measure for crops, but their ability to modify the tree microclimate might also alter tree growth and quality [22].

Even though several studies have been conducted on the effect of extreme weather events on grapevine growth [23,24] and physiology [24–26], as well as grape and wine quality [27–29], it is not clear yet what effects hailstorms have, directly or indirectly, on grapevine physiology and performance.

In light of this uncertainty, we studied the effects of natural hail on some important Thompson seedless grapevine leaf physiological parameters and determined its impact on vegetative growth and vine performance (yield and fruit composition).

2. Materials and Methods

2.1. Plant Material and Experimental Layout

The trial was undertaken at a commercial table grape vineyard located in Laliotis, a municipal unit of Kiato (northeastern Peloponnese, Greece, Figure 1), over the 2015 and 2016 growing seasons. The vineyard (38°01′10.9′′ N, 22°40′41.4′′ E, elevation 405 m a.s.l., silty clay loam soil) was a 15-year-old planting of *V. vinifera* L. cv. Thompson seedless grafted onto 110R rootstock. Plant distances were 1.20 m within the row and 2.80 m between rows (2976 vines/ha), trained to a Y-trellis system with three to four canes/vine and three to four spurs/vine. A 20-min hailstorm took place on 18 June 2015 (the 169th day of the year, DOY), four weeks after full bloom (known as development stage EL-33 [30]); it was preceded by heavy rain (36 mm/h) and accompanied by wind velocities up to 78 km/h and hailstone diameters of 25–30 mm. The hailstorm hit only one part of the vineyard, so this situation was very favorable for analyzing hailstorm effects on grapevine plants. So, two treatments were compared: control vines (ND, hail-damaged) and hail-damaged (HD) vines, the canopy of which was visibly damaged by the hailstorm. The vines were drip–irrigated at 3500–4000 m^3/ha, while fertilization, pest control, and cultural practices (berry thinning, leaf removal, shoot thinning, and shoot trimming) were conducted according to local practices. Furthermore, the following climatic data were recorded by a weather station (Wireless Vantage Pro2^{TM}, with a 24-h Fan Aspirated Radiation Shield, Davis Instruments, Canada USA) situated close to the trial site: monthly mean air temperature in 2015 and during 2005–2014, monthly rainfall in 2015, and average monthly rainfall during a recent decade (2005–2014).

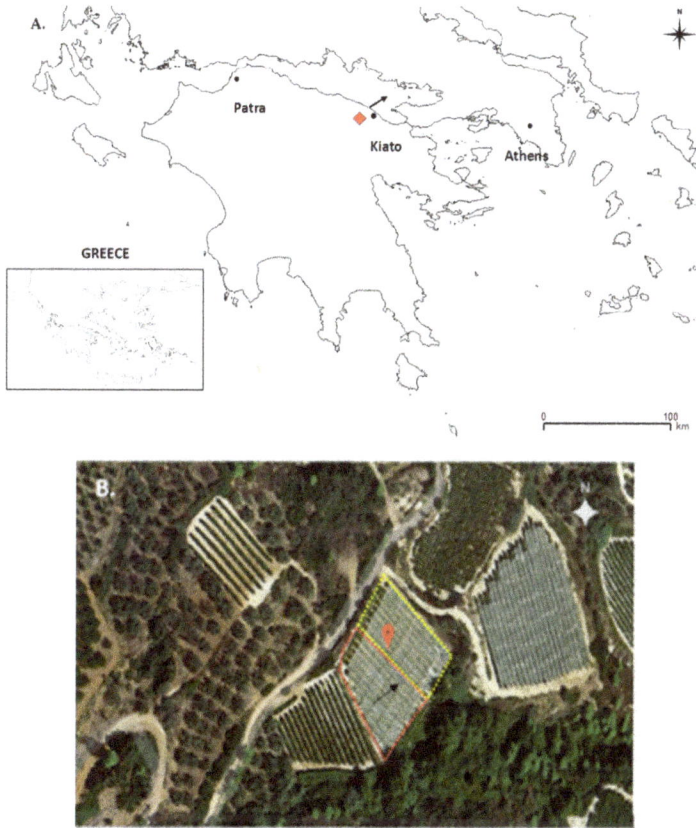

Figure 1. (**A**) Study area map of the Thompson seedless vineyard in the Gulf of Corinth (◆ northeastern Peloponnese, Greece). (**B**) The red and yellow dotted lines indicate the approximate areas of the vineyard that were affected or unaffected by the hailstorm, respectively. Black arrowheads indicate the approximate direction of the hailstorm on 18 June 2015.

The experiment was a randomized block design with two treatments—control (ND) and HD vines—in five replications. Each plot consisted of three grapevines (experimental unit), so there were 15 grapevines in each treatment.

Immediately after the hailstorm, visual effects of falling hailstones were observed. These included:

(a)   leaves still attached to grapevines with holes punched through them, creating hail injury;
(b)   partial defoliation to primary and lateral shoots;
(c)   partial injury or total cracking of berries;
(d)   partial stem and bark bruising;
(e)   primary and lateral shoots with partial hail injury on the epidermis.

All of these hail injuries were observed only on sides facing the direction of the hailstorm (i.e., to the south and east). No shoot was removed and any shoot topping was observed because the grapevines had already been trimmed by the time of the hailstorm.

No infection of pathogens to retained leaves, shoots, or bunches was observed due to the special nutrition and plant protection schedule applied before the hailstorm, immediately afterwards, and until leaf fall.

2.2. Leaf Gas-Exchange Measurements

Two days after the natural hailstorm (DOY 171), the single leaf assimilation rate (Pn), stomatal conductance (g_s), and leaf transpiration rate (E) of intact and hail-damaged leaves were measured simultaneously. The readings were repeated at DOY 186, 231, 247 (harvest), and 320 (two weeks before leaf fall) with a LC Pro+ portable photosynthesis system (ADC Bioscientific Ltd., Hoddesdon, UK). One shoot per vine (five shoots per treatment) was chosen and the readings were taken at the 3rd (basal) and 10th (medial) well-exposed leaves from the base of the main shoot and at the 3rd young leaf of the lateral shoot in the morning between 10 a.m. and 12 p.m. on sunny days under saturating photosynthetic active radiation (PAR > 1500 μmol photons $m^{-2}s^{-1}$). The readings of hail-damaged leaves were taken in leaf areas immediately adjacent to the hail injury. Concurrently, on the same leaves, intrinsic water use efficiency (WUE_i) was derived as the Pn-to-g_s ratio.

2.3. Vegetative Data, Productive Traits, and Fruit Composition

Two days after the hailstorm, the total leaf area from the primary and lateral canes and the corresponding total leaf area (m^2) from each were estimated by removing 10 canes from 10 randomly chosen vines per treatment and measuring the real leaf area (LA) with a portable leaf area meter LI-3000 (Li-Cor Biosciences, Lincoln, NE, USA); the same procedure was used at harvest in order to determine vegetative growth. Concurrently, the total number of nodes per vine, the diameter of the primary canes at the 3rd and 10th internode, and the diameter of the lateral canes at the 3rd internode were registered.

At the end of December 2015, the one-year-old pruning weight was recorded for all vines and the yield-to-pruning-weight ratio was calculated. Vine balance was also assessed by calculating the total leaf-area-to-yield ratio (vine basis) in both treatments.

At harvest, performed on 5 September 2015 (DOY 248), when the sugar accumulation on ND vines reached 20 degrees Brix (°Bx), all experimental vines were individually hand-picked and yield per vine was measured at the same time as the total number of bunches. Thereafter, the bunches were immediately weighed, and the total number of damaged and undamaged berries per bunch was counted. The bunch length and width were also recorded. The bunch compactness index was estimated as the bunch-weight-to-(bunch length)^2 ratio, according to Tello and Ibáñez [31].

Concurrently, five samples of 200 berries per treatment were randomly collected. From each sample, 175 berries were used in order to determine the following berry characteristics: berry weight, berry length and diameter (using an electronic digital caliper), and force pedicel detachment (using a digital dynamometer (PCE Italia s.r.l., Capannori, Italy)—expressed in Newton (N).

After these measurements, the same berry samples were crushed and the must was obtained in order to determine the following berry quality characteristics. Total soluble solids (TSS) were measured using a digital hand-held "pocket" refractometer PAL (Atago Co., Ltd., Tokyo, Japan) and expressed in °Bx at 20 °C. A digital HI-2002 Edge pH meter (Hanna Instruments, Rhode Island, USA) was used to measure must pH, and values were expressed in pH units. Titratable acidity (TA) was determined by titration of grape juice with a 0.1 N sodium hydroxide (NaOH) solution in the presence of a bromothymol blue indicator and expressed as tartaric acid percent (%). Also, the maturity index was calculated as the ratio TSS/TA. The remaining berries (25 berries per treatment) were frozen at −20 °C, and after a few days, total skin phenolic contents were determined as described by Slinkard and Singleton [32], and their concentration was expressed as milligrams per kilo of fresh berry weight (mg/Kg).

2.4. Carbohydrate and Nitrogen Storage in Above-Ground Permanent Vine Organs

At the end of December 2015, the soluble sugars and starch concentrations in primary canes (3rd and 10th internode) and lateral canes (3rd internode) were determined on seven replicates per treatment according to Morris (1948) [33] and Loewus (1952) [34] and expressed in mg g^{-1} of dry weight (DW).

2.5. Shoot Fertility and Blind Buds

In spring 2016, during the phenological phase of visible inflorescences, seven uniform vines per treatment were selected and the bud fertility index (number of bunches/shoot) was evaluated. At the same time, the number of blind buds per vine was determined in the same vines.

2.6. Statistical Analysis

All data were processed by a two-way analysis of variance using the SigmaStat software package (Systat Software, Inc. San Jose, California, USA). Treatment comparison was performed by *t*-test at $p < 0.05$ and $p < 0.01$. All traits are shown as mean ± standard error.

3. Results and Discussion

3.1. Meteorological Data

The average daily temperature during vegetative and reproductive growth until harvest was always higher compared with the same period in the recent decade (2005–2014), whereas the rainfall from February to October 2015 exceeded the average monthly rainfall as compared with the period 2005–2014 (Figure 2). The amount of rainfall in June 2015 was registered in concomitance with the hailstorm and represented only 5% of the total rainfall (692 mm) of that year (Figure 2).

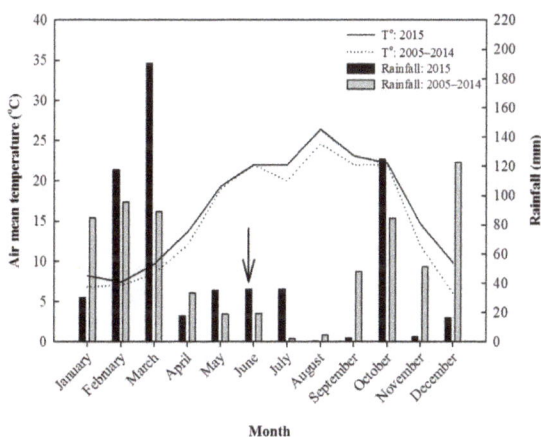

Figure 2. Seasonal trends (January–December) of monthly mean air temperature, monthly rainfall recorded in 2015, and average monthly temperature and rainfall in 2005–2014 close to the trial site. The arrow indicates the time of the hailstorm.

3.2. Effect of Natural Hail on Vegetative Growth

Hail impact differed strongly among the plant species and was related to their vegetative characteristics: plant height, growth form, leaf traits, and stem type [35]. For our experiment, total leaf area just after the hailstorm is shown in Table 1. The data indicate that this natural phenomenon removed 1.40 and 0.41 m^2 of leaf area/vine on primary and lateral shoots of HD vines, respectively, causing a total leaf area of HD vine reduction of 1.81 m^2/vine compared with ND vines (Table 1). This led to a 16.4% loss of the whole leaf area of HD vines, as demonstrated by the defoliated and scarred leaves caused by the hailstones (Figure 3). Such damage has been reported in other crops [36], and hail damage can be related to the architectural features of a plant species [35].

At harvest, the shoot number was very similar for both ND and HD vines, and despite the hail injury on the phloem (Figure 4) of the primary and lateral shoots of HD vines, there was no statistically significant difference in cane diameter between the two treatments (Table 1). Nevertheless, the lateral shoot number for HD vines was higher (31%) compared with ND vines and led to 8.6% more leaf area compared with the same vines just after hail (Table 1). This indicates the response of grapevines to natural hail (i.e., to stimulate the buds of primary shoots to outbreak and to develop new lateral shoots). Moreover, since no reports have been found in the scientific literature in which the effects of natural hail on grapevines were studied, it is not wrong to compare the effects of natural hail with those obtained by defoliation. In fact, the ability of vines to increase their lateral leaf area with defoliation (status derived from natural hail, in our case) is already known [37,38]. The total leaf area of HD vines was about 8.5% less compared with ND ones (Table 1) at harvest due to a lower leaf area of primary canes caused by hail damage (Table 1). These results are in accordance with other reports on crops such as maize [9].

Table 1. Vegetative growth (mean ± SE, *n* = 10) just after hail and at harvest in non-hail-damaged (ND) and hail-damaged (HD) field-grown Thompson seedless vines.

Parameters	ND	HD	Signif. [a]
Just after hail (DOY 169)			
Total leaf area from primary canes (m^2/vine)	8.90 ± 0.33	7.50 ± 0.18	*
Total leaf area from lateral canes (m^2/vine)	2.11 ± 0.12	1.70 ± 0.10	*
Total leaf area (m^2/vine)	11.01 ± 0.28	9.20 ± 0.17	*
At harvest (DOY 248)			
Shoots/vine	12.7 ± 0.05	13.0 ± 0.08	ns
Lateral shoots/vine	9.43 ± 0.04	12.34 ± 0.06	*
Cane diameter at 3rd internode (mm)	10.37 ± 0.29	10.73 ± 0.13	ns
Cane diameter at 10th internode (mm)	7.02 ± 0.39	7.15 ± 0.40	ns
Lateral cane diameter at 3rd internode (mm)	5.70 ± 0.22	5.65 ± 0.13	ns
Total leaf area from primary canes (m^2/vine)	10.97 ± 0.21	10.02 ± 0.28	*
Total leaf area from lateral shoots (m^2/vine)	3.13 ± 0.09	3.40 ± 0.07	*
Total leaf area (m^2/vine)	14.38 ± 0.28	13.16 ± 0.37	*

[a] Significant differences (Signif.) are indicated: *, $p < 0.05$; ns, (not significant) according to *t*-tests.

Figure 3. Undamaged leaves (**left**) and scarred basal leaves (**right**) of field-grown Thompson seedless vines. Picture was taken at harvest during the leaf gas exchange measurements; red ovals indicate the exact point where hailstones hit the leaf during the natural hail event that occurred four weeks after full bloom.

3.3. Effect of Natural Hail on Leaf Gas Exchange

Two days after natural hail (DOY 171), the leaf assimilation rate was not affected by natural hail in the basal or medial leaves (Figure 5A–C). Readings taken on lateral leaves essentially showed similar photosynthetic activity between treatments, although the lateral leaves from HD vines were developed after the hailstorm (Figure 5C).

Figure 4. Morphology of canes of field-grown Thompson seedless vines during winter pruning with evident callus tissues after being damaged by natural hail the previous season.

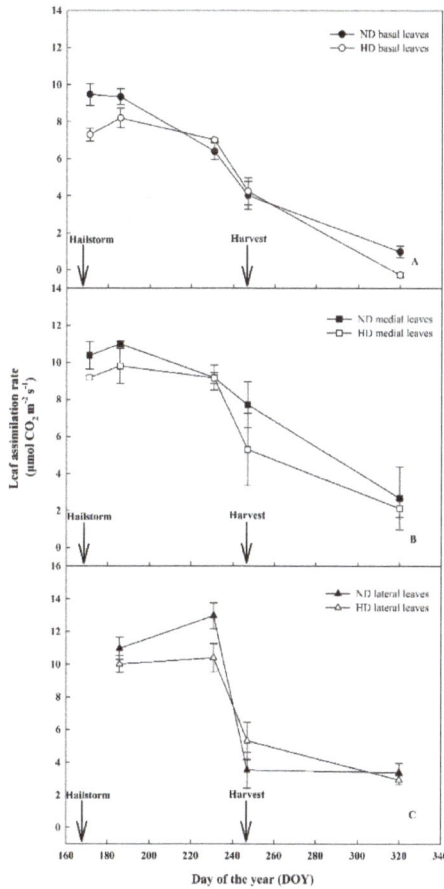

Figure 5. Seasonal changes in leaf assimilation rate (Pn) recorded in 2015 at different node positions along the main shoots: (**A**) in basal leaves, at the 3rd node; (**B**) medial leaves, at the 10th node; and the lateral shoots (**C**) (medial leaves, 3rd node) of non-hail-damaged (ND) and hail-damaged (HD) field-grown Thompson seedless vines. The hailstorm corresponded to day of the year (DOY) 169 in 2015, whereas harvest corresponded to DOY 248. Arrowheads indicate the time of hailstorm and harvest. Data are means ($n = 5$) ± standard errors represented by bars.

Even though no significant differences were recorded for E between ND and HD basal leaves (Figure 6A), the medial and lateral leaves from HD vines significantly increased E, with a maximum of 67% and 37%, respectively, during the first 20–25 days after the hailstorm (Figure 6B,C). Sixty days after the hailstorm, these HD leaves showed E values similar to ND leaves up to values of approximately 3.0 mmol H_2O m^{-2} s^{-1} (Figure 6B,C). The initial increase in transpiration of hail-injured leaves has also been documented in the literature for hail simulations [6], but in our case, it was more persistent. Perhaps due to a limited capacity of vine leaves to rapidly injure (e.g., by lignification at the edges of wounded leaf tissue) or to further abiotic stress (e.g., high-temperature stress), this abiotic stress situation was able to delay the recovery of evapotranspiration [26]. On the other hand, no significant differences of instantaneous WUE$_i$ were registered between treatments (Figure 7A–C).

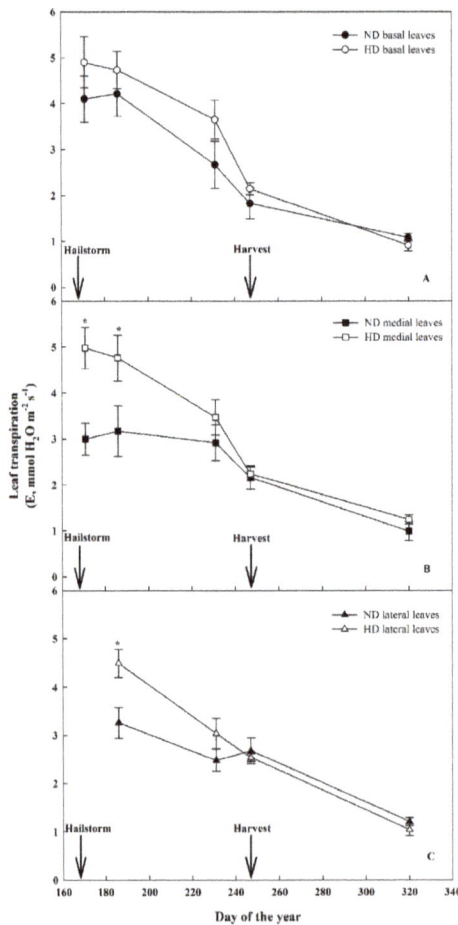

Figure 6. Seasonal changes in leaf transpiration rate (E) recorded in 2015 at different node positions along the main shoots: (**A**) in basal leaves, at the 3rd node; (**B**) medial leaves, at the 10th node; and the lateral shoots (**C**) (medial leaves, 3rd node) of non-hail-damaged (ND) and hail-damaged (HD) field-grown Thompson seedless vines. The hailstorm corresponded to day of the year (DOY) 169 in 2015, whereas harvest corresponded to DOY 248. Arrowheads indicate the time of the hailstorm and harvest. Data are means ($n = 5$) ± standard errors represented by bars. Asterisks indicate significant differences according to *t*-test, $p < 0.05$.

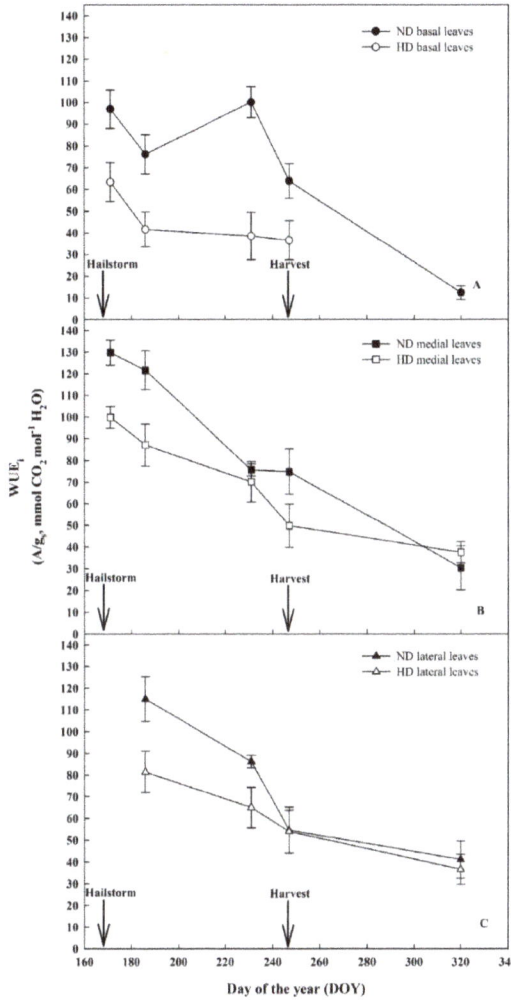

Figure 7. Seasonal changes in intrinsic water use efficiency (WUE$_i$) recorded in 2015 at different node positions along the main shoots: (**A**) in basal leaves, at the 3rd node; (**B**) medial leaves, at the 10th node; and the lateral shoots (**C**) (medial leaves, 3rd node) of non-hail-damaged (ND) and hail-damaged (HD) field-grown Thompson seedless vines. The hailstorm corresponded to day of the year (DOY) 169 in 2015, whereas harvest corresponded to DOY 248. Arrowheads indicate the time of the hailstorm and harvest. Data are means (*n* = 5) ± standard errors represented by bars.

These results are in accordance with previous works on the effects of mechanical stress caused by, for example, hailstorms [6,39,40] or an invasion of insect herbivory [41] on leaf gas exchange. As reported by Tartachnyk and Blanke [4] for mechanically hail-injured four-year-old apple trees (*Malus domestica* Borkh. cv. "Golden Delicious") and Aldea et al. [40] for soybean plants (*Glycine max* L., cv. Pioneer 93B15) damaged by Japanese beetles (*Popillia japonica*) and corn earworm caterpillars (*Helicoverpa zea* Bodie), the stress induced by hailstorms and insect herbivory, respectively, increased the transpiration but did not have a significant effect on the rate of net photosynthesis of damaged leaves.

3.4. Effect of Natural Hail on Productive Traits

Starting from a uniform bunch number per vine, the yield, bunch weight, and berry weight were affected by natural hail (Table 2). At harvest, in HD vines, the yield per vine and bunch weight were significantly reduced by about 39% and 29%, respectively, as compared with ND treatment. These reductions were attributable to a significantly lower number of undamaged berries per bunch (up to 44% less than ND), and consequently, bunch width and compactness were reduced in HD vines (Table 2).

Table 2. Yield components and bunch and berry characteristics (mean ± SE, $n = 20$) at harvest in non-hail-damaged (ND) and hail-damaged (HD) field-grown Thompson seedless vines.

Parameters	ND	HD	Signif. [a]
Bunches (number/vine)	15.0 ± 1.33	13.2 ± 2.65	ns
Yield (kg/vine)	13.10 ± 0.33	8.20 ± 0.65	*
Bunch weight (g)	874.10 ± 107.61	620.20 ± 117.79	*
Bunch length (cm)	21.8 ± 1.6	20.9 ± 1.2	ns
Bunch width (cm)	17.2 ± 0.7	14.6 ± 0.8	*
Bunch compactness index (g/(cm)^2)	1.89 ± 0.13	1.40 ± 0.19	*
Undamaged berries per bunch (n)	167.0 ± 3.2 [a]	93.0 ± 3.1	**
Undamaged berry weight (g)	5.07 ± 0.51	5.19 ± 0.61	ns
Hail-damaged berries per bunch (n)	- [†]	45.0 ± 1.8	-
Hail-damaged berry fresh weight (g)	-	3.3 ± 0.02	-
Undamaged berry length (mm)	27.11 ± 0.44	26.2 ± 0.66	ns
Undamaged berry diameter (mm)	17.88 ± 0.31	17.57 ± 0.42	ns
Force pedicel detachment (N)	1.6 ± 0.22	1.1 ± 0.12	ns

[a] Significant differences (Signif.) are indicated: *, $p < 0.05$; **, $p < 0.01$; ns, (not significant) according to t-tests. [†] Not detected.

Nevertheless, berry weight, width, diameter, and force pedicel detachment of undamaged berries from HD vines were unaffected by natural hail (Table 2), although the lack of response in terms of berry size on HD vines could simply be attributed to physiological berry growth compensation (fewer undamaged berries per bunch and a higher sink demand of undamaged berries compared with hail-damaged ones, on the same bunch, can lead to greater growth). It has to be noted that four weeks after full bloom (in concomitance with natural hail), this natural phenomenon was not effective enough to cause a severe source limitation, for example, on leaf gas exchange (Figure 5) during this stage of cell division, which in turn, could lead to a considerable reduction in final berry size. Indeed, Vasconcelos and Castagnoli [42] demonstrated that leaf removal, four weeks after bloom, had no impact on yield components, even though leaf removal can affect more or less the yield components in relation to the cultivar, severity, and time of treatment [43]. In our experimental conditions, natural hailstones were able to injure the epidermal cell of the berry and cause scarring (Figure 8), which probably led to suberification of the tissue, which in turn might have increased the weight loss of hail-damaged berries (Table 2). These hail-damaged berries should obviously be removed at harvest with appropriate clipping scissors in order to obtain marketable grapes—a process that increases the production cost of the crop.

3.5. Effect of Natural Hail on Fruit Composition

Natural hail occurring four weeks after full bloom increased TSS (+1.0 °Bx) at harvest (Table 3). The accumulation of sugar in the berries seemed to depend not only on the amount of yield (Table 2) but also the available active leaf area during the period between veraison and harvest. Indeed, during this period, HD vines presented a canopy composed of almost-young leaves, capable of fixing a sufficient amount of carbon, against ND vines composed only of old leaves for sugar accumulation in berries. This positive effect of lateral shoots on grape quality has already been demonstrated [44]. Several studies have shown that sugar accumulation is not affected [44,45] or even slightly increased [46–48] after reducing leaf area, similar to what happened in our study on HD vines (Table 1).

Figure 8. Cluster morphology of field-grown Thompson seedless HD (**left**) and ND (**right**) vines at harvest. Generally, the berries from HD vines presented unidirectional dorsal scars generated by natural hail that occurred four weeks after full bloom.

Table 3. Berry quality composition (mean ± SE, $n = 5$) at harvest in non-hail-damaged (ND) and hail-damaged (HD) field-grown Thompson seedless vines.

Parameters	ND	HD	Signif. [a]
Total soluble solids (TSS, °Bx)	20.0 ± 1.13	21.0 ± 0.85	*
Total acidity (TA, %)	0.39 ± 0.01	0.37 ± 0.02	ns
Must pH	4.27 ± 0.08	4.37 ± 0.10	ns
Maturity index (TSS/TA)	51.1 ± 0.05	57.0 ± 0.19	ns
Total phenolics (mg/kg)	1900.7 ± 39.2	975.5 ± 15.4	*

[a] Significant differences (Signif.) are indicated: *, $p < 0.05$; ns, (not significant), $p > 0.05$ according to t-tests.

Moreover, there were no significant differences in TA, juice pH, and maturity index among treatments (Table 3). These quality parameters at harvest were optimal and able to ensure a high-quality end-product, according to consumer preference [49]. The lack effect of natural hail on these qualitative parameters might be due to the minimal microclimate modifications around the cluster zone in our experimental conditions, since it is generally accepted that increasing cluster exposure to sunlight (perhaps indirectly caused by natural hail in our case, i.e., leaf gaps generated by hailstones) decreases juice acid content [46,50].

However, natural hail affected the total phenolic levels at harvest with a fold change of ~2 between HD and ND berries (Table 3). Abiotic constraints are known to exert a negative effect on secondary metabolites, such as phenolic compounds [51]. Also, in our study, a competitive relationship between primary and secondary metabolites in berries was registered (Table 3), confirming other results reported previously [52,53]. In HD berries, the impact of natural hail might have inhibited phenolic biosynthesis and/or promoted phenolic degradation. Indeed, the biosynthesis and degradation of phenolic compounds seem to be under the control of enzymes that strongly reduce their activity under source limitations [45,48].

3.6. Effect of Natural Hail on Pruning Data, Vine Balance Indices, and Carbohydrate and Nitrogen Storage in Permanent Vine Organs and Shoot Fertility

The pruning weight did not differ between treatments (Table 4). Conversely, the yield-to-pruning-weight ratio was affected by natural hail and decreased more than 37% on HD vines, while the leaf-to-fruit ratio increased by about 0.50 m^2 kg^{-1} in HD vines as compared with ND ones (Table 4).

Table 4. Vine balance indices, cane wood reserves, and bud fertility (mean ± SE, $n = 7$) in non-hail-damaged (ND) and hail-damaged (HD) Thompson seedless grapevines.

Parameters	ND	HD	Signif. [a]
Pruning weight (kg/vine)	2.99 ± 1.64	3.03 ± 0.13	ns
Yield-to-pruning-weight ratio (kg/kg)	4.38 ± 0.04	2.71 ± 0.13	**
Leaf-to-fruit ratio (m^2/kg)	1.10 ± 0.15	1.60 ± 0.09	*
Soluble solids of primary canes at 3rd internode (mg/g DW)	167.88 ± 16.98	155.79 ± 9.04	ns
Starch of primary canes at 3rd internode (mg/g DW)	40.62 ± 14.09	50.82 ± 11.58	ns
Soluble solids of primary canes at 10th internode (mg/g DW)	147.06 ± 3.63	151.66 ± 7.12	ns
Starch of primary canes at 10th internode (mg/g DW)	35.96 ± 2.49	35.58 ± 7.61	ns
Soluble solids of lateral canes at 3rd internode (mg/g DW)	119.19 ± 7.66	146.36 ± 17.95	ns
Starch of canes of lateral canes at 3rd internode (mg/g DW)	22.48 ± 10.25	39.67 ± 5.47	ns
‡ Bud fertility index	1.55 ± 0.20	2.08 ± 0.40	ns
‡ Blind buds/vine	4.00 ± 1.73	5.67 ± 2.40	ns

[a] Significant differences are indicated: *, $p < 0.05$; **, $p < 0.01$; ns, (not significant). ‡ Evaluated in spring 2016 and expressed as number of clusters per shoot per node position on the cane.

Analyses of the amount of total carbohydrates (total soluble sugars, reducing sugars, and sucrose) and insoluble (starch) carbohydrates at pruning (Figure 7) stored in the canes (primary and laterals) showed no differences between the two treatments, and consequently, no incidence to bud fertility or blind buds on HD vines was registered one year after the natural hail (Table 4). These results underline that the extent of this natural phenomenon was not large enough to cause mechanical damage to the shoots carrying the buds or latent bud differentiation interference, something that can be registered after a hailstorm [54] or during the year following defoliation, when lower vigor has been noted [55] as well as flower abscission [56] or lower bud fruitfulness. Moreover, lateral shoots were net exporters of carbohydrates, providing assimilates to support their own growth and sending the surplus to the main shoot, which contributed to berry ripening (Table 3).

4. Conclusions

The present study is the first attempt at quantifying natural hail effects on field-grown table grapes. Based on the results, the natural hailstorm caused an alteration in the source–sink relationship in Thompson seedless (*V. vinifera* L.) grapevines due to diminished leaf area induced by the hailstorm, which in turn led to lighter, less compact bunches and reduced the total phenolics of the berries. Moreover, this phenomenon implicates physiological and vegetative responses that can bring the vines to an acceptable maturity index without interfering with wood reserves or bud fertility in the following season.

Further investigation of these effects and in the presence of greater hail-damage impacts could be useful for better understanding the responses of grapevines under extreme abiotic conditions.

Author Contributions: Conceived, designed and supervised the experiment: D.G.P. Performed the analyses and collected the data: D.G.P., K.B., E.X., I.D., and. D.M. Wrote the manuscript: D.G.P. Reviewed the manuscript: D.G.P., K.B., and A.P. All authors have read and approved the manuscript.

Acknowledgments: We would like to thank the grapevine grower Georgios V. Petoumenos for hosting the vineyard trial, Evangelos Tyrlis for providing the meteorological data.

Conflicts of Interest: The authors declare no conflict of interest.

Agronomy **2019**, *9*, 197

References

1. Logothetis, B. The development of the vine and of viticulture in Greece based on *archaeological findings* in the area. In *Epistimoniki Epetiris tis Geoponikis kai Dasologikis Scholis*; University of Thessaloniki: Thessaloniki, Greece, 1970; Volume 13, pp. 167–249, (In Greek with English summary).
2. Staurakas, D.E. *Ampelography*, 2nd ed.; Ziti Press: Thessaloniki, Greece, 2015; pp. 588–594, ISBN 978-960-456-241-1.
3. Sioutas, M.; Meaden, G.; Webb, J. Hail frequency distribution and intensity in Northern Greece. *Atmos. Res.* **2007**, *93*, 526–533. [CrossRef]
4. Papagiannaki, K.; Lagouvardos, K.; Kotroni, V. A database of high-impact weather events in Greece: A descriptive impact analysis for the period 2001–2011. *Nat. Hazards Earth Syst. Sci.* **2013**, *13*, 727–736. [CrossRef]
5. Papagiannaki, K.; Lagouvardos, K.; Kotroni, V.; Papagiannakis, G. Agricultural losses related to frost events: Use of the 850 hPa level temperature as an explanatory variable of the damage cost. *Nat. Hazards Earth Syst. Sci.* **2014**, *14*, 2375–2386. [CrossRef]
6. Tartachnyk, I.; Blanke, M.M. Effect of mechanically-hail simulated on photosynthesis, dark respiration and transpiration of apple leaves. *Environ. Exp. Bot.* **2002**, *48*, 169–175. [CrossRef]
7. Wells, L.D.; MacManus, P.S. Effects of simulated hail events and subsequent fungicide applications on cranberry fruit rot incidence and yield. *Plant Dis.* **2013**, *97*, 1207–1211. [CrossRef] [PubMed]
8. Jalali, A.H. Potato (*Solanum tuberosum* L.) yield response to simulated hail damage. *Arch. Agron. Soil Sci.* **2013**, *59*, 981–987. [CrossRef]
9. Miya, S.P.; Modi, A.T.; Mabhaudhi, T. Interactive effects of simulated hail damage and plant density on maize seed quality. *Seed Sci. Technol.* **2017**, *45*, 100–111. [CrossRef]
10. Anda, A.; Burucs, Z.; Lõke, Z.; Decsi, É.K. Effects of Hail on Evapotranspiration and Plant Temperature of Maize. *J. Agron. Crop Sci.* **2002**, *188*, 335–341. [CrossRef]
11. Robertson, A.E.; Munkwold, G.P.; Hurburg, C.R.; Ensley, S. Effects of Natural Hail Damage on Ear Rots, Mycotoxins, and Grain Quality Characteristics of Corn. *Environ. Entomol.* **2011**, *103*, 1193–1199. [CrossRef]
12. Yue, Y.; Zhou, L.; Zhu, A.-X.; Ye, X. Vulnerability of cotton subjected to hail damage. *PLoS ONE* **2019**, *14*, e0210787. [CrossRef] [PubMed]
13. Baldwin, J.G.; Bleasdale, G.E.; Cadman, R.S.; Keens, I.L. Comparison of trellises and pruning levels for Sultana vines in the Murray Valley. *J. Exp. Agric. Anim. Husb.* **1979**, *19*, 634–640. [CrossRef]
14. Williams, L.E.; Ayars, J.E. Water use of Thompson Seedless grapevines as affected by the application of gibberellic acid (GA3) and trunk girdling-Practices to increase berry size. *Agric. For. Meteorol.* **2005**, *129*, 85–94. [CrossRef]
15. Warusavitharana, A.J.; Tambe, T.B.; Kshirsagar, D.B. Effect of cytokinins and brassinosteroid with gibberellic acid on yield and quality of Thompson Seedless grapes. *Acta Hortic.* **2008**, *785*, 217–224. [CrossRef]
16. Abu-Zahra, T.R. Berry size of Thompson seedless as influenced by the application of Gibberellic acid and cane girdling. *Pak. J. Bot.* **2010**, *42*, 1755–1760.
17. Webb, L.; Whiting, J.; Watt, A.; Hill, T.; Wigg, F.; Dunn, G.; Needs, S.; Barlow, E.W.R. Managing grapevines through severe heat: A survey of growers after the 2009 summer heatwave in south-eastern. *Aust. J. Wine Res.* **2010**, *21*, 147–165. [CrossRef]
18. Flexas, J.; Bota, J.; Escalona, J.M.; Sampol, B.; Medrano, H. Effects of drought on photosynthesis in grapevines under field conditions: An evaluation of stomatal and mesophyll limitations. *Funct. Plant Biol.* **2002**, *29*, 461–471. [CrossRef]
19. Mosedale, J.R.; Wilson, R.J.; Maclean, I.M.D. Climate Change and Crop Exposure to Adverse Weather: Changes to Frost Risk and Grapevine Flowering Conditions. *PLoS ONE* **2015**, *10*. [CrossRef]
20. Morales, M.; Toro, G.; Riquelme, A.; Sellés, G.; Pinto, M.; Ferreyra, R. Effect of different rootstocks on photosynthesis and nutritional response of grapevines cultivar 'Sultanina' under flooding stress. *Acta Hortic.* **2014**, *1045*, 123–131. [CrossRef]
21. Takahashi, K.; Kuranaka, M.; Miyagawa, A. The effects of wind on grapevine growth: Windbreaks for vineyards. *Bull. Shimane Agric. Exp. Stn.* **1976**, *14*, 39–83.
22. Manja, K.; Aoun, M. The use of nets for tree fruit crops and their impact on the production: A review. *Sci. Hortic.* **2019**, *246*, 110–122. [CrossRef]

23. Keller, M.; Romero, P.; Gohil, H.; Smithyman, R.P.; Riley, W.R.; Casassa, L.F.; Harbertson, J.F. Deficit irrigation alters grapevine growth, physiology, and fruit microclimate. *Am. J. Enol. Vitic.* **2016**, *67*, 426–435. [CrossRef]
24. Hendrickson, L.; Ball, M.C.; Wood, J.T.; Chow, W.S.; Furbank, R.T. Low temperature effects on photosynthesis and growth of grapevine. *Plant Cell Environ.* **2004**, *27*, 795–809. [CrossRef]
25. Petoumenou, G.D. Growth Rate, Morpho-Physiological and Ultrastructural Behaviour in Vitis Vinifera L. under Limited and Unlimited Conditions. Ph.D. Thesis, Università degli Studi di Perugia, Perugia, Italy, 2007.
26. Greer, D.H.; Weedon, M.M. The impact of high temperatures on *Vitis vinifera* cv. Semillon grapevine performance and berry ripening. *Front. Plant Sci.* **2013**, *4*, 1–9. [CrossRef]
27. Orduña, R. Climate change associated effects on grape and wine quality and production. *Food Res. Int.* **2010**, *43*, 1844–1855. [CrossRef]
28. Mori, K.; Goto-Yamamoto, N.; Kitayama, M.; Hashizume, K. Loss of anthocyanins in red-wine grape under high temperature. *J. Exp. Bot.* **2007**, *58*, 1935–1945. [CrossRef]
29. Sweetman, C.; Sadras, V.O.; Hancock, R.D.; Soole, K.L.; Ford, C.M. Metabolic effects of elevated temperature on organic acid degradation in ripening *Vitis vinifera* fruit. *J. Exp. Bot.* **2014**, *65*, 5975–5988. [CrossRef]
30. Lorentz, D.H.; Eichhorn, K.W.; Bleiholder, H.; Klose, R.; Meier, U.; Weber, E. Growth Stages of the Grapevine: Phenological growth stages of the grapevine (*Vitis vinifera* L. ssp. vinifera)-Codes and descriptions according to the extended BBCH scale. *Aust. J. Grape Wine Res.* **1995**, *1*, 100–103. [CrossRef]
31. Tello, J.; Ibáñez, J. Evaluation of indexes for the quantitative and objective estimation of grapevine bunch compactness. *Vitis* **2014**, *53*, 9–16.
32. Slinkard, K.; Singleton, V.L. Total Phenol Analysis: Automation and Comparison with Manual Methods. *Am. J. Enol. Vitic.* **1977**, *28*, 49–55.
33. Morris, D.L. Quantitative determination of carbohydrates with Dreywood's anthrone reagent. *Science* **1948**, *107*, 254–255. [CrossRef]
34. Loewus, F.A. Improvement in anthrone method for determination of carbohydrates. *Anal. Chem.* **1952**, *24*, 219. [CrossRef]
35. Fernandes, G.W.; Oki, Y.; Sales, N.M.; Quintini, A.V.; Freitas, C.; Caires, T.B. Hailstorm impact across plant taxa: Leaf fall in a mountain environment. *Neotrop. Biol. Conserv.* **2012**, *7*, 8–15.
36. Vescovo, L.; Gianelle, D.; Dalponte, M.; Miglietta, F.; Carotenuto, F.; Torresan, C. Hail defoliation assessment in corn (*Zea mays* L.) using airborne LiDAR. *Field Crop. Res.* **2016**, *196*, 426–437. [CrossRef]
37. Candolfi-Vasconcelos, M.C.; Koblet, W. Yield, fruit-quality, bud fertility and starch reserves of the wood as a function of leaf removal in *Vitis vinifera*-Evidence of compensation and stress recovering. *Vitis* **1990**, *29*, 199–221.
38. Caspari, H.W.; Lang, A.; Alspach, P. Effects of girdling and leaf removal on fruit set and vegetative growth in grape. *Am. J. Enol. Vitic.* **1998**, *49*, 359–366.
39. Tartachnyk, I.; Blanke, M.M. Temperature, evapotranspiration and primary photochemical responses of apple leaves to hail. *J. Plant Physiol.* **2008**, *165*, 1847–1852. [CrossRef]
40. Aldea, M.; Hamilton, J.G.; Resti, J.P.; Zangelr, A.R.; Byerenbaur, M.R.; Delucia, E.H. Indirect effects of insect herbivory on leaf gas exchange in soybean. *Plant Cell Environ.* **2005**, *28*, 402–411. [CrossRef]
41. Kerchev, P.I.; Fenton, B.; Foyer, C.H.; Hancock, R.D. Plant responses to insect herbivory: Interactions between photosynthesis, reactive oxygen species and hormonal signalling pathways. *Plant Cell Environ.* **2012**, *35*, 441–453. [CrossRef]
42. Vasconcelos, M.C.; Castagnoli, S. Leaf canopy structure and wine performance. *Am. J. Enol. Vitic.* **2000**, *51*, 390–396.
43. Tello, J.; Ibáñez, J. What we know about grapevine bunch compactness? A state-of-the-art review. *Aust. J. Wine Res.* **2018**, *24*, 6–23. [CrossRef]
44. Sabbatini, P.; Howell, G.S. Effects of early defoliation on yield, fruit composition, and harvest season cluster rot complex of grapevines. *HortScience* **2010**, *45*, 1804–1808. [CrossRef]
45. Pastore, C.; Zenoni, S.; Fasoli, M.; Pezzotti, M.; Tornielli, G.B.; Filippetti, I. Selective defoliation affects plant growth, fruit transcriptional ripening program and flavonoid metabolism in grapevine. *BMC Plant Biol.* **2013**, *13*, 30. [CrossRef]
46. Poni, S.; Casalini, L.; Bernizzoni, F.; Civardi, S.; Intrieri, C. Effects of early defoliation on shoot photosynthesis, yield components, and grape composition. *Am J. Enol. Vitic.* **2006**, *57*, 397–407.

47. Poni, S.; Bernizzoni, F.; Civardi, S. The effect of early leaf removal on whole-canopy gas exchange and vine performance of *Vitis vinifera* L. "Sangiovese". *Vitis* **2008**, *47*, 1–6.

48. Pastore, C.; Zenoni, S.; Tornielli, G.B.; Allegro, G.; Dal Santo, S.; Valentini, G.; Intrieri, C.; Pezzotti, M.; Filippetti, I. Increasing the source/sink ratio in *Vitis vinifera* (cv. Sangiovese) induces extensive transcriptome reprogramming and modifies berry ripening. *BMC Genomics* **2011**, *12*, 631. [CrossRef]

49. Ranganna, S. *Maturity Indeces and Quality Criteria. Grapes in Handbook of Analysis and Quality Control for Fruit and Vegetable Products*, 2nd ed.; Tata McGraw-Hill Publishing Company Ltd.: New Delhi, India, 1986; p. 171, ISBN 007-451-851-8.

50. Poni, S.; Bernizzoni, F.; Civardi, S.; Libelli, N. Effects of pre-bloom leaf removal on growth of berry tissues and must composition in two red *Vitis vinifera* L. cultivars. *Aust. J. Grape Wine Res.* **2009**, *15*, 185–193. [CrossRef]

51. Mirás-Avalos, J.M.; Intrigliolo, D.S. Grape Composition under Abiotic Constrains: Water Stress and Salinity. *Front. Plant Sci.* **2017**, *8*, 851. [CrossRef]

52. Sadras, V.O.; Moran, M.A. Elevated temperature decouples anthocyanins and sugars in berries of Shiraz and Cabernet Franc. *Aust. J. Grape Wine Res.* **2012**, *18*, 115–122. [CrossRef]

53. Wu, B.H.; Niu, N.; Li, J.H.; Li, S.H. Leaf:fruit ratio affects the proteomic profile of grape berry skins. *J. Am. Soc. Hortic. Sci.* **2013**, *138*, 416–427. [CrossRef]

54. May, P. The value of an estimate of fruiting potential in the 'Sultana'. *Vitis* **1961**, *3*, 15–26.

55. Palliotti, A.; Gardi, T.; Berrios, J.G.; Civardi, S.; Poni, S. Early source limitation as a tool for yield control and wine quality improvement in a high-yielding red *Vitis vinifera* L. cultivar. *Sci. Hortic.* **2012**, *145*, 10–16. [CrossRef]

56. Lohitnavy, N.; Bastian, S.; Collins, C. Early leaf removal increases flower abscission in *Vitis vinifera*, Semillon. *Vitis* **2010**, *49*, 51–53.

![agronomy logo] *agronomy*

MDPI

Article

Grapevine Phenology of cv. *Touriga Franca* and *Touriga Nacional* in the Douro Wine Region: Modelling and Climate Change Projections

Ricardo Costa [1], Helder Fraga [1], André Fonseca [1], Iñaki García de Cortázar-Atauri [2], Maria C. Val [3], Cristina Carlos [3], Samuel Reis [1] and João A. Santos [1,*]

[1] Centre for the Research and Technology of Agro-Environmental and Biological Sciences, CITAB, Universidade de Trás-os-Montes e Alto Douro, UTAD, 5000-801 Vila Real, Portugal; rjncosta@utad.pt (R.C.); hfraga@utad.pt (H.F.); andref@utad.pt (A.F.); sreis_1992@hotmail.com (S.R.)

[2] INRA, US1116 AgroClim, 84914 Avignon, France; inaki.garciadecortazar@inra.fr

[3] ADVID, Associação para o Desenvolvimento da Viticultura Duriense, Parque de Ciência e Tecnologia de Vila Real–Régia Douro Park, 5000-033 Vila Real, Portugal; carmo.val@advid.pt (M.C.V.); cristina.carlos@advid.pt (C.C.)

* Correspondence: jsantos@utad.pt; Tel.: +351-259350729

Received: 3 April 2019; Accepted: 23 April 2019; Published: 25 April 2019

Abstract: Projections of grapevine phenophases under future climate change scenarios are strategic decision support tools for viticulturists and wine producers. Several phenological models are tested for budburst, flowering, and veraison and for two main grapevine varieties (cv. *Touriga Franca* and *Touriga Nacional*) growing in the Douro Demarcated Region. Four forcing models (Growing degree-days, Richardson, Sigmoid, and Wang) and three dormancy models (Bidabe, Smoothed Utah and Chuine), with different parameterizations and combinations, are used. New datasets, combing phenology with weather station data, widespread over the Douro wine region, were used for this purpose. The eight best performing models and parameterizations were selected for each phenophase and variety, based on performance metrics. For both cultivars, results revealed moderate performances ($0.4 < R^2 < 0.7$) for budburst, while high performances ($R^2 > 0.7$) were found for flowering and veraison, particularly when Growing degree-days or Sigmoid models are used, respectively. Climate change projections were based on a two-member climate model ensemble from the EURO-CORDEX project under RCP4.5. Projections depicted an anticipation of phenophase timings by 6, 8 or 10–12 days until the end of the century for budburst, flowering, and veraison, respectively. The inter-model variability is of approximately 2–4 days for flowering and veraison and 4–6 days for budburst. These results establish grounds for the implementation of a decision support system for monitoring and short-term prediction of grapevine phenology, thus promoting a more efficient viticulture.

Keywords: grapevine; phenology; phenology modelling platform; *Touriga Franca*; *Touriga Nacional*; climate change; RCP4.5; EURO-CORDEX; Douro wine region; Portugal

1. Introduction

Viticulture is one of the most important socioeconomic sectors in Portugal. From a global perspective, Portugal is currently the 11th-largest wine producer and the 9th wine exporter in the world [1]. Portugal accounts for approximately 194,000 ha of vineyard area and with a wine production around 6.7×10^6 hL [1]. Almost half of this production is exported, accounting for nearly 2% of national export income [2]. The country is divided into 14 viticultural regions (12 in the mainland and 2 in the islands), with 25 Denominations of Origin (DOs). Douro/Porto, Vinhos-Verdes, Alentejo, and Lisboa are the main regions in terms of wine production. Although the Alentejo wine region is the most productive [3], the Douro/Porto wine region has the largest vineyard area and is the main region in terms of production, accounting for ca. 25% of all wine produced in Portugal [2].

The Douro/Porto wine region, also known as Douro Demarcated Region (DDR), has unique features, where vineyards are traditionally grown on terraces over steep slopes along the mountainous Douro valley and in an almost monocultural system, resulting in a UNESCO (United Nations Educational, Scientific and Cultural Organization) human heritage vineyard landscape. A large number of grapevine varieties (*Vitis vinifera* L.) are grown in the region, being many of them native. The complex physiographic and pedoclimatic characteristics of the DDR not only provide a wide range of terroirs for grapevine growth [4], but also make wine production very costly in terms of labour force and vineyard management [5]. Given the heterogeneous conditions in which grapevines are grown within the DDR, it is imperative to better understand the climatic influences on grapevine development in order to improve vineyard management in such a diverse region. In effect, monitoring the phenological phases and timings are some of the most useful information for planning the necessary viticultural practices in each developmental stage. This knowledge may promote more adjusted cultural practices, eventually optimizing production and quality at harvest.

It is known that grapevine development and growth are mainly driven by meteorological and climatic factors [6]. Therefore, factors such as temperature, solar radiation and precipitation may have a great influence on phenology, yields, and grape berry quality and attributes [7,8]. In particular, temperature is a fundamental factor in controlling grapevine phenological development and ripening [9–13]. Phenology models have been developed for a wide range of species, including grapevines, and using observational data from many different countries and regions [14]. Several previous studies have been performed in order to model the main phenological stages of grapevines, e.g., enabling a classification of varieties for technical proposes [15], to predict phenophases or to assess the impact of climate change on phenology [16].

Many different models have been developed during the last 60 years to simulate grapevine phenology. The modelling of budbreak was explored by several authors assuming different hypothesis [17–22]: taking or not into account the dormancy period, interactions between dormancy and post-dormancy periods, starting at a fixed date (e.g., January 1st, February 15th, March 15th). These models were tested for many different varieties under different climatic conditions. For flowering and veraison, the conventional linear growing degree-days model (GDD) was initially proposed by Amerine and Winkler [23]. This simple temperature accumulation model uses daily mean temperatures, with a fixed base temperature of 10 °C, and onsets temperature accumulation at April 1st. Others models have been calibrated and tested to simulate flowering and veraison, such as the linear model proposed by Duchene, et al. [24], with a formulation similar to the Winkler model, and the curvilinear model of Wang and Engel, later adapted by García de Cortázar-Atauri, et al. [25]. The latter model was improved by incorporating an optimal temperature (25 to 30 °C) and an upper critical threshold temperature (40 °C). The GFV model (Grapevine Flowering and Veraison model) was proposed by Parker, et al. [26] and Parker, et al. [27]. This model is similar to GDD, but with a 0 °C base temperature and starting temperature accumulation on March 1st. This last model was developed to classify different grapevine varieties within them. Finally, Molitor, et al. [28] developed a high-resolution model to simulate all the phenological stages described in the BBCH (*Biologische Bundesanstalt, Bundessortenamt und CHemische Industrie*) scale using a capped model. As such, the performance of several phenology models should be tested at a regional level, as there is no best model for all varieties and conditions.

Phenological models may also be coupled with climate change scenarios to project the potential impacts of future climates on phenological timings and grapevine development [21,24,25,29–33]. Projected increases in temperature are expected to drive earlier development stages and, as a result, a general advancement of grapevine phenology [24,34–37]. Those changes will affect the ripening period, commonly occurring in the warmest part of summer, and may lead to aromas and acidity losses [38]. Consequently, currently planted varieties, which are grown under quite specific conditions today, may no longer thrive in the same place under modified environmental conditions in the future [39]. Therefore, it is critical to understand to what extent temperature may influence the timings of the reproductive and vegetative cycles, as well as identifying varietal differences in phenology and maturity [26]. While several studies have assessed the phenological timings in many regions worldwide, a comprehensive study of phenology in the DDR was not undertaken so far, due to the overall lack of phenological and meteorological data. Hence, the present study attempts to overcome this drawback, by presenting the first grapevine phenological study for the DDR.

Along the previous lines, the aims of the present study are twofold: 1) to identify suitable phenological models for three main phenological timings (budburst, flowering, and veraison) of two leading grapevine varieties (cv. *Touriga Franca* and cv. *Touriga Nacional*) currently grown in the DDR; 2) to assess the climate change impacts on these phenological timings and varieties. The datasets and methodologies are described in Section 2. The results are presented in Section 3. Section 4 will provide a discussion and a summary of the main conclusions.

2. Materials and Methods

2.1. Study Area and Datasets

The present study targets the Douro/Porto Wine Region (DDR), located in Northern Portugal (Figure 1a). The DDR is characterized by Mediterranean-type climates, with typically dry and warm summers, though a wide range of mesoclimates and microclimates do exist throughout the region, mostly driven by the multifaceted orography and distance to the North Atlantic. Important east–west gradients in both temperatures and precipitation can be found within the region, and are typified by its 3 sub-regions: from *Baixo-Corgo* (westernmost sector) to *Cima-Corgo* (middle sector) and to *Douro Superior* (easternmost sector) (Figure 1b). This spatial variability is further enhanced by important differences in soil properties, though soils are mostly originated from schist bedrocks [40].

In the present study, phenological data recorded at 12 vineyards ("Quintas") in the DDR, with data available over the period 2014–2017 (4 years), were used (Figure 1b). The phenological timings were recorded through observations based on the BBCH scale adapted to grapevines and when at least 50% of a pre-defined set of homogenous plants reach the corresponding stage. Budburst (BBCH 07), flowering (BBCH 65) and veraison (BBCH 81) were considered herein [41]. Data for two main grapevines varieties grown in Portugal were selected: cv. *Touriga Franca* (TF) and cv. *Touriga Nacional* (TN). As information for other varieties (cv. *Tempranillo*, cv. *Tinta Amarela* and cv. *Moscatel-Galego*) was indeed very limited and inconsistent among vineyards, they were not taken into consideration. Furthermore, in order to enhance the statistical robustness of the selected models for a given grapevine variety, the phenological timings for the same variety for all sites (and terroirs) were considered for model calibration.

Owing to the relatively small sample size of the DDR dataset for model calibration (12 sites × 4 years = 48 records), the same phenological timings for TF and TN from two other Portuguese wine regions were also used: Lisboa (central-west) and Vinhos Verdes (north-west) (Figure 1a). Phenological records were available for the following locations (Table 1): "Dois Portos—DP" (Lisboa Wine Region), with 20 years of data within the full period of 1990–2014, and "*Estação Vitivinícola Amândio Galhano—EVAG*" (Vinhos Verdes Wine Region), with records over the period of 2005–2009 (5 years). Therefore, the total sample size is extended to 73 records for each phenological timing and variety, thus enabling a higher statistical robustness in model calibration. A summary of the entire phenology dataset is provided in Table 1.

For the DDR, daily mean temperature records from the meteorological stations installed at the different vineyard sites and over the period of 2014–2017 (4 years) were used (Figure 1b). For the other two wine regions (Vinhos Verdes-EVAG and Lisboa-DP), the local temperature time series were estimated from the nearest-point of a very-high resolution (~1 km) gridded daily temperature dataset for Portugal [42]. Phenology and temperature data for the Douro Wine Region were provided by the "*Associação para o Desenvolvimento da Viticultura Duriense—ADVID*", after collecting information among their members, while phenology data for Lisboa-DP and Vinhos Verdes-EVAG were provided by the *"Instituto Nacional de Investigação Agrária e Veterinária"*, INIAV, and by EVAG, respectively.

Table 1. Available time periods of the phenological data used in the present study (budburst, flowering and veraison dates) by wine region (Douro, Lisboa and Vinhos-Verdes) and for each grapevine variety (cv. *Touriga Franca* and cv. *Touriga Nacional*).

Wine Region	Latitude (°N)	Longitude (°W)	Altitude (meters)	Variety (cv.)	Available Time Period		
					Budburst [1]	Flowering [2]	Veraison [3]
Douro/Porto	41°14′24″ N–41°02′24″ N	7°01′48″ W–°47′24″ W	85–588	*Touriga Franca*	2014–2017	2014–2017	2014–2017
Douro/Porto	41°14′24″ N–41°02′24″ N	7°01′48″ W–7°47′24″ W	85–588	*Touriga Nacional*	2014–2017	2014–2017	2014–2017
Lisboa	39°02′24″ N	9°10′48″ W	85	*Touriga Franca*	1995–2014	1995–2014	1995–2014
Lisboa	39°02′24″ N	9°10′48″ W	85	*Touriga Nacional*	1990–2000; 2006–2014	1990–2000; 2006–2014	1990–2000; 2006–2014
Vinhos Verdes	41°48′36″ N	8°24′36″ W	70	*Touriga Franca*	2005–2009	2005–2009	2005–2009
Vinhos Verdes	41°48′36″ N	8°24′36″ W	70	*Touriga Nacional*	2005–2009	2005–2009	2005–2009

[1] BBCH 07, [2] BBCH 65, [3] BBCH 81.

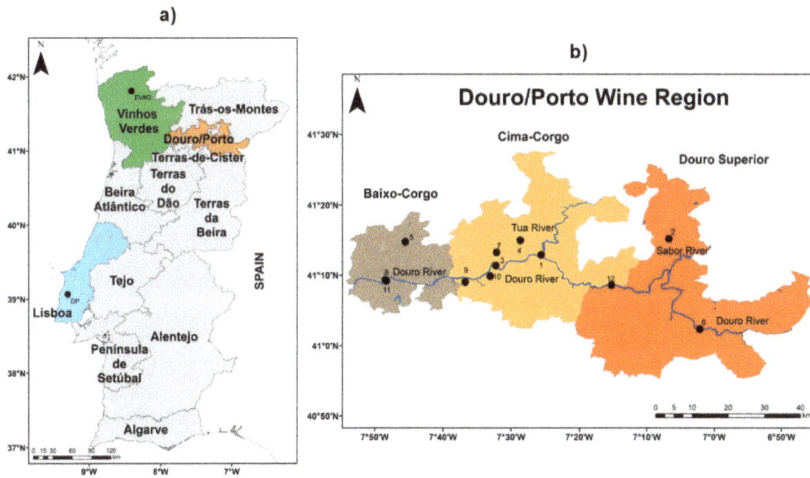

Figure 1. (**a**) Map of mainland Portugal with wine regions' boundaries. The three wine regions providing phenological data to the present study are grey shaded (Vinhos-Verdes, Lisboa, and Douro/Porto). The location of the vineyard sites EVAG, in Vinhos-Verdes wine region, and DP, in Lisboa wine region, are also outlined. (**b**) Douro/Porto wine region and its sub-regions (*Baixo-Corgo*, *Cima-Corgo*, and *Douro Superior*), along with the geographical locations of the 12 vineyard sites. The main rivers are also plotted, including the Douro River.

2.2. Phenological Models

Phenological models were selected based on previous studies, which have already shown moderate-to-high performance in simulating grapevine phenological timings in other regions [19,26,43]. Two main groups of phenological models were used: chilling models (Bidabe, Chuine, and Smoothed Utah), used to simulate the dormancy phase; and the forcing models (GDD, Richardson, Sigmoid, and Wang) applied to simulate thermal accumulation during the three phenophases: budburst, flowering, and veraison (Table 2).

In the case of budburst, two approaches were tested: sequential models that allow integrating a dormancy phase (described using chilling models) and models without dormancy period (starting at a fixed date—January 1st—and only using forcing models). For sequential models, dormancy was herein calculated starting from August 1st of the previous year, as proposed by García de Cortázar-Atauri, Brisson and Gaudillere [19]. For flowering and veraison only the forcing models were applied. Table 2 describes the tested models and their relevant bibliographic references, while Table 3 describes their corresponding mathematical formulations and tested parameters.

Table 2. Phenological models used in the present study, along with their relevant bibliographic references and pre-defined starting dates.

Type of Model	Model Name	Reference	Starting Date
Chilling model–Dormancy	Bidabe	[44]	
	Chuine	[45]	August 1st of previous year
	Smoothed Utah (SU)	[46]	
Forcing model, Post-Dormancy, Flowering, Veraison	Growing degree-days (GDD)	[47]	
	Richardson	[48]	January 1st, previous
	Sigmoid	[49]	phenological stage
	Wang	[50]	

Table 3. Grapevine phenology models selected in the present study, accompanied by their transfer function equations as a function of daily mean (T_d), maximum (T_{max}) and minimum (T_{min}) temperatures. The corresponding parameters are also outlined.

Model	Equation	Parameters *
Bidabe	$f_{Bidabe} = Q_{10}^{-T_{min}} + Q_{10}^{-T_{max}}$	$Q_{10} \in [0, 5]$
Chuine	$f_{Chuine} = 1/(1 + \exp[a(T_d - c)^2 + b(T_d - c)])$	a, b, c
GDD	$f_{GDD} = Max(0; T_d - T_b)$	T_b
Richardson	$f_{Richardson} = Max(Min(T_d - T_{low}, T_{high} - T_{low}), 0)$	T_{low}, T_{high}
Sigmoid	$f_{Sigmoid} = 1/(1 + \exp[e(T_d - d)])$	e, d
Smoothed Utah	$f_{Smoothed\ Utah} = \dfrac{1}{1+e^{-4\frac{T_d-T_{m1}}{T_{opt}-T_{m1}}}}$; If $T_d > T_{m1}$; $\dfrac{0.5\,(T_d-T_{opt})^2}{(T_{m1}-T_{opt})^2}$; If $T_{m1} < T_d < T_{opt}$; $1-(1-min)\dfrac{(T_d-T_{opt})^2}{2(T_{n2}-T_{opt})^2}$; If $T_{opt} < T_d < T_{n2}$; $min + \dfrac{1-min}{1+e^{-4\frac{T_{n2}-T_d}{T_{n2}-T_{opt}}}}$; If $T_{n2} < T_d$	$T_{m1}, T_{opt}, T_{n2}, min$
Wang	$f_{Wang} = Max[(2(T_d - T_m)^\alpha (T_{opt} - T_m)^\alpha - \left(\frac{T_d-T_m}{T_{opt}-T_m}\right)^{2\alpha}, 0]$; with $\alpha = \ln(2)/\ln\left(\frac{T_M-T_m}{T_{opt}-T_m}\right)$	T_m, T_M, T_{opt}

* T_{opt}—optimum temperature; T_b—base temperature; Q_{10}—Bidabe parameter; a, b, c—Chuine constant parameters; T_{low}—Richardson lower plateau temperature; T_{high}—Richardson upper plateau temperature; e, d—Sigmoid constant parameters; T_{m1}, T_{n2}, min—Smoothed Utah parameters; T_m—Wang lower threshold temperature; T_M—Wang upper threshold temperature.

For all these models, the parameters were fixed at specific thresholds, some of them defined in the literature taking into account previous research: $T_m = 0\ °C$, $T_M = 40\ °C$ and $T_{opt} = 25–26\ °C$ for Wang, $T_{low} = 5\ °C$ and $T_{high} = 20–25\ °C$ for Richardson, and $T_b = 0\ °C$ for GDD [19,20,26,43]. Besides these default models, the freely-adjusted best-fit parameters for each of these four models were also selected in the present study, as their performances can be significantly better than using the default parameters. The best eight performing models for each phenological phase were selected based on the performance metrics explained below. The models were considered for each of the three phenophases: dormancy to budburst (D-B), budburst to flowering (B-F), and flowering to veraison (F-V). The calibration of different models was performed independently for each phenophase and for the two selected grapevines varieties (TF and TN).

2.3. Modelling Tools and Performance Verification

In this study, the Phenological Modelling Platform, PMP [14], was used to test and calibrate different phenological models applied to grapevines. PMP is an intuitive platform that allows users to apply and test different phenological models for several purposes. In this platform, climatic and phenological data were used in order to calibrate the models, i.e., estimate the best-fit model parameters, for a specific location and a pre-defined time period. PMP estimates best-fit parameters following an iterative optimization procedure, based on the simulated annealing algorithm of Metropolis, et al. [51].

For each specific location, the input data used were daily mean (T_d), maximum (T_{max}) and minimum (T_{min}) temperatures, latitude and observed phenological timings for budburst, flowering, and veraison (in days of the year, DOY). The model performance is assessed herein by four commonly used metrics: the determination coefficient (R^2), root-mean-square error (RMSE), Nash–Sutcliffe coefficient of efficiency (EF) and the Akaike Information Criterion (AIC). Their definitions are as follows:

$$RMSE = \sqrt{\frac{\sum_{i=1}^{n}(O_i - P_i)^2}{n}}, \tag{1}$$

$$EF = 1 - \frac{\sum_{i=1}^{n}(O_i - P_i)^2}{\sum_{i=1}^{n}\left(O_i - \overline{O}_i\right)^2}, \tag{2}$$

$$AIC = n \times \ln\left(\frac{\sum_{i=1}^{n}(O_i - P_i)^2}{n}\right) + 2k + \left(\frac{2k(k+1)}{n-k-1}\right) \tag{3}$$

where O_i represents the observed phenological dates, $\overline{O}_i$ represents the average observed value, P_i represents simulated phenological dates, n is the sample size, and k is the number of parameters.

EF varies between $-\infty$ and +1, +1 corresponds to a perfect fit, 0 corresponds to the performance of the null model (average), while a negative value corresponds to a worse prediction than the null model. The *AIC* takes into account the number of model parameters, with the lowest value corresponding to a model that represents the highest variance ratio in the observed dataset with the fewest parameters (parsimony principle). Further, in order to take into consideration model overfitting, a leave-one-out cross-validation scheme was applied to every simulation, following the same methodology as described by Chuine, et al. [52], and the RMSE metric was accordingly adapted to Root Mean Square Error of Prediction (RMSEP), defined as:

$$RMSEP = \sqrt{\frac{\sum_{i=1}^{n}\left(O_i - \hat{P}_i\right)^2}{n}} \tag{4}$$

where $\hat{P}_i$ is the predicted value obtained from a leave-one-out cross-validation approach [53].

2.4. Future Climate Projections

PMP also allows running simulations using climatic data generated from different anthropogenic forcing scenarios and climate model experiments, applying the previously calibrated phenology models. For the simulations under future climates, the daily temperature time series generated by a two-member ensemble of climate model chains (Table 4), produced within the framework of the EURO-CORDEX (Coordinated Downscaling Experiment—European Domain) project, were used [54]. A distribution-based scaling method was applied as a bias correction method by EURO-CORDEX [55]. The nearest-point time series for each vineyard site was extracted from the original grid (~12 km spacing). Data for the future period of 2020–2100, under the Representative Concentration Pathway 4.5 (RCP4.5), was selected. This scenario corresponds to an intermediate/moderate anthropogenic (greenhouse gas) forcing and corresponds to an increase of the global mean temperature slightly above 2 °C, thus marginally fulfilling the Paris Agreements.

Table 4. List of the two EURO-CORDEX climate model chains selected for the present study. For each model chain the global climate model (GCM) and the regional climate model (RCM) are outlined, along with their abbreviation and relevant bibliographic reference.

GCM	RCM	Abbreviation
CNRM-CERFACS-CNRM-CM5	SMHI-RCA4	CNRMSMHI
MPI-M-MPI-ESM-LR	CLMcom-CCLM4-8-17	MPICLM

Although the model calibration was undertaken using data from three wine regions, which enabled an improvement of the statistical robustness of the model fits, the future projections were only developed for the Douro wine region, as this study was never carried out previously. The outputs for the eight selected phenology models and for each phenological phase were applied to each climate model output, thus obtaining 12 sites × 3 phases × 8 phenology models × 2 climate models (576 simulations). In the future simulations, the D-B phase started at August 1st (with dormancy) or January 1st (without dormancy, Sigmoid model), as previously described. As in the observational period, for the other two phenophases (B-F and F-V) the starting date was set at the previous phenological stage (budburst or flowering), but now using the average time series of simulated budburst or flowering dates by the corresponding eight phenology models.

3. Results

3.1. Model Performance Verification

As previously mentioned, the sample size of the observed phenology dataset used for model calibration is of 73 records for each phenological timing and variety. For TF, the minimum/mean/maximum DOY is 61/77/98 for budburst (from March 2nd to April 8th), 112/138/158 for flowering (from April 22nd to June 7th) and 177/201/222 for veraison (from June 26th to August 10th). For TN, the observed minimum/mean/maximum DOY is 64/78/100 (from March 5th to April 10th) for budburst, 112/140/158 for flowering (from April 22nd to June 7th) and 170/213/235 for veraison (from June 19th to August 23rd). Despite the large spread in the dates of each phenophase, these results show that phenophase timings are commonly earlier for TF than for TN, particularly for veraison. In addition, the correlation coefficients between the time series for DP and EVAG, in their overlapping period (2005–2009), are very high (>0.80), thus showing a strong coherency in the inter-annual variability of the phenophase timings, despite the regional differences. No correlations can be computed with the DDR, as only 2014 overlaps with DP.

The best eight models for each corresponding phenophase are shown in Table 5; Table 6 for TF and TN, respectively. For the D-B phenophase, seven sequential models and one forcing model were chosen. The results show that for D-B, the RMSEP is of 6–7 days for TF, while it reaches values of almost 8 days for TN (Tables 5 and 6). The EF varies from 0.35 to 0.41 (TF) or from 0.15 to 0.23 (TN). The corresponding coefficients of determination (R^2) show values ranging from 0.54 to 0.64 (TF) or from 0.41 to 0.48 (TN). It is important to highlight that most of the models chosen present different sets of parameters, showing very different temperature responses. As the sample size of observed data does not allow choosing a specific model with high statistical confidence, it is important to maintain this diversity in order to cover a wide range of possibilities. Further, there is no clear difference between the experiments with or without dormancy simulation. In fact, the use of a sequential model does not seem to improve over the single-model even if the number of model parameters is higher in the sequential-model [19]. The AIC values hint at this conclusion, revealing higher values in the sequential models compared to the Sigmoid model. Overall, the model performances for D-B are thus moderate and the best performances are obtained from the Sigmoid model, for both TF and TN, or, alternatively, from the Bidabe + Wang (TF) or Bidabe + GDD (TN) sequential models.

In the case of B-F (Tables 5 and 6), the RMSEP is of 5–6 days (TF and TN), i.e., less one day than for the D-B phase. The EF also presents much higher values than in the previous phenophase, between 0.60 to 0.70 in TF and from 0.61 to 0.69 in TN. The R^2 are also significantly higher, varying from 0.78 to 0.87 (0.78 to 0.84) in TF (TN). The AIC values are within the range 159–172 (170–180) in TF (TN). Among the three phenophases, the best results are achieved for F-V (Tables 5 and 6), with RMSEP of 4–5 days in TF and 5–6 days in TN. The EF varies from 0.76 to 0.84 in TF and from 0.83 to 0.86 in TN. The AIC shows relatively low values, from 141 to 158, in TF, and from 151 to 165, in TN.

The errors between simulated and observed dates for TF, over the period from 1995 to 2017, show relatively high values for D-B (up to ca. 20 days), while they are mostly lower than 10 days for B-F and F-V (Figure 2). As expected, the errors are much more scattered for D-B than for the other two phenophases, with an error compensation between the periods of 1998–2003 and 2005–2009. There is also some non-stationarity in errors, which may be attributed to non-homogeneities in phenophase and/or climate data. Similar considerations can be made for TN (Figure S1).

Table 5. List of the selected phenological models for cv. *Touriga Franca* and for each phenophase (D-B, B-F, and F-V), accompanied by the corresponding model performance parameters (Root Mean Square Error of Prediction (RMSEP), Nash–Sutcliffe coefficient of efficiency (EF), determination coefficient (R^2), and Akaike Information Criterion (AIC)). Models are sorted according to their model performance based on AIC.

Phenophase	Model	Description	RMSEP	EF	R^2	AIC
		With dormancy				
Dormancy-Budburst	Bidabe + Wang	$Q_{10} = 0.9, T_m = 5.5, T_M = 34.3, T_{opt} = 14.7$	6.4	0.36	0.62	185.1
	Bidabe + Sigmoid	$Q_{10} = 1.0, d = -22.4, e = 7.4$	6.5	0.35	0.58	186.5
	SU + Sigmoid	$T_{m1} = -13.4, T_{opt} = 29.2, T_{n2} = 44.6, min = -0.5, d = -39.9, e = 7.5$	6.2	0.41	0.64	187.4
	SU + GDD	$T_{m1} = -2.8, T_{opt} = 29.2, T_{n2} = 36.6, min = -0.3, T_b = 0$	6.5	0.35	0.59	190.1
	SU + Wang	$T_{m1} = -32.2, T_{opt} = 29.3, T_{n2} = 35.7, min = -0.3, T_m = 5.3, T_M = 34.8, T_{opt} = 15.1$	6.4	0.37	0.63	190.4
	Chuine + Sigmoid	$a = 0.3, b = -21.1, c = -2.5, d = -40, e = 7.5$	6.4	0.38	0.57	192.3
	Bidabe + Richardson	$Q_{10} = 0.9, T_{low} = 0.2, T_{high} = 43.2$	6.5	0.35	0.54	201.4
		Without dormancy				
	Sigmoid	$d = -40, e = 7.5$	6.4	0.38	0.61	181.6
	GDD	$T_b = 6.6$	5.1	0.69	0.85	159.1
	Sigmoid	$d = -0.2, e = 16.9$	5.1	0.70	0.86	159.4
	Richardson	$T_{low} = 6.5, T_{high} = 36.9$	5.1	0.69	0.85	161.1
Budburst-Flowering	Wang	$T_m = 0, T_M = 40, T_{opt} = 29.1$	5.1	0.69	0.86	161.9
	Richardson	$T_{low} = 5, T_{high} = 25$	5.2	0.68	0.84	162.6
	Richardson	$T_{low} = 5, T_{high} = 20$	5.3	0.67	0.83	164.1
	GDD	$T_b = 0$	5.9	0.60	0.78	171.1
	Wang	$T_m = 0, T_M = 31.6, T_{opt} = 25.5$	5.6	0.62	0.87	172.3
	Sigmoid	$d = 0.0, e = 13.0$	4.3	0.83	0.92	141.1
	Richardson	$T_{low} = 0.0, T_{high} = 21.5$	4.2	0.83	0.91	142.1
Flowering-Veraison	Wang	$T_m = 0, T_M = 40, T_{opt} = 25.3$	4.2	0.84	0.92	142.6
	Wang	$T_m = 0, T_M = 40, T_{opt} = 25$	4.2	0.84	0.92	143.1
	Wang	$T_m = 0, T_M = 40, T_{opt} = 26$	4.2	0.83	0.91	144.2
	Richardson	$T_{low} = 5, T_{high} = 20$	4.3	0.82	0.91	144.9
	Wang	$T_m = 0, T_M = 36.6, T_{opt} = 25.6$	4.4	0.82	0.91	148.9
	GDD	$T_b = 0$	5.1	0.76	0.90	157.6

Table 6. The same as Table 5, but for cv. *Touriga Nacional*.

Phenophase	Model	Description	Model Performance			
			RMSEP	EF	R^2	AIC
		With dormancy				
Dormancy-Budburst	Bidabe + GDD	$Q_{10} = 0.9$, $T_b = 0$	7.8	0.18	0.43	190.5
	Bidabe + Wang	$Q_{10} = 0.7$, $T_m = -36.7$, $T_M = 34$, $T_{opt} = 18$	7.7	0.20	0.44	193.7
	Bidabe + Richardson	$Q_{10} = 0.9$, $T_{low} = 0$, $T_{high} = 37.3$	7.8	0.18	0.42	193.9
	SU + Sigmoid	$T_{m1} = -34.1$, $T_{opt} = 29.9$, $T_{n2} = 43$, $min = -0.2$, $d = -39.7$, $e = 7.5$	7.6	0.23	0.48	196.0
	SU + GDD	$T_{m1} = -31.1$, $T_{opt} = 29.5$, $T_{n2} = 32.1$, $min = -0.1$, $T_b = 0$	8.0	0.15	0.41	197.8
	Bidabe + Sigmoid	$Q_{10} = 1$, $d = -38.3$, $e = 7.5$	7.8	0.19	0.41	197.9
	SU + Wang	$T_{m1} = -31.1$, $T_{opt} = 30$, $T_{n2} = 32.2$, $min = -0.4$, $T_m = -16.1$, $T_M = 25.1$, $T_{opt} = 16.4$	7.9	0.17	0.42	201.0
		Without dormancy				
Budburst-Flowering	Sigmoid	$d = -40$, $e = 7.5$	7.2	0.20	0.42	190.9
	GDD	$T_b = 7.0$	5.4	0.68	0.83	169.6
	Sigmoid	$d = -0.3$, $e = 14.4$	5.4	0.69	0.84	170.5
	Richardson	$T_{low} = 7.0$, $T_{high} = 20.8$	5.4	0.69	0.83	171.1
	Wang	$T_m = 0$, $T_M = 40$, $T_{opt} = 29.3$	5.3	0.69	0.83	172.5
	Richardson	$T_{low} = 5$, $T_{high} = 25$	5.7	0.67	0.82	173.6
	Richardson	$T_{low} = 5$, $T_{high} = 20$	5.8	0.67	0.82	173.7
	Wang	$T_m = 0$, $T_M = 26.5$, $T_{opt} = 21.8$	5.3	0.66	0.84	177.3
	GDD	$T_b = 0$	6.8	0.61	0.78	180.1
Flowering-Veraison	Sigmoid	$d = -0.4$, $e = 14.4$	6.0	0.85	0.93	151.3
	Richardson	$T_{low} = 0$, $T_{high} = 23.7$	5.6	0.86	0.93	151.6
	GDD	$T_b = 0$	5.6	0.85	0.92	154.8
	Wang	$T_m = 0$, $T_M = 40$, $T_{opt} = 26.9$	5.7	0.86	0.93	155.7
	Wang	$T_m = 0$, $T_M = 40$, $T_{opt} = 25$	6.2	0.85	0.91	161.0
	Richardson	$T_{low} = 5$, $T_{high} = 20$	6.1	0.83	0.92	161.7
	Richardson	$T_{low} = 5$, $T_{high} = 25$	5.9	0.83	0.91	161.8
	Wang	$T_m = 0$, $T_M = 43.2$, $T_{opt} = 29.9$	5.9	0.83	0.91	164.9

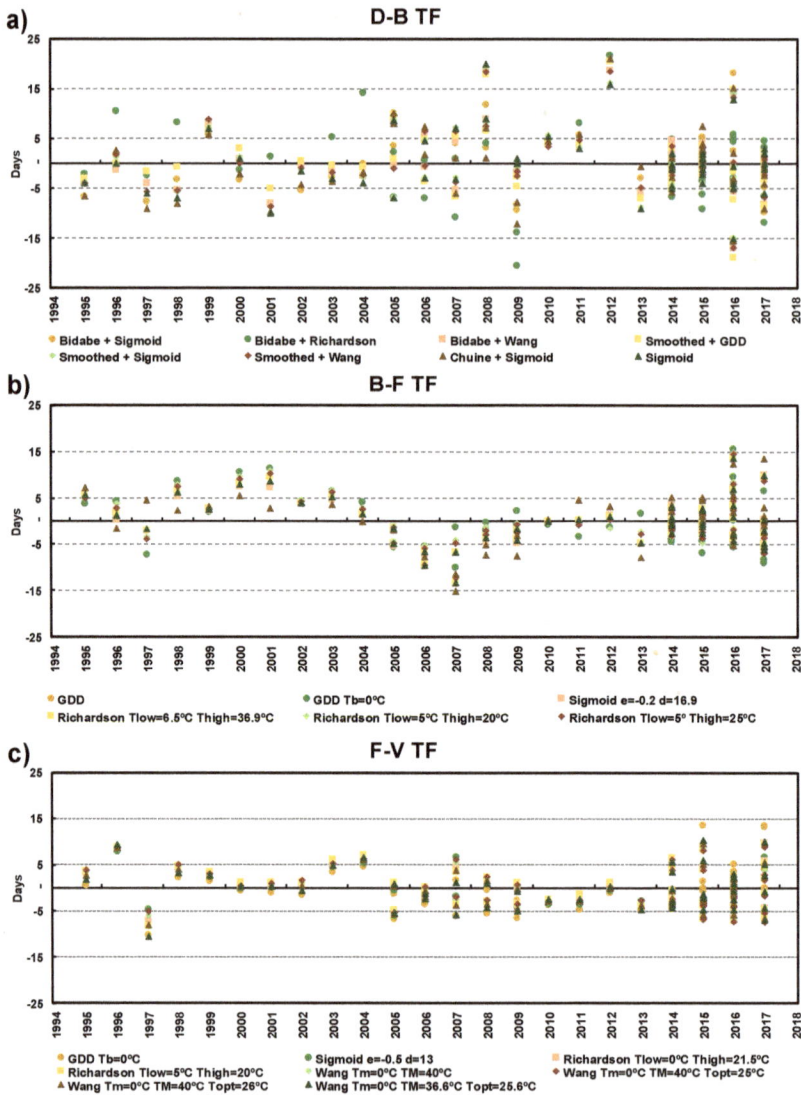

Figure 2. Chronograms of the departures between simulated and observed phenological timings of each phenophase on cv. *Touriga Franca*: (**a**) Dormancy–Budburst (D-B); (**b**) Budburst–Flowering (B-F), and (**c**) Flowering–Veraison (F-V), obtained from the eight selected models (cf. legends) and assembled for the 14 vineyard sites (Figure 1).

3.2. Future Projections

The simulated temperatures from the ensemble of two climate models under RCP4.5 were performed for 8 phenological models and for 12 vineyard sites in the DDR. In order to avoid an excessively large number of figures, only the mean curves over the 12 sites in the DDR are presented in Figure 3; Figure 4. The corresponding panels for each climate model chain are presented separately in Figures S2–S5. Apart from minor differences, site-specific figures are very similar (not shown). For all models, a gradual anticipation of the three main phenological phases (downward long-term trends) is clear for both TF (Figure 3) and TN (Figure 4), despite not being linear and revealing noteworthy fluctuations. Analysing the chronograms for both TF and TN, it is evident that the eight selected models for each phenophase render coherent results (low inter-model ranges), mostly for flowering and veraison. In general, the inter-model range is of nearly 4–6 days for budburst and 2–4 days for flowering and veraison. The correlations between D-B phase and January–April mean temperature, between B-F phase and May–June mean temperature and between F-V phase and July–August mean temperature are clear, thus highlighting that the projected changes are largely driven by changes in mean temperatures over specific time periods. However, some differences between the curves of temperature and mean phenophase timings are found. This can be explained by the fact that the temperature curves correspond to multi-month averages, while the phenological timings are determined by 1) the accumulation of daily temperatures over longer time periods and 2) by the preceding phenophase timings. Furthermore, apart from GDD, model responses are non-linear functions of temperature.

In the case of D-B for TF, a mean anticipation of 6 days is expected from 2020 to 2100, with important inter-decadal variability, without a clear trend until 2070, followed by a more pronounced long-term decrease from 2070 to 2100 (Figure 3a). The Smoothed-Wang and the Smoothed-Sigmoid are the closest models to the mean. For B-F in TF, a clear anticipation is expected after 2040, despite the delay in the period of 2060–2070 (Figure 3b), which can be attributed to a cooler period in the future climate (Figure 3b). Overall, a mean anticipation of 8 days in flowering is projected until 2100. The Sigmoid ($d = -0.2$, $e = 16.9$) and the Wang ($T_m = 0\,°C$, $T_M = 40\,°C$, $T_{opt} = 29.1\,°C$) models show the closest correspondence with the mean. For F-V, an anticipation of 8–10 days over the full period is projected (Figure 3c), with analogous temporal variability to B-F (Figure 3b). The Richardson ($T_{low} = 0\,°C$, $T_{high} = 21.5\,°C$) and the Wang ($T_m = 0\,°C$, $T_M = 40\,°C$, $T_{opt} = 26\,°C$) models are very close to the mean.

Similar considerations can be made for TN (Figure 4). For D-B of TN, the Smoothed-GDD and Smoothed-Wang are the closest to the mean. For B-F of TN, the Richardson ($T_{low} = 7\,°C$, $T_{high} = 20.8\,°C$) and the Sigmoid ($d = -0.3$, $e = 14.4$) models show the most central behaviour. For F-V of TN, the Richardson ($T_{low} = 0\,°C$, $T_{high} = 23.7\,°C$) and the Wang ($T_m = 0\,°C$, $T_M = 40\,°C$, $T_{opt} = 26.9\,°C$) models are the closest to the mean. In addition, the anticipation of veraison for TN is more pronounced than for TF (8 to 12 days). These anticipations in phenological timings are in clear agreement with other studies worldwide [21,32,34]. These shifts towards earlier timings may result in changes to the currently established cultural practices and potentially affect the wine characteristics of the Douro.

3.3. Inter-Model Spread

Regarding the inter-model variability, some differences can be highlighted. For D-B of TF (Figure 3a), all the selected models show very similar results, with the exception of the sequential models Bidabe-Richardson and Smoothed-GDD, which reveal some significant departures from the model average. These two models simulate systematically earlier budburst timings than the other models (Figure 3a), despite their strong agreement, as expected taking into account that their transfer functions are very similar (Table 5). Further, the Sigmoid, Bidabe-Sigmoid, Smoothed-Sigmoid, and Chuine-Sigmoid models are in close agreement, as the second model is the same, also suggesting that dormancy does not play a key role in the final outcome. Similarly, the Bidabe-Wang and Smoothed-Wang also reveal similar behaviour throughout the future period. For B-F of TF (Figure 3b), the GDD (Tb = $0\,°C$) model shows a tendency to have later timings. Conversely, the Wang ($T_m = 0\,°C$, $T_M = 31.6\,°C$, $T_{opt} = 25.5\,°C$) model reveal earlier timings throughout the analysed period. For F-V of TF (Figure 3c), the selected models highlight a very strong coherency, apart from some light discrepancies found for GDD ($T_b = 0\,°C$), with earlier timings.

For D-B in TN (Figure 4a), the Bidabe-Wang presents the earliest phenological timings throughout the study time period. For B-F in TN (Figure 4b), the GDD (Tb = $0\,°C$) model presents later phenological timings, while the Wang ($T_m = 0\,°C$, $T_M = 26.5\,°C$, $T_{opt} = 21.8\,°C$) model shows earlier phenological timings along all the study period. The other models show very strong agreement between them and with values closer to the mean. However, it should be stated that models showing higher agreement are not necessarily more realistic in reproducing timings under future conditions. Lastly, regarding F-V in TN (Figure 4c), the Wang ($T_m = 0$ °C, $T_M = 40$ °C, $T_{opt} = 25$ °C) shows the later timings, while the GDD ($Tb = 0$ °C) shows the earlier timings.

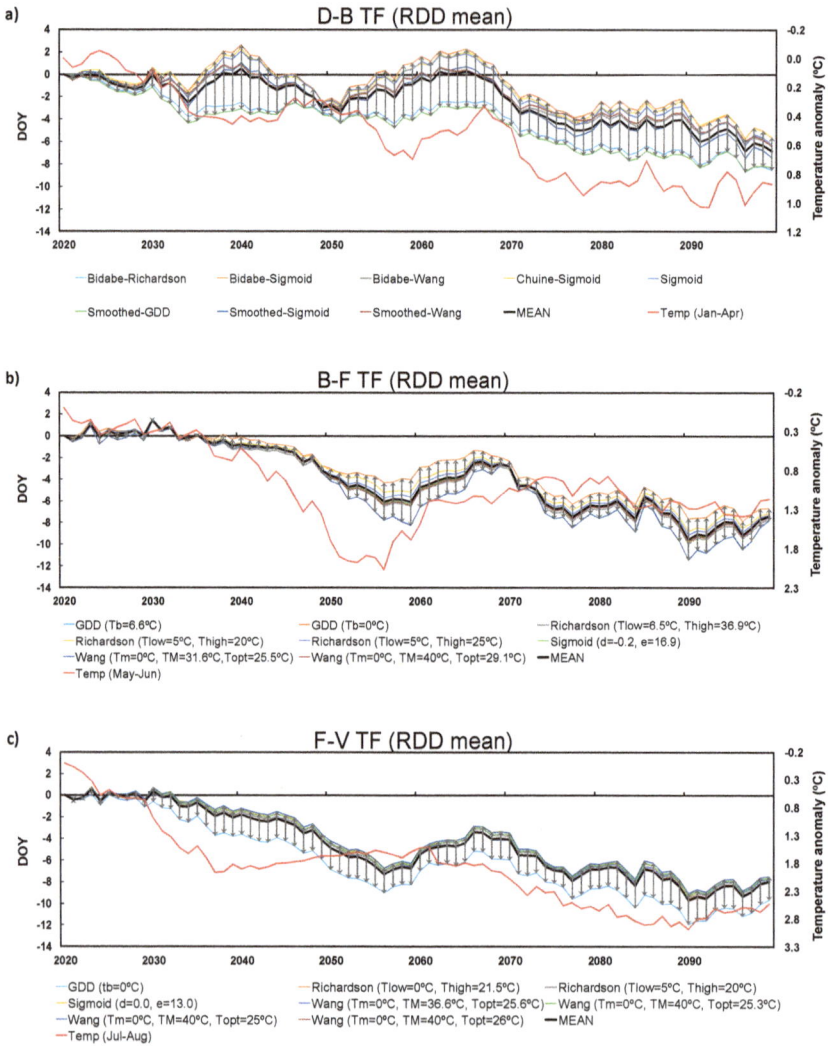

Figure 3. Chronograms (2020–2100, RCP4.5) of the phenological timings (days of the year, DOY) of (**a**) Dormancy–Budburst (D-B); (**b**) Budburst–Flowering (B-F) and (**c**) Flowering–Veraison (F-V), and for cv. *Touriga Franca* (TF), obtained from the eight selected models (cf. legends), averaged for all climate models and the 12 sites in the Douro Demarcated Region (DDR) (Figure 1). The average maximum–minimum ranges for each phenophase (among the different phenology models) are also pointed out (values in days within boxes in the right-upper corner of each panel). The red curves correspond to the (**a**) January–April, (**b**) May–June, and (**c**) July–August mean temperatures (temperature scale inverted).

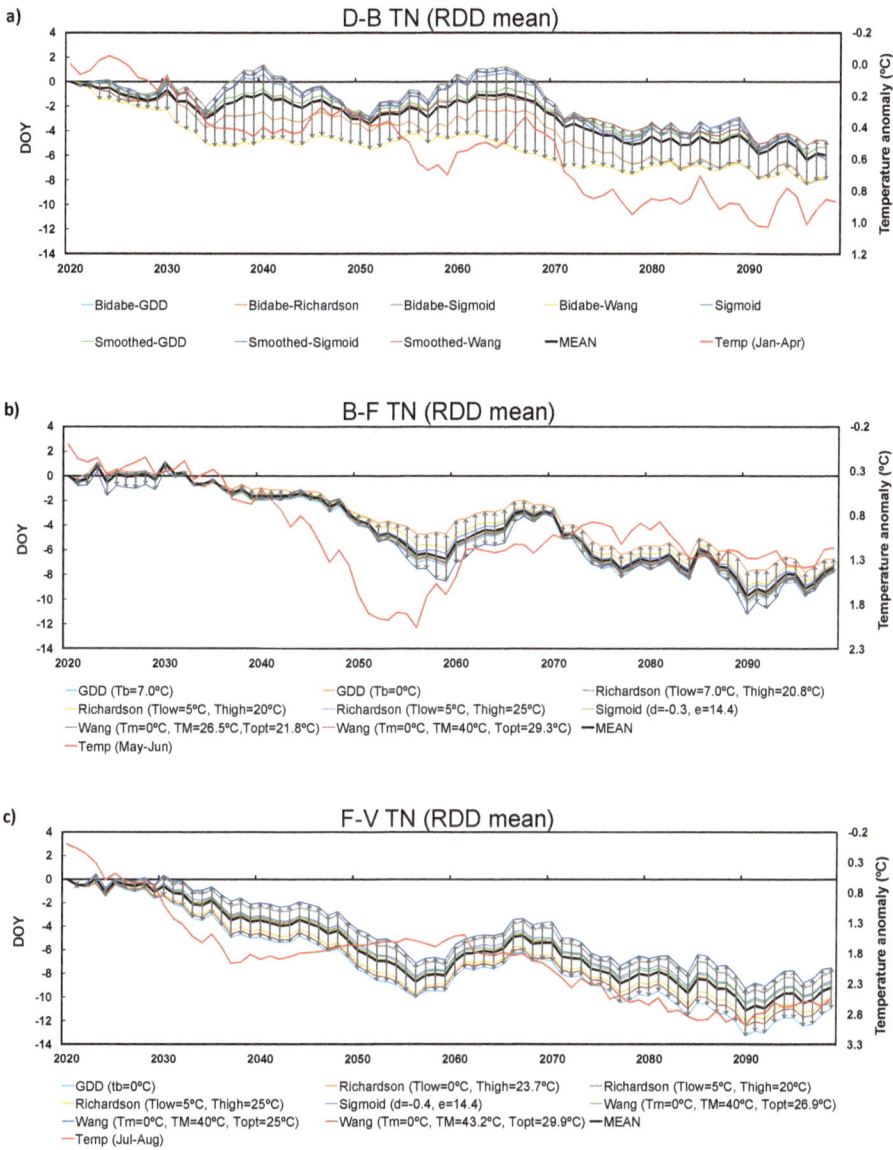

Figure 4. Chronograms (2020–2100, RCP4.5) of the phenological timings (days of the year, DOY) of (**a**) Dormancy–Budburst (D-B); (**b**) Budburst–Flowering (B-F) and (**c**) Flowering–Veraison (F-V), and for cv. *Touriga Nacional* (TF), obtained from the eight selected models (cf. legends), averaged for all climate models and the 12 sites in the Douro Demarcated Region (DDR) (Figure 1). The average maximum–minimum ranges for each phenophase (among the different phenology models) are also pointed out (values in days within boxes in the right-upper corner of each panel). The red curves correspond to the (**a**) January–April, (**b**) May–June, and (**c**) July–August mean temperatures (temperature scale inverted).

4. Discussion and Conclusions

Phenological models are a key tool for viticulturists, as they provide valuable information for management and planning in vineyards [7]. The accurate prediction phenological timings promotes best practices and the timely implementation of suitable measures to optimize grapevine yields and grape berry quality [56–59]. Moreover, future projections of potential changes in the phenological timings may also deliver important information for medium-to-long term planning [60]. Despite the aforementioned high economic value of the DDR, no previous studies were conducted for this region on phenological modelling and simulation. Unique datasets, combining phenological timings with weather station records, in the Douro wine region were used herein. The weather stations used were installed at the vineyard locations, which is of utmost importance for a proper phenological modelling [61]. Furthermore, two of the most important grapevine varieties cultivated in the Douro wine region (TF and TN) were selected (cv. *Touriga Franca*—TF and cv. *Touriga Nacional*—TN). Despite the relatively large number of sites in the Douro valley (12 *Quintas*), data is only available over a maximum period of 4 years. This dataset was thereby complemented with data from two other Portuguese wine regions (*Lisboa* and *Vinhos Verdes*) and for the same grapevine varieties (TF and TN). This approach led to more robust results, as the sample sizes for model calibration were significantly enlarged.

It was herein demonstrated that some specific phenological models are particularly suitable to the simulation of the two selected grapevine varieties. For all phenophases, specific parameter values were chosen within each model (e.g., Tb in GDD), either by selecting reference values, recommended from previous studies, or best-fit values. Although the recommended values did not always lead to the best performances in our study, no definite conclusions can be withdrawn, as only two varieties were selected herein and the corresponding sample sizes are relatively low. As a whole, the present study allowed the identification of the best phenological models (a set of eight models per phenophase), thus warranting their application both in an operational mode (real-time monitoring and short-term prediction) and in future climate change projections (long-term prediction). Nevertheless, more data will allow improving the selection of the best model and the accuracy of the predictions.

In general terms, the results highlight that the phenological models are better in simulating flowering or veraison than budburst. This may be due to the fact that dormancy and other relevant physiological processes are not sufficiently understood, but might be important drivers of budburst occurrence. As also described by García de Cortázar-Atauri, Brisson and Gaudillere [19], no clear improvements were found concerning the incorporation of a dormancy period in the model structure. Moreover, the model performances are generally higher for TF than for TN, which may suggest that the latter is slightly less sensitive to thermal forcing, though this needs to be properly addressed by further research, including field experiments. Overall, the Bidabe-Wang and Sigmoid models showed the best results for budburst in TF, while the Bidabe-GDD and Sigmoid models present the highest performances in TN. For flowering, in both grapevine varieties (TF and TN), the very simple GDD model, with a base temperature of ca. 7 °C, revealed the best performances. Lastly, the Sigmoid model provided the best performances for veraison in both varieties. Hence, the GDD linear model is not always the best approach to model and simulate grapevine phenology, as the phenology–temperature relationship may be of non-linear nature. The present study results are comparable, in terms of model performance, to other studies in other winemaking regions worldwide [19,26,27,43,62–65]. Although the aforementioned models provide the best approaches, it is important to bear in mind that they are bound to this particular phenology dataset.

Climate change projections of the phenophase timings, based on a two-member ensemble of state-of-the-art climate model chains, hinted at anticipations of the phenophase timings up to 10–12 days until the end of the twentieth first century, though with some dissimilarities amongst phenophases and varieties. In addition, there is a strong coherency between phenological models, as the results do not change significantly with the model choice, apart from a few exceptions, thus highlighting the robustness of the projections. Nonetheless, different grapevines varieties exhibit greater phenological sensitivity to temperature changes than others, which limits the extrapolation of the results found herein to other locations and varieties. These projections for the Douro are in agreement with observed trends worldwide, such as for Alsace (France), [Jones [66]] for the California (USA) and Bock, et al. [67] for Franconia (Germany). These shifts towards earlier phenophase onsets can potentially result in changes to the currently established wine characteristics and typicity. In effect, the expected warming may result in unbalanced wines, with high alcoholic content and excessively low acidity and altered colour and aroma [68,69]. These impacts can be partially overcome through the adoption of suitable adaptation measures, such as selection of grapevine varieties that tend to produce berries with lower sugar contents and higher acidity levels in warm climates. Suitable selections of variety-clone-rootstock combinations are among the most effective medium-to-long term measures to cope with climate change. The cultivation of late maturation cultivars is also key to mitigate the anticipation of phenological stages in the vineyards. Short term measures, like irrigation, application of sun screens, changes in training systems, soil management and mulching application, may also be envisioned by producers and winemakers.

The present study aims at providing some initial insights that can be of use not only to future research, but also to regional stakeholders. In the near future, it is expected that the outcomes of this study will be implemented on a web platform, maintained by ADVID, which will deliver information on phenological timings to regional wine companies and grounded on a network of operational weather stations. Future research should also envision the improvement of the current models, such as by extending the phenological datasets with new data from other regions in Portugal or in other countries where the selected varieties also grow. Other grapevine varieties may also be considered in future studies, but there is a critical lack of information for them, particularly on the Douro native/autochthonous varieties, which may considerably hamper their simulation.

Supplementary Materials: The following are available online at http://www.mdpi.com/2073-4395/9/4/210/s1, Figure S1: Chronograms of the departures between simulated and observed phenological timings of each phenophase on cv. *Touriga Nacional*: (a) Dormancy–Budburst (D-B); (b) Budburst–Flowering (B-F), and (c) Flowering–Veraison (F-V), obtained from the eight selected models (cf. legends) and assembled for the 14 vineyard sites (Figure 1). Figure S2. Chronograms (2020–2100, RCP4.5) of the phenological timings (days of the year, DOY) of (a) Dormancy–Budburst (D-B); (b) Budburst–Flowering (B-F); and (c) Flowering–Veraison (F-V), and for cv. *Touriga Franca* (TF), obtained from the eight selected models (cf. legends), averaged for the 12 sites in the DDR (Figure 1) and for the climate model CNRM-SHMI. The average maximum-minimum ranges for each phenophase (among the different phenology models) are also pointed out by vertical arrows. Figure S3. Chronograms (2020–2100, RCP4.5) of the phenological timings (days of the year, DOY) of (a) Dormancy–Budburst (D-B); (b) Budburst–Flowering (B-F); and (c) Flowering–Veraison (F-V), and for cv. *Touriga Nacional* (TN), obtained from the eight selected models (cf. legends), averaged for the 12 sites in the DDR (Figure 1) and for the climate model CNRM-SHMI. The average maximum-minimum ranges for each phenophase (among the different phenology models) are also pointed out by vertical arrows. Figure S4. As for Figure S2, but for the climate model MPI-CLM. Figure S5. As for Figure S3, but for the climate model MPI-CLM.

Author Contributions: Conceptualization, H.F., I.G.d.C.-A., and J.A.S.; Data curation, R.C., A.F., M.C.V., C.C., and S.R.; Formal analysis, R.C., H.F., and S.R.; Funding acquisition, J.A.S.; Investigation, R.C., H.F., A.F., M.C.V., C.C., S.R., and J.A.S.; Methodology, R.C., A.F., I.G.d.C.-A., and J.A.S.; Project administration, J.A.S.; Resources, A.F., I.G.d.C.-A., and J.A.S.; Software, R.C., H.F., A.F., I.G.d.C.-A., and S.R.; Supervision, H.F., I.G.d.C.-A., and J.A.S.; Validation, I.G.d.C.-A.; Visualization, S.R.; Writing—original draft, R.C.; Writing—review and editing, H.F., A.F., I.G.d.C.-A., M.C.V., C.C., S.R., and J.A.S.

Funding: This research and the APC were funded by the European Union's Horizon 2020 Research and Innovation Programme, grant number 810176.

Acknowledgments: This study was supported by the Clim4Vitis project—"Climate change impact mitigation for European viticulture: knowledge transfer for an integrated approach", funded by European Union's Horizon 2020 Research and Innovation Programme, under grant agreement n° 810176. This work was also supported by National Funds by FCT—Portuguese Foundation for Science and Technology, under the project UID/AGR/04033/2019. FCT scholarship SFRH/BPD/119461/2016 and contract CEECIND/00447/2017 are acknowledged. The authors also acknowledge the FCT scholarship given to R.C., PD/BD/135326/2017, under the Doctoral Programme 'Agricultural Production Chains—from fork to farm' (PD/00122/2012). Climate data was provided by ADVID, SOGRAPE vinhos SA, Quinta do Noval, and Direcção Regional de Agricultura e Pescas do Norte (DRAPN). Phenological data from DDR was obtained through the support of ADVID members.

Conflicts of Interest: The authors declare no conflict of interest.

References

1. OIV. *State of the Vitiviniculture World Market*; OIV: Paris, France, 2018; p. 14.
2. IVV. *Vinhos e Aguardentes de Portugal, Anuário 2017*; Ministério da Agricultura, do Desenvolvimento Rural e das Pescas, Instituto da Vinha e do Vinho: Lisboa, Portugal, 2017; p. 224.
3. Fraga, H.; Santos, J.A. Vineyard mulching as a climate change adaptation measure: Future simulations for Alentejo, Portugal. *Agric. Syst.* **2018**, *164*, 107–115. [CrossRef]
4. Fraga, H.; Costa, R.; Santos, J.A. Multivariate clustering of viticultural terroirs in the douro winemaking region. *Cienc. Tec. Vitivinic.* **2017**, *32*, 142–153. [CrossRef]
5. Fraga, H.; Santos, J.A. Daily prediction of seasonal grapevine production in the Douro wine region based on favourable meteorological conditions. *Austr. J. Grape Wine Res.* **2017**, *23*, 296–304. [CrossRef]
6. Keller, M. Phenology and growth cycle. In *The Science of Grapevines*, 2nd ed.; Keller, M., Ed.; Academic Press: San Diego, CA, USA, 2015; Chapter 2; pp. 59–99. [CrossRef]
7. Moriondo, M.; Ferrise, R.; Trombi, G.; Brilli, L.; Dibari, C.; Bindi, M. Modelling olive trees and grapevines in a changing climate. *Environ. Model. Softw.* **2015**, *72*, 387–401. [CrossRef]
8. Andreoli, V.; Cassardo, C.; La Iacona, T.; Spanna, F. Description and preliminary simulations with the Italian vineyard integrated numerical model for estimating physiological values (IVINE). *Agronomy* **2019**, *9*, 94. [CrossRef]
9. Winkler, A.J. *General Viticulture*; University of California Press: Berkly, CA, USA, 1974.
10. Huglin, P. *Nouveau Mode D'évaluation des Possibilités Héliothermiques d'un Milieu Viticole. COMPTES RENDUS de L'académie D'AGRICULTURE*; Académie d'agriculture de France: Paris, France, 1978.
11. Jones, G.V.; Davis, R.E. Climate influences on grapevine phenology, grape composition, and wine production and quality for Bordeaux, France. *Am. J. Enol. Viticult.* **2000**, *51*, 249–261.
12. Jones, G.V.; Duchêne, E.; Tomasi, D.; Yuste, J.; Braslavska, O.; Schultz, H.R.; Martinez, C.; Boso, S.; Langellier, F.; Perruchot, C.; et al. Changes in European winegrape phenology and relationships with climate. In Proceedings of the XIV GESCO Symposium, Geisenheim, Germany, 23–26 August 2005.
13. Van Leeuwen, C.; Garnier, C.; Agut, C.; Baculat, B.; Barbeau, G.; Besnard, E.; Bois, B.; Boursiquot, J.-M.; Chuine, I.; Dessup, T.; et al. Heat requirements for grapevine varieties is essential information to adapt plant material in a changing climate. In Proceedings of the 7th International Terroir Congress(Agroscope Changins-Wädenswil: Switzerland), Changins, Switzerland, 19–23 May 2008; pp. 222–227.
14. Chuine, I.; García de Cortázar-Atauri, I.; Kramer, K.; Hänninen, H. Plant development models. In *Phenology: An Integrative Environmental Science*; Springer: Berlin, Germany, 2013; pp. 275–293.
15. Orlandi, F.; Bonofiglio, T.; Aguilera, F.; Fornaciari, M. Phenological characteristics of different winegrape cultivars in Central Italy. *Vitis* **2015**, *54*, 129–136.
16. Quenol, H.; García de Cortázar-Atauri, I.; Bois, B.; Sturman, A.; Bonnardot, V.; Le Roux, R. Which climatic modeling to assess climate change impacts on vineyards? *Oeno One* **2017**, *51*, 91–97. [CrossRef]
17. Pouget, R. Le débourrement des bourgeons de la vigne: Méthode de prévision et principes d'établissement d'une échelle de précocité de débourrement. *Connaiss. Vigne Vin* **1988**, *22*, 105–123. [CrossRef]
18. Riou, C.; Carbonneau, A.; Becker, N.; Caló, A.; Costacurta, A.; Castro, R.; Pinto, P.A.; Carneiro, L.C.; Lopes, C.; Clímaco, P.; et al. *Le Determinisme Climatique de la Maturation du Raisin: Application au Zonage de la Teneur en Sucre Dans la Communauté Européenne*; Office des Publications Officielles des Communautés Européennes: Luxembourg, 1994; p. 319.
19. García de Cortázar-Atauri, I.; Brisson, N.; Gaudillere, J. Performance of several models for predicting budburst date of grapevine (Vitis vinifera L.). *Int. J. Biometeorol.* **2009**, *53*, 317–326. [CrossRef]
20. Nendel, C. Grapevine bud break prediction for cool winter climates. *Int. J. Biometeorol.* **2010**, *54*, 231–241. [CrossRef] [PubMed]
21. Caffarra, A.; Eccel, E. Projecting the impacts of climate change on the phenology of grapevine in a mountain area. *Austr. J. Grape Wine Res.* **2011**, *17*, 5–61. [CrossRef]
22. Fila, G.; Gardiman, M.; Belvini, P.; Meggio, F.; Pitacco, A. A comparison of different modelling solutions for studying grapevine phenology under present and future climate scenarios. *Agr. Forest Meteorol.* **2014**, *195*, 192–205. [CrossRef]
23. Amerine, M.; Winkler, A. Composition and quality of musts and wines of California grapes. *Hilgardia* **1944**, *15*, 493–675. [CrossRef]

24. Duchene, E.; Huard, F.; Dumas, V.; Schneider, C.; Merdinoglu, D. The challenge of adapting grapevine varieties to climate change. *Clim. Res.* **2010**, *41*, 193–204. [CrossRef]
25. García de Cortázar-Atauri, I.; Chuine, I.; Donatelli, M.; Parker, A.; Van Leeuwen, C. A curvilinear process-based phenological model to study impacts of climatic change on grapevine (Vitis vinifera L.). *Proc. Agro* **2010**, *11*, 907–908.
26. Parker, A.K.; García de Cortázar-Atauri, I.; van Leeuwen, C.; Chuine, I. General phenological model to characterise the timing of flowering and veraison of Vitis vinifera L. *Austr. J. Grape Wine Res.* **2011**, *17*, 206–216. [CrossRef]
27. Parker, A.; García de Cortázar-Atauri, I.; Chuine, I.; Barbeau, G.; Bois, B.; Boursiquot, J.-M.; Cahurel, J.-Y.; Claverie, M.; Dufourcq, T.; Gény, L.; et al. Classification of varieties for their timing of flowering and veraison using a modelling approach: A case study for the grapevine species Vitis vinifera L. *Agric. For. Meteorol.* **2013**, *180*, 249–264. [CrossRef]
28. Molitor, D.; Junk, J.; Evers, D.; Hoffmann, L.; Beyer, M. A High-resolution cumulative degree day-based model to simulate phenological development of grapevine. *Am. J. Enol. Viticult.* **2014**, *65*, 72–80. [CrossRef]
29. García de Cortázar-Atauri, I. Adaptation du modèle STICS à la vigne (Vitis vinifera L.): Utilisation dans le cadre d'une étude d'impact du changement climatique à l'échelle de la France. Ph.D. Thesis, École nationale supérieure agronomique, Montpellier, France, 2006.
30. Pieri, P. Climate change and grapevines: Main impacts. In *Climate Change, Agriculture and Forests in France: Simulations of the Impacts on the Main Species. The Green Book of the CLIMATOR Project (2007–2010)*; ADEME: Angers, France, 2010; pp. 213–223.
31. Cuccia, C.; Bois, B.; Richard, Y.; Parker, A.K.; García de Cortázar-Atauri, I.; Van Leeuwen, C.; Castel, T. Phenological Model Performance to Warmer Conditions: Application to Pinot Noir in Burgundy. *J. Int. Sci. Vigne Vin* **2014**, *48*, 169–178. [CrossRef]
32. Molitor, D.; Caffarra, A.; Sinigoj, P.; Pertot, I.; Hoffmann, L.; Junk, J. Late frost damage risk for viticulture under future climate conditions: A case study for the Luxembourgish winegrowing region. *Austr. J. Grape Wine Res.* **2014**, *20*, 160–168. [CrossRef]
33. Moriondo, M.; Bindi, M. Impact of climate change on the phenology of typical mediterranean crops. *Italian J. Agrometeorol.* **2007**, *3*, 5–12.
34. Webb, L.B.; Whetton, P.H.; Barlow, E.W.R. Modelled impact of future climate change on the phenology of winegrapes in Australia. *Austr. J. Grape Wine Res.* **2007**, *13*, 165–175. [CrossRef]
35. Petrie, P.R.; Sadras, V.O. Advancement of grapevine maturity in Australia between 1993 and 2006: Putative causes, magnitude of trends and viticultural consequences. *Austr. J. Grape Wine Res.* **2008**, *14*, 33–45. [CrossRef]
36. Fraga, H.; García de Cortázar-Atauri, I.; Malheiro, A.C.; Santos, J.A. Modelling climate change impacts on viticultural yield, phenology and stress conditions in Europe. *Glob. Chang. Biol.* **2016**, *22*, 3774–3788. [CrossRef]
37. Vivin, P.; Lebon, E.; Dai, Z.; Duchêne, E.; Marguerit, E.; García de Cortázar-Atauri, I.; Zhu, J.; Simonneau, T.; Van Leeuwen, C.; Delrot, S. Combining ecophysiological models and genetic analysis: A promising way to dissect complex adaptive traits in grapevine. *Oeno One* **2017**, *51*, 181. [CrossRef]
38. Van Leeuwen, C.; Seguin, G. The concept of terroir in viticulture. *J. Wine Res.* **2006**, *17*, 1–10. [CrossRef]
39. Wolkovich, E.M.; García de Cortázar-Atauri, I.; Morales-Castilla, I.; Nicholas, K.A.; Lacombe, T. From Pinot to Xinomavro in the world's future wine-growing regions. *Nat. Clim. Chang.* **2018**, *8*, 29–37. [CrossRef]
40. Magalhães, N. *Tratado de Viticultura: A Videira, a Vinha e o Terroir*; Chaves Ferreira: Lisboa, Portugal, 2008; p. 605.
41. Lorenz, D.H.; Eichhorn, K.W.; Bleiholder, H.; Klose, R.; Meier, U.; Weber, E. Growth Stages of the Grapevine: Phenological growth stages of the grapevine (Vitis vinifera L. ssp. vinifera)—Codes and descriptions according to the extended BBCH scale. *Austr. J. Grape Wine Res.* **1995**, *1*, 100–103. [CrossRef]
42. Fonseca, A.; Santos, J. High-resolution temperature datasets in Portugal from a geostatistical approach: Variability and extremes. *J. Appl. Meteorol. Climatol.* **2018**, *57*, 627–644. [CrossRef]
43. Caffarra, A.; Eccel, E. Increasing the robustness of phenological models for Vitis vinifera cv. Chardonnay. *Int. J. Biometeorol.* **2010**, *54*, 255–267. [CrossRef]

44. Bidabe, B. Action of temperature on evolution of buds in apple and comparison of methods used for estimating flowering time. *Ann. Physiol. Veg.* **1967**, *9*, 65.

45. Chuine, I. A unified model for budburst of trees. *J. Theor. Biol.* **2000**, *207*, 337–347. [CrossRef]

46. Bonhomme, M.; Rageau, R.; Lacointe, A. Optimization of endodormancy release models, using series of endodormancy release data collected in france. *Acta Hortic.* **2010**, *872*, 51–60. [CrossRef]

47. De Réaumur, M. Observations du thermometre, faites à Paris pendant l'annee 1735 comparŕeés avec celles qui ont été faites sous la ligne à l' Ile de France, à Alger et en quelques-unes de nos îles de l'Amérique. *Memoires de l'Academie Royale des Sciences* **1753**, 545–576.

48. Richardson, E. A model for estimating the completion of rest for 'Redhaven' and 'Elberta' peach trees. *HortScience* **1974**, *9*, 331–332.

49. Hänninen, H. Modelling bud dormancy release in trees from cool and temperate regions. *Acta Forestalia Fennica* **1990**, *213*, 1–47. [CrossRef]

50. Wang, E.L.; Engel, T. Simulation of phenological development of wheat crops. *Agric. Syst.* **1998**, *58*, 1–24. [CrossRef]

51. Metropolis, N.; Rosenbluth, A.W.; Rosenbluth, M.N.; Teller, A.H.; Teller, E. Equation of state calculations by fast computing machines. *J. Chem. Phys.* **1953**, *21*, 1087–1092. [CrossRef]

52. Chuine, I.; Bonhomme, M.; Legave, J.M.; García de Cortzar-Atauri, I.; Charrier, G.; Lacointe, A.; Améglio, T. Can phenological models predict tree phenology accurately in the future? The unrevealed hurdle of endodormancy break. *Glob. Chang. Biol.* **2016**, *22*, 3444–3460. [CrossRef]

53. Wallach, D.; Makowski, D.; Jones, J.W. *Working with Dynamic Crop Models: Evaluation, Analysis, Parameterization, and Applications*; Elsevier: Amsterdam, The Netherlands, 2006.

54. Jacob, D.; Petersen, J.; Eggert, B.; Alias, A.; Christensen, O.B.; Bouwer, L.M.; Braun, A.; Colette, A.; Déqué, M.; Georgievski, G.; et al. EURO-CORDEX: New high-resolution climate change projections for European impact research. *Reg. Environ. Chang.* **2014**, *14*, 563–578. [CrossRef]

55. Yang, W.; Andreasson, J.; Graham, L.P.; Olsson, J.; Rosberg, J.; Wetterhall, F. Distribution-based scaling to improve usability of regional climate model projections for hydrological climate change impacts studies. *Hydrol. Res.* **2010**, *41*, 211–229. [CrossRef]

56. Ramos, M.C. Projection of phenology response to climate change in rainfed vineyards in north-east Spain. *Agr. Forest Meteorol.* **2017**, *247*, 104–115. [CrossRef]

57. Fraga, H.; Costa, R.; Moutinho-Pereira, J.; Correia, C.M.; Dinis, L.-T.; Gonçalves, I.; Silvestre, J.; Eiras-Dias, J.; Malheiro, A.C.; Santos, J.A. Modeling phenology, water status, and yield components of three portuguese grapevines using the STICS crop model. *Am. J. Enol. Viticult.* **2015**, *66*, 482–491. [CrossRef]

58. Fraga, H.; Amraoui, M.; Malheiro, A.C.; Moutinho-Pereira, J.; Eiras-Dias, J.; Silvestre, J.; Santos, J.A. Examining the relationship between the enhanced vegetation index and grapevine phenology. *Eur. J. Remote Sens.* **2014**, *47*, 753–771. [CrossRef]

59. Cola, G.; Mariani, L.; Salinari, F.; Civardi, S.; Bernizzoni, F.; Gatti, M.; Poni, S. Description and testing of a weather-based model for predicting phenology, canopy development and source–sink balance in Vitis vinifera L. cv. Barbera. *Agric. Forest Meteorol.* **2014**, *184*, 117–136. [CrossRef]

60. Tomasi, D.; Jones, G.V.; Giust, M.; Lovat, L.; Gaiotti, F. Grapevine phenology and climate change: Relationships and trends in the Veneto Region of Italy for 1964–2009. *Am. J. Enol. Viticult.* **2011**, *62*, 329–339. [CrossRef]

61. Fraga, H.; Santos, J.A.; Moutinho-Pereira, J.; Carlos, C.; Silvestre, J.; Eiras-Dias, J.; Mota, T.; Malheiro, A.C. Statistical modelling of grapevine phenology in Portuguese wine regions: Observed trends and climate change projections. *J. Agr. Sci. Camb.* **2016**, *154*, 795–811. [CrossRef]

62. Bindi, M.; Miglietta, F.; Gozzini, B.; Orlandini, S.; Seghi, L. A simple model for simulation of growth and development in grapevine (Vitis vinifera L.). 2. Model validation. *Vitis* **1997**, *36*, 73–76.

63. Ramos, M.C.; Jones, G.V.; Yuste, J. Phenology of Tempranillo and Cabernet-Sauvignon varieties cultivated in the Ribera del Duero DO: Observed variability and predictions under climate change scenarios. *Oeno One* **2018**, *52*, 31–44. [CrossRef]

64. Oliveira, M. Calculation of budbreak and flowering base temperatures for vitis vinifera cv. touriga francesa in the Douro Region of Portugal. *Am. J. Enol. Viticult.* **1998**, *49*, 74–78.

65. Moncur, M.W.; Rattigan, K.; Mackenzie, D.H.; McIntyre, G.N. Base temperatures for budbreak and leaf appearance of grapevines. *Am. J. Enol. Viticult.* **1989**, *40*, 21–26.

66. Jones, G.V. Climate change in the western united states grape growing regions. *Proceedings of the Seventh International Symposium on Grapevine Physiology and Biotechnology. Acta Horticult.* **2005**, *689*, 41–59. [CrossRef]

67. Bock, A.; Sparks, T.; Estrella, N.; Menzel, A. Changes in the phenology and composition of wine from Franconia, Germany. *Clim. Res.* **2011**, *50*, 69–81. [CrossRef]

68. Jones, G.V.; White, M.A.; Cooper, O.; Storchmann, K. Climate change and global wine quality. *Clim. Chang.* **2005**, *73*, 319–343. [CrossRef]

69. Malheiro, A.C.; Campos, R.; Fraga, H.; Eiras-Dias, J.; Silvestre, J.; Santos, J.A. Winegrape phenology and temperature relationships in the Lisbon Wine Region, Portugal. *J. Int. Sci. Vigne Vin.* **2013**, *47*, 287–299. [CrossRef]

![agronomy logo]

MDPI

Article

Description and Preliminary Simulations with the Italian Vineyard Integrated Numerical Model for Estimating Physiological Values (IVINE)

Valentina Andreoli [1,*], Claudio Cassardo [2,3], Tiziana La Iacona [4] and Federico Spanna [4]

[1] Department of Physics, University of Torino "Alma Universitas Taurinorum", 10125 Torino, Italy
[2] Department of Physics and NatRisk Center, University of Torino "Alma Universitas Taurinorum", 10125 Torino, Italy; claudio.cassardo@unito.it
[3] College of Environmental Science and Engineering, Ewha Womans University, Seoul 03760, Korea
[4] Phytosanitary Sector, Regione Piemonte, 10144 Torino, Italy; federico.spanna@regione.piemonte.it (T.L.I.); tiziana.laiacona@mail.regione.piemonte.it (F.S.)
* Correspondence: valentina.andreoli@unito.it; Tel.: +39-011-670-7438

Received: 23 December 2018; Accepted: 14 February 2019; Published: 18 February 2019

Abstract: The numerical crop growth model Italian Vineyard Integrated Numerical model for Estimating physiological values (IVINE) was developed in order to evaluate environmental forcing effects on vine growth. The IVINE model simulates vine growth processes with parameterizations, allowing the understanding of plant conditions at a vineyard scale. It requires a set of meteorology data and soil water status as boundary conditions. The primary model outputs are main phenological stages, leaf development, yield, and sugar concentration. The model requires setting some variety information depending on the cultivar: At present, IVINE is optimized for *Vitis vinifera* L. Nebbiolo, a variety grown mostly in the Piedmont region (northwestern Italy). In order to evaluate the model accuracy, IVINE was validated using experimental observations gathered in Piedmontese vineyards, showing performances similar or slightly better than those of other widely used crop models. The results of a sensitivity analysis performed to highlight the effects of the variations of air temperature and soil water potential input variables on IVINE outputs showed that most phenological stages anticipated with increasing temperatures, while berry sugar content saturated at about 25.5 °Bx. Long-term (60 years, in the period 1950–2009) simulations performed over a Piedmontese subregion showed statistically significant variations of most IVINE output variables, with larger time trend slopes referring to the most recent 30-year period (1980–2009), thus confirming that ongoing climate change started influencing Piedmontese vineyards in 1980.

Keywords: viticulture; crop model; phenology; physiological processes; climate; micrometeorology; microclimate; climate change

---

1. Introduction

Grapevines are strongly dependent on environmental conditions, and several factors can influence their quality and productivity: Weather, climate, soil fertility, and management practices, among others. An increase in temperature has an important impact on crop growth and yield [1].

There is an increasing interest in the use of crop growth models as tools to assess climate variability and change in crop yields and quality [2–4]. Crop growth models can help to evaluate interactions between cultivar, the environment, and management strategies, and provide an instrument to understand complex plant processes and how they are influenced by pedoclimatic and management conditions. Crop growth models are currently employed at a regional scale for agricultural (yield or quality assessments) or environmental applications (crop water requirements, nitrate leaching) and as a tool to support the process of decision-making and planning in agriculture [5–9].

In particular, the study described in Costa et al. [10], related to the application of crop modeling to Portuguese viticulture, provided a review of research on grapevine models as key decision-supporting systems under current and future climatic conditions.

Many crop growth models operate on a daily time step and simulate the evolution of variables of agronomic interest through daily accumulation. Weather conditions are the input data that drive the crop models and have a noticeable effect on yield and other model outputs. Thus, these kinds of data need to be described accurately.

Since 1960, the Wageningen group has developed crop growth models of varied degrees of complexity for different purposes [11]. For example, the generic model BACROS was developed and improved between the 1960s and 1970s: This modeling approach was used by the modeling group [12,13], while the generic crop model SUCROS, developed in the 1980s [14], represented the basis of most recent Wageningen group crop models, such as WOFOST, ORYZA, INTERCOM, and LINTUL [15–17].

Some crop growth models are adaptable to various crops and can simulate crop growth and plant development, as well as water and nitrogen balances: This is the case of some Wageningen models, as well as STICS [18,19], developed since 1995 at the French National Institute for Agricultural Research (INRA). The STICS model is driven by daily climatic data and simulates crop growth, soil water, and nitrogen balance. It is adaptable to various crops by the use of generic parameters relevant for most crops and by the introduction of physiology and management formalization, chosen for each crop.

Specific crop growth models have also been developed to simulate grapevine growth and development. Among all models, we can mention a simple model for the simulation of growth and yield of a grapevine, specifically the Sangiovese vine [20]; the source-sink model developed to simulate the seasonal carbon supply and partition among reproductive and vegetative parts of a vine [21]; a model predicting phenology, leaf area development, and yield [22], and finally a decision-supporting system for sustainable management of vineyards and real-time monitoring [9]. Furthermore, a model for predicting daily carbon balance and dry matter accumulation in grapevines has been implemented [23].

In addition, the biophysical grape berry growth module described in Reference [24] has been developed and integrated with the whole-plant functional–structural model GrapevineXL and calibrated on two famous international varieties.

The generic crop model STICS has been adapted for grapevines and evaluated for different vineyards and cultivars in France [25]. Its ability to represent phenology, biomass production, yield, and soil water content has been studied for Portuguese grapevines and vineyards located in Chile and France [26,27].

Generally, crop growth models include specific modules calculating the occurrence of phenological stages that can also be used as stand-alone routines. Several models predicting the bud-burst date of a grapevine have been tested and compared [28]: The results of this study showed that calculation of dormancy break, provided by the BRIN model, is not a critical factor for improving the prediction of a bud-burst date under current climatic conditions, but it could become important in future climates. Models simulating the timing of flowering and veraison of grapevines have been tested, and a general phenological model (the spring warming model named the Grapevine Flowering Veraison model (GFV)) was developed and optimized [29], showing the best results in predicting flowering and veraison dates for different varieties.

Finally, grapevine phenology has recently been studied in connection with climate change by means of grape harvest dates used to reconstruct past climate [30] and phenological data of different cultivars in the Veneto region from a long-term collection [31]. The results showed that models used to relate temperature to grape harvest dates can be accurate, but both types of methodologies (linear regression and process-based phenological models) can induce some biases in temperature reconstruction. Grapevine phenology was influenced by the observed warming in the Veneto region: Flowering, veraison, and harvest dates were anticipated during the examined period (1964–2009).

From this overview, it is clear that there have been several studies on crop modeling applied to vineyards. However, few models have been specifically developed for studying grapevines [32], often deepening only certain aspects of crop growth, and few of them could evaluate water balance in vineyards. For this reason, we decided to develop a new crop model instead of adapting and implementing other existing models. The aim of this paper is to present the crop growth model Italian Vineyards Integrated Numerical model for Estimating physiological values (IVINE) [33,34]. IVINE is able to simulate a wide set of phenological and physiological parameters for vineyards using physically based equations for processes such as water balance and photosynthesis, and empirical equations for others. Since our intention was to study *Vitis vinifera* L. Nebbiolo, of which there are few studies in the literature and very few applications using crop models (none of them complete), we calibrated IVINE for cv. Nebbiolo. This cultivar usually is characterized by a large interval of time between the flowering and harvest stages, larger than for other more widespread and studied cultivars. Thus, the model calibration required particular attention.

Here, we would like to develop a grapevine growth model, based on previously described methods, studying the effects of climate change on phenology and yield in the northwestern Italian region of Piedmont. In fact, crop models can be applied to study vineyard complex agroecosystems and multilevel environments. However, before examining the consequences of future climate change on the vineyard environment, it is necessary to verify if and how much the selected crop model is able to provide an adequate representation of these processes in the present and recent climatic conditions.

The paper, after the model description, contains three sections. The first one is dedicated to IVINE validation with field observations. The second one presents a sensitivity analysis on the most important variables (air temperature and water potential) among the IVINE inputs (which also include air relative humidity, wind speed, global solar radiation, photosynthetically active radiation, atmospheric pressure, and soil temperature). The third describes long-term simulations (60 years) carried out over a specific wine area in the northwestern Italian region of Piedmont (Langhe, Roero, and Monferrato), famous for cv. Nebbiolo.

2. Materials and Methods

2.1. The IVINE Model

The numerical model IVINE is a crop growth model created to simulate physiological and phenological vineyard conditions. The model requires a set of meteorological data and vineyard and soil information. It runs on daily steps, and phenological phases dictate the timing of different model routines.

The required boundary conditions, provided during the simulation, are hourly data: Air temperature and relative humidity, solar global radiation, photosynthetically active radiation, soil temperature, soil water content, wind speed and direction, and atmospheric pressure.

Other data required as inputs (about vineyard and soil characteristics) are geographic information (latitude, longitude, and elevation), soil hydrology, variety characteristics, and vineyard management information. Soil parameters (the b-power parameter [35], hydraulic conductivity, soil porosity, wilting point, field capacity, saturated soil water potential, and soil thermal capacity) are required and can be evaluated according to empirical equations [35] by means of organic matter and sand and clay soil percentages, if available, or according to the U.S. Department of Agriculture soil textural classes [36–38]. We are aware of the approximations introduced by such kind of parameterizations, but we think that an even larger error could be produced by the large variations in soil parameters within the same soil class. Unfortunately, in the absence of specific measurements at a local scale, we think that this kind of error cannot be reduced. The presence of a steep slope on terrain has [32] a direct effect on air temperature, solar radiation, and soil status (temperature and moisture [39]). IVINE does not consider explicitly this parameter in its equations, but terrain slope information can be implicitly given to IVINE by selecting an accurate set of boundary conditions.

The IVINE model also requires the setting of some experimental parameters that depend on the cultivar (plant density, thermal thresholds, sugar content threshold at harvest, mean number of clusters per plant, and mean number of berries per cluster) and the site (soil layers number, texture, and depth). The following data about vineyard management are also required: The date and the severity of trimming and thinning (in case they are not available, IVINE prescribes fixed values at fixed dates). At present, the model is optimized for cv. Nebbiolo, since in Piedmont the most famous wines are produced from this cultivar.

The main model outputs are timing of the main phenological stages, leaf development, yield, and berry sugar concentration.

The occurrence of main simulated phenological stages (expressed in Julian days (JDs), used instead of calendar dates to represent the latter by integer values, starting from 1 on January 1st and ending with 365 or 366 on December 31st, and restarting the count at the beginning of each year) are dormancy break, bud-burst, flowering, fruit-set, beginning of ripening, veraison, and harvest: Their simulations use some thermal thresholds and the berry sugar concentration.

The phenological phase of dormancy break is simulated by means of chilling units (Cu) [28,40,41]: Its calculation starts on August 1st (a date close to the period in which the highest annual temperature is usually observed), and the phenological stage occurs when a critical amount of chilling units (100 Cu) is reached. Chilling units (Equation (1)) are calculated by means of maximum and minimum daily temperatures (T_x and T_n) and a parameter Q, set equal to 2.17 [28], while n refers to days [28]:

$$Cu = Q^{-Tx(n)/10} + Q^{-Tn(n)/10} \tag{1}$$

The postdormancy time period is calculated from the dormancy break using a sum of hourly temperatures $T_r(h,n)$, called growing degree hours or GDH, defined in Equation (2) and obtained by the method of Richardson [42,43], used in the BRIN model [28]. If not available, hourly temperatures are derived as in Section S1. The calculation stops when a threshold value equal to 8050 GDH (derived from the IVINE calibration) is reached:

$$GDH = \sum_h T_r(h, n) \tag{2}$$

The phenological phase of flowering (fruit-set) is simulated by means of growing degree-days GDD (Equation (3)) [44]: Its calculation starts from zero at bud-burst (flowering) and stops when an appropriate critical amount of GDD is reached (370 GDD and 50 GDD, respectively). GDDs are calculated through mean daily air temperature and a base temperature (set to 10 °C for cv. Nebbiolo):

$$GDD_n = \sum_n (T_{av}(n) - T_{base}(n)) \tag{3}$$

with the assumption that, when the mean daily temperature $T_{av}(n)$ is lower than T_{base}, GDD_n is set equal to 0 for that day.

The calculation of the beginning of ripening, veraison, and harvest occurs by means of amounts of GDDs (Equation (3)) and critical thresholds of berry sugar content. These thresholds were set in the calibration to 10 °Bx for the beginning of ripening, 12.5 °Bx for veraison, and 25 °Bx for the harvest, which are specific to cv. Nebbiolo.

The leaf area index (LAI, m^2 m^{-2}) is calculated as a measure of plant development [45,46],

$$LAI = \Delta_I \times F_T \times DENS \times I_W \tag{4}$$

using some functions and coefficients detailed in Supplementary Materials, Section S2 (Equations (S3)–(S5)). In more detail, leaf expansion is simulated by IVINE in terms of LAI modulating its value according to the phenological phases: The simulation of leaf development starts at bud-burst and stops at veraison and, from October 1st, leaf senescence is considered by IVINE by imposing a decreasing linear trend. Additional corrections are performed by taking into account the eventual vine trimming carried out in the vineyard.

The value of berry sugar content BSC (°Bx), considered a good indicator of maturity and quality, is evaluated by

$$BSC = \sigma_{Brix} BSC_{max} \tag{5}$$

in which BSC_{max} is the maximum berry sugar content (e.g., its value at the harvest), imposed equal to 25.5 °Bx for cv. Nebbiolo. The function σ_{Brix} is a normalized number, lower than 1, which is parameterized by means of a double sigmoid curve, a function of thermal time and of cultivar sugar content value at harvest [47,48], whose calculation starts from the phenological stage of flowering (see Supplementary Materials, Section S3).

The yield (kg vine^{-1}) is simulated by means of a photosynthetic process, starting from the flowering stage with the following equation [22],

$$Yield = 5.5 \frac{DM_{cluster}}{D\_P} \tag{6}$$

where $DM_{cluster}$ is the dry matter accumulation into vine clusters (see Supplementary Materials, Section S4), D_P the plant density, and 5.5 an empirical coefficient [22].

Other IVINE outputs are listed in Section S5.

2.2. Input Data

To feed IVINE, hourly data of atmospheric and soil variables are required (Section 2.1). Since sufficiently long series of meteorological or agrometeorological data to perform climatological analyses in that zone do not exist, external climatic databases of meteorological observations reconstructed by models and/or measurements were considered. All data but soil variables were directly extracted from the archive of the gridded database GLDAS2.0 (Global Land Data Assimilation System version 2.0 [49]). GLDAS is a global archive created by NASA (Goddard Earth Sciences Data and Information Services Center [50]), whose purpose is to assemble data observed from satellites and ground-based and surface models. Its version 2.0 (GLDAS2.0) contains data from 1948 to 2010 with a spatial resolution of 0.25° in longitude and latitude (about 25 km in the Piedmont region) and a temporal resolution of three hours. GLDAS2.0 data were then interpolated at an hourly rate.

Despite the GLDAS2.0 database also containing soil parameters produced by simulations performed using a land surface model (the NOAH model), we did not use such values. We ran instead the land surface model University of Torino model of land Process Interaction with Atmosphere (UTOPIA) [51], driven by atmospheric GLDAS2.0 data, to recalculate soil variables. The reason for this choice was derived by the results of an analysis [52,53] in which we demonstrated that UTOPIA soil variables (soil temperature and soil moisture) were proven to be closer than GLDAS2.0 ones to the observations carried out during a 3-year experimental campaign [54] carried out in the same area examined in this paper, in particular concerning the highest and lowest values of soil temperature and moisture. The GLDAS2.0 and UTOPIA hourly data used as inputs for IVINE covered a period of 60 years, from 1950 to 2009. Their domain was represented by an area of 15 grid points that included the Langhe and Monferrato wine regions (Figure 1 and Table 1) of Piedmont, characterized by different elevations varying from 95 to 623 m a.s.l. and two different soil textures (loam and clay loam).

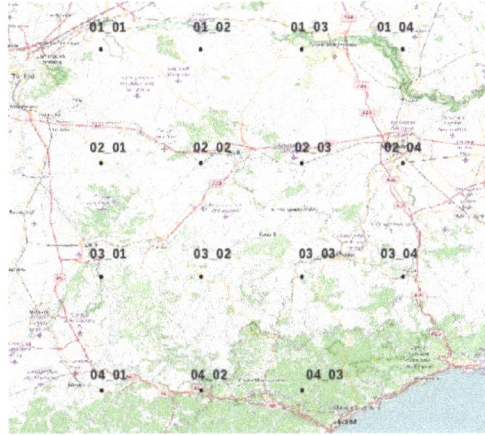

Figure 1. Location of the 15 grid points over the Piedmontese territory.

Table 1. Coordinates of grid points (longitude and latitude, in degrees and decimals, °E and °N, respectively, elevation in m a.s.l.) and soil texture of grid points considered in the study (the code refers to the [36] classification, also used by the US Department of Agriculture).

Grid Points	Coordinates	Elevation (m a.s.l.)	Soil Texture
01_01	7.875, 45.125	269	Clay loam-8
01_02	8.125, 45.125	207	Clay loam-8
01_03	8.375, 45.125	154	Clay loam-8
01_04	8.625, 45.125	95	Clay loam-8
02_01	7.875, 44.875	257	Loam-5
02_02	8.125, 44.875	181	Loam-5
02_03	8.375, 44.875	153	Clay loam-8
02_04	8.625, 44.875	107	Clay loam-8
03_01	7.875, 44.625	294	Loam-5
03_02	8.125, 44.625	416	Loam-5
03_03	8.375, 44.625	322	Loam-5
03_04	8.625, 44.625	342	Loam-5
04_01	7.875, 44.375	605	Loam-5
04_02	8.125, 44.375	623	Loam-5
04_03	8.375, 44.375	402	Loam-5

2.3. Model Validation

A comparison between IVINE outputs and measurements collected during some field experiments carried out in some Piedmontese vineyards was performed. The variables measured were the timing of some phenological stages, the vine *LAI*, the berry weight, and the sugar content.

Measurements were carried out from 2004 to 2010 on the cv. Nebbiolo in three different experimental vineyards located within the most famous wine regions in Piedmont (Langhe, Roero, and Monferrato): Castiglione Falletto (44°37′ N; 7°59′ E; 275 m a.s.l.), Fubine (44°58′ N; 8°26′ E; 200 m a.s.l.), and Castagnito (44°45′ N; 8°01′ E, 300 m a.s.l.). For the first two sites, input data required by IVINE were collected from sensors installed within the vineyards and from regional meteorological stations of Quargnento and Serralunga d'Alba. For the Castagnito site, input data were collected from the regional meteorological station of Castellinaldo and from the global archive GLDAS2.0.

Regarding phenological phases, observations performed in the experimental sites reported the BBCH stage (BBCH means Bundesanstalt, Bundessortenamt and CHemical industry; it is the German scale used to identify the phenological development stages of a plant) achieved at the date of the survey, based on the complete list of BBCH stages [55]. Surveys were performed during the

2008–2010 vegetative seasons at Castiglione Falletto, and during the 2008–2009 seasons at Fubine. IVINE simulations instead returned the dates (in Julian days) in which some BBCH stages occurred (to be precise, the beginning of bud break, BBCH 7; flowering, BBCH 65; fruit set, BBCH 71; veraison, BBCH 83 and °Bx ≥ 12.5; harvest, BBCH 89 and °Bx ≥ 25; Reference [55]). Despite the attempt to make the surveys in proximity to the beginning of the IVINE calculated stages, sometimes the achieved stage was not in the list of those evaluated by IVINE, making a direct comparison difficult.

Regarding the seasonal evolution of the leaf area index (LAI: $m^2_{leaf\ area}\ m^{-2}_{soil\ area}$), available measurements performed every 15–20 days refer to the period May–October 2009 in Castiglione Falletto and Fubine. LAI was estimated by comparing the radiation above the top of the vegetation to the one intercepted by the canopy using a solarimeter bar placed within the canopy, selecting the minimum value of radiation of the bar and comparing it to the radiation above the vegetation (see more details in Reference [56]). The vines were generally about 0.5 m thick, and measurements were carried out mostly during the central hours of the day.

Berry weight and sugar content measurements were measured at Castagnito approximately every 10 days in the periods July–harvest of 2004–2005 and July–September of 2006–2007. For each measurement, 200 berries were collected and weighed, and the juice obtained by their pressing was analyzed to determine their sugar concentration (°Bx).

2.4. Sensitivity Analysis

To understand the importance of the input data and their role in determining the IVINE output values, a sensitivity analysis test was carried out on main input variables.

Among input data, air temperature and soil water potential were chosen as the primary parameters to be investigated, in order to assess their relevance to model behavior. The impact of input variability was evaluated on the following output variables: Phenological phases, berry sugar content, leaf development, and yield.

One year of input data was the period selected for carrying out this kind of analysis: Since the choice of period and site were meaningless, we arbitrarily chose the last year of the selected time period (1950–2009) and one specific grid point (the one labeled as 03_01, whose details are listed in Table 1). The reason for choosing 2009 was to have the simulation output data in the same temporal period in which the IVINE model was calibrated for cv. Nebbiolo, while the reason for selecting the 03_01 grid point was that it was located at an intermediate elevation and its soil type (loam) was more common in the area.

The values of input temperature were varied in nine scenarios of simulation by summing to all air input temperatures a fixed value ΔT_{air} respectively equal to −2.0, −1.5, −1.0, −0.5, 0.0, +0.5, +1.0, +1.5, +2.0, where the value ΔT_{air} = 0 °C corresponds to no change in input temperature (control run). The values of soil water potential were varied in seven scenarios of simulation by summing to all input values a fixed value $\Delta \Psi$ respectively equal to −3.0, −2.0, −1.0, 0.0, 1.0, 2.0, 3.0 m (note that 1 m of hydraulic head roughly corresponds to 0.01 MPa of suction), where the value $\Delta \Psi$ = 0 m corresponds to no change in input soil water potential (control run). Since this variable has two realistic limiting thresholds, e.g., the wilting point and the field capacity, at each step it was checked that the modified values stayed between those thresholds.

2.5. Long-Term Simulations and Statistical Analysis

The time trend of each variable and dependence of phenological phases and physiological variables on elevation and soil type were examined for each grid point. Then, results of simulations performed in grid points with different values of soil type and elevation were analyzed and compared, in order to highlight the effects of such variables.

As a general premise in evaluating the results, it is necessary to consider, in the following analysis, that the IVINE model was calibrated for cv. Nebbiolo through comparisons with data recorded in the last 10 years, and thus this calibration (see details in Section S6) is representative of the standard practices currently performed. When current values are compared to those referring to the beginning of the simulation, the latter should be interpreted as the values of a plant raised similarly to the current plants, but 60 years before. Thus, the output variations can be considered to be the consequence of changes in the input data, e.g., they can highlight more efficiently the effects of climate change. For the same reason, this approach could not take into account the evolution of the change of vineyard methodologies in the analyzed time, in which vine grower standards have certainly changed.

A statistical test related to the significance of linear regression slopes over the whole period was performed on all IVINE output variables. In the paper, only those for the phenological phases, berry sugar content, maximum annual value of leaf area index, and yield are shown. In all cases, the selected test was the Cox–Stuart test, and the significance level was chosen at 95% (p-values ≤ 0.05).

3. Results

3.1. Model Validation

3.1.1. Phenological Stages

Since the in-field visits were not continuous, and IVINE does not simulate all BBCH stages, not always was there a correspondence between simulation and observations. For instance, during the first visit at Castiglione Falletto in 2008 (109th Julian day, e.g., April 18th), the achieved stage was the BBCH 11, while the last stage simulated by the model at that date was the BBCH 7 (on April 5th). Thus, the difference in the simulation of the BBCH 7 stage was certainly lower than 15 days. Considering these unavoidable discrepancies, looking at Tables 2 and 3 the typical error of IVINE in predicting the occurrence of phenological stages could be considered in the interval of 5–10 days (underestimation) in Castiglione Falletto (Table 2), and 0–5 days (overestimation) in Fubine (Table 3).

Table 2. Comparison between occurrence of simulated phenological stages (with their associated BBCH stages) and BBCH achieved stages in the Castiglione Falletto site.

Phenological Stage, Castiglione Falletto	Year	Simulated		Achieved	
		Julian Day	BBCH Stage	Julian Day	BBCH Stage
Bud-break	2008	96	7	109	11
Flowering	2008	165	65	161	63
Veraison	2008	236	83	223	81
Harvest	2008	300	89	289	89
Bud-break	2009	101	7	112	13
Flowering	2009	148	65	145	61
Veraison	2009	220	83	213	81
Harvest	2009	262	89	279	89
Flowering	2010	155	65	155	63
Veraison	2010	218	83	207	81
Harvest	2010	285	89	286	89

Table 3. Comparison between occurrence of simulated phenological stages (with their associated BBCH stages) and BBCH achieved stage in the Fubine site.

Phenological Stage, Fubine	Year	Simulated		Achieved	
		Julian Day	BBCH Stage	Julian Day	BBCH Stage
Bud-break	2008	117	7	120	17
Flowering	2008	172	65	148	60
Fruit-set	2008	175	71	171	73
Veraison	2008	244	83	240	83
Fruit-set	2009	164	71	160	73–75
Beginning of ripening	2009	230	81	224	81–83
Veraison	2009	237	83	224	81–83

3.1.2. Leaf Area Index (LAI)

In the simulations, available data about vine trimming (date and amount of trimming) were given to IVINE and were evident in the results. Comparisons show the quite good performances in Castiglione Falletto (Figure 2), with an overestimation after JD 240 (e.g., the beginning of September). On the contrary, in Fubine (not shown), IVINE underestimated the *LAI* by about 1 m^2 m^{-2} during spring (from March to June), but the growth trend was similar to the observed one: In the second part of the season, there was an overestimation (after JD 240) similar to that of Castiglione Falletto. In both cases, IVINE was able (also at Fubine, even if with an initial delay) to simulate well the potential growth of the leaf surface (until it was artificially reduced), while it had difficulties in capturing the slow decrease of *LAI* in the later part of the season.

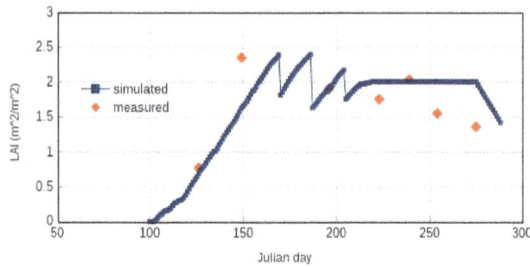

Figure 2. Comparison between simulated and measured leaf area index (*LAI*) in the Castiglione Falletto site.

3.1.3. Berry Growth

The IVINE model was able to simulate the evolution of berry growth during all examined seasons (Figure 3). The simulated values were generally well reproduced in the first part of the season, with an overestimation in July 2007 (Figure 3b), and were generally underestimated starting from about mid-August, with departures variable in the three years: Small in 2004, 2006, and 2007 (0.1–0.2 g), and larger in 2005 (0.3–0.4 g, Figure 3a). In all simulations, a "jump" of 0.1–0.2 g was present in JD 220: This was the effect of the cluster thinning that, in the absence of recorded information, was imposed on JD 220 of each year, with an intensity of 1 cluster/vine.

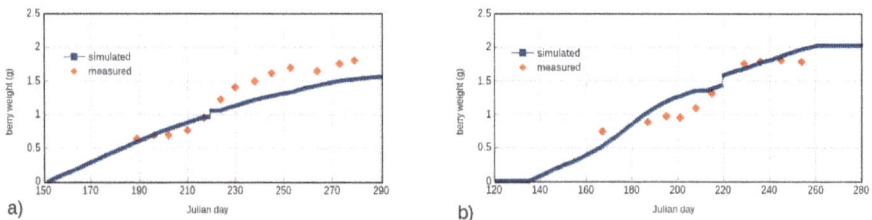

Figure 3. Comparison between simulated and measured berry weight (g) in the Castagnito site during 2005 (**a**), and in the Castagnito site during 2007 (**b**).

3.1.4. Berry Sugar Content

Regarding the berry sugar content, its trend simulated by IVINE was well reproduced during the whole season and in all years (Figure 4), with minor overestimations (always lower or equal to 2 °Bx) observed mainly between mid-July and mid-August. The simulated sugar content resulted close to the observed values in the central and final part of all seasons, while it was overestimated during the earlier part.

To quantify the above-mentioned intercomparisons, mean absolute error (MAE) between simulations and observations was calculated and averaged every year for *LAI*, berry weight, and berry sugar content (standard deviations refer to the annual average). The values are listed in Table 4.

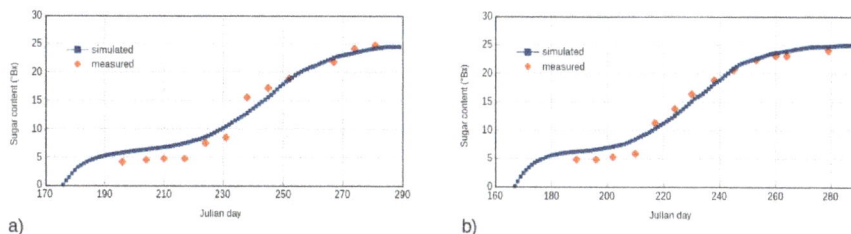

Figure 4. Comparison between simulated and measured berry sugar content (°Bx) in the Castagnito site during 2004 (**a**) and 2005 (**b**).

Table 4. Comparison between simulated and measured values of *LAI*, berry weight, and sugar content at Fubine. MAE: Mean absolute error.

Year	Experimental Site	Variable	Average MAE	Standard Deviation
2004	Castagnito	Berry weight (g)	0.15	0.11
2005	Castagnito	Berry weight (g)	0.19	0.12
2006	Castagnito	Berry weight (g)	0.16	0.08
2007	Castagnito	Berry weight (g)	0.16	0.1
2004	Castagnito	Sugar content (°Bx)	1.5	0.85
2005	Castagnito	Sugar content (°Bx)	1.12	0.7
2006	Castagnito	Sugar content (°Bx)	1.99	1.15
2007	Castagnito	Sugar content (°Bx)	1.51	0.74
2009	Castiglione Falletto	Leaf Area Index (m^2/m^2)	0.31	0.3
2009	Fubine	Leaf Area Index (m^2/m^2)	0.68	0.47

Based on those intercomparisons, performed in some experimental sites in Piemonte wine regions, we could conclude that IVINE seemed able to well represent the evolution of phenological phases and physiological parameters for cv. Nebbiolo and in that region.

3.2. Sensitivity Analysis

The sensitivity analysis was done on air temperature and soil water potential, and in this section the main results are reported.

Figure 5a shows the results of sensitivity analysis on the date of the flowering stage (expressed in Julian days) as a function of the variation in air temperature (ΔT_{air}). The graph clearly shows the effect of increasing temperature. The flowering stage tended to anticipate for higher values of air temperature. The anticipation was about 8 days for 1 °C of air temperature increment and, as expected, varied almost linearly in the range ±2 °C of ΔT_{air}, without signs of thresholds or saturation.

Other phenological phases showed similar behaviors related to the sensitivity analysis, varying almost linearly with ΔT_{air} and showing negative trends, except for the dormancy break, which occurred later with increasing ΔT_{air}. Since occurrence of all spring phenological stages but dormancy break anticipated, and dormancy break postponed, with increasing air temperature, the overall effect was a shortening of the period in which the vines prepared for the future vegetative season.

The effects of temperature variations were also analyzed for berry sugar content, evaluated on the 287th JD (corresponding to October 13th or 14th) (Figure 5b). The different simulations show that the sugar content increased with increasing air temperatures, but in this case the behavior was not linear. Around $\Delta T_{air} = 0$ °C, the rate of variation of the sugar content was about 1.1 °Bx °C^{-1}. As expected from Equation (5), Equations (S6), and (S7), above $\Delta T_{air} = 1$ °C, the value of sugar content stabilized at the quasi-asymptotic value of about 25.5 °Bx.

Other phenological phases showed similar behaviors related to the sensitivity analysis, varying almost linearly with ΔT_{air} and showing negative trends, except for the dormancy break, which occurred later with increasing ΔT_{air}. Since occurrence of all spring phenological stages but dormancy break anticipated, and dormancy break postponed, with increasing air temperature, the overall effect was a shortening of the period in which the vines prepared for the future vegetative season.

The effects of temperature variations were also analyzed for berry sugar content, evaluated on the 287th JD (corresponding to October 13th or 14th) (Figure 5b). The different simulations show that the sugar content increased with increasing air temperatures, but in this case the behavior was not linear. Around $\Delta T_{air} = 0$ °C, the rate of variation of the sugar content was about 1.1 °Bx °C^{-1}. As expected from Equation (5), Equations (S6), and (S7), above $\Delta T_{air} = 1$ °C, the value of sugar content stabilized at the quasi-asymptotic value of about 25.5 °Bx.

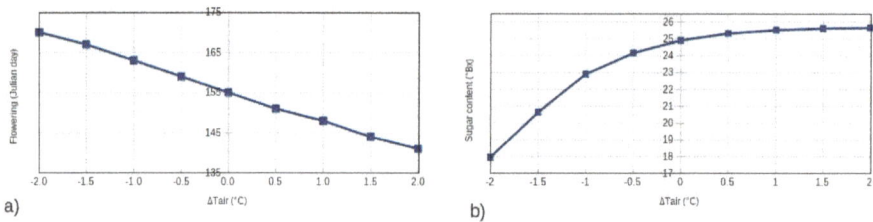

a)   b)

Figure 5. Sensitivity to changes of air temperature on the date of the flowering phase (expressed in Julian days) (**a**) and on the berry sugar content (in °Bx) evaluated at the 287th Julian day (corresponding to October 14th) (**b**). ΔT_{air} is the difference between the input temperature and the actual temperature record.

Figure 6a shows the results of sensitivity analysis related to the soil water potential on the maximum value of the *LAI* reached during the vegetative season as a function of $\Delta\Psi$. The graph shows that the value of *LAI* increased not linearly with increasing $\Delta\Psi$. Given the relation between Ψ and soil moisture ($\Psi = \Psi_s\, q^{-b}$, Ψ_s being the suction for saturated soil, q the soil saturation ratio, and b a coefficient, and b and Ψ_s depending on the soil texture [36]), the abscissae of Figure 6 can also be interpreted as a (nonlinear) soil moisture scale, with the lowest values on the left and the highest values on the right.

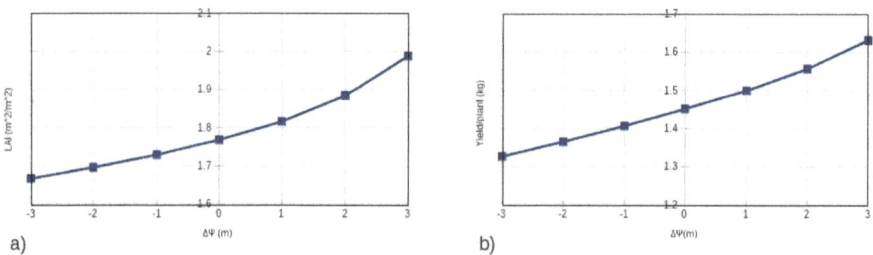

a)   b)

Figure 6. Sensitivity to changes in soil water potential on the maximum *LAI* (expressed in m^2 m^{-2}) (**a**) and on the yield/vine (expressed in kg) (**b**). $\Delta\Psi$ is the difference between the input soil water potential and the actual soil water potential record. Note that 1 m of water potential (hydraulic head) corresponds to about 0.01 MPa of suction.

With $\Delta\Psi = 0$ m, the variation rate of *LAI* was about 0.04 m^2 m^{-2} for 1 m of soil water potential increment. When $\Delta\Psi > 0$ (<0), the rate was larger (smaller).

The effects of soil moisture variation on the yield were also studied (Figure 6b). The sensitivity analysis highlighted a nonlinear positive trend of the yield with increasing soil moisture. The yield variation was about 0.04 kg vine^{-1} for 1 m of soil water potential, in the range of $\Delta\Psi = 0$ m. As in the case of *LAI*, the rate of growth of the yield vine^{-1} increased (decreased) when $\Delta\Psi > 0$ (<0).

3.3. Long-Term Simulations

Due to the impossibility of showing here all results (60 years of simulations carried out on the 15 grid points selected in GLDAS2.0, and 10 relevant variables to show), we decided to comment on figures showing time trends on groups of three grid points with the same soil texture and different elevations, and groups of two grid points at similar elevations and with different soil textures. We did not consider, in our study, the effect of exposition, since the horizontal resolution of the GLDAS2.0 database was too poor to highlight such kind of differences.

3.3.1. Effect of Elevation

Generally, the occurrence of all phenological stages showed the same trend: Thus, among all of them, the flowering stage was selected to show the results obtained in this study.

Figure 7a shows the evolution of the flowering date simulated by IVINE in the 60 years (1950–2009) in three sites characterized by elevations differing by approximately 400 m (from the lowest to the highest point). The effect of elevation was evident and seemed to remain constant along the entire analyzed period. Note that the flowering dates at the highest point since 2000 were in the same range as those near 1950 at the intermediate point. This result was in agreement with those relative to the sensitivity experiment on temperature, considering that, usually, temperature decreases about 0.6 °C for every 100 m of elevation.

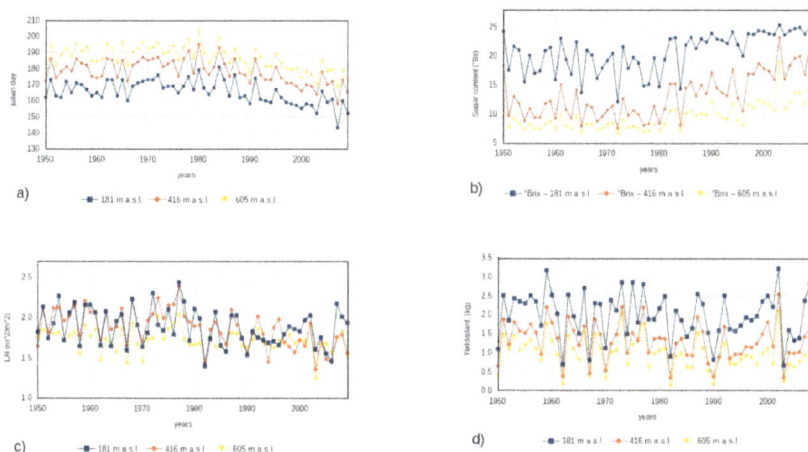

Figure 7. Flowering date (expressed in Julian days) (**a**), berry sugar content (**b**) at the date of 287th Julian day (expressed in °Bx) (**c**), *LAI* maximum value (expressed in m^2 m^{-2}), and yield per vine (expressed in kg) (**d**), simulated by IVINE in three grid points characterized by different elevation. Cv.: Nebbiolo.

Looking at the entire period, the largest variations occurred starting in 1980, which was the year showing the latest flowering date, while the earliest flowering date was observed in 2007. The large anomaly of 2007 was justified by the large positive thermal anomaly during the previous winter and spring over a large portion of Western Europe, more pronounced over northwestern Italy [57]. The trend over the total period evidenced in the simulation was negative, and accounted for about 21 days of anticipation of the flowering stage in the last 30 years.

In Figure 7b, values of berry sugar content simulated at the date of the 287th JD of each year, in the same three sites previously selected, are shown. The effect of elevation was evident also in this case, but, differently from the flowering stage, the difference of sugar content between the lowest elevation site (181 m a.s.l.) and the intermediate one (416 m a.s.l., i.e., 235 m higher) was much larger than the difference between the intermediate elevation site and the highest one (605 m a.s.l., i.e., 189 m higher), especially in the years with the lowest berry sugar content. There were no evident clear trends in the first 30 years in each site, while after 1980 increasing trends were visible, larger for the intermediate elevation site.

The difference in the trends of the two extreme points, evaluated over the whole period, showed a decrease of 2 °Bx for each 100 m of elevation gained. There was also evidence of a significant trend starting from about the 1980s, when interannual variability seemed to decrease in the highest site, with the exception of the year 2003, in which an exceptionally hot and dry summer [58] stimulated IVINE to estimate the highest sugar content of the whole period. Looking at the values at the various altitudes, it is also visible that the vineyards located at the intermediate site had initially very low values of sugar content, comparable to those of the highest elevation site, but starting in 1980 these values increased, almost equaling those of the lowest elevation site at the beginning of the simulations.

Regarding the maximum value of *LAI*, the simulations performed at different elevations are shown in Figure 7c. This variable was related to the vigor of the grapevine, and thus a large value indicated a larger number of leaves per plant, or larger leaves. *LAI* generally increased with warmer temperatures (but could be limited by too hot temperatures, too far from optimal), while it could be limited by too low soil moisture content (e.g., when soil moisture in the root zone approached the wilting point).

At first glance, the effects of elevation appeared less evident than for the previously examined variables. We could notice also for this variable a partition in two subperiods: In the first 30 years, *LAI* maximum values did not vary appreciably, while starting in 1980 there was a decreasing trend at all elevations. In the first period, the lowest and intermediate grid points showed similar values, while starting in 1990 the values of the intermediate grid point appeared more similar to those of the highest grid point. We think that both temperature and soil moisture values, which determine the *LAI* value (Equation (4), Equations (S3)–(S5)), could explain such behaviors, as previously stated. Another evident feature was the decrease of the interannual variability of simulations results since 1980.

The interannual variability of the yield per vine (Figure 7d) was very high, masking any visual trend, but we saw some more stable years in the periods 1953–1958 and 1993–2001. The largest yield at all elevations was observed in 2002, while curiously the lowest yield occurred one year later (2003) at the lowest elevation, and in 1962 at the highest one, and in both years at the intermediate elevation. The effect of elevation was evident among the three grid points, the lowest (highest) one showing the largest (smallest) yield/vine. Differently from the case of the *LAI*, here the grid point at the intermediate level showed yields more similar to those at the highest elevation during the entire analyzed period.

3.3.2. Effect of Soil Texture

The following figures show the time trends of the simulations of the same variables previously shown, but referring to two grid points located at very similar elevations but with different soil textures.

Figure 8a shows the Julian days of the flowering stage. Loam soil exhibited slightly anticipated stages with respect to clay loam soil, with differences generally of 1–2 days, which were not significant. Both soils evidenced a clear decreasing trend starting in 1980.

a)

b)

c)

d)

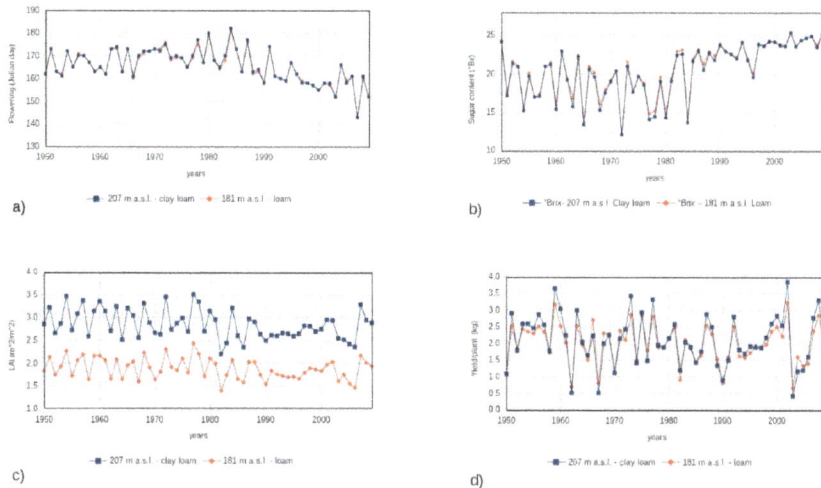

Figure 8. Flowering date (expressed in Julian days) (**a**), berry sugar content at the date of 287th Julian day (expressed in °Bx) (**b**), *LAI* maximum value (expressed in m^2 m^{-2}) (**c**), and yield per vine (expressed in kg) (**d**), simulated by IVINE in two grid points characterized by different soil texture. Cv.: Nebbiolo.

Figure 8b shows the time trend of the berry sugar content, evaluated on the 287th JD, for the two soil textures. Loam-type soil sometimes exhibited higher values of sugar content with respect to clay loam-type, the differences being limited to 0–3 °Bx, larger in the first 35 years of the simulation. Starting in 1984, the interannual variability of the sugar content dropped to its minimum values, and a clear increasing trend was present. During this period, only two years showed values larger than 25 °Bx: 2003 and 2009 (the last year of the simulation). A value of 18 °Bx, which can be associated with the occurrence of the berry softening phenological stage (BBCH 85) in current vineyard management for cv. Nebbiolo, was reached 18 times in the period 1950–1979 and 28 times in the period 1980–2009. The last time at which the 18 °Bx threshold was not reached was the year 1984.

In Figure 8c, the results of the simulations of *LAI* are reported. Differently from the previous figures, here the effect of different soil texture is evident: The grid point characterized by a clay loam-type soil showed the largest values of *LAI* during the whole period, with a systematic shift of about 1 m^2 m^{-2} above the values for loam soil. This was mainly caused by the different soil moistures in the two soil types (the soil saturation ratio is larger in loam soil due to its higher hydraulic conductivity). Values starting in 1980 were lower than in the previous period, almost equaling the minima during 1950–1979 (about 2.5 m^2 m^{-2} for clay loam soil), and interannual variability was very low during the period 1989–2002, perhaps due to the reduced effect of the combined variation of meteo-climatic parameters on maximum *LAI*.

The yield per vine (Figure 8d) evidenced that the values associated with the clay loam soil type were, on average, higher than those associated with loam soil due to higher values of soil water potential (in absolute value): In fact, since saturated soil water potential was more negative for clay loam soil (code 8), we expected higher soil moisture in such soils, and then higher yields. In this case, the differences were small but discernible (less than 0.5 kg vine^{-1}), and seemed larger when yield was larger. As already noted for the *LAI*, the period 1989–2002 was characterized by increasing yields with an extremely small interannual variability, due to the combined variations of meteo-climatic variables.

As a general conclusion for this section, the values belonging to simulations carried out on different soil types at an almost-the-same elevation showed that, compared to the elevation, the soil type played a smaller role. The differences were very low and could be slightly correlated with the soil moisture content, which was higher for clay loam soil, since saturated soil water potential and porosity are higher (in absolute value) in clay loam soil.

3.4. Slopes of Regression Trends

The following tables (Tables 5 and 6) contain the slope of the time trends obtained from the linear regression of the simulations over the whole time period, from 1950 to 2009 (Table 5), and over the most recent 30 years, from 1980 to 2009 (Table 6). As a preliminary note, we say that the time trends of the period 1980–2009, related to almost all output variables considered, resulted as statistically significant (see Section S7) and different with respect to the time trend of the whole time period. This result highlighted a sensitivity of IVINE to its input data starting in 1980, and it could be assumed to be a sign that climate change started to have significant effects on Piedmontese vineyards starting in 1980.

Table 5. Linear regression slopes evaluated for the variables discussed in the text over the whole 60-year period, 1950–2009, in three grid points with the same soil texture and different elevations.

Variable/Elevation	181 m a.s.l.	416 m a.s.l.	605 m a.s.l.
Flowering Stage (JD year^{-1})	−0.2	−0.2	−0.2
Berry Sugar Content (°Bx year^{-1})	0.1	0.1	0.1
LAI Maximum Value (m^2 m^{-2} year^{-1})	−0.004	−0.009	−0.003
Yield (kg year^{-1})	−0.004	−0.005	−0.005

Table 6. Linear regression slopes evaluated for the variables discussed in the text over the most recent 30-year period, 1980–2009, in three grid points with the same soil texture and different elevations.

Variable/Elevation	181 m a.s.l.	416 m a.s.l.	605 m a.s.l.
Flowering Stage (JD year^{-1})	−0.7	−0.7	−0.7
Berry Sugar Content (°Bx year^{-1})	0.2	0.3	0.2
LAI Maximum Value (m^2 m^{-2} year^{-1})	−0.001	−0.008	0.000
Yield (kg year^{-1})	0.007	0.008	0.004

The slope coefficients of the flowering phenological stage (as well as those of all other phases, not analyzed in this paper) were negative in both considered periods (1950–2009 and 1980–2009) and for all analyzed grid points. Those related to the most recent period were larger and evidenced a quickly decreasing trend (about three weeks of anticipation in 30 years).

The same consideration was valid for the berry sugar content, but with positive slope. Considering the most recent 30 years, the positive increasing trend corresponded to an increase of about 6–9 °Bx, which turned out to be quite consistent.

Due to the results of our sensitivity analysis for berry sugar content, we were expecting that this trend would slow down or even stop if temperatures continued to increase, since it was assumed to "saturate". However, it became quite unusual to have particularly low values with higher temperatures.

The *LAI*, as already observed in commenting on Figures 7c and 8c, showed a quite large interannual variability that masked the slopes. The largest negative slope was observed at the intermediate elevation grid point. The lowest grid point in recent times showed a very small negative slope. The highest grid point also did not show any slope in the most recent period. Even considering the most negative slope (intermediate elevation, most recent period), the total *LAI* decrease in 30 years was less than 0.25 m^2 m^{-2}.

Regarding the yield, its variations considering the whole 60-year period or the most recent 30-year period appeared opposite in sign and similar in amplitude. Recent slopes were positive and larger in the lowest and intermediate elevation grid points, with cumulative values of about 200–250 g of increment in 30 years, but this growing rate was limited by the quite low values recorded in the period 2003–2006 (Figure 7d).

The results of the statistical significance of linear regression slopes are listed in Table S1.

The linear regression slopes of flowering, fruit-set, beginning of ripening, and dormancy break stages resulted as significant for all grid points. Concerning the bud-break and veraison stages, most of their linear regression slopes were significant. Regarding the harvest stage, the linear regression slopes resulted as significant only for five grid points.

Regarding physiological outputs, berry sugar content and *LAI* maximum value linear regression slopes were always statistically significant (but in two grid points for *LAI*). Linear regression slopes of yield/vine instead resulted as statistically significant only at three grid points, due to its very large interannual variability.

4. Discussion

We validated IVINE by searching observational datasets gathered in field measurements carried out in cv. Nebbiolo vineyards. Despite this vine being widespread in the Piedmont region, there are not so many data suitable for checking IVINE reliability in reproducing the physio-phenological variables mentioned in the paper. Due to the impossibility of finding in the literature similar studies relative to cv. Nebbiolo and performed using other crop models, we compared our results to other recent results obtained for other red wine varieties in zones with climates not too different from the Piedmontese one.

Due to the missing contemporaneity between the experimental phenological stages (observed by two of us during twice-monthly visits) and those calculated by IVINE, there were some difficulties in validating the phenological stage occurrence, since sometimes it was impossible to reconstruct the exact day of some stages. Considering this problem, based on the results presented, we could evaluate the typical error of IVINE in predicting the occurrence of phenological stages in the interval −5–+10 days (see Tables 2 and 3). These values generally agreed with those found in the literature. For instance, Cola et al. (2014) [22], who analyzed the performance of their model on a vineyard of cv. Barbera in Italy, found a mean value of MAE of 0.7 BBCHs and yearly MAEs of 0.6–1.1 BBCHs, roughly corresponding to about 5–10 Julian days, similar to our values. Fraga et al. (2015) [27], who studied Portuguese red grapevines using the STICS model, found for the flowering phase a MAE generally lower than one week, with annual differences up to 13–17 days, and a higher accuracy for the harvest stage, with an overall MAE of a few days and yearly differences ranging from −7 to +2 days. Valdés-Gómez et al. (2009) [26], examining vineyards in France and Chile, found differences ranging from −6 to +1 days for the flowering stage and −4 to +4 days for the harvest stage, with an accuracy slightly superior to our values.

The MAEs for the IVINE simulations of berry weight were in the range 0.15–0.19 g. These values appeared lower than those of Mirás-Avalos et al. (2018) [59], who found for Tempranillo grapevines in Spain an MAE of about 0.47 g of dry mass per berry, and also those of Valdés-Gómez et al. (2009) [26], who found for French and Chilean sites MAE values of 0.1–0.5 g.

Regarding *LAI*, our MAEs were in the range 0.31–0.68 m^2 m^{-2}, and appeared to be slightly smaller than those obtained by Valdés-Gómez et al. (2009) [26] in their study, where the *LAI* absolute bias was 0.5 m^2 m^{-2}, with individual deviations larger than 1 m^2 m^{-2}.

In conclusion, the analysis of the validation experiments showed that IVINE performed at least similarly to other published crop models, even if it was impossible to find simulations for the same variety. Phenological phases seemed to be the variables that were less accurately predicted. In our opinion, the use of functions related only to air temperature, with specific thermal thresholds, at least for the first phases, could introduce some approximations that could cause such disagreements. In further studies, the influence of other variables on phenological phases could be taken into consideration.

This was also the main reason why the air temperature was the most sensible IVINE variable among its inputs, and explained why phenological stages anticipated almost linearly with increasing temperature, differently from other outputs. On the contrary, the quasiasymptotic threshold of about

25.5 °Bx, shown by the berry sugar content during sensitivity to air temperature, was an effect of the double-sigmoid curve (and the related parameter's choice) used in the parameterization of such variables (Equation (5), Equations (S6), and (S7)). Further investigations to be carried out in warmer climates and with other varieties could confirm if such a parameterization, adopted for IVINE, could be valid also in warmer climates.

The simulation of the harvest stage deserves a special consideration. IVINE simulates the harvest stage when the berry sugar content reaches a value of 25 °Bx: This value represents the harvest threshold in the current vineyard management practices for cv. Nebbiolo-making. Simulation results in other periods showed the behavior of plants raised using current practices, but with a past climate. They evidenced that this stage was rarely, or never, reached during the first 40 years of the analyzed period, particularly for grid points characterized by high elevations. This result could also be interpreted in another way. If 40 years ago it was impossible to produce Nebbiolo wine with the actual management practices in most of the Piedmont region, now it has become possible, mostly due to temperature increments connected with climate change.

Regarding the long-term simulations, the relevant decrease of yield (per plant) with elevation could be explained by considering that it was related to vegetation photosynthesis, which depends on radiation, temperature, and soil moisture. Among these effects, in our simulations the temperature was the factor changing most effectively in higher elevations. On the one hand, temperature delayed the flowering phase by more than 20 days, on average, considering our highest and lowest grid point (Figure 7a), thus postponing the growth of the berries. On the other hand, the photosynthetic activity was related indirectly to the temperature, and the difference mattered even if the quote difference between the highest and lowest grid points was only 424 m.

The linear regression slopes of most pheno-physiological variables examined during the long-term simulation showed that, for most of these variables, they were significant for most or all grid points, the only exception being the harvest stage (already discussed above) and yield. For the latter, the large interannual variability in each site was noticeable. The anticipation of the phenological phases looked numerically similar to the values reported by Tomasi et al. [31] for a study on a shorter time period in another northern Italy wine region (Veneto), but with different cultivars.

We considered the possible effects of volcanic eruptions on those data: The three volcanic eruptions with some discernible effect on global mean surface temperature in that period were in 1963 (Mt. Agung), 1982 (El Chichón), and 1991 (Mt. Pinatubo) [60]. These eruptions did not cause, in the analyzed region, temperature variations larger than those associated with natural interannual variability, and thus did not have discernible effects on the vineyard variables examined here.

The IVINE crop model, created to study some physio-phenological processes in vineyards on the basis of some micrometeorological and soil observations, and calibrated for cv. Nebbiolo, was able to give a realistic representation of such processes with quantitative data. IVINE could be applied as an instrument that gives to the winegrower some additional data useful for vineyard management. The results of our long-term simulation also showed that IVINE, once adequately calibrated also for other cultivars, could also be used, as we did, to show the effects of climate change on the variables affecting wine production. The use of regional climate model simulations, instead of past and recent observations, as inputs for IVINE could also allow for seeing the expected effects of future climate change on wine variables.

5. Conclusions

The crop growth model IVINE was developed to simulate grapevine phenological and physiological processes related to environmental conditions. It requires a set of meteorological and soil data as boundary conditions. The main model outputs are main phenological phases, leaf development, yield, and sugar concentration. At present, the model has been optimized and validated only for cv. Nebbiolo.

IVINE was validated using the data of phenological phases (bud break, flowering, fruit-set, beginning of ripening, veraison, and harvest) and physiological parameters (*LAI*, berry weight, and berry sugar content) recently observed in some Piedmontese vineyards. The model results were accurate in representing both the time trend and the numerical values of pheno-physiological variables related to the specific type of vine cultivar, with accuracies similar to, and in some cases higher than, those of other recent studies.

A sensitivity analysis was performed on air temperature and soil water potential inputs due to their relevance in the model equations, with other variables having minor effects. With increasing temperatures, phenological stages anticipated almost linearly (about 8 day $°C^{-1}$), while the berry sugar content showed a nonlinear increase (about 1.1 $°Bx °C^{-1}$), tending to stabilize its values at the quasiasymptotic threshold of about 25.5 $°Bx$. With increasing soil water potential, the *LAI* showed a nonlinear incremental rate (about 0.04 $m^2 m^{-2} m^{-1}$), and the yield showed a not linear positive trend (about 0.04 $kg m^{-1}$).

Long-term simulations, driven by GLDAS2.0 climatological atmospheric data and UTOPIA land surface model soil variables, were performed, running IVINE over 15 grid points for 60 years (1950–2009) within a selected Piedmontese area prone to viticulture. The results indicated significant trends of almost all variables related to physiology and phenology, combined with (for most variables) a reduction in interannual variability, particularly evident for berry sugar content in recent years. These results seem to indicate a strong influence of climate change, at least since 1980, after which almost all variable trends consistently increased.

The future perspective of this project will be the optimization of the IVINE crop model for other grapevine varieties and the execution of simulations in different regions of Italy or the world under current, past, or future climates.

Supplementary Materials: The following are available online at http://www.mdpi.com/2073-4395/9/2/94/s1, Section S1: Equations to Extrapolate Hourly Temperatures, Section S2: Parameterization of LAI, Section S3: Parameterization of Berry Sugar Content, Section S4: Parameterization of Yield, Section S5: Other Outputs of IVINE, Section S6: IVINE Calibration, Section S7: Slopes of Regression Trends of All IVINE Outputs; Table S1: Linear regression slopes of the main IVINE variables, evaluated over the full 60-year period. Phenological stage regression slopes are expressed in JD $year^{-1}$, sugar content in $°Bx year^{-1}$, *LAI* maximum value in $m^2 m^{-2} year^{-1}$, and yield in $kg year^{-1}$. Bold values represent statistically significant trends. For some phenological phases (veraison and harvest), the stage was not reached during several years, and thus the trend was not evaluated (we have put the *** symbol in such cases).

Author Contributions: V.A. and C.C. have contributed in equal measure to the research article. T.L.I. and F.S. have managed data acquisition and checking.

Funding: This research was partially funded by "MACSUR—Modelling European Agriculture for Food Security with Climate Change, a FACCE JPI knowledge hub", funded for an Italian partnership by the Italian Ministry of Agricultural Food and Forestry Policies (D.M. 24064/7303/15). The Lagrange Project—CRT Foundation/Isi Foundation and IWAY s.r.l. supported two grants of one year for developing the project.

Acknowledgments: We acknowledge Silvia Cavalletto, Silvia Ferrarese, Silvia Guidoni, Massimiliano Manfrin, Elena Mania, and the team of IWAY s.r.l. for useful discussions about agronomic and instrumental details, and Hualan Rui for help in managing the GLDAS dataset. We also acknowledge the scientific team that made available the GLDAS dataset in both versions for research purposes.

Conflicts of Interest: The authors declare no conflicts of interest.

References

1. Olesen, J.E.; Trnka, M.; Kersebaum, K.C.; Skjelvag, A.O.; Seguin, B.; Peltonen-Sainio, B.; Rossi, F.; Koryra, J.; Micale, F. Impacts and adaptation of European crop production systems to climate change. *Eur. J. Agron.* **2011**, *34*, 96–112. [CrossRef]
2. Challinor, A.J.; Simelton, E.S.; Fraser, E.; Hemming, D.; Collins, M. Increased crop failure due to the climate change: Assessing adaptation options using models and socio-economic data for wheat in China. *Env. Res. Lett.* **2010**, *5*, 034012. [CrossRef]

3. Olesen, J.E.; Bindi, M. Consequences of climate change for European agricultural productivity, land use and policy. *Eur. J. Agron.* **2002**, *16*, 239–262. [CrossRef]

4. Rotter, R.R.P.; Palosuo, T.; Kersebaum, K.K.C.; Angulo, C.; Bindi, M.; Ewert, F.; Ferrise, R.; Hlavinka, P.; Moriondo, M.; Nendel, C.; et al. Simulation of spring barley yield in different climatic zones of Northern and Central Europe. A comparison of nine crop models. *Field Crops Res.* **2012**, *133*, 23–36. [CrossRef]

5. Brouwer, F.; Van Ittersum, M. *Environmental and Agricultural Modelling. Integrated Approach for Policy Impact Assessment*; Springer: Dordrecht, The Netherlands, 2010.

6. Ewert, F.; Angulo, C.; Rumbaur, C.; Lock, R.; Enders, A.; Adenauer, M.; Heckelei, T.; Van Ittersum, M.K.; Wolf, J.; Rotter, R. *(AgriAdapt) Project of the Research Program Climate Change and Spatial Planning. Scenario Development and Assessment of the Potential Impacts of Climate and Market Changes on Crops in Europe*; Climate Changes Spatial Planning Programme: Wageningen, The Netherlands, 2011; p. 49.

7. Basso, B.; Cammarano, D.; Sartori, L. A strategic and tactical management approach to select optimal N fertilizer rates for wheat in a spatially variable field. *Eur. J. Agron.* **2011**, *35*, 215–222. [CrossRef]

8. Thorp, K.R.; Dejong, K.C.; Kaleita, A.L.; Batchelor, W.D.; Paz, J.O. Methodology for the use of DSSAT models for precision agriculture decision support. *Comp. Elect. Agric.* **2008**, *64*, 276–285. [CrossRef]

9. Rossi, V.; Salinari, F.; Poni, S.; Caffi, T.; Bettati, T. Addressing the implementation problem in agricultural decision support systems: The example of vite.net. *Comp. Elect. Agric.* **2014**, *2014*, 88–99. [CrossRef]

10. Costa, R.; Fraga, H.; Malheiro, A.C.; Santos, J.A. Application of crop modelling to portuguese viticulture: Implementation and added-values for strategic planning. *Cienc. Tec. Vitiv.* **2015**, *30*, 29–42. [CrossRef]

11. Van Ittersum, M.K.; Leffelaar, P.A.; Van Keulen, H.; Kropff, M.J.; Bastiaans, L.; Goudriaan, J. On approaches and applications of the Wageningen crop models. *Eur. J. Agron.* **2003**, *18*, 201–234. [CrossRef]

12. De Wit, C.T.; Goudriaan, J.; van Laar, H.H.; Penning de Vries, F.W.T.; Rabbinge, R.; van Keulen, H.; Louwerse, W.; Sibma, L.; de Jonge, C. *Simulation of Assimilation, Respiration and Transpiration of Crops (Simulation Monographs)*; Pudoc, Center for Agricultural Publishing and Documentation: Wageningen, The Netherlands, 1978.

13. Goudriaan, J. *Crop Micrometeorology: A Simulation Study. Simulation Monographs*; Pudoc, Center for Agricultural Publishing and Documentation: Wageningen, The Netherlands, 1977.

14. Van Keulen, H.; Van Heemst, H.D.J. *Crop Response to the Supply of Macronutrients*; Centre for Agricultural Publishing and Documentation, Pudoc: Wageningen, The Netherlands, 1982; p. 46.

15. Van Keulen, H.; Wolf, J. *Modelling of Agricultural Production: Weather, Soils and Crops*; Centre for Agricultural Publishing and Documentation, Pudoc: Wageningen, The Netherlands, 1986; p. 46.

16. Penning de Vries, F.W.T.; Jansen, D.; Ten Berge, H.F.M.; Bakema, A. *Simulation of Ecophysiological Processes of Growth in Several Annual Crops*; Pudoc/IRRI: Wageningen, The Netherlands; Los Banos, Philippines, 1989; Volumn 29.

17. Boumann, B.A.M.; Kropff, M.J.; Tuong, T.P.; Wopereis, M.C.S.; Ten Berge, H.F.M.; Van Laar, H. *ORYZA2000: Modelling Lowland Rice*; International Rice Research Institute/University and Research Centre: Los Banos, Philippines; Wageningen, The Netherlands, 2001.

18. Brisson, N.; Mary, B.; Ripoche, D.; Jeuffroy, M.H.; Ruget, F.; Nicoullaud, B.; Gate, P.; Devienne-Barret, F.; Antonioletti, R.; Durr, C.; et al. STICS: A generic model for the simulation of crops and their water and nitrogen balances. I. Theory and parameterization applied to wheat and corn. *Agronomie* **1998**, *18*, 311–346. [CrossRef]

19. Brisson, N.; Gary, C.; Justes, E.; Roche, R.; Mary, B.; Ripoche, D.; Zimmer, D.; Sierra, J.; Bertuzzi, P.; Burger, P.; et al. An overview of the crop model STICS. *Eur. J. Agron.* **2003**, *18*, 309–332. [CrossRef]

20. Bindi, M.; Miglietta, F.; Gozzini, B.; Orlandini, S.; Seghi, L. A simple model for simulation of growth and development in grapevine (*Vitis vinifera* L.). I. Model description. *Vitis* **1997**, *36*, 67–71.

21. Vivin, P.; Castelan, M.; Gaudillere, J.P. A source/sink model to simulate seasonal allocation of carbon in grapevine. *Acta Hort.* **2002**, *584*, 43–56. [CrossRef]

22. Cola, G.; Mariani, L.; Salinari, F.; Civardi, S.; Bernizzoni, F.; Gatti, M.; Poni, S. Description and testing of a weather-base model for predicting phenology, canopy development and source-sink balance in *Vitis vinifera* L. cv. Barbera. *Agric. For. Meteorol.* **2014**, *184*, 117–136. [CrossRef]

23. Poni, S.; Palliotti, A.; Bernizzoni, F. Calibration and Evaluation of a STELLA Software-based Daily CO2 Balance Model in *Vitis vinifera* L. *J. Amer. Soc. Hort. Sci.* **2006**, *131*, 273–283. [CrossRef]

24. Zhu, J.; Génard, M.; Poni, S.; Gambetta, G.A.; Vivin, P.; Vercambre, G.; Trought, M.C.T.; Ollat, N.; Delrot, S.; Dai, Z. Modelling grape growth in relation to whole-plant carbon and water fluxes. *J. Exp. Bot.* **2018**. [CrossRef] [PubMed]

25. Garcia de Cortazar Atauri, I. Adapatation du Model STICS a la Vigne (*Vitis vinifera* L.). Utilisations dans le Cadre D'une Etude D'impact du Changement Climatique a L'echelle de la France. Ph.D. Thesis, Ecole Nationale Superiore Agronomique de Montpellier, Ecole Doctarale, Montpellier, France, 2006.

26. Valdes-Gomez, H.; Celette, F.; Garcia de Cortazar Atauri, I.; Jara-Rojas, F.; Gary Ortega-Farias, C.C. Modelling soil water content and grapevine growth and development with the STICS crop-soil model under two different water management strategies. *J. Int. Sci. Vigne. Vin.* **2009**, *43*, 13–28. [CrossRef]

27. Fraga, H.; Costa, R.; Moutinho-Pereira, J.; Correia, C.M.; Dinis, L.T.; Goncalves, I.; Silvestre, J.; Eiras-Dias, J.; Malheiro, A.C.; Santos, J. Modeling phenology, water status, and yield components of three portuguese grapevines using the STICS crop model. *Am. J. Enol. Vitic.* **2015**, *66*, 482–491. [CrossRef]

28. Garcia de Cortazar Atauri, I.; Brisson, N.; Gaudillere, J.P. Performance of several models for predicting budburst date of grapevine (*Vitis vinifera* L.). *Int. J. Biometeorol.* **2009**, *53*, 317–326. [CrossRef] [PubMed]

29. Parker, A.K.; Garcia de Cortazar Atauri, I.; Van Leeuwen, C.; Chuine, I. General phenological model to characterise the timing of flowering and veraison of *Vitis vinifera* L. *Aust. J. Grape Wine Res.* **2011**, *17*, 206–216. [CrossRef]

30. Garcia de Cortazar Atauri, I.; Daux, V.; Garnier, E.; Yiou, P.; Viovy, N.; Seguin, B.; Boursiquot, J.M.; van Leeuwen, C.; Parker, A.K.; Chuine, I. Climate reconstructions from grape harvest dates: Methodology and uncertainties. *Holocene* **2010**, *20*, 1–10. [CrossRef]

31. Tomasi, D.; Jones, G.; Giust, M.; Lovat, L.; Gaiotti, F. Grapevine phenology and climate change: Relationships and trends in the Veneto region of Italy for 1964–2009. *Am. J. Enol. Vitic.* **2011**, *62*, 329–339. [CrossRef]

32. Brillante, L.; Mathieu, O.; Lévêque, J.; Bois, B. Ecophysiological modeling of grapevine water stress in Burgundy terroirs by a machine-learning. *Front. Plant Sci.* **2016**, *7*, 796. [CrossRef] [PubMed]

33. Andreoli, V.; Cassardo, C.; Cavalletto, S.; Ferrarese, S.; Guidoni, S.; Mania, E.; Spanna, F. *Representation of Micrometeorological and Physiological Parameters with Numerical Models Influencing the Vineyard Ecosystem: The Case of Piemonte (Italy)*; European Geophysics Union: Vienna, Austria, 2017; Volumn 19.

34. Cassardo, C.; Andreoli, V.; Spanna, F.; Ferrarese, S.; Picco, S. *Climatology of Phenological and other Micrometeorological Variables Parameters in Regional Vineyard Ecosystem in Piedmont (Italy)*; European Geophysics Union: Vienna, Austria, 2018; Volumn 20.

35. Saxton, K.E.; Rawls, W.J. Soil Water Characteristic Estimate by Texture and Organic Matter for Hydrologic Solutions. *Soil Sci. Soc. Am. J.* **2006**, *70*, 1569–1578. [CrossRef]

36. Clapp, R.B.; Hornberger, G.M. Empirical Equations for Some Soil Hydraulic Properties. *Water Resour. Res.* **1978**, *14*, 601–604. [CrossRef]

37. McCumber, M.C.; Pielke, R.A. Simulation of the effects of surface fluxes of heat and moisture in a mesoscale numerical model: 1. *J. Geophys. Res.* **1981**, *86*, 9929–9938. [CrossRef]

38. Tremback, C.J.; Kessler, R.L. *A Surface Temperature and Moisture Parametrization for Use in Mesoscale Numerical Models, 7th Conference on Numerical Weather Prediction*; American Meteorological Society: Boston, MA, USA, 1985.

39. Brillante, L.; Mathieu, O.; Lévêque, J.; Van Leeuwen, C.; Bois, B. Water status and must composition in grapevine cv. Chardonnay with different soils and topography and a mini meta-analysis of the δ13C/water potentials correlation. *J. Sci. Food Agric.* **2017**, *98*, 691–697. [CrossRef] [PubMed]

40. Bidabe, B. Controle de l'epoque de floraison du pommier per une nouvelle conception de l'action de temperature. *Comptes Rendus Séances L'Acad. D'Agric. Fr.* **1965**, *49*, 934–945.

41. Bidabe, B. *L'action des Temperatures sur l'Evolution des Bourgeons de L'entrée en Dormance à la Floraison*; Société Pomologique de France: Lyon, France, 1965.

42. Richardson, E.A.; Seeley, S.D.; Walker, R.D. A model for estimating the completion of rest for Redhaven and Elberta peach trees. *HortScience* **1974**, *9*, 331–332.

43. Richardson, E.A.; Seeley, S.D.; Walker, R.D.; Ashcroft, G. Pheno-climatology of spring peach bud development. *HortScience* **1975**, *10*, 236–237.

44. Winkler, A.J.; Cook, J.A.; Kliewer, W.M.; Lider, L.A. *General Viticulture*; Cerruti, L., Ed.; University of California Press: Oakland, CA, USA, 1974.

45. Singels, A.; de Jager, J.M. Refinement and validation of the PUTU wheat crop growth model. 2. Leaf area expansion. *Afr. J. Plant Soil* **1991**, *8*, 67–72. [CrossRef]
46. Brisson, N. *Notices Concepts et Formalismes STICS Version 5.0*; INRA-Bioclimatologie Avignon: Avignon, France, 2002.
47. Letchov, G.; Roichev, V. Growth kinetics of grape berry density (*Vitis vinifera* L. 'Black Corinth'). *Vitis* **2017**, *56*, 155–159.
48. Garcia de Cortazar Atauri, I.; Brisson, N.; Ollat, N.; Jacquet, O.; Payan, J.-C. Asynchronous dynamics of grapevine (*Vitis vinifera*) maturation: Experimental study for a modelling approach. *J. Int. Sci. Vigne. Vin.* **2009**, *43*, 83–97. [CrossRef]
49. Rui, H.; Beaudoing, H. *README Document for NASA GLDAS Version 2 Data Products*; Goddart Earth Sciences Data and Information Services Center (GES DISC): Greenbelt, MD, USA, 2018.
50. Rodell, M.; Houser, P.R.; Jambor, U.; Gottschalck, J.; Mitchell, K.; Meng, C.-J.; Arsenault, K.; Cosgrove, B.; Radakovich, J.; Bosilovich, M.; et al. The Global Land Data Assimilation System. *Bull. Am. Meteor. Soc.* **2004**, *85*, 381–394. [CrossRef]
51. Cassardo, C. *The University of Torino Model of Land Process Interaction with Atmosphere (UTOPIA) Version 2015*; CCCPR/SSRC, Ewha Womans University: Seoul, Korea, 2015; p. 80.
52. Lazzarato, F. Reconstruction of Parameters Useful to Define the Microclimate of Piedmontese Vineyards Using Simulations (in Italian). Master's Thesis, Department of Physics, University of Torino, Torino, Italy, 2017.
53. Picco, S. Climatology of Piedmontese Wine-Vocated Zones, with Particular Attention to Phenological and Physiological Parameters (In Italian). Master's Thesis, Department of Physics, University of Torino, Torino, Italy, 2017.
54. Andreoli, V.; Bertoni, D.; Cassardo, C.; Ferrarese, S.; Francone, C.; Spanna, F. Analysis of micrometeorological conditions in Piedmontese vineyards. *Ital. J. Agrometeorol.* **2018**, *2018*, 27–40. [CrossRef]
55. Bleiholder, H.; Weber, E.; Feller, C.; Hess, M.; Wicke, H.; Meier, U.; van den Boom, T.; Lancashire, P.D.; Buhr, L.; Hack, H.; et al. *Growth Stages of Mono-and Dicotyledonous Plants*; BBCH Monograph, Federal Biological Research Centre for Agriculture and Forestry: Berlin and Braunschweig, Germany, 2001.
56. Francone, C. Study of the Atmospheric Boundary Layer Processes over Sloping Terrain Covered by Sparse Canopy. Ph.D. Thesis, Dottorato in Fluidodinamica, Politecnico di Torino, Torino, Italy, 2011–2012.
57. Paesano, G. *Health and Environment Report 2008. The State of Environmental Components. Climate*; Agenzia Regionale Protezione Ambientale: Piemonte, Italy, 2008; pp. 149–169.
58. Cassardo, C.; Mercalli, L.; Cat Berro, D. Characteristics of the summer 2003 heat wave in Piedmont, Italy, and its effects on water resources. *J. Korean Meteorol. Soc.* **2007**, *43*, 195–221.
59. Mirás-Avalos, J.M.; Uriarte, D.; Lakso, A.N.; Intrigliolo, D.S. Modeling grapevine performance with 'VitiSim', a weather-based carbon balance model: Water status and climate change scenarios. *Sci. Hortic.* **2018**, *240*, 561–571. [CrossRef]
60. Bindoff, N.L.; Stott, P.A.; AchutaRao, K.M.; Allen, M.R.; Gillett, N.; Gutzler, D.; Hansingo, K.; Hegerl, G.; Hu, Y.; Jain, S.; et al. 2013: Detection and Attribution of Climate Change: From Global to Regional. In *Climate Change 2013: The Physical Science Basis. Contribution of Working Group I to the Fifth Assessment Report of the Intergovernmental Panel on Climate Change*; Stocker, T.F., Qin, D., Plattner, G.-K., Tignor, M., Allen, S.K., Boschung, J., Nauels, A., Xia, Y., Bex, V., Midgley, P.M., Eds.; Cambridge University Press: Cambridge, UK; New York, NY, USA, 2013.

agronomy

MDPI

Article

Modelling Approach for Predicting the Impact of Changing Temperature Conditions on Grapevine Canopy Architectures

Dominik Schmidt [1,*], Christopher Bahr [2], Matthias Friedel [2] and Katrin Kahlen [3]

[1] Department of Modeling and Systems Analysis, Hochschule Geisenheim University, Von-Lade-Str. 1, 65366 Geisenheim, Germany

[2] Department of General and Organic Viticulture, Hochschule Geisenheim University, Von-Lade-Str. 1, 65366 Geisenheim, Germany

[3] Department of Vegetable Crops, Hochschule Geisenheim University, Von-Lade-Str. 1, 65366 Geisenheim, Germany

* Correspondence: dominik.schmidt@hs-gm.de; Tel.: +49-6722-502-79734

Received: 29 June 2019; Accepted: 31 July 2019; Published: 3 August 2019

Abstract: Future climatic conditions might have severe effects on grapevine architecture, which will be highly relevant for vineyard management decisions on shoot positioning, pruning or cutting. This study was designed to help gaining insight into how, in particular, increasing temperatures might affect grapevine canopies. We developed a functional-structural model for Riesling, *Virtual Riesling*, based on digitised data of real plants and a comprehensive state-of-the-art data analysis. The model accounts for the variability in temperature-sensitive morphological processes, such as bud break and appearance rates. Our simulation study using historical weather data revealed significant effects of the thermal time course over the year on bud burst of the cane and on primary shoots. High variabilities in these events affect canopy growth and leaf area distribution. This report shows that *Virtual Riesling* can be useful in assessing the significance of changing temperatures for grapevine architecture and thereby considering management techniques such as vertical shoot positioning. Further developments of *Virtual Riesling* might support the knowledge gain for developing necessary adaptations in future vineyard management and, thus, facilitate future work on climate change research using functional-structural model approaches.

Keywords: grapevine; Virtual Riesling; climate change; temperature; plant architecture; crop management; modelling

---

1. Introduction

Climate change will affect traditional forms of viticulture from multiple perspectives. Temperature increases in wine growing regions with a cool to moderate climate have already advanced the onset of ripening by two weeks in the past 20 years and are expected to advance phenological development by another two weeks in the near future [1]. An earlier ripening period will expose ripening grapes to higher temperatures and lead to a higher degree of alcohol, a lower concentration of organic acids and to changes in the aroma composition of wines. This may ultimately lead to a loss of typicality of regional wine styles. Elevated temperatures during shoot growth and ripening will also lead to an increased pressure of infection of *Plasmopara viticola* [2] and *Botrytis cinerea* in most European regions [3]. An earlier bud break will also increase the risk of late frost damages in grapevines. Apart from accelerating the phenological development of grapevines, increasing temperatures will also affect grapevine growth and shoot architecture [4]. Developmental rates of leaf primordia, unfolded and fully expanded leaves are constant when expressed in thermal time as observed for naturally varying field and greenhouse conditions [5,6]. For example, appearance rates linearly relate to temperature. However, even this stable program might be affected by the trophic state and the water status of the vine [7]. In addition, growth rates and durations of Grenache Noir are not stable when expressed in thermal time [5] and different grapevine cultivars might respond differently to changes in temperatures, as shown by Luchaire et al. [6]. In the growth chamber experiment of Buttrose [8], which covered a wide range of constant temperatures from 15 °C to 30 °C, Riesling reached maximal shoot lengths at 30 °C, whereas the maximal node number was reached at lower temperatures (25 °C). At even lower temperatures (<20 °C), relatively more dry weight was distributed to the leaves than at higher temperatures, and the lengths of lateral shoots sharply increased at extreme temperatures (>30 °C). However, it is not known how this knowledge transfers to fully grown vines in the field. The temperature responsiveness of developmental growth processes may also be summarised by common Arrhenius-like functions [9]. The increasing phase between base temperature and below optimal temperatures is in agreement with the thermal-based program, whereas responses to extreme, higher temperature conditions are non-linear with different pattern. Overall, these complex patterns with both thermal-stable and different organ-specific temperature responses will challenge the predictability of grapevine architecture under future environmental conditions and the wine industry will have to cope with these challenges [10]. In the long run, it may be necessary to either move production of traditional cultivars to cooler vineyards, e.g., at higher elevation, or adapt the cultivar profile of existing wine regions. In the medium term, however, the production of typical wines from traditional cultivars could be maintained if vineyard management techniques are adapted [11]. Developing necessary adaptations in vineyard management will require enormous scientific, experimental and practical efforts. Modelling approaches, which consider plant architecture explicitly, might support and facilitate these approaches [10,12–14]. The class of functional-structural plant models explicitly combines plant architecture and plant functioning. They have proven useful for analysing feedback processes between plant architecture and physiological processes, if local environmental conditions are the key process drivers [15–18]. A milestone for modelling plant architecture of grapevine is the work of Louarn et al. [19] and Louarn et al. [20] on *TopVine*. Their statistical approach for a static architectural model of grapevine was based on digitised real plants. This approach allowed integrating inter-plant variability. The variability was mimicked for basal positions of the shoot, parameters for the spatial paths of the shoots and leaf azimuth and elevation angles, whereas each shoot has the same leaf area, which is a model input, and the length of each sub-unit is the same. *TopVine* was used to simulate light-sensitive differences in canopy structure variability within and between cultivar × training system pairs for Grenache Noir and Syrah [20]. A similar study of Iandolino et al. [21] aimed at simplifying simulations of grapevine canopy reconstruction. Random sample measurements in the field and in the lab were used to parameterise allometric relationships, which serve as input for *YPLANT*, a statistical plant generator. *YPLANT* reconstructs static plants with option for high variability and estimations of light distributions within the virtual canopy. Recent modelling studies, which also aim at including knowledge on grapevine architecture, either are developments of the *TopVine*-approach or are based on greenhouse grapevine fruiting cuttings trained to one shoot axis as model plant ([22], *GrapevineXL*). Prieto et al. [23] modelled intra-canopy variability of gas exchange by considering leaf nitrogen content and local acclimation to radiation in grapevine. Here, the static model for one digitised plant from the *TopVine* study was used and adapted to match the leaf size of Syrah. Garin et al. [24] used an architectural dataset from *TopVine* to set up a first dynamic grapevine model used to analyse the development of powdery mildew within the virtual grapevine canopy. The dynamic approach was implemented based on *L-Py*, a programming language for Lindenmayer-systems. One of the latest functional-structural models for grapevine was designed to simulate berry quality based on carbon and water fluxes (e.g., [22,25,26], for *GrapevineXL*). This was achieved by linking a biomechanical gas exchange model and a complex water status model to local plant architectural conditions. Here, the model for plant architecture was simplified and descriptive to mimic the conditions of greenhouse grown grapevine fruiting cuttings of

Cabernet Sauvignon. Just recently, the leaf-based functional-structural plant model *HydroShoot* was published [27], focusing on simulations of leaf gas-exchange rates in complex canopies by coupling hydraulic, energy and exchange modules. It was exemplarily evaluated using static mock-ups of virtual grapevine canopies to study plant-scale gas-exchange rates and leaf-scale temperature and water potential in response to canopy architecture.

In summary, grapevine architecture models could be promising tools to predict the impact of rising temperatures on canopy structure of grapevines, and thus provide the basis for the simulation of canopy management techniques under future climatic conditions. However, this requires an appropriate consideration of the variability in the grapevine canopy architecture, modelling approaches for the temperature-responsiveness of morphological processes and the sensitivity of the virtual crop to management measures such as shoot positioning. With this study, we wanted to highlight possible effects of temperature conditions on architectural traits of growing grapevine canopies using historical weather data. The aim was to show that a simple, descriptive dynamic three-dimensional model for grapevine architecture can be used to mimic effects of changing temperatures on canopy growth in a typical Riesling vineyard.

2. Materials and Methods

First, we developed a simple, descriptive, but dynamic three-dimensional model for grapevine architecture, which is able to simulate the natural variability of a growing canopy in a typical Riesling vineyard (Rheingau, Germany) including the responsiveness to shoot positioning. Data of digitised grapevines were used for model conceptualisation and parameterisation. Second, this model was extended to cope with the temperature-sensitivity of morphological processes such as bud break and organ growth. Third, we ran simulations studies for Riesling growth and development under different historical temperatures to assess the integrated temperature-effect on canopy level and compared canopy architectures under different climates.

2.1. Field Site

Vines for digitisation were grown in the VineyardFACE field experiment located at Hochschule Geisenheim University in the Rheingau region, Germany (49°59' N, 7°57' E). The region has a moderate oceanic climate with an average temperature of 10.5 °C and annual rainfall of 543 mm. The experimental vineyard was planted in 2012 with *Vitis vinifera* L. cv. Riesling (clone 198-30 Gm) grafted to rootstock SO4 (clone 47 Gm) with at a vine spacing of 0.9 m and a row spacing of 1.8 m and a north–south row orientation. The soil at the field site is characterised as a sandy loam. Soil management consisted of alternating row of open soil and a grass mixture cover crop. The cover crop was mulched several times during the vegetation period. Integrated plant protection was carried out according to the code of good practice. Vines were cane pruned and trained to a single Guyot vertical shoot positioning system (VSP) with one cane pruned to approximately ten nodes (Figure 1). After bud break, shoot number was adjusted to eight shoots per vine. Vineyard area was 0.5 ha (5000 m^2), within which a free air CO_2 (FACE) enrichment system was installed. Details of the FACE system have been published elsewhere [28,29].

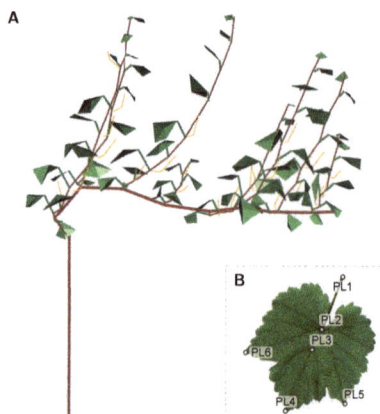

Figure 1. (**A**) Exemplary early-stage reconstructed digitised vine showing vine structure with single Guyot vertical shoot positioning system (VSP) with one cane pruned to approximately eight nodes (green: leaves; yellow: flowers; brown: cane/shoot); and (**B**) Riesling leaf digitisation scheme.

2.2. Plant Digitisation

All shoots from three grapevines were digitised at three developmental stages on 3–4 May 2018, 17 May 2018 and 7–8 June 2018, corresponding approximately to E-L stages 12 (5 leaves expanded), 15 (8 leaves expanded), and 26 (cap fall complete), respectively (see Coombe [30] for E-L Stages). The three-dimensional structure of the canopy was recorded using an electromagnetic 3D-digisiter ("Fastrak", Polhemus, Colchester, U.S.). Kahlen and Stützel [31] and Schmidt and Kahlen [32] described the general digitisation procedure. The specific digitisation protocol applied to the grapevines took into account the topological relationships between main organs, and was designed to allow the reconstruction of the guyot cane, primary shoots, lateral shoots, leaves and flowers. For digitisation, the transmitter, which generates the measurement sphere and includes the origin of the Cartesian coordinate system, was placed at a distance of 1 m to the canopy at cane level, with the x-, y- and z-axes pointing towards the canopy, parallel to the row and vertically, respectively. All metallic objects within a 3 m radius of the transmitter were removed to avoid inferences with the electromagnetic sphere. Digitisation started at the cane level, at which all nodes were digitised with one point per node, taken in the axil of each node, irrespectively of bearing a shoot or not. On a node bearing a shoot, the shoot was digitised, before digitisation continued along the cane. For shoot digitisation, all primary nodes of the shoot, beginning at the base and moving towards the apex, were recorded with one point at the bud axil. For each node, all leaves, flowers and lateral shoots emanating from the node were digitised according to the following sequences. Flowers were digitised with three points: PF1 was set at the branching point of the flower towards the shoot apex. PF2 was set at the first branching point of the flower itself towards the flower tip. PF3 was the tip of the flower itself. Leaves were digitised with six points. PL1 was set at the petiole base towards the shoot apex (Figure 1). PL2 was the leaf base, recorded at the adaxial side of the leaf. PL3 was the joining point of the midrib and the veins spanning the central lobe of the leaf, taken on the adaxial side. PL4 was the tip of the mid rip. PL5 and PL6 were the tips of the veins spanning the middle lobes to the left and the right side (seen from leaf base to tip) of the central lobe, respectively. Lateral shoots were digitised in the same way as primary shoots. Apical nodes of primary or lateral shoots with less than 1 cm of length and leaves with less than 3 cm of primary vein length were not digitised.

2.3. Weather Data

Weather data for model parameterisation during the 2018 season were collected from a weather station located at the experimental site. Historical weather data from 1927 to 2018 (91 years) were provided by Germany's National Meteorological Service (DWD) from a station located approximately 200 m from the experimental site (Station-ID: 1580, 49°59′ N, 7°57′ E, 110.2 m above NN).

We used daily mean T_m (°C), minimum T_{min} (°C) and maximum T_{max} (°C) temperatures as estimated by the respective DWD standards. This includes slight adaptations of the methodology; for instance, between 1935 and 1986, T_m was calculated based on three measurements per day (7, 14, and 21 MOZ) as $T_m = (T_7 + T_{14} + T_{21} \cdot 2)/4$, while since 2006 at least 21 hourly measurements were used in an arithmetic mean calculation. This slightly different estimation throughout the historic time frame should not be of any concern for our simulation results, as these data were exemplarily used to compare different climates.

Selection of candidate years followed the principle of using some of the most distinct years. As a first basis, we categorised by annual mean temperature ($\bar{T}_{m,anno}$, °C), selecting the year with the lowest (1940, $\bar{T}_{m,anno} = 8.3$ °C) and the highest (2018, $\bar{T}_{m,anno} = 12.4$ °C) mean temperature. Grouping years in blocks of 10 based on $\bar{T}_{m,anno}$ we selected the year 1987, as the coldest year of the second block, i.e., the eleventh coldest year, with $\bar{T}_{m,anno} = 9.2$ °C. In addition, we selected 2014 as the year with the earliest predicted bud break (day of year (DOY) 99 ; see Section 2.4.1), 1979 with the latest bud break (DOY 135) and, finally, 2017, as the immediate predecessor of the measurement year 2018. More details to specific ranks of $\bar{T}_{m,anno}$ and bud break BB (DOY) for the selected years are provided in Table 1 and Figure 2.

Table 1. Selected years with ranking position (1–91) for annual average of daily mean air temperature ($\bar{T}_{m,anno}$) and estimated bud break date (BB) ([1] - BB for 2018 is based on phenological observations instead of thermal time data; see Section 2.4.1).

Year	Air Temperature Ranking	$\bar{T}_{m,anno}$ (°C)	Bud Break Ranking	BB (DOY)
1940	1	8.3	11	122
1979	15	9.4	91	135
1987	11	9.2	2	120
2014	90	12.0	1	99
2017	83	11.3	46	101
2018	91	12.4	51	111 [1]

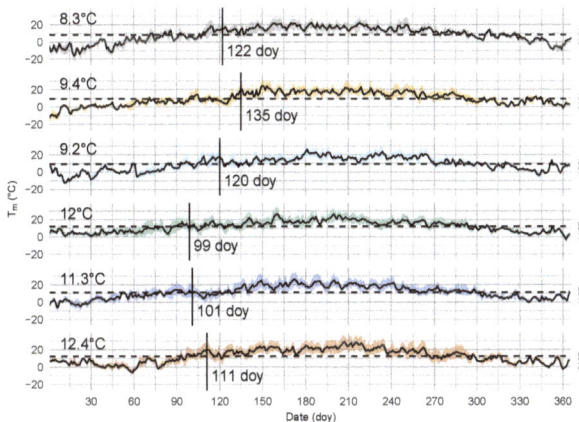

Figure 2. Annual cycle of mean air temperature (T_m) for selected years (1940, 1979, 1987, 2014, 2017, and 2018) with their respective bud break date (vertical bar) and annual mean temperature ($T_{m,anno}$; dashed line). Ribbon range illustrates daily temperature range (min, max).

The years selected for the simulation study were 1940, 1987, 2014, 2017 and 2018. We only used the 2018 on-site measurements for model parameterisation (thermal time estimations), while the simulations solely used the historical weather data.

2.4. Data Analysis

Data analyses were conducted within R (v3.6.0) [33] using the package *rstanarm* (v2.17.4) [34] for Bayesian analysis and *lme4* (v1.1-21) [35] for (non-)linear mixed models.

2.4.1. Bud Break

Phenological data were gathered during 2018 starting at DOY 109 (19 April) following the modified E-L-stage scoring by Coombe [30], where Stage 4 represents the bud break. For practical reasons and with the aim to only estimate average bud burst, shoots were selected totally at random, hence no information on bud position on cane was available. To estimate the variability of bud break with respect to thermal time (THT ,°C d), thermal time was estimated following the approach of Schultz [36] as

$$\text{THT} = \sum_{i=1}^{n} \frac{(T_{max,i} - T_{min,i})}{2} - T_b,$$ (1)

with a base temperature of $T_b = 10\,°C$ and depending on daily maximum T_{max} and minimum T_{min} temperatures only; and n stands for days considered in THT summation. Applying a Bayesian linear mixed model to predict the most probable THT-value for the E-L-Stage 4, i.e., the bud break date, while controlling for replications (blocks) and plants (repeated measures) the posterior predictive bud break date was estimated to DOY 111 (21 April). For considering variability around predicted bud break dates with respect to THT, we used the estimated posterior predictive standard deviation of $\sigma_{THT,BB} = 12.38\,°C\,d$.

Prediction of bud break dates for all other years were realised applying a model from Nendel [37]. Nendel [37] used the single triangle algorithm from Zalom and Goodell [38] for calculation of degree-days D (°C d),

$$D = \begin{cases} 0 & \text{for } T_0 \geq T_{max} \\ \left(\frac{T_{max}-T_0}{2}\right) \cdot \left(\frac{T_{max}-T_0}{T_{max}-T_{min}}\right) & \text{for } T_{min} < T_0 < T_{max} \\ T_m - T_0 & \text{for } T_0 \leq T_{min} \end{cases},$$ (2)

depending on the daily maximum T_{max}, mean T_m, and minimum temperature T_{min}. For our predictions, we set T_0 to 5.9 °C, the estimated threshold temperature for Riesling bud break in Germany [37]. Beginning with 1 March the degree-days, D, were summed until the respective bud-break threshold $D_{BB} = (186.1 \pm 24.7)\,°C\,d$ was reached [37]. For the virtual plant simulations, we used the average date within the date range as the bud break date (BB, DOY). This information was used together with the in situ estimated standard deviation ($\sigma_{THT,BB}$) to model bud break and its natural variability.

For the bud break prediction for years different from 2018 we always used thermal time (THT) estimated by Equation (1) starting with the respective bud break date.

2.4.2. Phytomer Appearance

To model phytomer appearance rate μ (phytomers/°Cd), we used the maximum rank for each digitised primary shoot and the corresponding thermal time of the measurement date (Figure 3). Using a linear mixed effect model controlling for data from the same plants and repeated measurements in time, we found $\mu = 0.0453$ phytomers/°Cd or in other words the necessary thermal time for the development of a new phytomer is $1/\mu = 22.0892\,°C\,d$. For the lateral shoots, we assumed equal development rates.

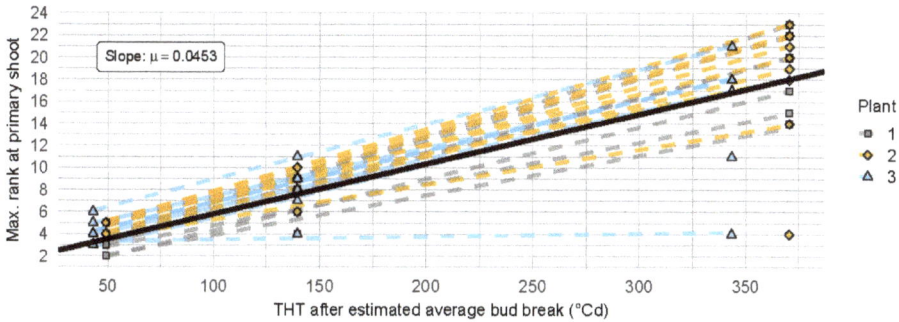

Figure 3. Estimation of phytomer appearance rate (μ) as the slope of linear mixed model fit.

2.4.3. Internodes

Phytomer appearance rate was used to estimate the thermal time age (THTage, °C d) of each node at the different measurement dates. This information was combined with the corresponding primary internode length IL_1 (cm), estimated as the Cartesian distance between two node coordinates, to fit asymptotic growth curves through the origin, following

$$IL_1(R, \text{THT}_{\text{age}}) = IL_{1,\max}(R) \cdot \left(1 - \exp\left(-\exp\left(k_{IL_1}\right) \cdot \text{THT}_{\text{age}}\right)\right),\tag{3}$$

with the growth constant $k_{IL_1} = -3.1812$ and an asymptotic value, i.e., the maximum internode length $IL_{1,\max}$, that was found to be dependent on the node's rank (R) at the primary shoot. We found the $IL_{1,\max}$ coefficients for the different ranks to follow a repetitive sequence for higher ranks, while for lower ranks ($R \leq 7$) a simple linear increase in $IL_{1,\max}$ with rank fitted the data well (Figure 4).

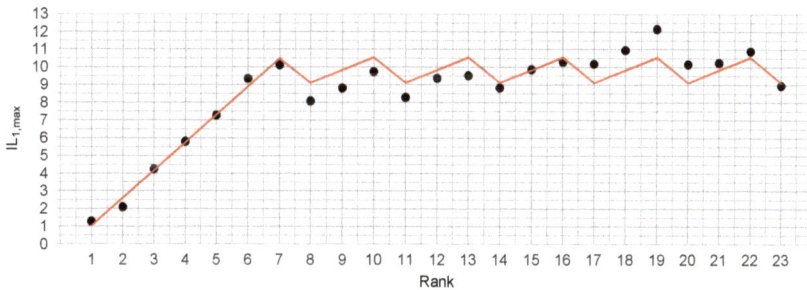

Figure 4. Estimated coefficients (dots) for rank wise maximum internode length ($IL_{1,\max}$) and model fit (red line, Equation (4)).

Hence, a step-function (Equation (4)) was used to model this systematic:

$$IL_{1,\max}(R) = \begin{cases} R \cdot a_1 + b_1 & R \leq 7 \\ ((R+1) \mod 3) \cdot a_2 + b_2 & \text{else} \end{cases},\tag{4}$$

with $a_1 = -0.5562$, $b_1 = 1.5729$ and $a_2 = 0.7212$, $b_2 = 9.0909$.

Model curves versus measurement data are shown in Figure 5A indicating good model performance even for higher ranks, where only sparse measurement data were available.

A similar approach was applied for lateral internode length IL_2 (cm), which is calculated as

$$IL_2(R, \text{THT}_{\text{age}}) = IL_{2,\max}(R) \cdot \left(1 - \exp\left(-\exp\left(k_{IL_2}\right) \cdot \text{THT}_{\text{age}}\right)\right),\tag{5}$$

with $k_{IL_2} = -3.9706$. In this case, the asymptotic values $IL_{2,max}$ showed no sign of a distinct repetitive sequence at higher levels, hence a asymptotic growth model was used to model rank dependency (Equation (6)),

$$IL_{2,max}(R) = b_3 + (a_3 - b_3) \cdot \exp\left(-\exp\left(k_{IL_{2,max}}\right) \cdot R\right),\qquad(6)$$

with the coefficients $k_{IL_{2,max}} = -0.1183$, $a_3 = -2.4822$ and $b_3 = 9.4788$. Model fit versus data is given in Figure 5B showing sparse data availability at ranks above four.

To account for variability in lateral shoots regarding appearance probability and time of a lateral bud break, locations (rank) of lateral shoots and their respective frequency at this location were extracted from the digitised plant data (Figure 6). The appearance probability $P_{app,2}$ (-) was calculated as the quotient of the number of lateral shoot at a specific rank and the total count of this rank ($n_{R,tot} = 50$), equal to the number of digitised shoots. As the development of lateral shoots was not yet completed at the final measurement date, we assumed a maximum appearance probability of 98 % for $R > 7$.

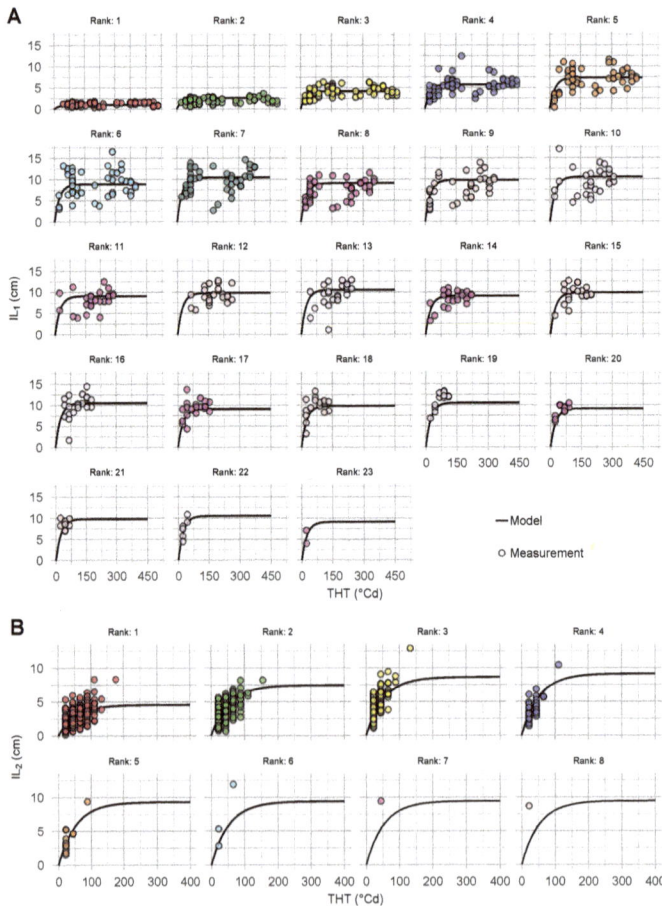

Figure 5. Model fit (solid line) and measurement data (circles) for internode length over thermal time: (**A**) internodes of primary shoots, where colours indicate different model coefficients, showing the repetitive pattern (equal colours) at ranks $R > 7$ (Equation (4)); and (**B**) internodes of lateral shoots, where colours indicate different model coefficients (ranks).

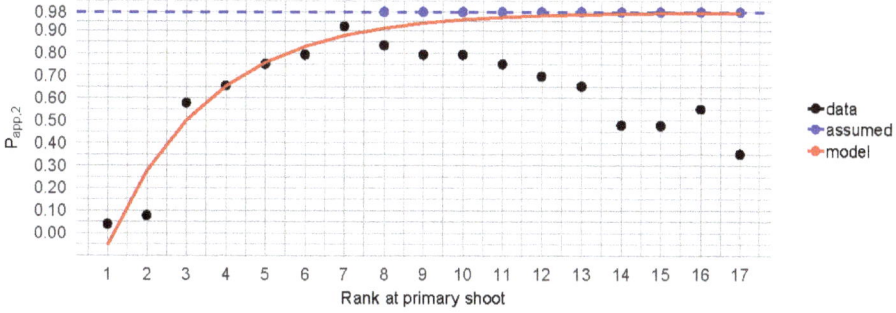

Figure 6. Estimated appearance probability for lateral shoots $P_{app,2}$ (black dots) at specific ranks and model fit (red line, Equation (4)). Blue dots represent assumed maximum probability for not yet fully developed ranks at the measurement dates.

Using this dataset, an asymptotic function was fitted describing the appearance probability at different ranks, as follows:

$$P_{app,2}(R) = p_0 + (p_1 - p_0) \cdot \exp(-\exp(k_{P_{app,2}}) \cdot R) \tag{7}$$

with $p_0 = 0.9800$, $p_1 = -0.5376$ and $k_{P_{app,2}} = -0.9585$.

Lateral bud break from a primary shoot's node was modelled as a thermal time (THT) after appearance of the corresponding rank (Equation (8)). Based on the previous analysis and on expert opinion, Ranks 1 and 2 were excluded from the analysis, i.e., lateral shoots only grow from Rank 3 upwards. Furthermore, an outlier with a negative THT was excluded, too, leaving 23 primary shoots with a total of 236 lateral shoots within the analysis.

A linear function was fitted to predict necessary post-appearance THT at which the lateral bud break occurs (THT$_{BB,2}$,°C d) at a specific rank. This accounts for the observation, that the first lateral shoots did not appear from the lowest ranks, even after excluding the first two ranks. Model fit was realised with a Bayesian mixed effect model to control for sampling structure, i.e., measurements from the same plant or the same shoot, and to estimate a posterior predictive standard deviation to be used as an uncertainty measure in the predictions of THT$_{BB,2}$. Best fit to the linear function,

$$\mathrm{THT}_{BB,2}(R) = q_0 + q_1 \cdot R \quad \text{for } R > 2, \tag{8}$$

was found for the coefficients $q_0 = 337.0334$ and $q_1 = -18.5344$. Average posterior predictive standard deviation of THT$_{BB,2}$ across ranks was estimated to $\sigma_{THT,BB,2} = 27.98$ °C d. Since this linear model would allow for negative THT$_{BB,2}$, i.e., lateral shoots would start growing from a bud not yet present, we limited the earliest thermal time passed after bud appearance to $\mathrm{THT}_{BB,2,min} = 3 \cdot \frac{1}{\mu} = 66.2676$ °C d. This represents the state that a lateral shoot appears at the earliest when three further buds are present at the primary shoot. It is based on data from lateral shoots growing from Rank 17, the highest rank a lateral shoot was already present during the experimental time frame (data not shown). Hence, implementation of this limitation follows:

$$\mathrm{THT}_{BB,2}(R) = \max(\mathrm{THT}_{BB,2,min}, q_0 + q_1 \cdot R) \quad \text{for } R > 2, \tag{9}$$

while this correction was applied twice—before and after sampling from a Gaussian normal distribution with mean THT$_{BB,2}$ and standard deviation $\sigma_{THT,BB,2}$ (Figure 7).

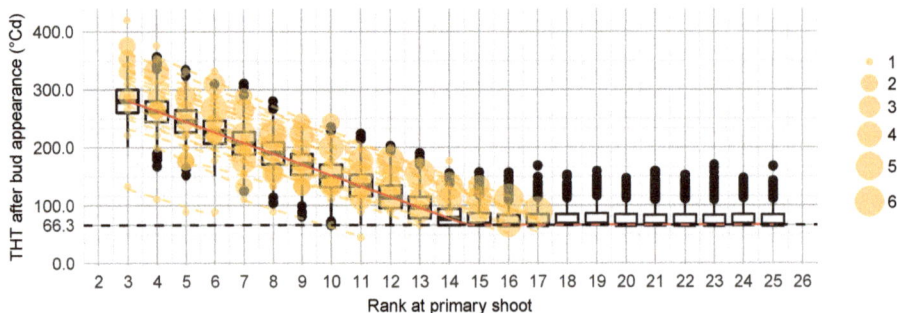

Figure 7. Data (orange) and model (red) to describe lateral bud break as thermal time (THT) after primary bud evolution. Box plots are results of sampling ($n = 1000$) from a Gaussian normal distribution around the mean ($\sigma_{THT,BB,2} = 27.98003$) after correcting for minimal $THT_{BB,2,min} = 66.2676229\,°C\,d$.

Cane internodes were initialised in the simulation scene using the maximum internode length of a full-grown primary shoot ($IL_{1,max}$), as this matches the applied training system, where one primary shoot was used as the cane in the next season.

2.4.4. Leaves

The landmark definition for digitising leaves already considered a model to estimate leaf area (LA, cm^2) utilising only the sum of the length's of the two secondary leaf veins (L_{2nd}, cm), e.g., as proposed by Beslic et al. [39] for a different variety of *Vitis vinifera* L. (cv. Blaufränkisch) (Equation (10)).

$$\sqrt{LA} = s_{LA} \cdot L_{2nd} .$$ (10)

The model was parameterised using two-dimensional vein measurements of 302 Riesling leaves and their respective leaf area (LI-3100C Area Meter, LICOR) provided by Döring et al. [40], leading to a scaling factor of $s_{LA} = 0.6869294$.

The sum of the length's of the two secondary leaf veins (L_{2nd}) in our digitised data is given by Equation (11):

$$L_{2nd} = \overline{PL2, PL5} + \overline{PL2, PL6} .$$ (11)

where $\overline{PL2, PL5}$ and $\overline{PL2, PL6}$ are measured as Cartesian distances between two landmarks (Figure 1).

Following this approach to describe leaf size, our leaf growth model was set up for the sum of the secondary leaf veins and using a similar approach as for the internode length. For the primary shoot leaves, $L_{2nd,1}$ is modelled as

$$L_{2nd,1}(R, THT_{age}) = L_{2nd,max,1}(R) \cdot \left(1 - \exp\left(-\exp\left(k_{L_{2nd,1}}\right) \cdot THT_{age}\right)\right) ,$$ (12)

with $L_{2nd,1}$ depending on rank R and thermal time age THT_{age}. The fixed growth coefficient estimated from our digitised plants was $k_{L_{2nd,1}} = -4.1819$ and the asymptotic values $L_{2nd,max,1}$ were found to be rank dependent, as follows:

$$L_{2nd,max,1}(R) = b_4 + (a_4 - b_4) \cdot \exp\left(-\exp\left(k_{L_{2nd,max,1}}\right) \cdot R\right) ,$$ (13)

with the coefficients $k_{L_{2nd,max,1}} = -0.8131$, $a_4 = 2.5499$ and $b_4 = 19.7196$.

The sum of the length of secondary veins on lateral shoot leaves ($L_{2nd,2}$) is modelled by

$$L_{2nd,2}(R, THT_{age}) = L_{2nd,max,2} \cdot \left(1 - \exp\left(-\exp\left(k_{L_{2nd,2}}\right) \cdot THT_{age}\right)\right) ,$$ (14)

with two constants $L_{2nd,max,2} = 11.9418$ and $k_{L_{2nd,2}} = -3.0008$. In this case, no clear dependence of $L_{2nd,max,2}$ on rank was found; hence, we assumed all leaves on lateral shoots can reach the same maximum size. It should be noted that this assumption was only based on measurements up to Rank 5; hence, measurements later in the seasons might have revealed a more diverse pattern.

Petiole length was found to be proportional to leaf size as sum of secondary veins ($L_{2nd,1}$) and best fit with a second-order polynomial model. Equations (15) and (16) are the results for primary (PL_1, cm) and lateral leaf petiole lengths (PL_2, cm) estimated with a mixed model controlling for repeated measurements and sampling structure.

$$PL_1(L_{2nd,1}) = a_5 \cdot L_{2nd,1} + b_5 \cdot (L_{2nd,1})^2 , \tag{15}$$

$$PL_2(L_{2nd,2}) = a_6 \cdot L_{2nd,2} + b_6 \cdot (L_{2nd,2})^2 , \tag{16}$$

with $a_5 = 0.3122$, $b_5 = 0.0060$, $a_6 = 0.3247$, and $b_6 = 0.0071$.

Petiole orientation in relation to a shoot follows an alternate-distichous phyllotaxis [36,41] where the petiole is assumed to grow out of the shoot at an angle of approximately 45°. In the model, this initial orientation is adjusted to further match the average horizontal angle found at primary and lateral shoots (Section 2.5.1). For primary shoots, this angle to the horizon was estimated to $\alpha_{Pet,1} = 40.0441°$ (upward) applying an intercept-only mixed effect model controlling for sampling structure and repeated measurements. In the same way, the average angle of lateral shoots was found to be $\alpha_{Pet,2} = 35.6239°$.

For primary leaf's midrib angle in relation to the horizon ($\alpha_{Mid,1}$) descriptive statistics indicated an interacting dependence on rank and size. This was transferred into a model equation as follows:

$$\alpha_{Mid,1}(R, L_{2nd,1}) = c_0 + c_1 \cdot L_{2nd,1} + c_2 \cdot R + c_3 \cdot L_{2nd,1} \cdot R , \tag{17}$$

with coefficient values of $c_0 = 50.2723$, $c_1 = -6.1390$, $c_2 = -5.4465$ and $c_3 = 0.4749$. Only measured once, we used an intercept only-model for lateral leaves leading to $\alpha_{Mid,2} = -25.0179°$.

In addition, we estimated a three-dimensional leaf shape by a generalised Procrustes analysis with the R-package *geomorph* (v.3.1.1) [42] based on data from 30 randomly selected leaves that were digitised with 18 instead of only 6 landmarks (data not shown). Thus far, the leaf's shape information and its horizontal orientation (midrib) were only used in visualisations and had no effect on the presented results. Hence, we also did not yet consider other leaf orientation angles, i.e. leaf roll angles around the midrib, that would become of interest as soon as plant–light interaction is introduced into the model.

2.5. Architectural Model of Riesling Canopy Growth

The architectural model was based on a Lindenmayer model approach [43]. To include variability of architectural traits not only point estimators (single parameters, such as mean data) but also distributions were considered in the developmental rules for organ appearance, orientation in space and growth rates.

2.5.1. Model Description

The functional-structural plant model for Riesling (*Virtual Riesling*) aims at simulating the dynamic growth of Riesling from bud break on the cane until end of flowering. Our modelling approach focusses on the architecture of the canopy, which is influenced by the growth behaviour of the vine, environmental conditions and vineyard management. The model was written in the programming language XL according to the formalism of relational growth grammars (RGG-formalism, [44]), which is a generalisation of Lindenmayer systems [43]; and we used the interactive modelling platform *GroIMP* (v.1.5) for model development (www.grogra.de/software/groimp) [44].

To simulate the three-dimensional architecture of Riesling, all main organs and their topological connections represent the aerial part of the plant. Rules for growth and development were extracted from digitised data (see Section 2.4) and applied in the model.

2.5.2. Scene

A virtual vineyard including three Riesling vines in a row, posts and horizontal wires built the initial set-up of a simulation. To match our experimental field conditions, the cane-supporting wires were placed in-between two posts at 70 cm above the ground. If more than one plant were initiated, the distance between trunks was $\Delta y_{trunk} = 90$ cm. The initial stage of each plant in the scene was represented by its trunk and cane. Grapevine genetics controls that winter buds on the cane were located in alternate-distichous phyllotaxis [41]. Thus, the model samples an alternation angle per bud based on a uniformly distributed random variable in an interval of 170° to 190° to introduce natural variability. At the beginning of a simulation, primary shoots will emerge from winter buds accounting for the variability of burst day (Section 2.4.1). Later on, lateral shoots evolve from buds on primary shoots. During an entire simulation, virtual plant management included events of shoot positioning (SP) similar to vineyard management practices. A system of two wires, each moving from one side to the posts, was implemented in the model to mimic SP. At each SP event, primary shoots that were long enough to be caught, were relocated between the wires. According to the practice at the experimental site, SP was applied successively at four different heights above the ground (90 cm, 110 cm, 140 cm and 170 cm). Two pairs of wires remained on the posts at the heights of 110 cm and 170 cm, after they were moved from the previous heights at 90 cm and 140 cm, respectively. A simulation ended when more than 90 % of shoots in the virtual stand reached a height of 2.3 m above the ground (Figure 8).

Figure 8. Exemplary end-of-simulation plant architecture including the scene of the *Virtual Riesling* vineyard section consisting of three plants between two posts and the shoot positioning wire pairs.

2.5.3. Shoot Development and Growth

Primary and lateral shoots were built based on phytomers, which differentiated from the apex of a shoot. In the model, a phytomere was the compound of an internode, an axillary bud and a leaf, consisting of a lamina (blade) and petiole.

Bud Break

The concept of bud break is described in Section 2.4.1. In brief, to mimic bud break variability of primary buds, we predicted an average day of bud burst on the cane for each year based on the annual temperature cycle and used a standard deviation for bud burst estimated from phenological data of 2018 ($\sigma_{THT,BB}$, see Section 2.4.1). Sampling a value of a normal distribution around zero with $\sigma_{THT,BB}$ the deviation in thermal time from the predicted average day of bud burst is used to estimate the day of primary bud break for each bud on the cane accordingly. The break of lateral buds on primary shoots depends on a rank-specific probability and a minimum threshold thermal time age (see Section 2.4.3). If a lateral bud is old enough, the chance of burst is checked once, to determine whether a new phytomer appears or the bud becomes inactive.

Phytomers

The phytomer appearance model multiplies the accumulated thermal time after burst until the current day by $\mu = 0.0453$ phytomers/$°Cd$ and divides it by the corresponding rank of the apex (see Section 2.4.3). If the result is equal to or greater than one, a new phytomer appears. At appearance, phytomers are oriented in the three-dimensional space according to the local orientation of the preceding apex. After phytomer emergence from the cane (primary shoot growth) or from the primary shoots (lateral shoot growth), the orientation of the new instance of apex is modified to mimic negative gravitropism. The local orientation of the apex may rotate upwards towards the direction of the vertical of the global coordinate system (z-axis). The rotation angle depends on a parameter for gravitropism strength with values between 0 and 1. The strength was parameterised to mimic the final shape of primary shoots observed on the digitised plants in the vineyard. Although primary shoots show an upwards orientation, they often do not point straight up. While some shoots grow upwards quickly, others show a clear incline to the cane. Moreover, shoots tend to grow upwards faster the more they are initially pointing downwards (Figure 1). To achieve this natural variability of shapes and growth behaviour, the strength of the negative gravitropism is adjusted to positions of phytomers and the orientation (local z-axis) of the apices. The more an apex points downwards, the stronger it should be rotated upwards. Therefore, the model estimates the apex's orientation as the angle to the *x-y*-plane. If this angle exceeds $-22.5°$, the strength increases from 0.5 to 1 with increasing absolute angle to the plane. From $-22.5°$ below the plane until an incline above the *x-y*-plane of 45°, the strength is fixed to 0.5. With an incline equal to or bigger than 45°, the strength is set to the weak value of 0.15 and stays constant.

Internodes

When a phytomer emerges from the apex, internodes might also be reoriented. While an entire shoot shows an upwards orientation, individual internodes do not consistently follow this trend and they show some variation in orientation from internode to internode. In the model, a slight rotation around the local *x*-axis (sampled from a uniform distribution between $-10°$ and $10°$) modifies the initial orientation of each internode. Internodes of primary and lateral shoots elongate according to Equations (3) and (5) with details described in Section 2.4.3.

Petioles and Leaves

Petioles first emerge in a fixed yaw-angle of 45° to their parent internode, in order to maintain the alternating leaf position, before they are adjusted to their final horizontal orientation. The petioles at the experimental site tended to emerge parallel to the global *x*-axis (data not shown), i.e., perpendicular to the row (*y*-axis), but this depends on the actual orientation of the shoots and can be influenced by the timing of shoot positioning managements with respect to the measurement dates. Hence, we only adjust the petioles horizontal orientation after initialisation. Therefore, the model first projects the petiole on the *x-y* plane, before adjusting the angle to the *z*-axis. On primary shoots petioles of digitised plants were slightly more inclined ($\alpha_{Pet,1} = 40.0441°$) than on lateral shoots ($\alpha_{Pet,2} = 35.6239°$). Petiole elongation was found to be proportional to leaf growth (see Section 2.4.4).

Along primary shoots leaves at the bottom tend to point downwards, while leaves at the top appear more horizontal (see Equation (17)). Leaf orientation is controlled via the midrib. The model adapts the angles in the same way as the horizontal petiole angles. A fixed midrib angle of $\alpha_{Mid,2} = -25.0179°$ and hence pointing downwards is used for leaves on lateral shoots. Leaf areas are modelled according to Equations (10), (12) and (14) depending on their order (primary or lateral shoot), rank and thermal time (see Section 2.4.4).

Shoot Positioning

The model performs shoot positioning automatically up to four times during a simulation. As soon as more than 90 % of primary shoots in the virtual canopy tower over one of the positioning heights, two wires from both sides bend every shoot they catch into the remaining gap. A wire is moved to its final position starting from below the final height on the post (approximately a movement of 45° upwards) to mimic this manual management practice. This controls whether a shoot is caught by the wires or if it is not affected by shoot positioning. A shoot that has not reached the necessary height for a SP event still might be caught from a wire at the next SP event. The wires get attached to the posts, and, therefore, the extent of the remaining gap between the wires equals the diameter of a post (10 cm). For SP, the model computes the relation between the number of shoots, that are high enough, and the total number of shoots. If more than 90 % reached or exceeded the corresponding wire height, the model initiates a shoot positioning event by determining the lengths of all shoots, but taking into account the internodes above the cane only. In addition, it subtracts 5 cm from this shoot's length, since a wire may slide upwards along the shoot during the positioning process and since the trajectory can lead to a too low position in the end. If the resulting value is larger than the height of the wire, the virtual shoot will be caught by the wire and is reoriented as follows: The wire will push the shoot into the centre and induce a reorientation of the shoot. Therefore, we included a joint-command at the position of each node. This command may change the orientation towards the direction of a defined reference point with a defined strength. The reference point is the position on the wire, where the shoot is tucked after the bend. To define its coordinates, the model extracts the y-coordinate (along the row) from the first node on the shoot, that is above the final height of the wire. The x- and z-coordinates of the reference point are taken from the wire's distance to the middle of the row and the height, respectively. All nodes of a shoot that lie below the reference point, minus a buffer of 5 cm, will be modified. The buffer guarantees that no node is chosen, which would lie above the reference point after the procedure. This is crucial to avoid a downwards reorientation of shoots. Older nodes close to the bottom of a shoot are assumed to be less flexible than younger nodes. Thus, the model modifies younger nodes with higher strength. Additionally, it excludes all nodes from reorientation, that are located more than 50 cm below the height of the current pair of wires used in the SP event. Using strength and reference point the model can apply the reorientation to all affected nodes. This is a cascade of position modifications, as the change at one node influences positioning of all following objects (Figure 9).

Avoid

To avoid that organs collide with each other or objects (posts, wires), we implemented a method where nodes change their direction, if obstacles are in their prospective path. Therefore, the model examines if other organs or objects lie within a cone of a certain inner angle and length. If so, the model randomly assigns a new direction within this cone excluding the direction of the detected obstacle (for details, see http://wwwuser.gwdg.de/~groimp/grogra.de/gallery/Technics/smart_line.html).

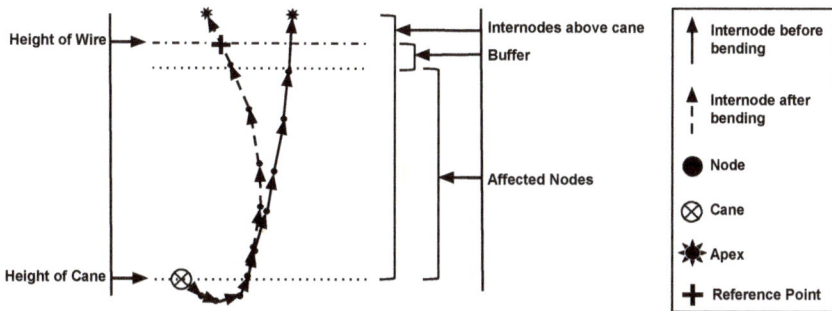

Figure 9. Schematic representation of reorientation of a shoot due to shoot positioning by tucking with a wire: shoot before bending (**right**); and shoot after bending (**left**).

2.5.4. Loop

Using a semi-stochastic model, i.e., some functions allow for observed variability, multiple simulations of a scenario year are necessary for comparisons across years. Hence, a loop function was implemented to run the model multiple times and automatically printing out internode and leaf data for external post-processing. The output includes the seed value of the random generator to be able to reproduce the exact same growth pattern, whenever needed. Currently, the model resets the scene for multiple simulations after each simulation is completed, i.e., 90 % of the primary shoots within the scene are higher than 2.3 m, and restarts with a new seed value.

2.6. Simulations

We set up 2000 simulations per year each with three plants in a single row. Those were run until the threshold for maximum height (foliage cut) was reached. Besides results from the final date we also kept result from DOY 173, the most frequent end date (653/2000) from the reference year 2018 (June 22nd). Effectiveness of sample size to estimate accurate means was tested following an estimation method from Byrne [45]. Using data on leaf area density variability from the final dates of the 2018 simulation with a confidence interval width of 5 % around the mean on a 99 % confidence level, a sample size of $n \geq 434$ was found to be sufficient.

2.7. Post-Processing

2.7.1. Descriptive Statistics

Descriptive statistics are based on all simulations results ($n = 2000$ per year) and were conducted using *R* (v3.6.0) [33] in combination with *ggplot2* (v3.1.1) [46].

2.7.2. Leaf Area Density

Leaf area density (LAD, $m^2 m^{-3}$) was estimated within a representative canopy section of the virtual canopy for each simulation. As a representative canopy section we cut out the central zone of the three-plant canopy. Boundaries of the zone were derived fixing the trunk-base of the central plant to the origin and using the trunk distance of $\Delta y_{trunk} = 0.9$ m.

Following the experimental approach of Schultz [47] LAD was estimated for cubes of $0.003\,375\,m^3$ ($a_{cube} = 0.15$ m). Hence, the length of the central zone is represented by six cubes while the volume filling process always started with a cube fixed within the bunch zone ($0.85\,m \geq z \geq 1.15\,m$) with the cube centre coordinates of $x = 0\,m$, $y = 0.5 \cdot a_{cube}$ and $z = 1\,m$. Hence, the figures in the results section are limited to the transformed canopy section where the origin matches the central coordinate of this initial cube. An exemplary division of the central zone is shown Figure 10.

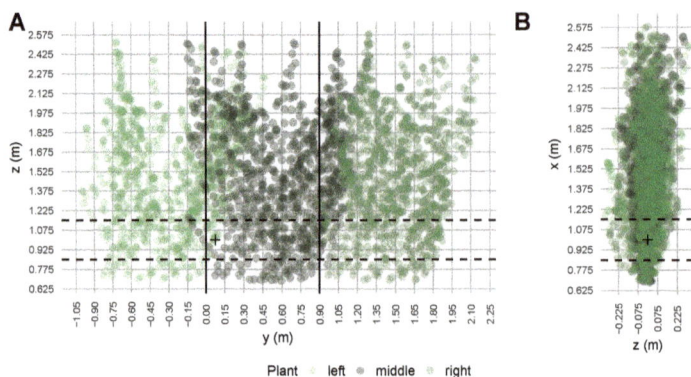

Figure 10. Canopy representation by leaf location points of three simulation plants. Calculations of leaf area density are conducted using the central zone (vertical solid lines, $\Delta y_{trunk} = 0.9$ m (trunk distance)), i.e., a representative canopy section divided into cubes of $0.003\,375$ m^3 ($a_{cube} = 0.15$ m) (dashed line: bunch zone (0.85 m $\geq z \geq 1.15$ m); +: centre point of initial cube): (**A**) side view; and (**B**) backward view.

3. Results and Discussion

3.1. Thermal Time Course

Selection of historical growth seasons was based on the fixed parameters annual mean temperature and bud break date aiming for a diversified architectural development between years. As grapevine growth was modelled in dependence of accumulated thermal time, we first compared the average thermal time course for all simulation years (Figure 11) based on average bud break and simulation end dates of the 2000 simulations per year. Figure 11 shows that daily THT contributions during the selected years differed in volatility and absolute values, leading to different simulation durations (see also Table 3).

For example, although 2017 was on average a warmer year, a cold period with no THT contribution led to longer simulations when compared to the on average colder years 1940 and 1979. In addition, Figure 11 confirms that simulation duration depended on total accumulated THT (approximately 500 °C d) (see also Figure 14). This was expected, as the end of simulation depended on the plants overall development (shoot height), which in turn depended on THT. As a consequence, this supports the necessity for conducting future climate predictions with this model to not only consider the average global warming, but also the effect on extreme temperature events.

In the following, we go into more details on how our *Virtual Riesling* vine model reacted to these different seasonal climates.

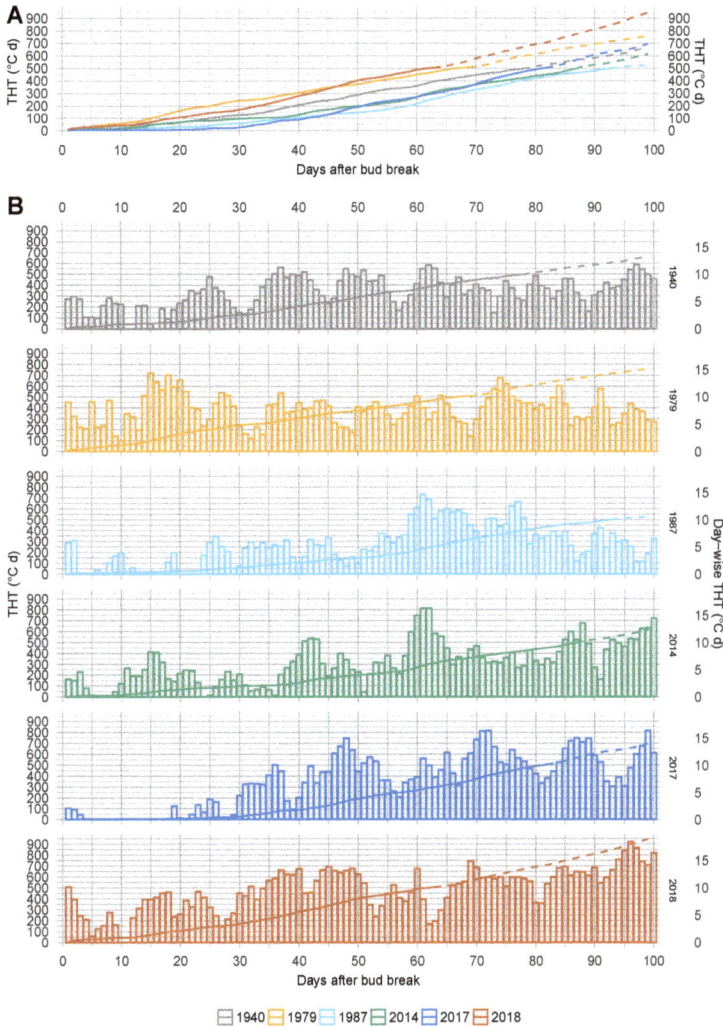

Figure 11. Thermal time (THT) development for simulations years: (**A**) direct comparison of thermal time (sum); and (**B**) year-wise plots including daily THT contributions. Start of dashed lines indicates average end of simulation date per year.

3.2. Bud Break

Figure 12 shows similar distributions around the estimated bud break along the cane within years (vertical profile), but high variability between years, i.e., multiple-peak years (1987) versus single-peak years (2018). The retrospectively estimated overall durations for bud break shows large differences from fewer than 20 days in 2018 up to 60 days in 2017 Table 2. Multiple-peak bud break years can be explained by slow and uneven temperature sum development during the bud break phase, as observed in 2017. Phases of mean temperatures below the threshold in the sensitive phase result in gaps between bud break events (e.g., DOYs 105–117 in 2017, Figure 11) and low average mean daily THT indicate long durations (Table 2). In contrast, distinct single-peak appearances of bud break days can be explained by a compact phase of high mean temperature around the estimated day of bud break (e.g., 1940 and Table 2). The simulated mean bud break data (Table 2) were in good accordance with historical data obtained from a long-term collection of Hochschule Geisenheim (e.g., http://rebschutz.hs-geisenheim.de/klima/witterung.php?Auswahl=Weinjahr). The data on interval estimates are, however, difficult to validate due to a lack of historical data of bud break variability. The earlier on-set of bud break in the more recent years is line with the observed shift towards warmer springs [48]. Considering the expected increased variability in weather events with more extreme events [49], we might speculate that the variability from single to multiple primary bud break events will be maintained, but that the general expectation interval for BB predictions has to be increased.

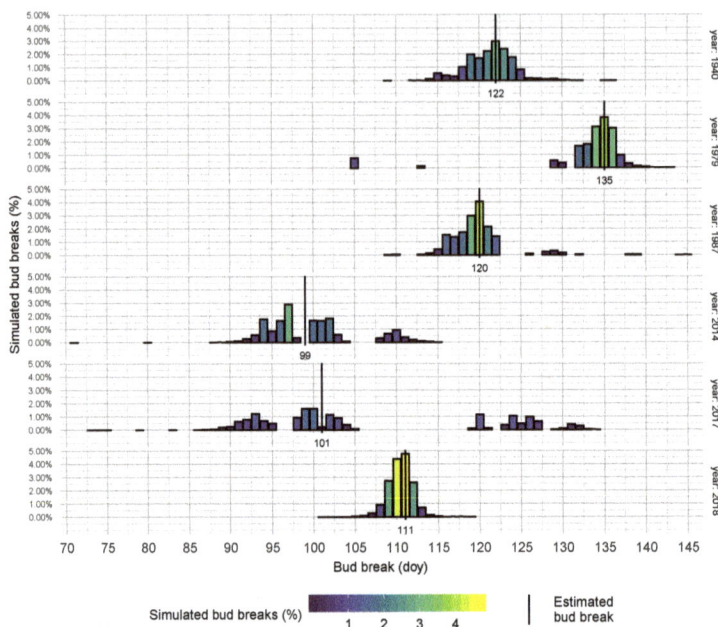

Figure 12. Distribution of simulated bud breaks at day of year (DOY) for all simulations ($n = 2000$ per year; 2000 simulations $\times$ 3 plants $\times$ 8 shoots = 48,000 simulated bud breaks per year).

Lateral bud break shows similar patterns over years and ranks (Figure 13). The latter differs significantly from the pattern of bud break for primary shoots. Bud break for lateral shoots clearly shifts in time from lower to higher ranks, which mimics the successive appearance of new phytomers on the primary shoots (see Section 2.4). This shift follows a linear trend for the time periods of lateral bud break with highest density (lighter colours on top of the scatter plots), almost without variability towards earlier DOY, but similar dispersion for phases later than those associated with highest densities. Thus, this linear trend marks an upper limit for the rank-dependent day of lateral bud break caused by the maximal established appearance rate (see Section 2.4). Earlier years show a similar slope in this linear trend, whereas 2017 and 2018 show steeper increases, and thus shorter overall durations. This is associated with the higher mean temperatures of more than 20 °C during the bud break phase in the latter years versus ca. 18 °C in earlier years (Table 2). The overall duration for lateral bud break between years varied between 35 in 2017 and 53 in 1979, which can be explained by low high to low average daily mean temperatures and mean daily THT in the corresponding years (Table 2).

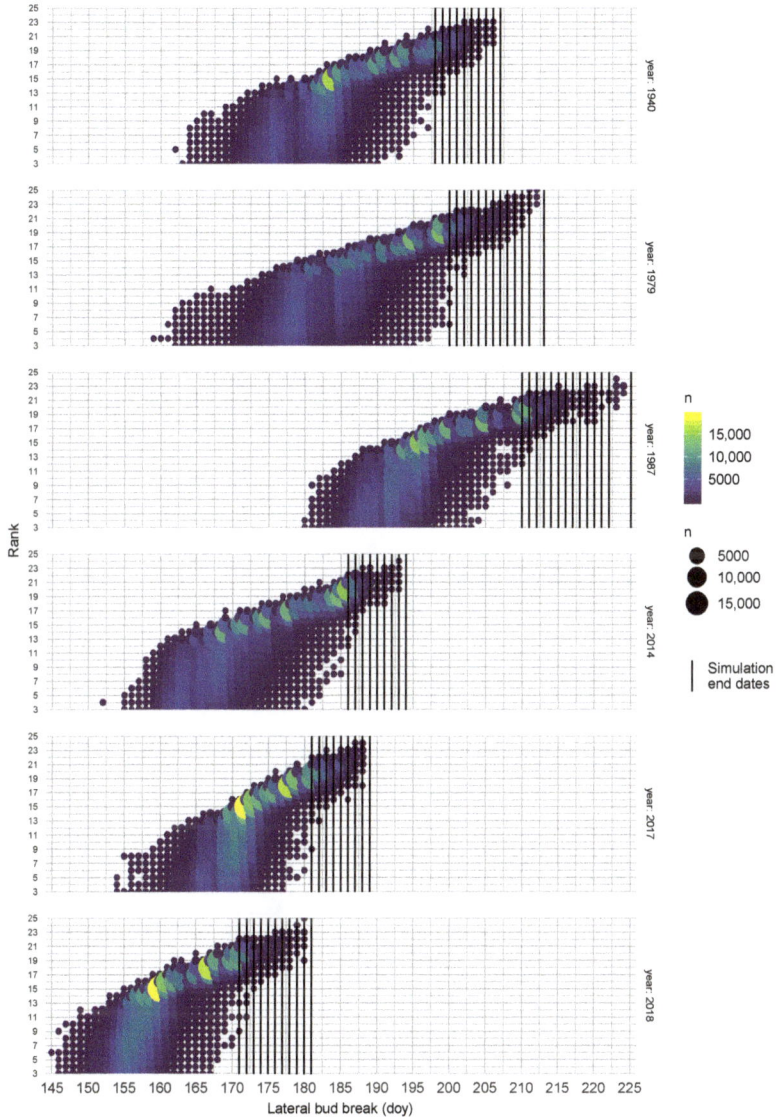

Figure 13. Lateral bud break counts of different ranks at day of year (DOY) for all simulations ($n = 2000$ per year).

Table 2. Bud break statistics per year: Range of Dates (first BB_{min} and last BB_{max}) and average daily mean temperature ($\bar{T}_m$).

	Year	BB_{min} (DOY)	BB_{max} (DOY)	$\bar{T}_m$ (°C)	Mean Daily THT (°C d)
Primary	1940	109	136	13.7 ± 2.3	3.8 ± 2.1
	1979	105	143	10.5 ± 24.6	2.2 ± 3.1
	1987	109	145	11.9 ± 23.4	2.5 ± 2.6
	2014	71	115	11.6 ± 23.2	2.3 ± 2.4
	2017	73	134	10.7 ± 22.5	1.5 ± 1.9
	2018	101	119	15.5 ± 22.7	5.4 ± 2.8
Lateral	1940	162	206	17.7 ± 22.4	7.7 ± 2.1
	1979	159	212	17.4 ± 22.5	7.4 ± 2.5
	1987	180	224	17.8 ± 23.5	7.9 ± 3.2
	2014	152	193	18.4 ± 23.0	8.5 ± 3.0
	2017	154	188	20.1 ± 23.0	10.1 ± 3.1
	2018	145	180	20.3 ± 22.8	10.2 ± 3.0

3.3. End of Simulation

As expected from the thermal-time based model concept, the different trends in temperatures over the simulation periods did not affect average accumulated thermal time in the different years (Figure 14), e.g., 520 °C d^{-1} for 2018. Small deviations of less than 12 °C d^{-1} cover the time for the appearance of just one new phytomer. In contrast, the range of thermal time for reaching the final height differs between 8 and 17 days (Figure 14). Mean DOY for reaching the termination criteria for a simulation describes a time interval of 40 days starting from DOY 174 (Table 3).

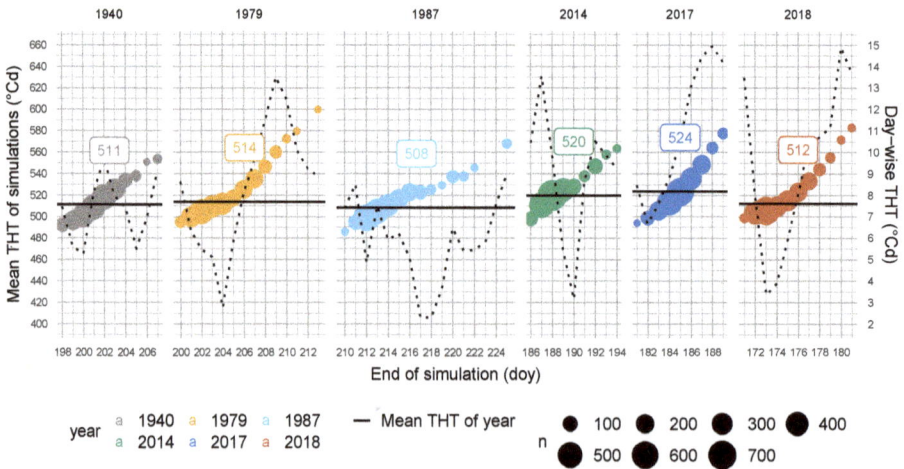

Figure 14. Variability in end of simulation date (DOY) with average accumulated THT per end-date and per year, plus daily THT contributions of adjacent dates.

Our simulations covered the growth period from primary to lateral bud break over the years. Note that the resulting simulation duration was the shortest for 2018 with just 63 days (Table 3). The longest duration was 50% larger, which means that for 1987 it was predicted to take 94 days until 90% of the shoots in the virtual stand reach the final height. Mean temperature during the simulation is a good predictor for duration of the considered growth period (linear model with $R^2 = 0.98$). However, the stable overall pattern did not predict the year-specific variability of the whole growth period over the years, which was also found by Kahlen and Chen [17]. This means that predicting future variability in bud break of grapevine will likely remain a challenging task and deserves further investigations.

Table 3. Average number of leaves in canopy zone ($\Delta y = 0.9$ m) per year with corresponding mean output date (90 % Shoot-Height $\geq$ 2.3 m), mean simulation duration and air temperature during simulations.

Year	Leaves in Canopy Zone (mean ± sd)	End of Simulation Date (mean DOY)	Simulation Duration (mean DOY)	Simulation T_m (mean ± sd)
1940	917 ± 108	201	79	16.5 ± 3.2
1979	967 ± 135	204	69	17.2 ± 2.9
1987	915 ± 128	214	94	15.0 ± 4.2
2014	978 ± 130	188	89	15.7 ± 3.8
2017	926 ± 113	185	84	15.8 ± 5.4
2018	917 ± 108	174	63	17.9 ± 3.6

3.4. Leaf Area

The variability in reaching the final heights (Figure 14) also affects the mean leaf area per shoot at the end of the simulations (Figure 15). The longer it takes until the final height is reached, the larger is the mean leaf area per shoot. Mean values cover a range from 0.5 m^2 to 1.3 m^2, which is in the range of typical values for grapevine in the production before the first hedging event. Schultz [36] reported a primary leaf area per shoot of 0.27 m^2 in the 1987 growing season for vines bearing 16 shoots. Plants simulated with *Virtual Riesling* bearing eight shoots per vine reach a primary shoot leaf area of 0.294 m^2 in our study (data not shown) at a comparable growth stage with 1987 weather data. Furthermore, Figure 14, on the one hand, shows the limitations of a standard cumulated thermal-time scheme, where more is always more, on the other hand, it illustrates the improvement due to consideration of variability in bud break events (Figures 12 and 13), which allow for reactions on short-term temperature events by altering temperature-related developmental spreads, i.e., time range of end of simulation dates. Figure 15 also shows the dispersion of leaf area over the simulations per output day (day of year). Even though the variability is almost stable over output date and year, it should be considered in further research on the impact of temperatures on grapevine architecture.

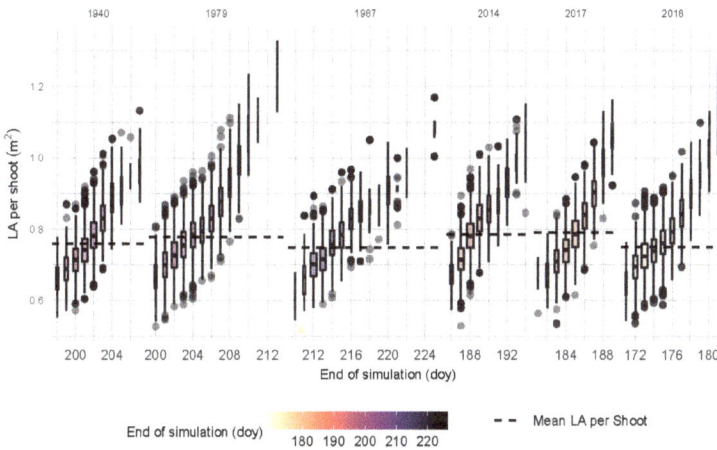

Figure 15. Total leaf area (LA) per shoot at the different end of simulation dates (DOY) for all years. Width of box plots indicate the number of simulations that ended at this date (total number of simulations per year: $n = 2000$).

3.5. Leaf Area Index

Leaf area index (LAI, $m^2\,m^{-2}$) was estimated for the representative canopy using the leaf area data from within $\Delta x = 1.8\,m$ (row distance) and $\Delta y = 0.9\,m$ (trunk distance).

The variability driven spread in reaching the final height translates from shoot leaf area to leaf area index (Figure 16). However, production typical average values are about 3.8 [50].

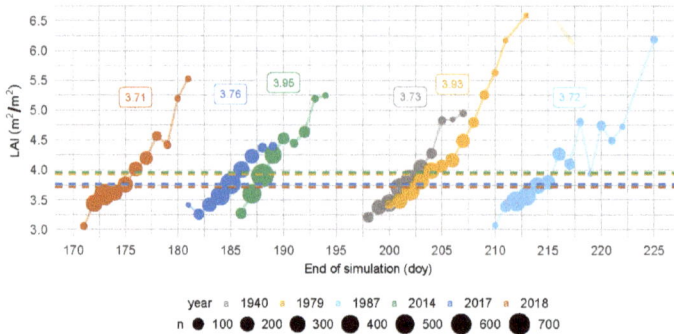

Figure 16. Leaf area index (LAI, $m^2\,m^{-2}$) averages per end of simulation (DOY) and year. Dashed lines and annotation indicate year-wise mean LAI.

3.6. Leaf Area Density

Leaf area density (LAD, $m^2\,m^{-3}$) is an important means to measure the distribution of leaf surface in space and identify locations of elevated density in the canopy [47]. Zones of high density have a higher disease infection pressure due to unfavourable microclimatic conditions [51,52] and contain leaves with a low intrinsic water use efficiency [53]. In addition, high LAD in the bunch zone may lead to a loss of grape quality [54]. LAD can locally reach values of $24\,m^2\,m^{-3}$ in dense canopies [55]. This was the case in our virtual grapevine population, in which the vertical distribution of LAD along the cane ($x = 0\,m$) was similar and seemingly high in all years (Figure 17). LADs of up to $30\,m^2\,m^{-3}$ were observed above the bunch zone in the canopy centre after the last shoot positioning event. Figure 17 shows the distribution of LAD in cubes of $0.003\,375\,m^3$ located along the cane (y-direction), perpendicular to the row (x-direction) as well as in the vertical direction (z) for all years. Note that LAD about $30\,m^2\,m^{-3}$ does reflect the average presence of circa five fully grown grapevine leaves in a single cube, whereas values of $0\,m^2\,m^{-3}$ indicate that the LAD is positive, but below $0.5\,m^2\,m^{-3}$. Such a low LAD might occur, if in one of ten simulations just one leaf would be positioned in a cube of interest. The high LAD compared to literature data can be explained by the following facts: First, the cubes for the LAD estimations in this study were placed in the middle of the cane, while Schultz [47] placed the cubes to the right and left sides of the cane. Second, a meticulous shoot positioning protocol was observed for the vines used for digitisation, leading to a relatively compact, but slim canopy. Third, as shoot primary leaf area was comparable to previously published data, the high LADs above the bunch zone, as well as the comparatively high LAIs may be explained by a strong side shoot growth. This is underlined by the fact that the main contributor to LAD above the bunch zone were indeed leaves from lateral shoots (Figure 18), while they were leaves from primary shoots in the study of Schultz [47]. This strong lateral shoot growth may on the one hand be related to the very vigorous shoots in our dataset, originating from vines with only eight shoots. On the other hand, it has been shown that, under warm climatic conditions around bud burst, the apical dominance of the shoots is reduced and lateral shoot growth promoted [4]. Thus, a larger dataset for model parameterisation might be necessary to give better estimates of lateral shoot growth under changing climatic conditions. The width of the resulting virtual canopies is represented by the LAD-values for different x-positions. Interestingly, the canopy in 1987 has a smaller width than in all other years (Figure 17). This could only be explained by the interplay of shoot positioning and the specific pattern of thermal time during the simulation in 1987. Here, on average over all simulations more shoots might have be caught at vertical positioning. Even though the overall width is larger, the differences may not be significant because of the quite low LAD for higher x-values. However, the high-resolution of space in cubes can be easily changed to different cubes sizes and thus adapted to specific analytical purposes.

To further analyse these LAD-values, we summarised the data over horizontal layers, i.e., cuboids covering the entire canopy width (Figure 18). The LAD is similar along the cane (y-direction) for all years (Figure 18). Figure 18B shows that layer-specific LAD values are in the range of up to $4\,\text{m}^2\,\text{m}^{-3}$ at the final simulation days and that this is not affected by the year. In contrast, if LAD data were compared for a specific DOY (here 173), the year and, therefore, temperature would significantly affect leaf area distribution within the canopy in height (z-direction). This has important effects for many viticultural aspects, such as radiation partitioning, cover crop growth, or soil water relations.

In a final coarsening step, LAD of the canopy was estimated depending only on height or width of the canopy using layers of 0.15 m thickness while differentiating between primary and lateral leaves' LAD (Figure 19). As discussed above, when comparing the end of simulation dates between years, the canopies are more or less similar. Figure 19 shows that the small differences in LAD between years area more related to lateral leaf growth than to primary leaves.

This can be explained by the fact that lateral shoots counts and internodes are rising nearly exponential at thermal times around $500\,°\text{C}\,\text{d}$, as more and more lateral shoots appear and grow, while the number of primary shoots remains constant. In other studies (e.g., [47]), primary leaves usually dominate LAD instead of lateral leaves, but foliage cut events during the season might have affected this ratio. This discrepancy between our simulations and other findings might be related to the fact, that lateral shoot growth parameterisation of our model was limited to one measurement date per plant. Hence, for instance, we assumed similar phytomer development rates for lateral and primary shoots and did not analyse any effects of possible different phytomer types, as indicated by Figure 4 and identified for other varieties, Grenache Noir and Syrah, by Louarn et al. [56] and Pallas et al. [57]. As a consequence, our model should benefit by a more extensive parameterisation based on digitisations in another, less extreme year and later in the season, ideally immediately before and after a foliage cut.

It is evident that other changing environmental factors, such as CO_2 concentration [28,58] and water availability [59,60] will also affect growth and plant architecture of grapevines. Especially growth constraints due to limited water ability drastically affect grapevine shoot architecture (internode length, side shoot growth, leaf size, [21,61,62]) and have to be incorporated into descriptive plant architecture models if climate change effects are to be realistically modelled. Water deficit effects on grapevine growth have not yet been integrated into the virtual Riesling, but will be the next step in model development. An additional factor of climate change is the rise of air CO_2 levels. Elevated air CO_2 concentration might help to alleviate water deficit effects on plant growth due to an increased intrinsic water use efficiency. In addition, changes in CO_2 levels will also affect plant architecture [63]. However, results for grapevine are sparse and sometimes contradictory. Greenhouse experiments have shown a reduced growth under elevated CO_2 [61,62], while field trials using FACE systems have shown the opposite results [28,64]. More research is thus necessary to get reliable field data for modelling CO_2 effects on grapevine growth. Modelling future vine growth and canopy characteristics will be of great advantage to vine growers with regards to planning canopy manipulation techniques, optimise plant protection, forecast future workforce demand and facilitate vineyard planning. A faster vegetative development, as seen in Geisenheim in the year 2018, will not only affect grape quality due to higher temperatures and a higher light interception of bunches during berry development, but also challenge viticulturists: As viticultural tasks such as suckering, shoot positioning and hedging will occur in a compressed time frame, workforce demand during the season will change. At the same time, there will be a need to carry out these tasks more accurately to avoid zones with extreme LAD, which favour the development of fungal diseases, especially under warmer conditions.

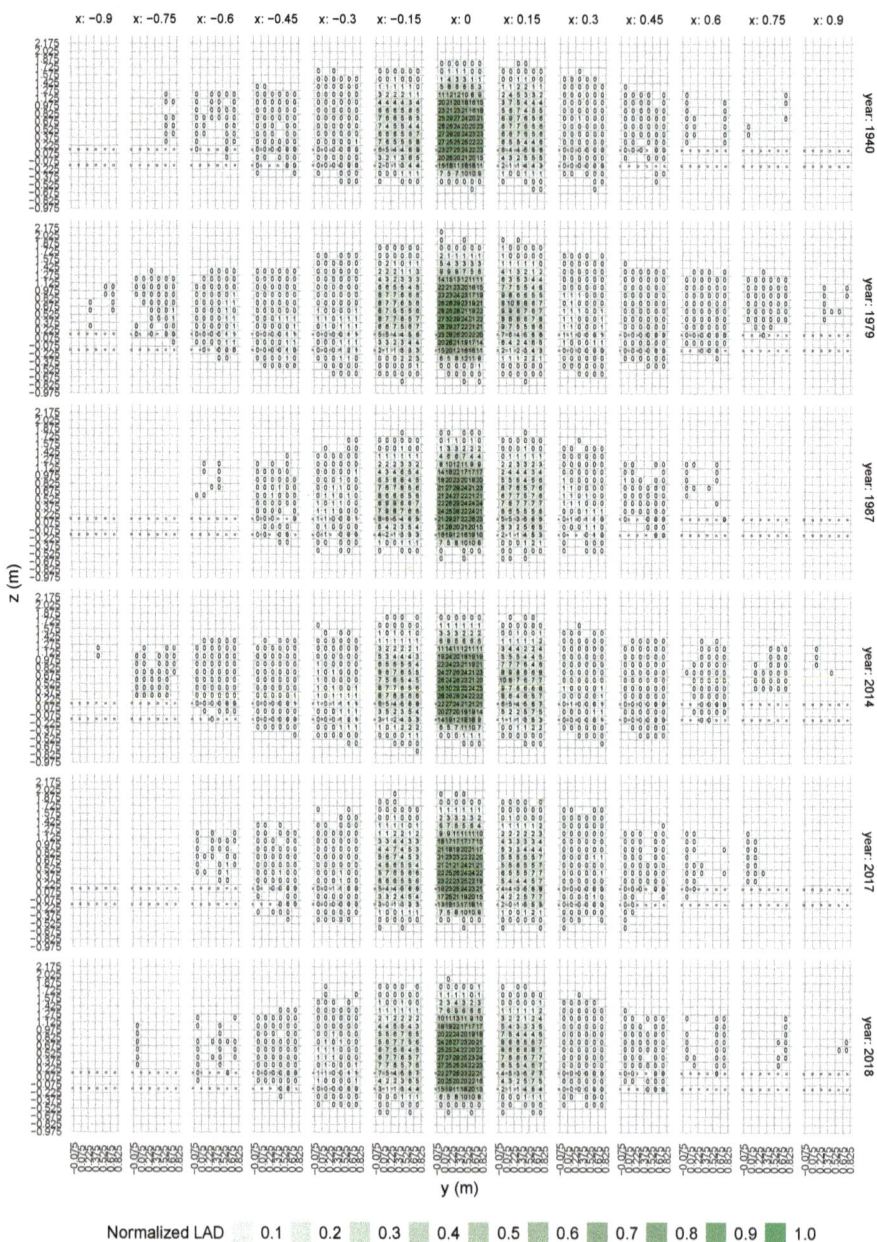

Figure 17. Leaf area density (LAD, m^2 m^{-3}) for cubes of 0.003 375 m^3 (15 cm × 15 cm × 15 cm) at final simulation dates (i.e., 90 % ≥ 2.3 m) per simulation year. Opacity: normalised LAD (LAD/ max(LAD); side view of canopy: cane/canopy aligned with *y*-axis; *x*: distance to cane; *y*: distance along cane/canopy; *z*: height relative to centre of bunch zone; dotted line: bunch zone.

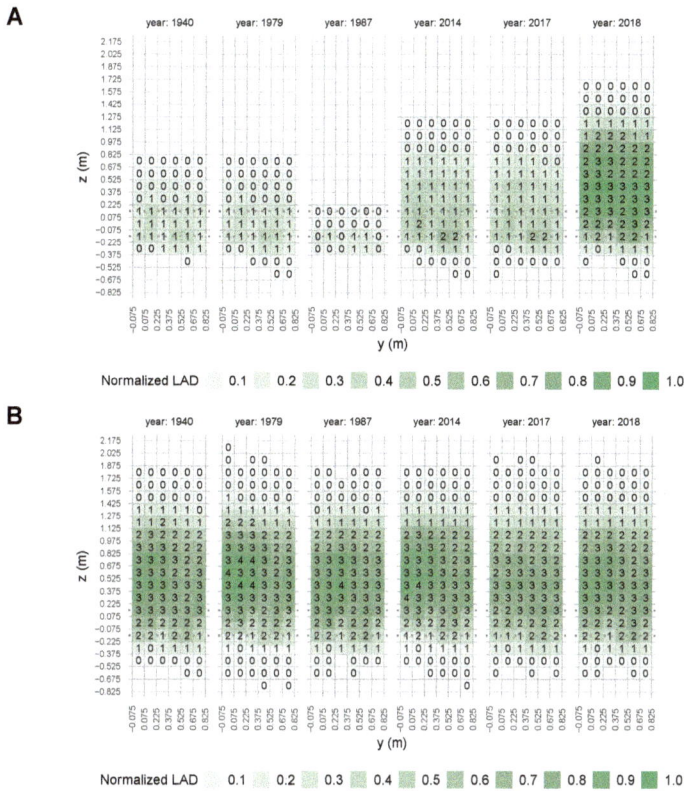

Figure 18. Differences in leaf area density (LAD, $m^2\,m^{-3}$) for cuboids of $0.050\,625\,m^3$ ($195\,cm \times 15\,cm \times 15\,cm$) caused by bud burst date and temperatures after bud burst: (**A**) at DOY 173 per year ($n_{2018} = 653$, $n_{else} = 2000$); and (**B**) for averages at the end of simulation per year ($n = 2000$). Opacity: normalised LAD (LAD/ max(LAD). Side view of canopy: cane/canopy aligned with y-axis; y: distance along cane/canopy; z: height relative to centre of bunch zone; dotted line: bunch zone.

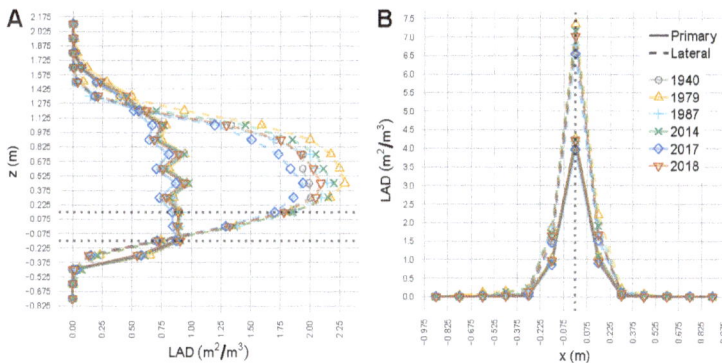

Figure 19. Leaf area density (LAD, $m^2\,m^{-3}$) of primary and lateral leaves within the canopy section for: (**A**) horizontal (z: height relative to centre of bunch zone; dotted: bunch zone); and (**B**) vertical (left and right of cane, dotted: cane-/x-axis) layers with a thickness $0.15\,m$.

127

4. Conclusions

We successfully developed a new functional-structural plant model *Virtual Riesling* based on digitised data of real plants grown in a vineyard and a comprehensive state-of-the-art data analysis. The model allows for the management technique "vertical shoot positioning" including bending of the shoots and subsequent reorientation of the canopy, the latter an intrinsic characteristic of the virtual canopy. We conducted an extensive simulation study based on *Virtual Riesling* using historical data of candidate years covering a 91-year timespan with some most distinct years. Simulation results reveal significant effects of the thermal time course over the year on primary and lateral bud breaks. High variabilities in these events affect canopy growth and leaf area distribution. All of them might have severe effects on crop management measures. The discussion of simulation output showed some discrepancies between reported data on leaf area characteristics such as the leaf area densities. We attributed this to the specific dataset used for model parameterisation. A more extensive dataset considering more plants and years with different temperature conditions as well as extended measurement periods should overcome this issue, when considered in future investigations. Despite the limitations, this report shows that a model such as *Virtual Riesling* could be used to assess the significance of changing temperatures for grapevine architecture and thereby considering management techniques such as vertical shoot positioning. Further developments of *Virtual Riesling* might support the knowledge gain for developing necessary adaptations in future vineyard management and, thus, facilitate future work on climate change research using functional-structural model approaches.

Author Contributions: Conceptualization, D.S., C.B., M.F. and K.K.; methodology, D.S., C.B., M.F. and K.K.; software, C.B.; validation, D.S., C.B., M.F. and K.K.; formal analysis, D.S.; investigation, D.S., M.F. and K.K.; resources, D.S., M.F. and K.K.; data curation, D.S. and C.B.; writing–original draft preparation, D.S., C.B., M.F. and K.K.; writing–review and editing, D.S., C.B., M.F. and K.K.; visualization, D.S. and C.B.; supervision, K.K.; project administration, D.S., M.F. and K.K.;

Funding: Funds for this study were provided by The Hessen State Ministry for Higher Education, Research and the Arts (HMWK).

Acknowledgments: The authors would like to thank the Geisenheim VineyardFACE team for facilitating the digitisation experiments and providing supplementary data.

Conflicts of Interest: The authors declare no conflict of interest.

References

1. van Leeuwen, C.; Darriet, P. The impact of climate change on viticulture and wine quality. *J. Wine Econ.* **2016**, *11*, 150–167. [CrossRef]
2. Francesca, S.; Simona, G.; Francesco Nicola, T.; Andrea, R.; Vittorio, R.; Federico, S.; Cynthia, R.; Maria Lodovica, G. Downy mildew (*Plasmopara viticola*) epidemics on grapevine under climate change. *Glob. Chang. Biol.* **2006**, *12*, 1299–1307. [CrossRef]
3. Bregaglio, S.; Donatelli, M.; Confalonieri, R. Fungal infections of rice, wheat, and grape in Europe in 2030–2050. *Agron. Sustain. Dev.* **2013**, *33*, 767–776. [CrossRef]
4. Keller, M.; Tarara, J.M. Warm spring temperatures induce persistent season-long changes in shoot development in grapevines. *Ann. Bot.* **2010**, *106*, 131–141. [CrossRef] [PubMed]
5. Lebon, E.; Pellegrino, A.; Tardieu, F.; Lecoeur, J. Shoot Development in Grapevine (*Vitis vinifera*) is Affected by the Modular Branching Pattern of the Stem and Intra- and Inter-shoot Trophic Competition. *Ann. Bot.* **2004**, *93*, 263–274, [CrossRef] [PubMed]
6. Luchaire, N.; Rienth, M.; Romieu, C.; Nehe, A.; Chatbanyong, R.; Houel, C.; Ageorges, A.; Gibon, Y.; Turc, O.; Muller, B.; et al. Microvine: A New Model to Study Grapevine Growth and Developmental Patterns and their Responses to Elevated Temperature. *Am. J. Enol. Vitic.* **2017**, *68*, 283–292, [CrossRef]
7. Pallas, B.; Loi, C.; Christophe, A.; Cournède, P.H.; Lecoeur, J. Comparison of three approaches to model grapevine organogenesis in conditions of fluctuating temperature, solar radiation and soil water content. *Ann. Bot.* **2010**, *107*, 729–745. [CrossRef] [PubMed]
8. Buttrose, M. Vegetative growth of grapevine varieties under controlled temperature and light intensity. *Vitis* **1969**, *8*, 280–285.

9.   Parent, B.; Tardieu, F. Temperature responses of developmental processes have not been affected by breeding in different ecological areas for 17 crop species. *New Phytol.* **2012**, *194*, 760–774. [CrossRef] [PubMed]

10. Schultz, H.R.; Stoll, M. Some critical issues in environmental physiology of grapevines: Future challenges and current limitations. *Aust. J. Grape Wine Res.* **2010**, *16*, 4–24. [CrossRef]

11. Van Leeuwen, C.; Destrac-Irvine, A. Modified grape composition under climate change conditions requires adaptations in the vineyard. *Oeno One* **2017**, *51*, 147–154. [CrossRef]

12. Vos, J.; Evers, J.B.; Buck-Sorlin, G.H.; Andrieu, B.; Chelle, M.; De Visser, P.H. Functional–structural plant modelling: A new versatile tool in crop science. *J. Exp. Bot.* **2009**, *61*, 2101–2115. [CrossRef] [PubMed]

13. Stützel, H.; Kahlen, K. Editorial: Virtual Plants: Modeling Plant Architecture in Changing Environments. *Front. Plant Sci.* **2016**, *7*, 1734, [CrossRef] [PubMed]

14. Evers, J.B.; Letort, V.; Renton, M.; Kang, M. Computational botany: Advancing plant science through functional–structural plant modelling. *Ann. Bot.* **2018**, *121*, 767–772, [CrossRef]

15. Chen, T.W.; Stützel, H.; Kahlen, K. High light aggravates functional limitations of cucumber canopy photosynthesis under salinity. *Ann. Bot.* **2017**, *121*, 797–807, [CrossRef] [PubMed]

16. Chen, T.W.; Nguyen, T.; Kahlen, K.; Stützel, H. High temperature and vapor pressure deficit aggravate architectural effects but ameliorate non-architectural effects of salinity on dry mass production of tomato. *Front. Plant Sci.* **2015**, *6*, 887. [CrossRef] [PubMed]

17. Kahlen, K.; Chen, T.W. Predicting Plant Performance Under Simultaneously Changing Environmental Conditions—The Interplay Between Temperature, Light, and Internode Growth. *Front. Plant Sci.* **2015**, *6*, 1130, [CrossRef] [PubMed]

18. Cieslak, M.; Seleznyova, A.N.; Hanan, J. A functional–structural kiwifruit vine model integrating architecture, carbon dynamics and effects of the environment. *Ann. Bot.* **2010**, *107*, 747–764, [CrossRef] [PubMed]

19. Louarn, G.; Lebon, E.; Lecoeur, J. "Top-vine", a topiary approach based architectural model to simulate vine canopy structure. In Proceedings of the XIV International GESCO Viticulture Congress, Geisenheim, Germany, 23–27 August 2005; Groupe d'Etude des Systemes de COnduite de la vigne (GESCO): Geisenheim, Germany, 2005; pp. 464–470.

20. Louarn, G.; Dauzat, J.; Lecoeur, J.; Lebon, E. Influence of trellis system and shoot positioning on light interception and distribution in two grapevine cultivars with different architectures: An original approach based on 3D canopy modelling. *Aust. J. Grape Wine Res.* **2008**, *14*, 143–152. [CrossRef]

21. Iandolino, A.; Pearcy, R.; Williams, L.E. Simulating three-dimensional grapevine canopies and modelling their light interception characteristics. *Aust. J. Grape Wine Res.* **2013**, *19*, 388–400. [CrossRef]

22. Zhu, J.; Gambetta, G.A.; Vivin, P.; Ollat, N.; Delrot, S.; Dai, Z.; Génard, M.; Vercambre, G. Growing grapes on a virtual plant. In Proceedings of the IEEE 2018 6th International Symposium on Plant Growth Modeling, Simulation, Visualization and Applications (PMA), Hefei, China, 4–8 November 2018 ; pp. 56–60.

23. Prieto, J.A.; Louarn, G.; Perez Pena, J.; Ojeda, H.; Simonneau, T.; Lebon, E. A leaf gas exchange model that accounts for intra-canopy variability by considering leaf nitrogen content and local acclimation to radiation in grapevine (*Vitis vinifera* L.). *Plant Cell Environ.* **2012**, *35*, 1313–1328. [CrossRef] [PubMed]

24. Garin, G.; Fournier, C.; Andrieu, B.; Houlès, V.; Robert, C.; Pradal, C. A modelling framework to simulate foliar fungal epidemics using functional–structural plant models. *Ann. Bot.* **2014**, *114*, 795–812. [CrossRef] [PubMed]

25. Zhu, J.; Dai, Z.; Vivin, P.; Gambetta, G.A.; Henke, M.; Peccoux, A.; Ollat, N.; Delrot, S. A 3-D functional–structural grapevine model that couples the dynamics of water transport with leaf gas exchange. *Ann. Bot.* **2017**, *121*, 833–848. [CrossRef] [PubMed]

26. Zhu, J.; Génard, M.; Poni, S.; Gambetta, G.A.; Vivin, P.; Vercambre, G.; Trought, M.C.T.; Ollat, N.; Delrot, S.; Dai, Z. Modelling grape growth in relation to whole-plant carbon and water fluxes. *J. Exp. Bot.* **2018**, *70*, 2505–2521.10.1093/jxb/ery367. [CrossRef] [PubMed]

27. Albasha, R.; Fournier, C.; Pradal, C.; Chelle, M.; Prieto, J.A.; Louarn, G.; Simonneau, T.; Lebon, E. HydroShoot: A functional-structural plant model for simulating hydraulic structure, gas and energy exchange dynamics of complex plant canopies under water deficit - application to grapevine (*Vitisvinifera* L.). *In Silico Plants* **2019**, [CrossRef]

28. Wohlfahrt, Y.; Smith, J.; Tittmann, S.; Honermeier, B.; Stoll, M. Primary productivity and physiological responses of Vitis vinifera L. cvs. under Free Air Carbon dioxide Enrichment (FACE). *Eur. J. Agron.* **2018**, *101*, 149–162, [CrossRef]

29. Schulze-Sylvester, M.; Reineke, A. Elevated CO_2 Levels Impact Fitness Traits of Vine Mealybug Planococcus ficus Signoret, but Not Its Parasitoid Leptomastix dactylopii Howard. *Agronomy* **2019**, *9*. [CrossRef]

30. Coombe, B. Growth stages of the grapevine: Adoption of a system for identifying grapevine growth stages. *Aust. J. Grape Wine Res.* **1995**, *1*, 104–110. [CrossRef]

31. Kahlen, K.; Stützel, H. Estimation of Geometric Attributes and Masses of Individual Cucumber Organs Using Three-dimensional Digitizing and Allometric Relationships. *J. Am. Soc. Hortic. Sci.* **2007**, *132*, 439–446. [CrossRef]

32. Schmidt, D.; Kahlen, K. Towards More Realistic Leaf Shapes in Functional-Structural Plant Models. *Symmetry* **2018**, *10*, 278. [CrossRef]

33. R Core Team. *R: A Language and Environment for Statistical Computing*; R Foundation for Statistical Computing: Vienna, Austria, 2019.

34. Goodrich, B.; Gabry, J.; Ali, I.; Brilleman, S. Rstanarm: Bayesian Applied regression Modeling via Stan., 2018. R Package Version 2.17.4. Available online: http://mc-stan.org/ (accessed on 29 June 2019).

35. Bates, D.; Mächler, M.; Bolker, B.; Walker, S. Fitting Linear Mixed-Effects Models Using lme4. *J. Stat. Softw.* **2015**, *67*, 1–48, [CrossRef]

36. Schultz, H.R. An empirical model for the simulation of leaf appearance and leaf area development of primary shoots of several grapevine (*Vitis vinifera* L.) canopy-systems. *Sci. Hortic.* **1992**, *52*, 179–200. [CrossRef]

37. Nendel, C. Grapevine bud break prediction for cool winter climates. *Int. J. Biometeorol.* **2010**, *54*, 231–241. [CrossRef]

38. Zalom, F.G.; Goodell, P.B. *Degree Days: The Calculation and Use of Heat Units in Pest Management*; University of California, Division of Agriculture and Natural Resources: Davis, CA, USA , 1983.

39. Beslic, Z.; Todic, S.; Tesic, D. Validation of non-destructive methodology of grapevine leaf area estimation on cv. Blaufränkisch (*Vitis vinifera* L.). *S. Afr. J. Enol. Vitic.* **2010**, *31*, 22–25. [CrossRef]

40. Döring, J.; Stöber, V.; Tittmann, S.; Kauer, R.; Stoll, M. Estimating leaf area and leaf area index in VSP trained grapevines under different management systems (*Vitis vinifera* cv. Riesling). In Proceedings of the 18th International Symposium GiESCO, Porto, Portugal, 7–11 July 2013.

41. Pratt, C. Vegetative anatomy of cultivated grapes—A review. *Am. J. Enol. Vitic.* **1974**, *25*, 131–150.

42. Adams, D.; Collyer, M.; Kaliontzopoulou, A. Geomorph: Software for Geometric Morphometric Analyses. R Package Version 3.1.0. Available online: https://cran.r-project.org/package=geomorph (accessed on 29 June 2019).

43. Prusinkiewicz, P.; Lindenmayer, A. *The Algorithmic Beauty of Plants*; Springer Science & Business Media: New York, NY, USA, 1990.

44. Kniemeyer, O. Design and Implementation of a Graph Grammar Based Language for Functional-Structural Plant Modelling. Ph.D. Thesis, Brandenburg University of Technology, Cottbus-Senftenberg, Germany, 2008.

45. Byrne, M. How many times should a stochastic model be run? An approach based on confidence intervals. In Proceedings of the 12th International Conference on Cognitive Modelling, University Park, PA, USA, 11 July 2013.; pp. 445–450.

46. Wickham, H. *ggplot2: Elegant Graphics for Data Analysis*; Springer-Verlag New York: New York, NY, USA, 2016.

47. Schultz, H. Grape canopy structure, light microclimate and photosynthesis. I. A two-dimensional model of the spatial distribution of surface area densities and leaf ages in two canopy systems. *Vitis* **1995**, *34*, 211–216.

48. Walther, G.R.; Post, E.; Convey, P.; Menzel, A.; Parmesan, C.; Beebee, T.J.; Fromentin, J.M.; Hoegh-Guldberg, O.; Bairlein, F. Ecological responses to recent climate change. *Nature* **2002**, *416*, 389. [CrossRef]

49. Diffenbaugh, N.S.; Singh, D.; Mankin, J.S. Unprecedented climate events: Historical changes, aspirational targets, and national commitments. *Sci. Adv.* **2018**, *4*. [CrossRef]

50. Siegfried, W.; Viret, O.; Huber, B.; Wohlhauser, R. Dosage of plant protection products adapted to leaf area index in viticulture. *Crop Prot.* **2007**, *26*, 73–82. [CrossRef]

51. Zahavi, T.; Reuveni, M.; Scheglov, D.; Lavee, S. Effect of grapevine training systems on development of powdery mildew. *Eur. J. Plant Pathol.* **2001**, *107*, 495–501. [CrossRef]

52. Austin, C.N.; Grove, G.G.; Meyers, J.M.; Wilcox, W.F. Powdery mildew severity as a function of canopy density: Associated impacts on sunlight penetration and spray coverage. *Am. J. Enol. Vitic.* **2011**, *62*, 23–31. [CrossRef]

53. Medrano, H.; Pou, A.; Tomás, M.; Martorell, S.; Gulias, J.; Flexas, J.; Escalona, J.M. Average daily light interception determines leaf water use efficiency among different canopy locations in grapevine. *Agric. Water Manag.* **2012**, *114*, 4–10, For a better use and distribution of water, [CrossRef]

54. Smart, R.; Dick, J.K.; Gravett, I.M.; Fisher, B. Canopy management to improve grape yield and wine quality-principles and practices. *S. Afr. J. Enol. Vitic.* **1990**, *11*, 3–17. [CrossRef]

55. Schultz, H.R. Extension of a Farquhar model for limitations of leaf photosynthesis induced by light environment, phenology and leaf age in grapevines (*Vitis vinifera* L. cvv. White Riesling and Zinfandel). *Funct. Plant Biol.* **2003**, *30*, 673–687. [CrossRef]

56. Louarn, G.; Guedon, Y.; Lecoeur, J.; Lebon, E. Quantitative Analysis of the Phenotypic Variability of Shoot Architecture in Two Grapevine (*Vitis vinifera*) Cultivars . *Ann. Bot.* **2007**, *99*, 425–437, [CrossRef]

57. Pallas, B.; Louarn, G.; Christophe, A.; Lebon, E.; Lecoeur, J. Influence of intra-shoot trophic competition on shoot development in two grapevine cultivars (*Vitis vinifera*). *Physiol. Plant.* **2008**, *134*, 49–63, [CrossRef]

58. Bindi, M.; Fibbi, L.; Miglietta, F. Free Air CO_2 Enrichment (FACE) of grapevine (*Vitis vinifera* L.): II. Growth and quality of grape and wine in response to elevated CO_2 concentrations. *Eur. J. Agron.* **2001**, *14*, 145–155. [CrossRef]

59. Escalona, J.; Flexas, J.; Medrano, H. Drought effects on water flow, photosynthesis and growth of potted grapevines. *Vitis* **2002**, *41*, 57–62.

60. Dry, P.; Loveys, B. Grapevine shoot growth and stomatal conductance are reduced when part of the root system is dried. *Vitis* **1999**, *38*, 151–156.

61. Kizildeniz, T.; Mekni, I.; Santesteban, H.; Pascual, I.; Morales, F.; Irigoyen, J.J. Effects of climate change including elevated CO_2 concentration, temperature and water deficit on growth, water status, and yield quality of grapevine (*Vitis vinifera* L.) cultivars. *Agric. Water Manag.* **2015**, *159*, 155–164. [CrossRef]

62. Martínez-Lüscher, J.; Kizildeniz, T.; Vučetić, V.; Dai, Z.; Luedeling, E.; van Leeuwen, C.; Gomès, E.; Pascual, I.; Irigoyen, J.J.; Morales, F.; et al. Sensitivity of grapevine phenology to water availability, temperature and CO2 concentration. *Front. Environ. Sci.* **2016**, *4*, 48.

63. Pritchard, S.G.; Rogers, H.H.; Prior, S.A.; Peterson, C.M. Elevated CO_2 and plant structure: A review. *Glob. Chang. Biol.* **1999**, *5*, 807–837. [CrossRef]

64. Bindi, M.; Raschi, A.; Lanini, M.; Miglietta, F.; Tognetti, R. Physiological and Yield Responses of Grapevine (*Vitis vinifera* L.) Exposed to Elevated CO_2 Concentrations in a Free Air CO_2 Enrichment (FACE). *J. Crop Improv.* **2005**, *13*, 345–359, [CrossRef]

agronomy

MDPI

Article

Variability among Young Table Grape Cultivars in Response to Water Deficit and Water Use Efficiency

Carolin Susanne Weiler [1,*], Nikolaus Merkt [2], Jens Hartung [3] and Simone Graeff-Hönninger [1]

[1] Institute for Crop Science, University of Hohenheim, Fruwirthstr. 23, 70599 Stuttgart, Germany; simone.graeff@uni-hohenheim.de

[2] Institute for Crop Science, Quality of Plant Products, University of Hohenheim, Emil-Wolff-Str. 25, 70599 Stuttgart, Germany; nikolaus.merkt@uni-hohenheim.de

[3] Institute for Crop Science, Biostatistics, University of Hohenheim, Fruwirthstr. 23, 70599 Stuttgart, Germany; moehring@uni-hohenheim.de

* Correspondence: carolin.weiler@uni-hohenheim.de

Received: 28 January 2019; Accepted: 11 March 2019; Published: 15 March 2019

Abstract: Climate change will lead to higher frequencies and durations of water limitations during the growing season, which may affect table grape yield. The aim of this experiment was to determine the variability among 3-year old table grape cultivars under the influence of prolonged water deficit during fruit development on gas exchange, growth, and water use efficiency. Six own rooted, potted table grape cultivars (cv. 'Muscat Bleu', 'Fanny', 'Nero', 'Palatina', 'Crimson Seedless' and 'Thompson Seedless') were subjected to three water deficit treatments (Control treatment with daily irrigation to 75% of available water capacity (AWC), moderate (50% AWC), and severe water deficit treatment (25% AWC)) for three consecutive years during vegetative growth/fruit development. Water deficit reduced assimilation, stomatal conductance, and transpiration, and increased water use efficiencies (WUE) with severity of water limitation. While leaf area and number of leaves were not affected by treatments in any of the tested cultivars, the response of specific leaf area to water deficit depended on the cultivar. Plant dry mass decreased with increasing water limitation. Overall, high variability of cultivars to gas exchange and water use efficiencies in response to water limitation was observed. 'Palatina' was the cultivar having a high productivity (high net assimilation) and low water use (low stomatal conductance) and the cultivar 'Fanny' was characterized by the highest amount of total annual dry mass as well as the highest total dry mass production per water supplied during the experiment (WUE_{DM}). Hence, 'Fanny' and 'Palatina' have shown to be cultivars able to cope with water limiting conditions and should be extensively tested in further studies.

Keywords: water limitation; dry mass partitioning; assimilation; intercellular CO_2; stomatal conductance; leaf water potential

---

1. Introduction

Climate change and the resulting alterations in temperature, precipitation as well as frequency and duration of extreme weather events, have a huge impact on crop production worldwide and will result in positive and negative changes in the quality and quantity of agricultural products [1]. Water will be one of the most limiting factors for agricultural crop production [2]. According to the IPCC [3], the central and southern part of Europe will have a higher risk of summer droughts due to increasing temperatures and annual precipitation decreases [3]. Additionally, more frequent and intense heat waves will occur all over Europe [3]. High temperatures and decreasing water availability might make Southern Europe unsuitable for wine as well as for table grapes, while northern and central Europe may offer better growing conditions. Increasing temperatures in northern and central Europe will result in an enlarged production area, which will continue to extend further north [4–6].

Climate conditions in regions from France and Germany will likely resemble to those located in the Mediterranean Basin [7]. Due to very high annual yields and high water requirements, table grape production has already been affected and will be more affected in the future by water shortages [8].

Adaptations of table grape production to changing environmental conditions are possible but will require additional irrigation, time-consuming breeding, or the selection of drought tolerant cultivars, which are able to cope with limited water availability. Until now, most research in the field of water limitation was done on vines and very few studies exist for table grapes, such as 'Crimson Seedless' [9,10] and 'Thompson Seedless' [11]. From our knowledge, no screening was done yet on the cultivars cultivated in Germany, especially with regard to their physiological and growth response to water deficit, their ability to use water efficiently, and their potential to grow under water limiting conditions in the future. Within several studies, grapevine cultivars showed a high variability to water limitation on leaf and on whole-plant level parameters. This was demonstrated under water-stress [12–14] and also under non-stressed conditions [15,16]. Screenings can be based on direct or indirect measurements for the determination of water limitation on the physiological level. Non-destructive gas exchange measurements on a single leaf are often used as an indicator for the detection of water stress in plants, as stomatal closure is one of the first adaptable plant responses to water limitation, and will result in limiting plant water losses [17,18]. While protecting plants against water loss, the closure of stomata will also reduce the amount of assimilated carbon [18], which can decrease yield and reduce the quality of table grapes. Furthermore, additional observations on the plant-level are important to evaluate the impact of water deficit on table grape cultivars, as grapevines adapt to water limitation by decreasing leaf area, reducing the number of leaves, and limiting growth rate [19,20]. Water use efficiency (WUE) can be calculated on a single-organ or on whole canopy scale. On leaf-scale, WUE can be distinguished between intrinsic WUE (WUE_i) and instantaneous WUE (WUE_{inst}). WUE_i represents the link between net assimilation of CO_2 (A_n) and the stomatal conductance of water (g_s) [21] and WUE_{inst} of A_n and transpiration (E). Both leaf-level WUE are used as parameters to characterize genetic as well as environmental effects [16,22,23]. Plant-level WUE is expressed as the accumulation of biomass per water lost/used [24,25] and shows the response of the plant during the growing season. In contrast to leaf-level WUE, plant-level WUE is not based on a single gas exchange measurement at a specific time and environmental conditions. The main objective of the present work was to determine the influence of water deficit on growth, physiology, and WUEs of six 3-year old table grape cultivars and to indentify possible cultivars able to cope with water limitation.

2. Materials and Methods

2.1. Plant Material and Treatments

The experiment was conducted from 2014 to 2016 on potted, own rooted table grapes in a greenhouse of the University of Hohenheim, Germany. Overall, six table grape cultivars ('Muscat Bleu', 'Nero', 'Fanny', 'Palatina', 'Crimson Seedless', and 'Thompson Seedless') subjected to three water deficit levels were tested with eight replications/plants per combination (six cultivars × three treatments × eight replications). For the current study, only data of 2016 was analyzed. For experimental setup, a non resolvable block design was chosen as it allows to cover a potential temperature gradient within the greenhouse.

The plant material of the table grape cultivars 'Thompson Seedless' and 'Crimson Seedless' originated from Israel (The Volcani Center, ARO, Bet-Dagan, Israel), while the other cultivars were obtained from Germany (Rebveredlung Kühner, Lauffen, Germany). One-bud cuttings of all cultivars were grown in sand, kept hydrated until they grew 4 to 6 leaves and developed a sufficient root. Twenty-four plants per cultivar were transplanted in 7-L pots with six kilograms of a loam, sand, and peat mixture (40:50:10, % per volume) in July 2014, with a maximum water holding capacity of 37.8%. During the consecutive three-year experiment, plants were kept at field capacity before and after stress treatment. Additionally, plants were fertilized biweekly with 1 g Hakapos® Blue (N 15% + P 10% + K 15% + Mg 2%) (CAMPO EXPERT, Münster, Germany) and 0.1 g Fetrilon[®1] Combi (BASF, Ludwigshafen, Germany). Treatments and experiment information (timeframe, no. of weeks of water deficit treatment, and BBCH) are summarized in Table 1. The first water deficit treatment started in 2014, after an establishment phase of 16 weeks. In the second year, water was limited during vegetative growth starting at an average shoot height of 60 cm and 6 to 8 leaves for 10 weeks. Furthermore, grapevines developing inflorescences were defruited before the treatments started. In 2016, table grapes were kept well-watered during flowering and water deficit treatments started at fruit set and ended at harvest. Over the entire three-year experiment, plants were maintained with only one shoot, attached to bamboo sticks.

Table 1. Characterization of water deficit treatments and experimental information from 2014 to 2016.

Water Deficit Treatment			Daily Irrigation to
Control			75% AWC
Moderate			50% AWC
Severe			25% AWC
Year	Timeframe	Weeks of Water Deficit	BBCH (at the Beginning of Water Deficit)
2014	22.9. – 29.10.	5.5	19
2015	12.5. – 21.7.	10	16–18
2016	15.6. – 16.9.	12	71

In 2014, one bud cuttings were planted. 2014 & 2015: Only vegetative growth. AWC was determined gravimetrically for each pot. AWC, available water content; BBCH, Biologische Bundesanstalt, Bundessortenamt und Chemische Industrie.

For determining the water usage of every plant during the imposition of water deficit, plant and soil water loss was measured gravimetrically on a daily basis using a platform scale (FKB 36K0.1, KERN, KERN & SOHN GmbH, Balingen, Germany) with a maximum range of 36 kg and 0.1 g accuracy. Control plants were irrigated daily to 75% available water content (AWC), moderate to 50% AWC, and severe deficit to 25% AWC. Before starting the treatment, AWC was determined for each pot/plant individually by flooding the pots after sunset to avoid transpiration losses. The excess water was able to drain overnight. Before sunrise, pots were weighed to get the maximum pot weight/field capacity. Wilting point was considered as the minimum weight of the pots. Therefore, all pots were dried out until a constant weight was reached and plants started wilting. Plants were rewatered and adjusted to the plant-pot specific weight. The following formula was used to calculate the individual pot weight for every plant in the treatments:

$$\text{Individual Pot Weight} = \text{PotMin} + (\text{PotMax} - \text{PotMin}) \cdot \text{Treatment} \qquad (1)$$

Within Formula (1), we used 0.75, 0.5, and 0.25 of the total available water content for the respective treatments (Control (75% AWC), moderate (50% AWC), and severe (25% AWC)).

During 2014 and 2015, pot weight was not adjusted to the increasing plant weight during the water deficit treatment. In 2016, due to additional bunch weight, pot weight for irrigation was modified by including bunch weights at veraison. Therefore, bunch weights were determined individually by a handheld scale and their weight was added to the corresponding pot's weight. Irrigation during water deficit treatment was applied daily, by the gravimetric determination of water used by each plant/pot and manually refilling to the plant specific weight, calculated with Formula (1).

During the water deficit treatment in 2016, temperature and relative humidity were measured in five-minute intervals using a datalogger (TGP-4500, Gemini Data Loggers, Chichester, UK). Mean temperature over the experimental period in 2016 was 21.8 °C and relative humidity was 63.9% (Figure 1). Vapour-pressure deficit was calculated based on measured values of temperature and relative humidity and ranged between 0.42–1.74 kPa.

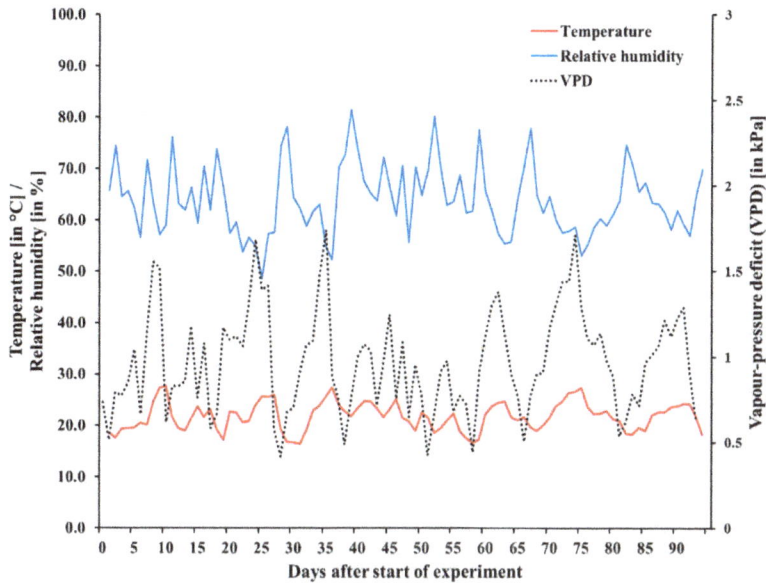

Figure 1. Daily mean values of air temperature, air humidity, and vapour pressure deficit in the open greenhouse during the experimental period (fruit set to harvest) in 2016.

2.2. Plant Water Status

The plant water status was estimated by measuring predawn leaf water potential (Ψ_{pd}) in two consecutive nights before harvest. Measurements were performed with a pressure chamber at harvest, according to the methodology developed by Scholander et al. [26] on one leaf per plant before dawn (03.00 to 06.00 a.m.).

2.3. Gas Exchange Measurements

Net assimilation (A_n), transpiration (E), stomatal conductance (g_s), and intercellular CO_2 (C_i) were measured using the portable gas exchange system GFS 3000 (Walz, Effeltrich, Germany) on one mid plant level leaf of every plant per cultivar–treatment combination. The system was equipped with the Basic System Package, including the Control Unit 3100-C, Standard Measuring Head 3010-S, and LED Light Source 3040-L (90% red and 10% blue light). Measurements were carried out within a timeframe of six days before harvest (10:00 a.m.–06:00 p.m.). Gas exchange was determined on an area of four cm^2 with a flowing rate of 750 µmol s^{-1} and impeller setting of 7. For the simulation of future climate conditions, a PPFD intensity of 1300 µmol $m^{-2}s^{-1}$, 400 ppm CO_2, a temperature of 30 °C, and relative humidity of 50% were configured as the chamber environment.

Instantaneous WUE was calculated by A_n/E and intrinsic WUE by A_n/g_s.

2.4. Plant Dry Weight and Leaf Area

The total leaf area (LA), dry mass (DM) of leaves, stems, and petioles were determined and the number of leaves were counted at harvest for each vine individually. Total leaf area was determined using an LI-3100C Area Meter (LI-COR, Lincoln, NE, USA). Dry mass was measured after drying at 60 °C until reaching constant weight. Specific leaf area (SLA) ($cm^2 g^{-1}$) was calculated as the ratio of LA and leaf dry mass and total dry mass water use efficiency (WUE_{DM}) as the ratio of total plant dry mass and water supplied during the experimental period (g L^{-1}).

2.5. Statistical Analysis

The physiological response of leaf-level gas exchange, WUE_i and WUE_{inst}, Ψ_{pd}, growth parameters, and WUE_{DM} of six table grape cultivars (cultivars: 6) subjected to three water deficit stress levels (treatments: 3) were analyzed using PROC MIXED (SAS version 9.2., SAS Institute Inc., Cary, NC, USA) with the following model:

$$y_{ijkl} = \mu + t_k + b_{kl} + \tau_i + \varphi_j + (\tau\varphi)_{ij} + e_{ijkl}, \tag{2}$$

where μ is the general effect, t_k and b_{kl} are random block effects for the kth table and the lth block on the kth table, respectively. τ_i, φ_j, and $(\tau\varphi)_{ij}$ corresponds to fixed main effects of the ith cultivar and jth water deficit treatment and their interaction effects, respectively. e_{ijkl} are the error effects of observations y_{ijkl}. Residuals were checked graphically for normality and homogeneity of variances. To reach normality and homogeneity of variances, data of E and g_s needed to be square-root transformed prior to analysis. Data of WUE_i, WUE_{inst}, LA, total DM, WUE_{DM}, and Ψ_{pd} were log-transformed. In case of a significant F-test, multiple comparisons for levels of the corresponding factor were done based on LSD ($\alpha = 0.05$). Significant differences were presented using a letter display created by the SAS macro %mult [27]. Within the letter display, capital letters show significant differences among cultivars in one or for all water deficit treatments. Lower case letters indicate significances among treatments in one cultivar or for all cultivars. If data needed transformation before analysis, statistical analysis are based on the transformed data. However, the same statistical analysis was conducted for transformed and non-transformed data. For the presentation of the results, transformed data were back-transformed (back-transformation: LOG: $y = e^x$; square-root: $y = x^2$). However, the corresponding letter display is based on previously transformed data.

3. Results

3.1. Plant Water Status

Predawn leaf water potential (Ψ_{pd}) showed significant interactions between cultivar and treatment. Ψ_{pd} values for the control treatment ranged between −0.2 to −0.36 MPa, for the moderate treatment between −0.2 to −0.69 MPa, and for the severe treatment between −0.25 to −1.10 MPa (Figure 2). For 'Fanny', all treatments differed significantly from each other and Ψ_{pd} decreased (−0.32 to −1.1 MPa) when water deficit intensified. Differences between the control and severe treatments were observed for 'Palatina' (−0.33 to −0.48 MPa) and 'Crimson Seedless' (−0.27 to −0.36 MPa), while no differences between treatments for 'Nero' and 'Thompson Seedless' were observed. When comparing cultivars within the treatments, 'Thompson Seedless' was the cultivar that differed the most from other cultivars and had the least negative Ψ_{pd}. In contrast, the most negative Ψ_{pd} was observed in 'Fanny' with a moderate (−0.69 MPa) and severe water deficit (−1.1 MPa).

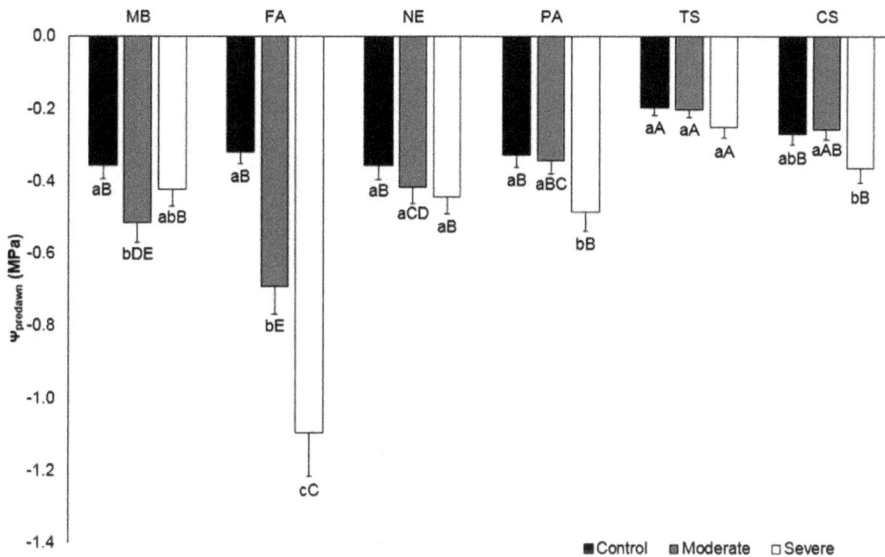

Figure 2. Median values of predawn leaf water potential (Ψ_{pd}) of six table grape cultivars subjected to three water deficit treatments at harvest in 2016. The data represent values of back transformed data. Treatments included: Control: daily irrigation to 75% of available water capacity (AWC), Moderate: daily irrigation to 50% of AWC, and Severe: daily irrigation to 25% of AWC. CS: Crimson Seedless, FA: Fanny, MB: Muscat Bleu, NE: Nero, PA: Palatina, TS: Thompson Seedless; Error bars represent standard errors; Values with identical letters indicate non-significant differences among cultivars (capital letters) and treatments (lower case letters) at $\alpha = 0.05$.

3.2. Gas Exchange Measurement

Gas exchange parameters differed significantly between cultivars and treatments (Table 2). 'Fanny' had the highest rate of all cultivars (10.96 µmol m^{-2}s^{-1}), while 'Muscat Bleu' (5.18 µmol m^{-2}s^{-1}), 'Thompson Seedless' (5.78 µmol m^{-2}s^{-1}), and 'Crimson Seedless' (4.16 µmol m^{-2}s^{-1}) were the cultivars with lowest assimilation rates. Similar results were observed for E and g_s, where 'Thompson Seedless', 'Crimson Seedless', and 'Muscat Bleu' had approximately 55 to 60% lower transpiration and 58 to 63% lower stomatal conductance in comparison to 'Fanny'. The highest C_i was found in 'Nero' with 245.48 µmol m^{-2}s^{-1}, whereas 'Palatina' (143.91 µmol m^{-2}s^{-1}) and 'Thompson Seedless' (161.58 µmol m^{-2}s^{-1}) had the smallest C_i values. WUE$_{inst}$ and WUE$_i$ were highest for 'Palatina' with 6.42 µmol CO_2 mmol^{-1} H_2O and 0.16 µmol CO_2 mmol^{-1} H_2O respectively, but did not differ significantly from 'Thompson Seedless', 'Fanny', and 'Muscat Bleu'. 'Nero', on the other hand, represented the least efficient cultivar at this development stage with 45% lower instantaneous and 47% lower intrinsic WUE compared to 'Palatina'. Among the treatments, all parameters differed significantly between the control and severe water deficit. Control vines had the highest E (1.85 mmol m^{-2}s^{-1}) and g_s (77.83 mmol m^{-2}s^{-1}), followed by moderately stressed plants (E = 1.51 mmol m^{-2}s^{-1}, g_s = 62.24 mmol m^{-2}s^{-1}), and the lowest values were found in plants treated with severe water deficit (E = 0.94 mmol m^{-2}s^{-1}, g_s = 37.6 mmol m^{-2}s^{-1}). For A_n, C_i, and both leaf-level WUEs, control and moderate treatments did not significantly differ from each other. Vines under severe water deficits had a 22 to 29% lower A_n and 22 to 26% lower C_i, while WUE$_i$ and WUE$_{inst}$ increased by approximately 22 to 30% and 19 to 26%, respectively.

Table 2. Mean values of net assimilation (A_n) and intercellular CO_2 (C_i), and median values of transpiration (E), stomatal conductance (g_s), and leaf-level water use efficiencies (intrinsic: WUE_i and instantaneous: WUE_{inst}) of six table grape cultivars subjected to three water deficit treatments.

Cultivar	Treatment	A_n (μmol m^{-2}s^{-1})			E (mmol m^{-2}s^{-1})			g_s (mmol m^{-2}s^{-1})			C_i (μmol m^{-2}s^{-1})			WUE_{inst} A/E			WUE_i A/g_s		
Muscat Bleu	Control	4.65	a		0.96	a		39.60	a		218.12	a		3.98	b		0.10	b	
	Moderate	5.57	a	C	1.07	b	C	43.84	b	C	203.69	a	ABC	4.76	b	ABC	0.12	b	ABC
	Severe	5.32	b		0.69	c		27.19	c		120.74	b		6.57	a		0.17	a	
Fanny	Control	12.30	a		3.28	a		148.08	a		247.20	a		3.74	b		0.08	b	
	Moderate	11.86	a	A	2.45	b	A	104.30	b	A	213.00	a	ABC	4.23	b	ABC	0.10	b	ABC
	Severe	8.71	b		1.32	c		53.11	c		149.88	b		6.84	a		0.17	a	
Nero	Control	10.67	a		3.14	a		135.63	a		258.41	a		3.25	b		0.08	b	
	Moderate	7.54	a	B	1.93	b	A	81.64	b	A	241.87	a	A	3.67	b	C	0.09	b	C
	Severe	7.14	b		1.77	c		75.26	c		236.15	b		3.69	a		0.09	a	
Palatina	Control	11.03	a		1.80	a		74.22	a		155.00	a		5.83	b		0.14	b	
	Moderate	9.89	a	AB	1.44	b	B	59.32	b	B	149.40	a	BC	6.72	b	AB	0.17	b	AB
	Severe	7.80	b		1.03	c		41.17	c		127.33	b		6.75	a		0.17	a	
Thompson Seedless	Control	6.64	a		1.29	a		51.29	a		175.67	a		5.15	b		0.13	b	
	Moderate	5.88	a	C	1.14	b	C	46.38	b	C	191.45	a	C	4.87	b	A	0.12	b	A
	Severe	4.81	b		0.71	c		27.05	c		117.62	b		7.11	a		0.19	a	
Crimson Seedless	Control	5.17	a		1.24	a		49.66	a		225.53	a		4.00	b		0.10	b	
	Moderate	5.19	a	C	1.25	b	C	48.36	b	C	220.30	a	AB	4.05	b	BC	0.10	b	BC
	Severe	2.12	b		0.44	c		15.98	c		192.44	b		4.51	a		0.12	a	
ANOVA																			
	Cultivar (C)	<0.0001			<0.0001			<0.0001			0.0361			0.0355			0.0324		
	Treatment (T)	0.0006			<0.0001			<0.0001			<0.0001			0.0001			<0.0001		
	C*T	0.5774			0.4113			0.3597			0.5151			0.4336			0.3549		

The data represent mean values (A_n, and C_i) and median values of back transformed data (E, g_s, WUE_{inst}, and WUE_i). Treatments included: Control: daily irrigation to 75% of available water capacity, Moderate: daily irrigation to 50% of available water capacity, and Severe: daily irrigation to 25% of available water capacity. Different letters indicate significant differences among cultivars (capital letters) and treatments (lower case letters) at $\alpha = 0.05$. ANOVA: *p*-values are given for the global F-test of the corresponding factor.

3.3. Growth Parameters, Dry Mass Partitioning, and Plant WUE

Leaf area differed significantly within the cultivars but was not affected by treatments. 'Crimson Seedless' was the cultivar that produced the highest LA (1921 cm^2), followed by 'Thompson Seedless' (1660 cm^2), 'Nero' (1213 cm^2), 'Palatina' (1211 cm^2), 'Fanny' (1168 cm^2), and lastly 'Muscat Bleu' (1116 cm^2) (Table 3). For SLA, significant interactions of treatment and cultivar were observed. Though 'Fanny', 'Palatina', and 'Thompson Seedless' showed no differences between treatments, SLA of 'Crimson Seedless' significantly decreased with increasing water limitation. Additionally, 'Nero' reached the highest SLA values under severe water deficit conditions (159.73 cm^2 g^{-1}). Within all levels of treatments, we observed the highest SLA for 'Fanny' and 'Crimson Seedless'. In contrast, the lowest values were found in 'Palatina' and 'Nero'. Differences between cultivars were determined by the number of leaves per plant (Table 3), where 'Crimson Seedless' produced the most leaves (20.5) while 'Fanny' and 'Muscat Bleu' only formed 12 and 9.8 leaves per plant, respectively. However, no differences between the water deficit levels were observed for any of the cultivars studied. Significant effects of treatment and cultivar were found for the total annual DM production and resulting WUE (Table 3). Among all cultivars, 'Fanny' had the highest values with a DM of 61.35 g and WUE of 0.08 g L^{-1}. 'Thompson Seedless' (30.12 g) produced the least amount of dry mass but did not differ significantly from 'Crimson Seedless' (30.45 g). Due to high water usage of both cultivars during the experiment, WUE was 55 to 58% lower than 'Fanny'. Besides cultivar, deficit treatment led to significant differences in both parameters. Severely stressed vines had 10 to 12% higher annual DM production as well as 19 to 33% higher WUE than the moderate treatment and the control.

Table 3. Mean values of specific leaf area (SLA) and number of leaves and median values of leaf area (LA), total dry mass (Total DM), and total dry mass water use efficiency (WUE$_{DM}$) of six table grape cultivars subjected to three water deficit treatments.

Cultivar	Treatment	LA (cm^2)		SLA (cm^2 g^{-1})			Total DM (g)			Number of Leaves		WUE$_{DM}$ (g L^{-1})		
Muscat Bleu	Control	1074.27		171.72	a	AB	32.54	a		9.88		0.04	c	
	Moderate	1151.48	C	152.17	b	C	35.95	a	CD	10.00	D	0.05	b	B
	Severe	1122.71		167.16	a	AB	36.08	b		9.38		0.06	a	
Fanny	Control	1191.18		168.62	a	B	60.33	a		12.75		0.07	c	
	Moderate	1228.93	C	171.35	a	A	70.98	a	A	11.63	C	0.09	b	A
	Severe	1087.03		175.67	a	A	53.89	b		11.63		0.09	a	
Nero	Control	1282.80		152.30	ab	C	43.45	a		16.38		0.05	c	
	Moderate	1229.79	C	144.03	b	C	41.43	a	B	16.25	B	0.05	b	B
	Severe	1130.14		159.73	a	BC	35.79	b		16.25		0.06	a	
Palatina	Control	1311.60		152.09	a	C	34.06	a		16.25		0.04	c	
	Moderate	1206.16	C	153.33	a	BC	37.52	a	BC	15.25	B	0.05	b	B
	Severe	1121.81		149.95	a	C	33.65	b		15.25		0.05	a	
Thompson Seedless	Control	1766.40		169.15	a	B	32.31	a		14.71		0.03	c	
	Moderate	1560.87	B	164.10	a	AB	29.46	a	E	14.88	B	0.03	b	C
	Severe	1658.88		161.83	a	B	29.37	b		15.13		0.05	a	
Crimson Seedless	Control	2185.06		180.90	a	A	34.90	a		21.13		0.03	c	
	Moderate	1836.10	A	174.25	ab	A	31.77	a	DE	19.25	A	0.03	b	C
	Severe	1767.82		165.01	b	AB	25.47	b		21.25		0.04	a	
ANOVA														
	Cultivar (C)	<0.0001		<0.0001			<0.0001			<0.0001		<0.0001		
	Treatment (T)	0.0633		0.0448			0.0324			0.8144		<0.0001		
	C*T	0.7734		0.0069			0.4589			0.9975		0.7317		

The data represent mean values (SLA and number of leaves) and median values of back transformed data (LA, Total DM, and WUE$_{DM}$). Treatments included: Control: daily irrigation to 75% of available water capacity, Moderate: daily irrigation to 50% of available water capacity, and Severe: daily irrigation to 25% of available water capacity. Different letters indicate significant differences among cultivars (capital letters) and treatments (lower case letters) at $\alpha = 0.05$; ANOVA: p-values are given for the global F-test of the corresponding factor.

Total annual dry mass production of vines and relative dry mass production of fruit, leaves, stem, and petioles are shown in Tables 3 and 4. Overall, we determined significant differences between cultivars regarding leaves, petioles, and fruit dry mass. Petioles were also affected by the water deficit treatments (Table 4). For stem dry mass, significant interactions between cultivar and treatment have been determined. 'Fanny' had the highest fruit dry mass and the lowest DM of leaves, stem, and petioles, while 'Crimson Seedless' had the highest dry mass of leaves, and petioles, but no plant of 'Crimson Seedless' produced fruit. Lowest dry mass of leaves and petioles were determined for 'Fanny' (6.9 g and 1.08 g).

Table 4. Mean values of annual dry mass production of leaf, stem, petioles, and fruit of six table grape cultivars subjected to three water deficit treatments.

Cultivar	Treatment	Leaf DM (g)		Stem DM (g)		Petioles DM (g)		Fruit DM (g)	
	Control	6.33		17.99a	AB	1.05	a		21.98	
Muscat Bleu	Moderate	7.66	D	18.06a	AB	1.04	a	D	14.81	B
	Severe	6.76		16.06a	A	1.03	b		11.58	
	Control	7.14		13.46a	C	1.14	a		35.48	
Fanny	Moderate	7.27	D	12.51a	C	1.15	a	D	57.62	A
	Severe	6.27		11.36a	C	0.93	b		37.45	
	Control	8.55		16.83a	B	1.33	a		27.08	
Nero	Moderate	9.12	C	16.31a	B	1.38	a	C	17.80	B
	Severe	7.27		14.98a	AB	1.09	b		14.62	
	Control	8.72		17.49a	B	1.63	a		17.55	
Palatina	Moderate	7.98	C	17.80a	AB	1.47	a	B	23.13	B
	Severe	7.69		15.20b	AB	1.33	b		16.74	
Thompson Seedless	Control	10.47		17.47a	B	1.61	a		18.51	
	Moderate	10.01	B	17.27a	AB	1.49	a	B	10.11	B
	Severe	10.39		16.54a	A	1.46	b		14.03	
	Control	12.11		20.03a	A	2.07	a		n.a.	
CrimsonSeedless	Moderate	11.07	A	18.96a	A	1.85	a	A	n.a.	n.a.
	Severe	10.90		13.06b	BC	1.62	b		n.a.	
ANOVA									
	Cultivar (C)	<0.0001		<0.0001		<0.0001			<0.0001	
	Treatment (T)	0.1221		<0.0001		0.0003			0.5901	
	C*T	0.6549		0.0191		0.6271			0.4693	

The data represents mean values of leaves, stem, petioles, and fruit dry mass. Treatments included: Control: daily irrigation to 75% of available water capacity, Moderate: daily irrigation to 50% of available water capacity, and Severe: daily irrigation to 25% of available water capacity; n.a.: not available; Different letters indicate significant differences among cultivars (capital letters) and treatments (lower case letters) at $\alpha = 0.05$; ANOVA: p-values are given for the global F-test of the corresponding factor.

4. Discussion

At the end of the water deficit treatments, cultivars showed a high variation in plant water potential when exposed to water limitation. According to Ojeda et al. [28], who defined four levels of water deficit, the cultivar suffering from the most severe water stress was 'Fanny' with about -1.1 MPa at the end of the treatment, while 'Thompson Seedless' had a stress level that ranged between none to weak stress (-0.2 to -0.25 MPa). Differential behaviors and responses of plant water potential to water deficit were described by Costa et al. [29] for the cultivars 'Aragonez' and 'Trincadeira'. In the study of Ojeda et al. [28], they also determined Ψ_{pd} continuously during the experiment and Ψ_{pd} of the stress treatment showed high variations. At some measurements, they could observe only a weak or non-existent stress level. In our study, only minor differences between the levels of water deficit were found for some cultivars. Based on the studies of Ezzahouani and Williams [30] and Wenter et al. [31], who determined decreasing Ψ_{pd} values towards the end of stress/growing season, plants with highest water limitation could have experienced a period with severe water stress (defined by Ojeda et al. [28]) in this study. In order to identify differences in the behavior of different cultivars to water limitation during the experimental period, additional measurements of water potential should be carried out before and during the experimental period.

Even though no clear results were found for Ψ_{pd}, gas exchange measurements and leaf-level WUE showed a definite reaction to water limitation. For g_s, rates decreased when the deficit intensified, which is in accordance with other studies on grapevines [13], table grapes [10], and rootstocks [32,33]. Since stomata closure is the first reaction to water limitations [34], g_s is often used as a non-destructive indicator to detect water stress. Therefore, water stress was classified into three levels. The first level of mild water stress is defined by g_s from 150 to 500 mmol H_2O m^{-2}s^{-1} ($\hat{=}$max. g_s), the second level of moderate water stress by g_s between 50 to 150 mmol H_2O m^{-2}s^{-1}, and the third level of severe water stress by $g_s < 50$ mmol H_2O m^{-2}s^{-1} [34,35]. According to these definitions, the control and the moderate treatment had a moderate water stress level at harvest, while vines subjected to severe water deficit had g_s values within the third level of water stress. As a consequence of increasing stomatal closure, we observed a downregulation of A_n when water deficit intensified. Previous studies observed similar results and determined a curvilinear relationship of A_n and g_s [34,35]. Furthermore, the range of A_n and g_s values are in agreement with studies by Chaves et al. [36] and Jara-Rojas et al. [37]. Decreasing C_i values with intensified water limitation, as observed in our study, imply stomatal limitations as the dominant factor for regulation at moderate stress [38,39], while the dominant factors for an upregulation of C_i, at the threshold value of g_s (50 mmol H_2O m^{-2}s^{-1}), are non-stomatal limitations [38]. As we could not determine increasing C_i at the threshold value within our study, stomatal closure may have led to decreasing g_s values. As a result of a higher decrease of g_s and E than A_n, both leaf-level WUEs increased with severity of water deficit. Medrano et al. [34] described similar results, where g_s decreased by 50% while A_n only decreased by 30% when the deficit progressed and led to higher WUE values when the water deficit intensified. Based on the observations for g_s, A_n, E, C_i, and leaf-level WUEs, our results indicate stomatal limitations as the limiting factor for lower A_n values in table grapes exposed to severe water limitation. The stomatal limitation could have been caused by increasing ABA concentration within xylem sap [33,40,41] and/or decrease of hydraulic conductance [33,42,43], as they are considered as main factors regulating stomatal conductance. Besides the effect of water deficit treatment, cultivar selection had a major influence on all gas exchange parameters and WUE$_{leaf}$. Variations and differences among grapevine cultivars in gas exchange under non-limiting and limiting water conditions were observed in several studies [13,44,45] and the response to water limitation is highly dependent on environmental conditions [45]. However, results obtained by gas exchange measurements could be overestimated, due to the possible occurrence of non-uniform closure of stomata (patchiness) in grapevines, when subjected to water deficit [46]. Furthermore, single leaf WUEs are limited due to high variability of measurements within the canopy, differences in leaf-response to the cumulative daily irradiance and leaf age, as young leaves have a higher gas exchange than older leaves [24]. Within our study, we determined differences among the cultivars for plant-level WUE$_{DM}$. The result of a high variability of cultivars are conform with other grapevine studies, comparing whole plant WUE (WUE$_{WP}$) of 19 cultivars under well-watered conditions in a glasshouse [47] or eight cultivars under well-watered and water-stressed conditions [12]. In contrast to our study, Palliotti et al. [44] could not find any differences in the measurement of whole plants WUE$_{Canopy}$ with regard to the response to higher water limitation. The comparison of results based on whole plants in relation to water stress is problematic and difficult, as the results are based, among other things, on gas exchange measurements of the canopy [44], biomass growth during the experiment [12], or, as in our study, on the total dry matter of the plant. When comparing leaf and plants WUEs, we observed increasing efficiencies with increasing water deficits, while in other studies no clear relationship was found between leaf and plant WUEs [24,44]. Medrano et al. [24] suggested the analysis of additional physiological parameters to reveal cultivar specific responses.

In most studies investigating the influence of water deficiency on plant growth and the adaptation of plants to the limited availability of soil water, it was observed that leaf area, dry matter, and number of leaves decreased in response to water limitation [19]. In our study, leaf area and the number of leaves were not negatively affected by the water deficit, in contrast to the results of Gomez-del-Campo et al. [20] where less leaf area was produced under water limiting conditions and the number of leaves was lower than under well-watered conditions. Plant growth, indicated by annual dry mass production, decreased, when the deficit was more severe which is in accordance with the study of Toumi et al. [19]. Within the study of Tardieu et al. [48], SLA was reported to decrease, if environmental conditions led to greater reduction of growth than on photosynthesis [48]. Therefore, it is used as a tool for the detection of changes in leaf structure [49,50]. Within our study, response of SLA depended on cultivar–treatment. Only SLA values of 'Crimson Seedless' decreased with increasing water limitation, indicating a higher influence of water deficit conditions on growth than on photosynthesis. While no clear behavior of the cultivars with regard to water deficit could be determined for SLA, differences could be determined for dry mass. A severe water deficit led to a lower dry mass. Since A_n rates and dry mass production have a close relationship [51], the reduced carbon assimilation, as an effect of closed stomata, could have led to a decrease in dry mass production in severely stressed plants within our study. Reductions of plant dry matter in case of a severe water deficit are in line with other studies [19,33] and indicated reduced plant growth due to a prolonged water deficit during fruit development and ripening. Cultivar differences, as they occurred in this study, were also observed by Gómez del Campo et al. [52,53], where cultivar selection and cultivar–irrigation interactions were the main factors influencing leaf area, number of leaves, SLA, and dry mass production [53–55].

5. Conclusions

Based on the obtained results, we identified gas exchange and water use efficiencies of table grapes to be affected by cultivar and by water deficit treatment. Since high productivity (high A_n) with low water loss (g_s) is a selection criterion for cultivation in water limiting environments, 'Palatina' could be a possible cultivar for cultivation under these environmental conditions. In addition. 'Fanny' appeared to be the cultivar least influenced by the deficit treatment. Hence, under changing climatic conditions with increasingly limited water availability during the growing period, 'Palatina' and 'Fanny' seem to be the most promising table grape cultivars of our study. However, further studies in the field under limited water conditions, as well as grafting on different rootstock, are necessary to confirm the ability of these cultivars to cope with water limitations.

Author Contributions: Conceptualization, C.S.W., N.M. and S.G.-H.; Data curation, C.S.W.; Formal analysis, C.S.W. and J.H.; Funding acquisition, C.S.W. and S.G.-H.; Investigation, C.S.W.; Methodology, C.S.W. and S.G.-H.; Project administration, C.S.W. and S.G.-H.; Resources, C.S.W., N.M. and S.G.-H.; Supervision, N.M. and S.G.-H.; Validation, C.S.W. and S.G.-H.; Visualization, C.S.W.; Writing – original draft, C.S.W.; Writing – review & editing, C.S.W., N.M., J.H. and S.G.-H.

Funding: This research was funded by the Anton & Petra Ehrmann-Stiftung Research Training Group "Water People Agriculture".

Acknowledgments: We want to thank Folkard Asch and his department for providing the gas-exchange measuring system "Walz GFS-3000" and especially Marc Schmierer for his help and support. This study was conducted within the framework of the Anton & Petra Ehrmann-Stiftung Research Training Group "Water People Agriculture" at the University of Hohenheim.

Conflicts of Interest: The authors declare no conflict of interest. The founding sponsors had no role in the design of the study; in the collection, analyses, or interpretation of data; in the writing of the manuscript, and in the decision to publish the results.

References

1. IPCC. *Climate Change 2014: Synthesis Report. Contribution of Working Groups I, II and III to the Fifth Assessment Report of the Intergovernmental Panel on Climate Change*; IPCC: Geneva, Switzerland, 2014.
2. Costa, J.M.; Ortuño, M.F.; Chaves, M.M. Deficit irrigation as a strategy to save water: Physiology and potential application to horticulture. *J. Integr. Plant Biol.* **2007**, *49*, 1421–1434. [CrossRef]
3. IPCC. *Climate Change 2014: Impacts, Adaptation, and Vulnerability. Part B: Regional Aspects. Contribution of Working Group II to the Fifth Assessment Report of the Intergovernmental Panel on Climate Change*; Cambridge University Press: Cambridge, UK; New York, NY, USA, 2014; ISBN 9781107683860.
4. Moriondo, M.; Bindi, M.; Fagarazzi, C.; Ferrise, R.; Trombi, G. Framework for high-resolution climate change impact assessment on grapevines at a regional scale. *Reg. Environ. Chang.* **2011**, *11*, 553–567. [CrossRef]
5. Fraga, H.; Santos, J.A.; Malheiro, A.C.; Oliveira, A.A.; Moutinho-Pereira, J.; Jones, G.V. Climatic suitability of Portuguese grapevine varieties and climate change adaptation. *Int. J. Climatol.* **2016**, *36*, 1–12. [CrossRef]
6. Fraga, H.; García de Cortázar Atauri, I.; Malheiro, A.C.; Santos, J.A. Modelling climate change impacts on viticultural yield, phenology and stress conditions in Europe. *Glob. Chang. Biol.* **2016**, *22*, 3774–3788. [CrossRef] [PubMed]

7.   Fraga, H.; Malheiro, A.C.; Moutinho-Pereira, J.; Santos, J.A. Future scenarios for viticultural zoning in Europe: Ensemble projections and uncertainties. *Int. J. Biometeorol.* **2013**, *57*, 909–925. [CrossRef] [PubMed]

8.   Permanhani, M.; Costa, J.M.; Conceição, M.A.F.; de Souza, R.T.; Vasconcellos, M.A.S.; Chaves, M.M. Deficit irrigation in table grape: Eco-physiological basis and potential use to save water and improve quality. *Theor. Exp. Plant Physiol.* **2016**, *28*, 85–108. [CrossRef]

9.   Pinillos, V.; Chiamolera, F.M.; Ortiz, J.F.; Hueso, J.J.; Cuevas, J. Post-veraison regulated deficit irrigation in "Crimson Seedless" tablegrape saves water and improves berry skin color. *Agric. Water Manag.* **2015**, *165*, 181–189. [CrossRef]

10. Conesa, M.R.R.; de la Rosa, J.M.M.; Domingo, R.; Bañon, S.; Pérez-Pastor, A. Changes induced by water stress on water relations, stomatal behaviour and morphology of table grapes (cv. Crimson Seedless) grown in pots. *Sci. Hortic.* **2016**, *202*, 9–16. [CrossRef]

11. Williams, L.E. Effects of applied water amounts at various fractions of evapotranspiration (ETc) on leaf gas exchange of Thompson Seedless grapevines. *Aust. J. Grape Wine Res.* **2012**, *18*, 100–108. [CrossRef]

12. Tomás, M.; Medrano, H.; Pou, A.; Escalona, J.M.; Martorell, S.; Ribas-Carbó, M.; Flexas, J. Water-use efficiency in grapevine cultivars grown under controlled conditions: Effects of water stress at the leaf and whole-plant level. *Aust. J. Grape Wine Res.* **2012**, *18*, 164–172. [CrossRef]

13. Bota, J.; Tomás, M.; Flexas, J.; Medrano, H.; Escalona, J.M. Differences among grapevine cultivars in their stomatal behavior and water use efficiency under progressive water stress. *Agric. Water Manag.* **2016**, *164*, 91–99. [CrossRef]

14. Bota, J.; Flexas, J.; Medrano, H. Genetic variability of photosynthesis and water use in Balearic grapevine cultivars. *Ann. Appl. Biol.* **2001**, *138*, 353–361. [CrossRef]

15. Tortosa, I.; Escalona, J.M.; Bota, J.; Tomas, M.; Hernandez, E.; Escudero, E.G.; Medrano, H. Exploring the genetic variability in water use efficiency: Evaluation of inter and intra cultivar genetic diversity in grapevines. *Plant Sci.* **2016**, *251*, 35–43. [CrossRef] [PubMed]

16. Tomás, M.; Medrano, H.; Escalona, J.M.; Martorell, S.; Pou, A.; Ribas-Carbó, M.; Flexas, J. Variability of water use efficiency in grapevines. *Environ. Exp. Bot.* **2014**, *103*, 148–157. [CrossRef]

17. Schulze, E. Carbon dioxide and water vapor exchane in repsponse to drought in the atmosphere and in the soil. *Annu. Rev. Plant Physiol.* **1986**, *37*, 247–274. [CrossRef]

18. Chaves, M.M. Effects of water deficits on carbon assimilation. *J. Exp. Bot.* **1991**, *42*, 1–16. [CrossRef]

19. Toumi, I.; M'Sehli, W.; Bourgou, S.; Jallouli, N.; Bensalem-Fnayou, A.; Ghorbel, A.; Mliki, A. Response of ungrafted and grafted grapevine cultivars and rootstocks (*Vitis* sp.) to water stress. *J. Int. Sci. Vigne Vin.* **2007**, *41*, 85–93. [CrossRef]

20. Gómez-del-Campo, M.; Ruiz, C.; Baeza, P.; Lissarrague, J.R. Drought adaptation strategies of four grapevine cultivars (*Vitis vinifera* L.): Modification of the properties of the leaf area. *J. Int. Sci. Vigne Vin.* **2003**, *37*, 131–143. [CrossRef]

21. Lavoie-Lamoureux, A.; Sacco, D.; Risse, P.-A.; Lovisolo, C. Factors influencing stomatal conductance in response to water availability in grapevine: A meta-analysis. *Physiol. Plant.* **2017**. [CrossRef]

22. Chaves, M.M.; Oliveira, M.M. Mechanisms underlying plant resilience to water deficits: Prospects for water-saving agriculture. *J. Exp. Bot.* **2004**, *55*, 2365–2384. [CrossRef]

23. Morison, J.I.L.; Baker, N.R.; Mullineaux, P.M.; Davies, W.J. Improving water use in crop production. *Philos. Trans. R. Soc. B Biol. Sci.* **2008**, *363*, 639–658. [CrossRef] [PubMed]

24. Medrano, H.; Tomás, M.; Martorell, S.; Flexas, J.; Hernández, E.; Rosselló, J.; Pou, A.; Escalona, J.-M.; Bota, J. From leaf to whole-plant water use efficiency (WUE) in complex canopies: Limitations of leaf WUE as a selection target. *Crop J.* **2015**, *3*, 220–228. [CrossRef]

25. Merli, M.C.; Gatti, M.; Galbignani, M.; Bernizzoni, F.; Magnanini, E.; Poni, S. Water use efficiency in Sangiovese grapes (*Vitis vinifera* L.) subjected to water stress before veraison: Different levels of assessment lead to different conclusions. *Funct. Plant Biol.* **2015**, *42*, 198–208. [CrossRef]

26. Scholander, P.F.; Hammel, H.T.; Bradstreet, E.D.; Hemmingsen, E.A. Sap Pressure in Vascular Plants: Negative hydrostatic pressure can be measured in plants. *Science* **1965**, *148*, 339–346. [CrossRef]

27. Piepho, H.P. A SAS macro for generating letter displays of pairwise mean comparisons. *Commun. Biometry Crop Sci.* **2012**, *7*, 4–13.

28. Ojeda, H.; Deloire, A.; Carbonneau, A. Influence of water deficits on grape berry growth. *Vitis* **2001**, *40*, 141–145.

29. Costa, J.M.; Ortuño, M.F.; Lopes, C.M.; Chaves, M.M. Grapevine varieties exhibiting differences in stomatal response to water deficit. *Funct. Plant Biol.* **2012**, *39*, 179–189. [CrossRef]

30. Ezzahouani, A.; Williams, L.E. Effect of irrigation amount and preharvest cutoff date on vine water status and productivity of Danlas grapevines. *Am. J. Enol. Vitic.* **2007**, *58*, 333–340.

31. Wenter, A.; Zanotelli, D.; Montagnani, L.; Tagliavini, M.; Andreotti, C. Effects of an early-summer drought stress on leaf photosynthesis, growth and yields of grapevine in mountain conditions. *Acta Hortic.* **2017**, 457–462. [CrossRef]

32. Meggio, F.; Prinsi, B.; Negri, A.S.; Simone Di Lorenzo, G.; Lucchini, G.; Pitacco, A.; Failla, O.; Scienza, A.; Cocucci, M.; Espen, L. Biochemical and physiological responses of two grapevine rootstock genotypes to drought and salt treatments. *Aust. J. Grape Wine Res.* **2014**, *20*, 310–323. [CrossRef]

33. Pou, A.; Flexas, J.; Alsina, M.D.M.; Bota, J.; Carambula, C.; De Herralde, F.; Galmés, J.; Lovisolo, C.; Jiménez, M.; Ribas-Carbó, M.; et al. Adjustments of water use efficiency by stomatal regulation during drought and recovery in the drought-adapted Vitis hybrid Richter-110 (*V. berlandieri* x *V. rupestris*). *Physiol. Plant.* **2008**, *134*, 313–323. [CrossRef] [PubMed]

34. Medrano, H.; Escalona, J.M.; Bota, J.; Gulías, J.; Flexas, J. Regulation of photosynthesis of C3 plants in response to progressive drought: Stomatal conductance as a reference parameter. *Ann. Bot.* **2002**, *89*, 895–905. [CrossRef] [PubMed]

35. Flexas, J.; Bota, J.; Escalona, J.-M.; Sampol, B.; Medrano, H. Effects of drought on photosynthesis in grapevines under field conditions: An evaluation of stomatal and mesophyll limitations. *Funct. Plant Biol.* **2002**, *29*, 461–471. [CrossRef]

36. Chaves, M.M.; Zarrouk, O.; Francisco, R.; Costa, J.M.; Santos, T.; Regalado, A.P.; Rodrigues, M.L.; Lopes, C.M. Grapevine under deficit irrigation: Hints from physiological and molecular data. *Ann. Bot.* **2010**, *105*, 661–676. [CrossRef] [PubMed]

37. Jara-Rojas, F.; Ortega-Farías, S.; Valdés-Gómez, H.; Acevedo-Opazo, C. Gas exchange relations of ungrafted grapevines (cv. Carménère) growing under irrigated field conditions. *S. Afr. J. Enol. Vitic.* **2015**, *36*, 231–242. [CrossRef]

38. Flexas, J.; Medrano, H. Drought-inhibition of photosynthesis in C3 plants: Stomatal and non-stomatal limitations revisited. *Ann. Bot.* **2002**, *89*, 183–189. [CrossRef] [PubMed]

39. Ghaderi, N.; Talai, A.R.; Ebadi, A.; Lessani, H. The physiological response of three Iranian grape cultivars to progressive drought stress. *J. Agric. Sci. Technol.* **2011**, *13*, 601–610.

40. Romero, P.; Fernández-Fernández, J.I.; Martinez-Cutillas, A. Physiological Thresholds for Efficient Regulated Deficit Irrigation Management in Winegrapes Under Semiarid Conditions: Soil-Plant-Water Relationships and Berry Composition. *Acta Hortic.* **2012**, 171–178. [CrossRef]

41. Tramontini, S.; Döring, J.; Vitali, M.; Ferrandino, A.; Stoll, M.; Lovisolo, C. Soil water-holding capacity mediates hydraulic and hormonal signals of near-isohydric and near-anisohydric Vitis cultivars in potted grapevines. *Funct. Plant Biol.* **2014**, *41*, 1119–1128. [CrossRef]

42. Schultz, H.R. Differences in hydraulic architecture account for near-isohydric and anisohydric behaviour of two eld-grown. *Plant Cell Environ.* **2003**, *26*, 1393–1406. [CrossRef]

43. Lovisolo, C.; Perrone, I.; Hartung, W.; Schubert, A. An abscisic acid-related reduced transpiration promotes gradual embolism repair when grapevines are rehydrated after drought. *New Phytol.* **2008**, *180*, 642–651. [CrossRef] [PubMed]

44. Palliotti, A.; Tombesi, S.; Frioni, T.; Famiani, F.; Silvestroni, O.; Zamboni, M.; Poni, S. Morpho-structural and physiological response of container-grown Sangiovese and Montepulciano cvv. (*Vitis vinifera*) to re-watering after a pre-veraison limiting water deficit. *Funct. Plant Biol.* **2014**, *41*, 634–647. [CrossRef]

45. Pou, A.; Medrano, H.; Tomàs, M.; Martorell, S.; Ribas-Carbó, M.; Flexas, J. Anisohydric behaviour in grapevines results in better performance under moderate water stress and recovery than isohydric behaviour. *Plant Soil* **2012**, *359*, 335–349. [CrossRef]

46. Maroco, J.P.; Rodrigues, M.L.; Lopes, C.; Chaves, M.M. Limitations to leaf photosynthesis in field-grown grapevine under drought—Metabolic and modelling approaches. *Funct. Plant Biol.* **2002**, *29*, 451–459. [CrossRef]

47. Gibberd, M.R.; Walker, R.R.; Blackmore, D.H.; Condon, A.G. Transpiration efficiency and carbon-isotope discrimination of grapevines grown under well-watered conditions in either glasshouse or vineyard. *Aust. J. Grape Wine Res.* **2001**, *7*, 110–117. [CrossRef]

48. Tardieu, F.; Granier, C.; Muller, B. Modelling Leaf Expansion in a Fluctuating Environment: Are Changes in Specific Leaf Area a Consequence of Changes in Expansion Rate? *New Phytol.* **1999**, *143*, 33–43. [CrossRef]

49. De Pinheiro Henriques, A.R.; Marcelis, L.F.M. Regulation of growth at steady-state nitrogen nutrition in lettuce (*Lactuca sativa* L.): Interactive effects of nitrogen and irradiance. *Ann. Bot.* **2000**, *86*, 1073–1080. [CrossRef]

50. Koundouras, S.; Tsialtas, I.T.; Zioziou, E.; Nikolaou, N. Rootstock effects on the adaptive strategies of grapevine (*Vitis vinifera* L. cv. Cabernet-Sauvignon) under contrasting water status: Leaf physiological and structural responses. *Agric. Ecosyst. Environ.* **2008**, *128*, 86–96. [CrossRef]

51. Peterson, R.B.; Zelitch, I. Relationship between Net CO_2 Assimilation and Dry Weight Accumulation in Field-Grown Tobacco. *Plant Physiol.* **1982**, *70*, 677–685. [CrossRef]

52. Gómez-del-Campo, M.; Ruiz, C.; Sotés, V.; Lissarrague, J.R. Consequences of Water Consumption in the Leaf Area and Dry Matter Partitioning in Four Grapevine Varieties. *Acta Hortic.* **1996**, 331–338. [CrossRef]

53. Gómez-del-Campo, M.; Baeza, P.; Ruiz, C.; Lissarrague, J.R. Effects of water stress on dry matter content and partitioning in four grapevine cultivars (*Vitis vinifera* L.). *J. Int. Sci. Vigne Vin* **2005**, *39*, 1–10. [CrossRef]

54. Mullins, M.G.; Bouquet, A.; Williams, L.E. *Biology of the Grapevine*; Cambridge University Press: Cambridge, UK, 1992.

55. Williams, L.E. Grape. In *Photoassimilate Distribution in Plants and Crops Source-Sink Relationship*, 1st ed.; Zamski, E., Ed.; CRC Press: New York, NY, USA, 1996; pp. 851–881. ISBN 9780824794408.

![agronomy logo] *agronomy*

MDPI

Article

Anti-Transpirant Effects on Vine Physiology, Berry and Wine Composition of cv. Aglianico (*Vitis vinifera* L.) Grown in South Italy

Claudio Di Vaio [1], Nadia Marallo [1], Rosario Di Lorenzo [2] and Antonino Pisciotta [2,*]

[1] University of Naples Federico II, Department of Agricultural Sciences, Via Università 100, 80055 Portici, Italy; claudio.divaio@unina.it (C.D.V.); nadia.marallo@hotmail.it (N.M.)
[2] University of Palermo, Department of Agricultural, Food and Forest Sciences, Viale delle Scienze, Ed 4, 90128 Palermo, Italy; rosario.dilorenzo@unipa.it
* Correspondence: antonino.pisciotta@unipa.it; Tel.: +39-091-23861217

Received: 20 February 2019; Accepted: 13 May 2019; Published: 14 May 2019

Abstract: In viticulture, global warming requires reconsideration of current production models. At the base of this need there are some emerging phenomena: modification of phenological phases; acceleration of the maturation process of grapes, with significant increases in the concentration of sugar musts; decoupling between technological grape maturity and phenolic maturity. The aim of our study was to evaluate the effect of a natural anti-transpirant on grapevine physiology, berry, and wine composition of Aglianico cultivar. For two years, Aglianico vines were treated at veraison with the anti-transpirant Vapor Gard and compared with a control sprayed with only water. A bunch thinning was also applied to both treatments. The effectiveness of Vapor Gard were assessed through measurements of net photosynthesis and transpiration and analyzing the vegetative, productive and qualitative parameters. The results demonstrate that the application of anti-transpirant reduced assimilation and transpiration rate, stomatal conductance, berry sugar accumulation, and wine alcohol content. No significant differences between treatments were observed for other berry and wine compositional parameters. This method may be a useful tool to reduce berry sugar content and to produce wines with a lower alcohol content.

Keywords: global warming; technological and phenolic ripeness; grape; wine; sensory analysis

---

1. Introduction

In the last 20 years, the acceleration of ripening in wine grapes has been extensively documented worldwide. An increase in carbon dioxide emissions and other greenhouse gases is altering the composition of the atmosphere. It is likely that most of the global warming since the mid-20th century has been due to increases in greenhouse gases from human activities [1]. World climate is changing and becoming warmer [2,3], with great effects on agricultural production, whose products are directly impacted by meteorological conditions. For example, by 2050, the projected increase in annual average temperature in grape-growing regions is estimated to range from 0.4 to 2.6 °C. For example, increases in annual average temperature between the present day and the year 2030 are expected to range from 0.2 to 1.1 °C in many of the Australian grape-growing regions [4]. A steady trend of increased warming is pushing traditional areas of grape-growing towards accelerated ripening [1], leading, in turn, to excessive sugar accumulation in the fruit and high alcohol in the wine.

Wine consumer preferences over the last decade are changing [1,5] towards lower-alcohol wines. The growing demand for wines with moderate alcohol content is leading to a reappraisal of current production systems as well as management techniques. Vineyard management practices are able to increase, stabilize, or slow maturation [6–10], and grapevine phenology is predominantly temperature-driven [11,12]. Matching the critical developmental stages of grapevines to a suitable climate is a fundamental factor in the planning of new vineyards where optimizing quality is a priority. McIntyre et al. described the timing of phenology in many grape varieties and the possibility of a "best fit" variety for a particular climate [13]. In a future climate change scenario, rising temperatures may change the timing of grape ripening and consequent harvest date and may affect grape quality and yield [4,14–18]. Therefore, the projected temperature increases could have a major impact on such phenological events in terms of winegrape production and quality across wine regions, especially as grapevine phenology varies with regions and varieties [19]. The impact in question could be positive or negative depending on the present climate of the region [20].

The alcohol content of wines is reported to be increasing worldwide. In Australia, during the period 1984–2004, the alcohol content rose from 12.3% to 13.9% in red wines and from 12.2% to 13.2% in whites [21]. Dokoozlian reported that the average sugar content of Cabernet Sauvignon musts increased from 21–22 °Brix in 1990 to 24–25 °Brix in 2008 in the Napa Valley [22]. This finding was supported by Vierra, who found that the average alcohol content of Napa Valley wine increased from 12.5% to 14.8% during the period 1971 to 2001 [23]. Duchene and Schneider also reported that the alcohol potential of Riesling produced in Alsace had increased by 2.5% over the previous 30 years due to higher temperatures during ripening [24]. Although all changes in phenological development have been well documented, perhaps the most striking is the advance of harvest time by more than a month. Ganichot compared harvest dates from 1945 to 2005 in Chateauneuf du Pape (France) and found that harvest time was getting earlier, advancing from early October in 1945 to early September in 2000 [25]. In recent years, the harvest date of Montepulciano, grown in Abruzzo, advanced by 14–15 days in the central part of the region and by 10 days when grown closer to the coast [26,27].

As a means to reduce sugar accumulation, numerous studies have considered agronomic practices that limit photosynthetic activity and increase competition between sink and source. The use of commercial products that reduce the transpiration rate, and hence photosynthesis, induces a variation in the metabolism of carbohydrate compounds and their translocation in the berries [6,28–30].

2. Materials and Methods

2.1. Experimental Site, Design, and Treatments

The trial was carried out in Benevento province (in Southern Italy) (lat. 41°15′32″ N, long. 14°35′54″ E) at an altitude of 300 m above sea level (a.s.l.). The experimental trial was conducted on a uniform clay-loamy soil type. The study was carried out over the 2013 and 2014 growing seasons in an Aglianico/110 Richter vineyard that is more than 10 years old. Vines were spaced with 2.40 m between rows and 1.40 m within a row, trained to a Vertical Shoot Position (VSP) system and pruned to a bilateral guyot with 30 nodes per vine (15 for each cane). The vineyard was in a dry condition during the two growing seasons. Pest management was carried out according to local standard practice. Daily minimum, maximum, and average air temperature (°C) and monthly rainfall (mm) data were recorded in both years and were taken from a weather station located in Guardia Sanframondi (BN), close to the vineyard. In total, 40 vines of Aglianico were selected: 20 vines were assigned to Vapor Gard® anti-transpirant treatment (VG), and 20 vines were used as an unsprayed control (C). At VG application time, in half of the vines of treatments VG and C, manual bunch-thinning (±BT) was applied at BBCH stage 81, decreasing the total bunches to 50%. Ten vines for each treatment were assigned in a completely randomized design throughout the vineyard.

Four treatments, finally, were compared: C ± BT for control vines, with and without bunch-thinning; and VG ± BT for vines treated with the anti-transpirant, with and without bunch-thinning. The anti-transpirant product used was Vapor Gard® (Intrachem Bio Italia, Grassobbio, Italy), a water-emulsifiable organic concentrate for use on plants, designed to reduce transpiration by forming a clear, soft, and flexible film that retards normal transpiration loss. Its active ingredient is di-1-p-menthene ($C_{20}H_{34}$), a terpenic polymer also known as pinolene. VG was prepared as a 2%

solution in water and stirred slowly to form an emulsion before treatment. All the leaves of the canopy located above the cluster area were sprayed at 0.336 L/vine rate using a portable pump. The abaxial surfaces of the leaves were wetted well in order to cover the stomatal pores [31]. The entire canopy of all VG vines was sprayed with Vapor Gard until run-off. The VG treatments were applied at veraison (BBCH stage 83–85), approximately one month before harvest.

2.2. Physiological Measurements

Measures of gas exchanges were carried out three days after anti-transpirant was sprayed, onto 10 mature (10–12 node position of the main shoot) and fully expanded leaves (in 10 vines, 1 per vine). Single-leaf gas exchange readings were taken at midday of clear days using a portable photosynthetic open-system (Li-6400, LICOR, Lincoln, NB, USA) featuring a broad leaf chamber (6.0 cm^2). PPFD incident on the leaves was always greater than 1 000 µmol m^{-2} s^{-1}. The CO2 inside leaf chamber was supplied by an external tank to obtain a flow rate of 360 µmol mol^{-1} air.

Assimilation rate (A), transpiration rate (E), and stomatal conductance (gs) were calculated from inlet and outlet CO_2 and H_2O relative concentrations. Intrinsic water-use efficiency (WUEi) was then derived as the A to gs ratio. Measurements were taken. gs was measured at midday using a non-steady state porometer (AP4, Delta-T Devices, Cambridge, UK). Measuring was done four times after VG application, until harvest.

2.3. Growth, Yield, and Grape Composition

Each year four repetitions of 50 berries (5 berries × 10 vines) were randomly collected on four calendar dates, from veraison to harvest (one before and three after VG and BT applications). The berries were randomly collected from different sections of the bunch (top, middle and bottom) and from sun exposed and non-sun-exposed bunch sides, to obtain grape maturity data and to determine the optimal harvest date. The berries were also weighed with a digital precision weighing scale (Acculab Sartorius Group ECON EC-411).

The 50 berries of the four different repetitions were manually crushed, and their juice was used to determine: soluble solids (°Brix), pH, and titratable acidity (TA). Total soluble solids (TSS) concentration was determined with a digital refractometer (Model L-R 01 Digital Refractometer, Maselli Misure S.p.a., 43100 Parma, Italy) on 2 mL of juice at 20 °C. Samples of 10 mL of juice were used for pH and TA measurements. pH was measured by a digital pH meter (Crison Instrument GLP 21 pH); TA was determined using the official method for TA determination, with 0.1 N NaOH to a pH 8.2 end-point, and was expressed as g L^{-1} of tartaric acid, phenolic maturity was determined according to Glories' method [32] and (expressed as mg L^{-1}).

Yield and bunch number per vine were determined at harvest time. At harvest, 100 kg of fruit per treatment was randomly harvested and transported to the laboratory. The bunches were collected from both sides of the vines and from shaded and non-shaded vine sections to avoid bias.

During winter, for each year of the trial, pruning weight per vine was also determined.

2.4. Microvinification and Wine Analysis

In 2013 and 2014, wines were made using microvinification techniques. At harvest, 100 kg of fruit per treatment were manually harvested in plastic boxes of 20 kg and transported to the experimental cellar to be microvinified.

For each treatment, two microvinifications were carried out. Grapes from each treatment were mechanically crushed, destemmed, transferred to fermentation containers. potassium metabisulphite was added to obtain a total SO2 level of about 35 mg L^{-1} and 20 g hL^{-1} of a commercial yeast strain (BCS 103 Springer Oenologie) was inoculated. Musts were fermented for 16 to 18 days on the skin and punched down twice daily, with the fermentation temperature ranging from 20 to 23 °C. After alcoholic fermentation, the wines were pressed at 0 °Brix and inoculated with 30 g hL^{-1} *Oenococcus oeni* (Lalvin Elios 1 MBR; Lallemand). After completion of malolactic fermentation, the samples were racked and transferred to glass bottles, and 50 mg L^{-1} of potassium metabisulphite was added. Two months later, the wines were racked again, bottled into 750 mL bottles, and then closed with cork stoppers. The wines were analyzed for alcohol, TA, pH, total phenol, and anthocyanin concentrations were determined with Foss (Wine Scan™ Auto, Hillerod, Denmark). All determinations on wines were carried out in duplicate yielding four repetitions per treatment.

2.5. Sensory Analysis

A quantitative sensory analysis (QDA) of the experimental wines was performed. Sensory analysis was carried out on wine products using the official method of the International Union of Oenologues, to describe the sensory profiles of wines. A panel of 12 judges composed of agri-food experts (seven males and five females between the ages of 22 and 55 years) were selected. All of the judges were experienced wine tasters, they were previously selected on the basis of their sensory abilities, trained in recognize and describe odors (chemical standards), and several wine typologies.

Samples of 30 mL of each wine were served at 10 °C in black tulip-shaped glasses, coded with random three-digit codes. Samples were evaluated in duplicate (two duplicate sessions). Each judge evaluated all the wines in each session and the wines were served according to a randomized service design. The judges were asked to focus on the perceived odor descriptors and rate the corresponding intensities ranging to 8–11 point scale. They were provided with a list of 27 taste/odor descriptors (the order was randomized among the judges).

2.6. Statistical Analysis

Analysis of variance (ANOVA) and mean separation by Duncan's multiple range test ($p < 0.05$) were performed using the statistical package XL-Stat Version, 2013 (New York, NY, USA).

3. Results and Discussion

From the trend of average monthly temperatures recorded at the farm in Guardia Sanframondi and the monthly rainfalls for the same area in 2013 and 2014 (Figure 1a), it was observed that minimum temperatures were 6.1, 5.1, and 5.9 °C, respectively during January, February, and December in the year 2013. Peak maximum temperatures were recorded during August (22.4 °C). The same trend was shown for the temperatures measured in the second year of study; however, the minimum temperatures in this year were higher, 7.7 and 9.3 °C in January and February, respectively, except for December (−1.5 °C), while the maximum temperatures seemed to remain quite similar to the prior year (21.3 °C, once again during August) (Figure 1b). In 2013 and 2014 at Guardia Sanframondi, there was a total rainfall of 2,037.2 and 1,734.8 mm, respectively.

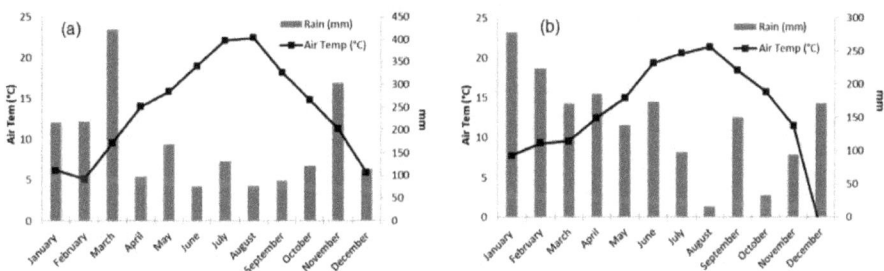

Figure 1. Monthly averages air temperature and monthly rainfall recorded in 2013 (**a**) and 2014 (**b**). The line indicates average monthly temperature, and the bars the monthly rain.

The rainiest months were March and November for the year 2013 (422.8 and 303 mm, respectively), and January and February (278 and 223.6 mm, respectively) for the year 2014.

The VG treatments were applied at veraison (BBCH stage 83–85), on 2 September 2013 and on 1 September 2014. In both years, from VGapplication to harvest time, we monitored gs. As reported in Figure 2, it is possible to see how these parameters evolved during the season from VG application to harvest time and to appreciate the significant differences in gs between treatments. The gs for the VG treatment was lower for VG-treated vines in the first 20 days after application (Figure 2).

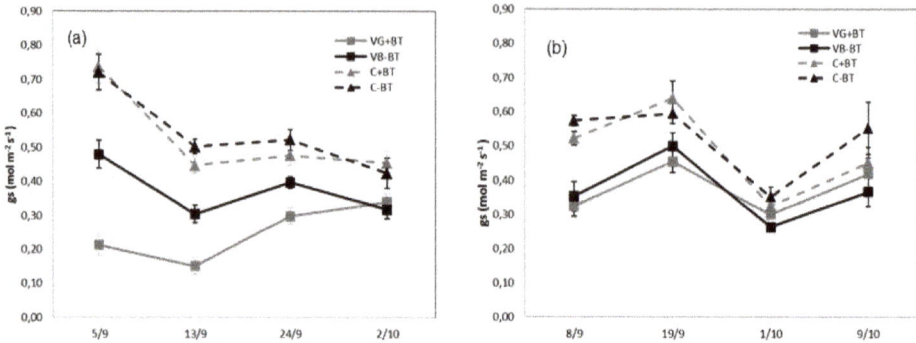

Figure 2. Stomatal conductance (gs) measured by porometry in control (C) and treated (VG = Vapor Gard anti-transpirant application; BT = 50% bunch-thinning) Aglianico vines in (**a**) 2013 and (**b**) 2014. Data are averages of 10 replicates ± SE.

Stomatal conductance (gs) was significantly reduced each year in the sprayed Aglianico vines as compared with C vines (Figure 2). In 2013, Aglianico vines showed less leaf conductance, amounting to 0.47 vs. 0.72 mol m^{-2} s^{-1} for VG -BTand C-BT vines, respectively, and 0.21 vs. 0.73 mol m^{-2} s^{-1} for VG+BT and C+BT vines, respectively, after 3 days of application (Figure 2a). We can observe the same trend for VG Aglianico vines in the year 2014 (Figure 2b). It is interesting also to describe the same trend between VG-BT vines and VG+BT- vines; VG+BT vine had less leaf conductance in both years.

A few days after VG treatment, the sprayed leaves showed a great reduction in A and E and an increase of WUEi (Figures 3–5) in both years (2013 and 2014). Leaf assimilation values in 2013 were: 17.4 and 26.6 μmol m^{-2} s^{-1} for VG and C, respectively (Figure 3a). Palliotti et al. and Brillante et al., reported similar observations [31,33]. There was more reduction in leaf assimilation for the BT treatment: 25.4 vs. 10.8 μmol m^{-2} s^{-1} for VG+ BT and C+ BT, respectively, during 2013 (Figure 3A). The same behavior was observed in 2014. When BT was combined with VG treatment, a reduction in leaf assimilation was recorded. In fact, the reduction was 34.7% and 57.6%, respectively, in VG -BT and VG+ BT in 2013, while in 2014 it was 62.4% and 45.3%, respectively (Figure 3b).

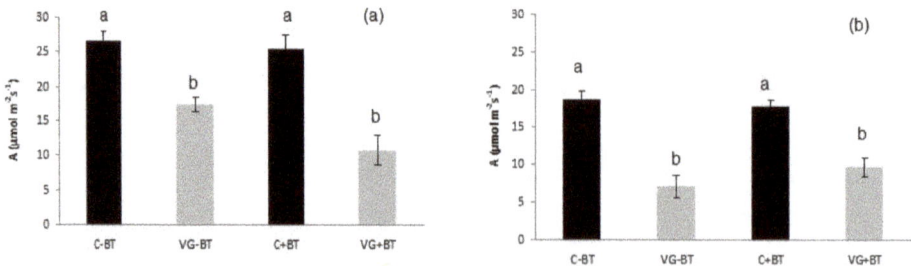

Figure 3. Assimilation rate (A) measured on fully expanded leaves in control (C) and treated (VG = Vapor Gard anti-transpirant application; BT = 50% bunch-thinning) Aglianico vines in (**a**) 2013 and (**b**) 2014. Data are averages of 10 replicates ± SE. The same letter indicates non-significant differences by Duncan's post hoc test ($p < 0.05$).

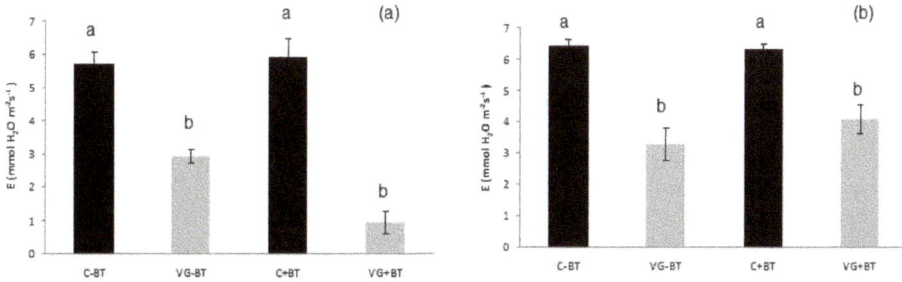

Figure 4. Transpiration rate (E) measured on fully expanded leaves in control (C) and treated (VG = Vapor Gard anti-transpirant application; BT = 50% bunch-thinning) Aglianico vines in (**a**) 2013 and (**b**) 2014. Data are averages of 10 replicates ± SE. The same letter indicates non-significant differences by Duncan's post hoc test ($p < 0.05$).

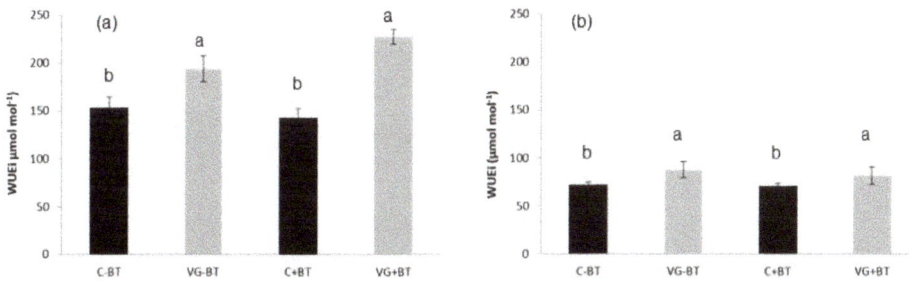

Figure 5. Intrinsic water-use efficiency (WUEi) calculated as A/gs measured on fully expanded leaves in control (C) and treated (VG = Vapor Gard anti-transpirant application; BT = 50% bunch-thinning) Aglianico vines in (**a**) 2013 and (**b**) 2014. Data are averages of 10 replicates ± SE. The same letter indicates non-significant differences by Duncan's post hoc test ($p < 0.05$).

No statistical difference was found in the C treatment when combined with BT.

There were significant differences in E between VG and C vines in 2013 and 2014. VG caused a 66.6% reduction in E after application in 2013, and a 42.2% reduction in 2014 compared to the control vines. These effects were the same when BT was also applied (Figure 4a). In 2013, E values were 5.70 mmol H_2O m^{-2} s^{-1} in the control and 2.91 mmol H_2O m^{-2} s^{-1} in the VG vines. The BT treatment also showed major differences in E: for the control with BT it was 5.92 mmol H_2O m^{-2} s^{-1}, and for the sprayed treatment it was 0.92 mmol H_2O m^{-2} s^{-1} (Figure 4A). The same results, with statistically significant differences between treated and control vines, were recorded in the year 2014 (Figure 4b). Independently of BT, the treated vines showed a lower E with respect to control vines (Figure 4).

These findings are in agreement with those of several researchers [31,33]. In fact, they described that the reduction of gs, A and E following VG spraying was accompanied by a marked reduction (from 60% to 70% compared to leaves of control vines) of substomatal CO_2 concentration (182 to 218 ppm in control leaves versus 112 to 165 ppm in VG-treated leaves); it is apparent that this behavior was linked to some physical impairment of stomatal opening and function.

The reverse trend was instead shown for WUEi, derived as the A to gs ratio. In 2013, WUEi measured 3 days after VG application was 153.46 μmol mol^{-1} in C and 193.97 μmol mol^{-1} in VG vines (Figure 5a). The BT treatment also showed the same trend for WUEi: for the control it was 142.51 μmol mol^{-1}, and for the sprayed treatment 227.57 μmol mol^{-1}. The same results, with statistically significant differences between treated and control vines, were recorded in 2014 for Aglianico: 72.51 vs. 87.92 μmol mol^{-1} for VG-BT vines, and 71.23 vs. 81.91 μmol mol^{-1} for VG+ BT vines (Figure 5b). After VG application, A and E rates again decreased, demonstrating the effectiveness of VG in rapidly reducing stomatal opening upon treatment. Thereafter, the capacity for carbon gain of VG-treated leaves remained limited for a period of 4 weeks until harvest, when gs again converged towards levels seen in C leaves. Conversely, at harvest, sprayed leaves still had lower E than control leaves. The depression of E after VG application resulted in a significant increase of WUEi in VG relative to C vines and was of similar duration, suggesting a lower amount of water consumed per carbon assimilated in VG relative to C vines, while both achieved a similar carbon gain to that reported in the literature [28,31].

These findings are comparable with those reported in the literature [30,33,34]. As reported by Palliotti et al., the decrease in E can be attributed to an increase in resistance to water transport related to the film-forming anti-transpirant [34]. Our study showed that after application, Aglianico plants were able to recover, although a reduced A compared to the control was still observed After treatment in the VG-sprayed leaves, a large reduction in leaf A and gs was observed, which continued over the following 60 days with peak reductions compared with C [30,33,34]. Post-veraison, the effect on stomatal closure was reduced in part, although E was lower than in the control even late in the season, in agreement with Palliotti et al. [34]. The depression of transpiration after VG application resulted in a significant increase in WUEi in VG- relative to C vines. Our results are confirmed by other studies: Sangiovese and Ciliegiolo leaves showed a smaller decrease in WUEi during the season in response to application of VG [31,33,34].

The significant improvement of intrinsic WUEi, from VG application until the final stage of ripening, indicates less water loss through stomata for a similar carbon gain. This behavior occurred because the limitation in gs of H_2O was proportionally higher than the depression of A [31].

The fact that the film-forming VG exerts a physical barrier to gas exchange, thus hampering the CO_2 entering the stomata and the water vapor leaving the stomata, was found almost 40 years ago on *Vicia faba* by Davenport et al., who also noted that under the transparent film the stomata were more open [35]. Scanning electron micrographs on bean plants confirmed these results [36]. Moreover, in peach, midday leaf water potential increased after anti-transpirant application as compared to unsprayed plants [35]. Thus, maintenance of a high moisture level of the leaf tissue in conjunction with possible effects of light reflectance might explain why treated leaves did not heat up significantly, in agreement with findings in a tropical plant using the same compound [37]. In terms of light reflectance, VG behaves differently than kaolin-based foliar reflectants, which have been proven to cause a significant reduction of leaf and/or berry temperature, especially under limiting water supply [37–39]. The significant improvement of intrinsic WUEi, extending from the time of VG application until the final stage of ripening, indicates less water loss through stomata for a similar carbon gain. This behavior occurred because the limitation in gs of H_2O was proportionally higher than the depression of A.

A significant source limitation following Vapor Gard spraying has been previously assessed in different species [34,40], including grapevine [34], and, quite remarkably, this source limitation is reached without modifying the vine leaf-to-fruit ratio or the cluster microclimate during ripening. The product, applied late in the season, has been effective in reducing the pace of sugar accumulation in the berry, as compared to control vines, scoring −1.2 °Brix at harvest and lowering the alcohol content in the resulting wines by −1% vol. It can be recommended as a valuable cultural practice in viticultural areas where berry ripening takes place early during the hottest part of the season [31,34].

From veraison to harvest, we monitored average berry weight (g), TSS, pH and TA for both years, 2013 and 2014. The experimental vines were individually and manually picked, in 2013 on 7 October, and in 2014 on 9 October. In Figures 6–9, it is possible to see how these parameters evolve during the season. In both years, despite some changes between the theses after the applications with VG, at harvest time no significant differences in Aglianico berry weight were observed (Figure 6 and Table 1) according to other authors [33,34].

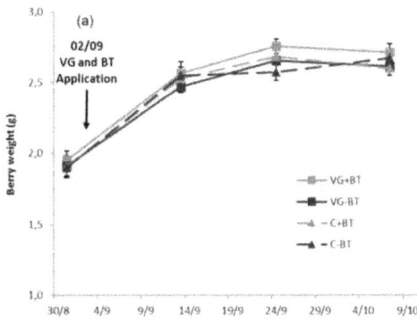

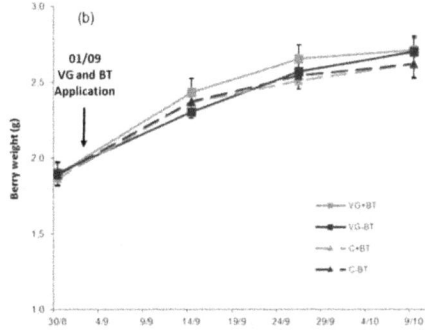

Figure 6. Berry weight measured in control (C) and treated (VG = Vapor Gard anti-transpirant application; BT = 50% bunch-thinning) Aglianico vines in (**a**) 2013 and (**b**) 2014. Data are averages of 4 repetitions ± SE.

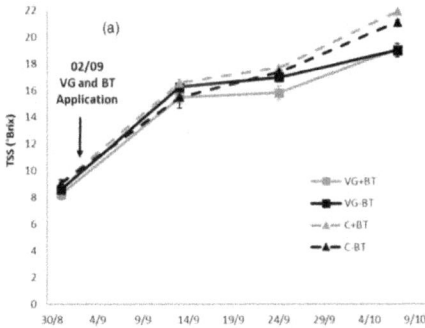

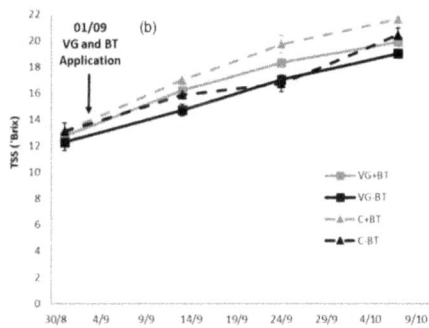

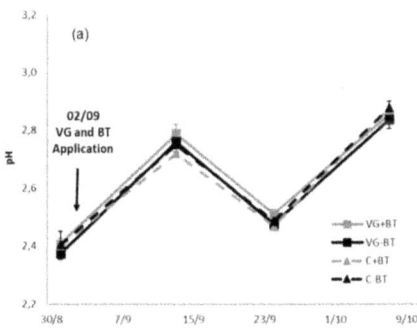

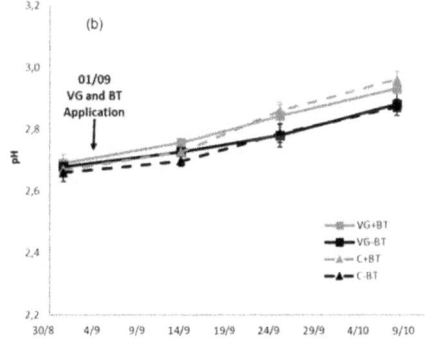

Figure 7. Total soluble solids (TSS) measured in control (C) and treated (VG = Vapor Gard anti-transpirant application; BT = 50% bunch-thinning) Aglianico vines in (**a**) 2013 and (**b**) 2014. Data are averages of 4 repetitions ± SE.

Figure 8. pH measured in control (C) and treated (VG = Vapor Gard anti-transpirant application; BT = 50% bunch-thinning) Aglianico vines in (**a**) 2013 and (**b**) 2014. Data are averages of 4 repetitions ± SE.

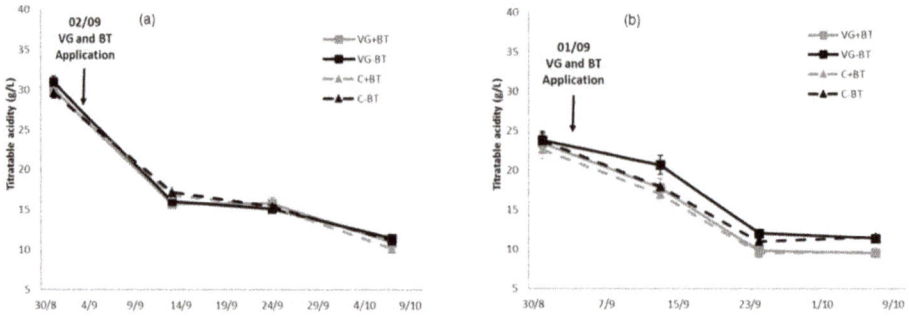

Figure 9. Titratable acidity (TA) measured in control (C) and treated (VG = Vapor Gard anti-transpirant application; BT = 50% bunch-thinning) Aglianico vines in (**a**) 2013 and (**b**) 2014. Data are averages of 4 repetitions ± SE.

Table 1. Yield components, bunch morphology and grape composition recorded in control (C) and treated (VG = Vapor Gard anti-transpirant application; BT = 50% bunch-thinning) Aglianico cultivar in 2013 and 2014 Data are averages of 10 replicates for yield and number of bunches per vine and averages of 4 repetitions for other parameters. For each parameter and for each year, row values with the same letter are not significantly different by Duncan's post hoc test ($p < 0.05$).

	2013				2014			
Parameter	C-BT	V G-BT	C+ BT	VG+ BT	C-BT	VG-BT	C+BT	VG+ BT
Yield/vine (kg)	7.6 b	8.5 b	6.2 a	5.4 a	7.6 b	7.1 b	4.8 a	4.6 a
Bunches/vine	24.8 b	27.3 b	14.9 a	11.6 a	21.3 b	20.0 b	11.5 a	11.9 a
Berry weight (g)	2.67 a	2.61 a	2.60 a	2.71 a	2.52 a	2.70 a	2.52 a	2.71 a
°Brix berry	21.1 a	19.0 b	21.9 a	19.1 b	20.4 bc	19.0 a	21.6 c	19.9 ab
Juice pH	2.88 a	2.84 a	2.87 a	2.85 a	2.85 a	2.84 a	2.96 a	2.93 a
Juice TA (g L^{-1} of tartaric acid)	11.17 ab	11.37 a	10.23 c	10.93 b	11.61 b	11.40 b	9.67 a	9.53 a

Sugar accumulation in the berry showed that, after VG treatment, the accumulation is slower according to other authors (Figure 7a,b) [33,34]. In both years, we observed less sugar accumulation at harvest time, 19.1 vs. 21.9 °Brix in VG+BT and C+BT, in 2013 and 19.9 vs. 21.6 °Brix, in 2014. We can observe the same trend for treatment without BT: 19.0 vs. 21.1 °Brix for VG-BT and C-BT, respectively (Table 1). After VG application, we found a difference of 2.8 °Brix for VG+BT vines, and 2.1 °Brix for VG-BT vines (Figure 7). These values are in agreement with those found in other works; the reduction in TSS in VG-treated vines may be linked to a reduction in canopy photosynthetic capacity and/or limitation in sugar translocation from leaves to berries [30,31,33,34,41].

As shown in Figure 8a,b and Table 1, during the growing season and at harvest, there were no significant differences between treatments in pH values. VG applications did not show significant changes in values of titratable acidity during the vegetative season, while BT, in particular at harvest, showed, in both years, a significantly lower titratable acidity (Figure 9a,b and Table 1).

In both years, as expected, BT vines had a lower yield and lower bunch number per vine than controls. VG applied at veraison did not affect yield per vine or average bunch weight (Table 1) [31].

Extractable anthocyanins (pH 1) differed significantly between the two treatments (VG and C vines): VG vines had more (1044 mg L^{-1}) than C vines (996 mg L^{-1}) in 2013 without BT treatment (Table 2); 1124 vs. 1224 mg L^{-1} was recorded for C and VG, respectively, in 2014. We observed the same results in both years for treatments with BT, while extractable anthocyanins (pH 3.2) and total phenolics (D.O.280) were similar between control and VG vines ±BT (Table 2) in both years, without statistically-significant differences.

Table 2. Total and extractable anthocyanins and total phenolics recorded in control (C) and treated (VG = Vapor Gard anti-transpirant application; BT = 50% bunch-thinning) Aglianico cultivar in 2013 and 2014. Data are averages of four repetitions. For each parameter and for each year, row values with the same letter are not significantly different by Duncan's post hoc test ($p < 0.05$).

Parameter	2013				2014			
	C-BT	V G-BT	C+ BT	V G+ BT	C-BT	V G-BT	C+BT	VG+ BT
Total anthocyanins (mg L^{-1})	996 a	1044 b	992 a	1108 b	1124 a	1224 b	1228 b	1476 c
Extractable anthocyanins (mg L^{-1})	902 a	912 a	910 a	923 a	928 a	952 a	964 a	904 a
Total phenolics OD	75.0 a	64.5 a	69.0 a	75.3 a	60.8 a	65.9 a	62.0 a	64.3 a

In the wines, a lower alcohol percentage was observed for both VG treatments (±BT)) (Table 3), particularly in 2013: 11% and 12.3% were recorded in VG-BT and C-BT, respectively, while 10.9% and 12.9% were recorded in VG+ BT and C+ BT, respectively. Similarly, statistical difference was found in the second year of study (2014): 11.0% vs. 12.5% (VG and C vines, respectively) for treatment - BT, and 10.6% vs. 12.7% (VG and C vines, respectively) for treatment + BT. Total phenolics and total anthocyanins did not show any statistical difference among treatments [31].

Table 3. Wine composition recorded in control (C) and treated (VG = Vapor Gard anti-transpirant application; BT = 50% bunch-thinning) Aglianico vines in 2013 and 2014. Data are averages of four repetitions. For each parameter and for each year, row values with the same letter are not significantly different by Duncan's post hoc test ($p < 0.05$).

Parameter	2013				2014			
	C-BT	V G-BT	C+ BT	VG+ BT	C-BT	V G-BT	C+ BT	VG+ BT
Alcohol (%)	12.3 b	11.0 a	12.9 b	10.9 a	12.5 b	11.0 a	12.7 b	10.6 a
Total anthocyanins (mg kg^{-1})	510 a	490 a	526 a	520 a	181 a	163 a	166 a	197 a
Total phenolics (mg kg^{-1})	1555 a	1467 a	1720 a	1601 a	1719 a	1779 a	1797 a	1814 a

Phenol composition is an important aspect of high-quality red wines. Phenols are responsible for astringency and bitterness [42] and play a role in color stability [43]. The phenolic profile of wine has been shown to be influenced by different viticultural practices [44–47] and different oenological techniques [47–49]. The variety [50], vintage [46,51] and region where the grapes are grown [47,48] all affect the phenolic composition of the wine. Anti-transpirant effects did not affect the total phenolic composition, demonstrating in this way that it is possible to conceive this method as a better way for reducing sugar and alcohol content without influencing the quality of the wine product.

The amounts of wine aroma components can be influenced by various factors, among others the environment (climate, soil), grape variety, degree of ripeness, fermentation conditions (pH, temperature, yeast flora), wine production (oenological methods, treatment substances), and ageing (bottle maturation) of the wine.

After sensory analysis of the wines produced in two years of study, it was possible to detect the typical notes of Aglianico in both 2013 and 2014. The wine products present a good intensity and persistency and also good body and harmony; we observed the same results for the second year of study (Figure 10). In Aglianico wine, we found notes of: phenol leather, good structure, acidity, and typicality. Red fruit notes were presented during the wine tasting in both 2013 and 2014 (Figure 11). No significant difference was shown between the wines produced by grapes treated with anti-transpirant and untreated grapes.

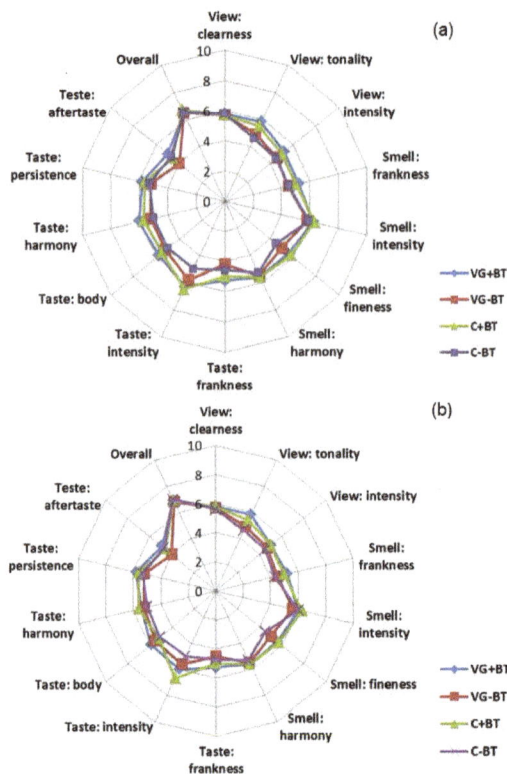

Figure 10. Attributes view, smell and taste scores of Aglianico wines obtained by microvinifications in 2013 (**a**) and 2014 (**b**).

The aroma of wine consists of 600 to 800 aroma compounds, of which especially those typical for the variety are already present in the grapes. There are significant varietal differences between the aromagrams ('fingerprint patterns'). Thus, the amount of some flavor compounds ('key substances') shows typical dependence on the variety. In particular, monoterpene compounds play an important role in the differentiation of wine varieties. We can show after this sensory analysis and wine tasting that the anti-transpirant product does not affect the wine notes and their characteristic structure.

Pruning weight was significantly reduced in each year in the VG-sprayed vines as compared with C vines (Figure 12). In 2013, pruning weight measured in VG-treated vines was 2.9 kg while in the control it was 3.8 kg. The BT treatment also showed differences in pruning weight: the control vine + BT reached values of 3.2 kg, and the sprayed vines 2.5 kg (Figure 12a). The same results, with statistically significant differences between VG and control vines, were recorded in 2014 (Figure 12b). Independently of BT, the VG vines showed a lower pruning weight with respect to control vines. Notably, lower pruning weight emphasizes that vine 'vigor' was restrained by VG to the benefit of the ripening process, suggesting that this compound could be considered for applications aimed at controlling vigor while avoiding or limiting the counteracting effect of a smaller source potential, according to Palliotti et al. [31,34].

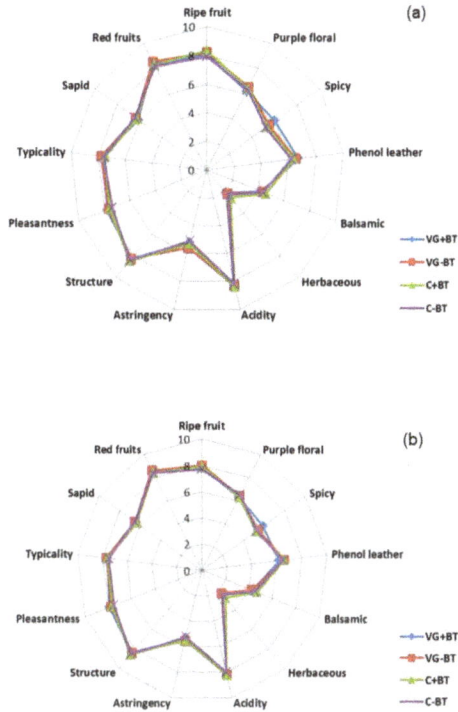

Figure 11. Sensory scores of Aglianico wines obtained by microvinifications in 2013 (**a**) and 2014 (**b**).

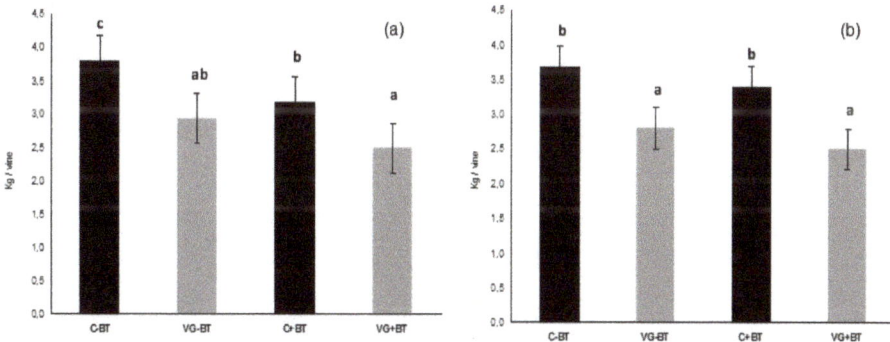

Figure 12. Pruning weight per vine measured in control (C) and treated (VG = Vapor Gard anti-transpirant application; BT = 50% bunch-thinning) Aglianico vines in 2013 (**a**) and 2014 (**b**). Data are averages of 10 replicates ± SE. The same letter indicates non-significant differences by Duncan's post hoc test ($p < 0.05$).

4. Conclusions

The application in post-veraison of the organic film-forming anti-transpirant is a suitable strategy to delay grape ripening. The method proved to be effective and easy to apply in order to hinder the sugaring of berries and to obtain wines with a lower alcohol percentage. Concurrently, this method had no other negative impact on phenolic compounds, organic acids, or pH in grapes and wines. Moreover, the anti-transpirant does not show adverse effects on the production per plant or on berry

Agronomy **2019**, *9*, 244

size for each vintage examined. The application of anti-transpirant leads to a reduction in stomatal conductance and A and an increase in WUEi in Mediterranean climatic conditions. To be effective in reducing the accumulation of TSS in the berries, the VG emulsion should be applied at the time of veraison and should completely wet the lower leaf surface where stomata are located. The effectivity of the product depends also on the concentration of preparation; in our case, the concentration of 2% has been shown to be very efficient. Another important aspect to consider is that applying the anti-transpirant product does not produce any differences in the notes and in the wine characteristics produced in both years of trial.

After the sensory analysis and wine tasting, no negative notes or unpleasant characteristics were detected in the wines produced. Finally, the reduction of sugar content in the berries and the reduction of alcohol content in the wines did not result in any negative qualitative or quantitative characteristics that could affect the final product.

Author Contributions: N.M. wrote the first draft of the manuscript and followed the agronomical and physiological measurements. R.D.L. and A.P. were involved in data analysis, data interpretation and writing. C.D.V. coordinated the whole project, provided the intellectual input, set up the experiment and corrected the manuscript.

Funding: This research received no external funding.

Acknowledgments: The authors wish to thank their colleague Stefano Barone (Associate Professor of Statistics at University of Palermo, Department of Agricultural, Food and Forest Sciences) for his review of the statistical aspects in the presentation of the manuscript.

Conflicts of Interest: The authors declare no conflict of interest.

References

1. Jones, G.V.; White, M.A.; Cooper, O.R.; Storchmann, K. Climate change and global wine quality. *Clim. Chang.* **2005**, *73*, 319–343. [CrossRef]
2. Hansen, J.; Sato, M.; Ruedy, R.; Lo, K.; Lea, D.W.; Medina-Elizade, M. Global temperature change. *Proc. Natl. Acad. Sci. USA* **2006**, *103*, 14288–14293. [CrossRef] [PubMed]
3. Tomasi, D.; Jones, G.V.; Giust, M.; Lovat, L.; Gaiotti, F. Grapevine phenology and climate change: Relationships and trends in the Veneto region of Italy for 1964–2009. *Am. J. Enol. Vitic.* **2011**, *62*, 329–339. [CrossRef]
4. Webb, L. The Impact of Greenhouse Gas-Induced Climate Change on the Australian Wine Industry. Ph.D. Thesis, School of Agriculture and Food Systems, University of Melbourne, Parkville, VIC, Australia, 2006; p. 277.
5. Schultz, H. Climate change and viticulture: A European perspective on climatology, carbon dioxide and UV-B effects. *Aust. J. Grape Wine Res.* **2000**, *6*, 2–12. [CrossRef]
6. Palliotti, A.; Silvestroni, O.; Leoni, F.; Poni, S. Maturazione dell'uva e gestione della chioma in *Vitis vinifera*: Processi e tecniche da riconsiderare in funzione del cambiamento del clima e delle nuove esigenze di mercato. *Italus Hortus* **2012**, *19*, 1–15.
7. Scafidi, P.; Barbagallo, M.G.; Pisciotta, A.; Mazza, M.; Downey, M.O. Defoliation of two-wire vertical trellis: Effect on grape quality. *New Zeal. J. Crop Hort. Sci.* **2017**, *46*, 18–38. [CrossRef]
8. Pisciotta, A.; Scafidi, P.; Di Lorenzo, R.; Barbagallo, M.G. Manual and mechanical leaf removal in the bunch zone (*Vitis vinifera* L. 'Nero d'Avola'): Effects on plant physiology, vegetative parameters, yield and grape quality in a warm area. *Acta Hortic.* **2013**, *978*, 285–292. [CrossRef]
9. Hunter, J.J.; Volschenk, C.G.; Novello, V.; Pisciotta, A.; Booyse, M.; Fouché, G.W. Integrative effects of vine water relations and grape ripeness level of *Vitis vinifera* L. cv. Shiraz/Richter 99. II. grape composition and wine quality. *S. Afr. J. Enol. Vitic.* **2014**, *35*, 359–374. [CrossRef]
10. Pisciotta, A.; Di Lorenzo, R.; Santalucia, G.; Barbagallo, M.G. Response of grapevine (Cabernet Sauvignon cv) to above ground and subsurface drip irrigation under arid conditions. *Agric. Water Manag.* **2018**, *197*, 122–131. [CrossRef]
11. Jones, G.V.; Davis, R.E. Climate influences on grapevine phenology, grape composition, and wine production and quality for Bordeaux, France. *Am. J. Enol. Vitic.* **2000**, *51*, 249–261.
12. Pearce, I.; Coombe, B.G. Grapevine Phenology. In *Viticulture Volume 1—Resources*; Dry, P., Coombe, B.G., Eds.; Winetitles: Adelaide, South Australia, 2004; pp. 150–166.

13. McIntyre, G.N.; Lider, L.A.; Ferrari, N.L. The chronological classification of grapevine phenology. *Am. J. Enol. Vitic.* **1982**, *33*, 80–85.

14. Haselgrove, L.; Botting, D.; Heeswijck, R.V.; Høj, P.B.; Dry, P.R.; Ford, C.; Land, P.G.I. Canopy microclimate and berry composition: The effect of bunch exposure on the phenolic composition of *Vitis vinifera* L cv. Shiraz grape berries. *Aust. J. Grape Wine Res.* **2000**, *6*, 141–149. [CrossRef]

15. Marais, J. Effect of grape temperature and yeast strain on Sauvignon blanc wine aroma composition and quality. *S. Afr. J. Enol. Vitic.* **2001**, *22*, 47–51. [CrossRef]

16. Marais, J.; Calitz, F.; Haasbroek, P.D. Relationship between microclimatic data, aroma component concentrations and wine quality parameters in the prediction of Sauvignon blanc wine quality. *S. Afr. J. Enol. Vitic.* **2001**, *22*, 22–26. [CrossRef]

17. Spayd, S.E.; Tarara, J.M.; Mee, D.L.; Ferguson, J.C. Separation of sunlight and temperature effects on the composition of *Vitis vinifera* cv. Merlot berries. *Am. J. Enol. Vitic.* **2002**, *53*, 171–182.

18. Walker, R.P.; Famiani, F. Organic acids in fruits: Metabolism, functions and contents. *Hort. Rev.* **2018**, *45*, 371–430.

19. Smart, R.E.; Alcorso, C.; Hornsby, D.A. A comparison of winegrape performance at the present limits of Australian viticultural climates Alice Springs and Hobart. *Aust. Grapegrow. Winemak.* **1980**, *184*, 28–30.

20. Dry, P.R. Climate change and the Australian grape and wine industry. *Aust. Grapegrow. Winemak.* **1988**, *300*, 14–15.

21. Godden, P.; Gishen, M. The trends in the composition of Australian wine from vintages 1984 to 2004. *Austral. New Zealand Wine Ind. J.* **2005**, *20*, 21.

22. Dokoozlian, N. Integrated canopy management: A twenty year evolution in California. In *Recent Advances in Grapevine Canopy Management*; University of California: Davis, CA, USA, 16 July 2009; pp. 43–52.

23. Vierra, G. Pretenders at the table—Are table wines no longer food friendly? *Wine Business Monthly* **2004**, *11*, 17–21.

24. Duchêne, E.; Schneider, C. Grapevine and climatic changes: A glance at the situation in Alsace. *Agron. Sustain. Dev.* **2005**, *25*, 93–99. [CrossRef]

25. Ganichot, B. Evolution de la date des vendanges dans les Côtes du Rhône méridionales. In *Actes des 6e Rencontres Rhodaniennes*; Institut Rhodanien: Orange, France, 2002; pp. 38–41.

26. Di Lena, B.; Silvestroni, O.; Mariani, L.; Parisi, S.; Antenucci, F. European climate variability effects on grapevine harvest date time series in the Abruzzi (Italy). *Acta Hortic.* **2012**, *931*, 63–69. [CrossRef]

27. Scaglione, G.; Graziani, G.; Federico, R.; Di Vaio, C.; Nadal, M.; Formato, A.; Di Lorenzo, R.; Ritieni, A. Effect of canopy managment techniques on the nutritional quality of the Montepulciano grapewine in Puglia (southern Italy). *OENO One* **2012**, *46*, 253–261. [CrossRef]

28. Carnevali, P.; Falcetti, M. Utilizzo di Antitraspiranti a Base di Pinolene per la Modulazione della Maturità Glucidica. In Proceedings of the 'IV Convegno Nazionale di Viticoltura Asti, Asti, Italy, 10–12 July 2012.

29. Mattii, G.; Lazzini, F.; Binda, C. Effetto di trattamenti con antitraspiranti sulla maturazione di uve Merlot. In Proceedings of the 'IV Convegno Nazionale di Viticoltura Asti, Asti, Italy, 10–12 July 2012.

30. Tittmann, S.; Stöber, V.; Bischoff-Schaefer, M.; Stoll, M. Application of anti-transpirant under greenhouse conditions of grapevines (*Vitis vinifera* cv. Riesling and cv. Müller-thurgau) reduce photosynthesis under greenhouse conditions. In Proceedings of the 18th International Symposium GiESCO, Porto, Portugal, 7–11 July 2013; Ciência e Técnica Vitivinícola, 28. pp. 276–282.

31. Palliotti, A.; Panara, F.; Famiani, F.; Sabbatini, P.; Howell, G.S.; Silvestroni, O.; Poni, S. Postveraison application of antitranspirant di-1-p-menthene to control sugar accumulation in Sangiovese grapevines. *Am. J. Enol. Vitic.* **2013**, *64*, 378–385. [CrossRef]

32. Glories, Y.; Agustin, M. Maturité phénolique du raisin, conséquences technologiques: Application aux millésimes 1991 et 1992. In *Actes du Colloque Journée Technique du CIVB*; CIVB: Bordeaux, France, 21 January 1993; pp. 56–61.

33. Brillante, L.; Belfiore, N.; Gaiotti, F.; Lovat, L.; Sansone, L.; Poni, S.; Tomasi, D. Comparing Kaolin and Pinolene to Improve Sustainable Grapevine Production during Drought. *PLoS ONE* **2016**, *11*, e0156631. [CrossRef]

34. Palliotti, A.; Poni, S.; Berrios, J.G.; Bernizzoni, F. Vine performance and grape composition as affected by early-season source limitation induced with anti-transpirants in two red *Vitis vinifera* L. cultivars. *Aust. J. Grape Wine Res.* **2010**, *16*, 426–433. [CrossRef]

35. Davenport, D.C.; Fisher, M.A.; Hagan, R.M. Some counteractive effects of antitranspirants. *Plant Physiol.* **1972**, *49*, 722–724. [CrossRef] [PubMed]
36. Iriti, M.; Picchi, V.; Rossoni, M.; Gomarasca, S.; Ludwig, N.; Gargano, M.; Faoro, F. Chitosan antitranspirant activity is due to abscisic acid-dependent stomatal closure. *Environ. Exp. Bot.* **2009**, *66*, 493–500. [CrossRef]
37. Moftah, A.E.; Al-Humaid, A.R.I. Effects of antitranspirants on water relations and photosynthetic rate of cultivated tropical plant (*Polianthes tuberosa* L.). *Pol. J. Ecol.* **2005**, *53*, 165–175.
38. Rosati, A. Physiological effects of kaolin particle film technology: A review. *Funct. Plant Sci. Biotech.* **2007**, *1*, 100–105.
39. Shellie, K.C.; King, B.A. Kaolin-based foliar reflectant and water deficit influence Malbec leaf and berry temperature, pigments, and photosynthesis. *Am. J. Enol. Vitic.* **2013**, *64*, 223–230. [CrossRef]
40. Francini, A.; Lorenzini, G.; Nali, C. The antitranspirant di-1-p-menthene, a potential chemical protectant of ozone damage to plants. *Water Air Soil Poll.* **2011**, *219*, 459–472. [CrossRef]
41. Palliotti, A.; Poni, S.; Petoumenou, D.; Vignaroli, S. Effects of modulated limitation of canopy photoassimilation using anti-traspirant on yield and grape composition. In Proceedings of the 'II Convegno Nazionale di Viticoltura', Marsala, Italia, 14–19 luglio 2008. *Italus Hortus* **2010**, *17*, 21–26.
42. Fischer, U.; Noble, A.C. The effect of ethanol, catechin concentration, and pH on sourness and bitterness of wine. *Am. J. Enol. Vitic.* **1994**, *45*, 6–10.
43. Robinson, W.B.; Weirs, L.D.; Bertino, J.J.; Mattick, L.R. The relation of anthocyanin composition to color stability of New York State wines. *Am. J. Enol. Vitic.* **1966**, *17*, 178–184.
44. Price, S.F.; Breen, P.J.; Valladao, M.; Watson, B.T. Cluster sun exposure and quercetin in Pinot noir grapes and wine. *Am. J. Enol. Vitic.* **1995**, *46*, 187–194.
45. Reynolds, A.G.; Price, S.F.; Wardle, D.A.; Watson, B.T. Fruit environment and crop level effects on Pinot noir. I. Vine performance and fruit composition in British Columbia. *Am. J. Enol. Vitic.* **1994**, *45*, 452–459.
46. Yokotsuka, K.; Nagao, A.; Nakazawa, K.; Sato, M. Changes in anthocyanins in berry skins of Merlot and Cabernet Sauvignon grapes grown in two soils modified with limestone or oyster shell versus a native soil over two years. *Am. J. Enol. Vitic.* **1999**, *50*, 1–12.
47. Zoecklein, B.W.; Fugelsang, K.C.; Gump, B.H.; Nury, F.S. *Wine Analysis and Production*; Chapman & Hall: New York, NY, USA, 1995; pp. 407–409.
48. Sims, C.A.; Bates, R.P. Effects of skin fermentation time on the phenols, anthocyanins, ellagic acid sediment, and sensory characteristics of a red *Vitis rotundifolia* wine. *Am. J. Enol. Vitic.* **1994**, *45*, 56–62.
49. Wightman, J.D.; Price, S.F.; Watson, B.T.; Wrolstad, R.E. Some effects of processing enzymes on anthocyanins and phenolics in Pinot noir and Cabernet Sauvignon wines. *Am. J. Enol. Vitic.* **1997**, *48*, 39–48.
50. Goldberg, D.M.; Karumanchiri, A.; Tsang, E.; Soleas, G.J. Catechin and epicatechin concentrations of red wines: Regional and cultivar-related differences. *Am. J. Enol. Vitic.* **1998**, *49*, 23–34.
51. Brossaud, F.; Cheynier, V.; Asselin, C.; Moutounet, M. Flavonoid compositional differences of grapes among site test plantings of Cabernet franc. *Am. J. Enol. Vitic.* **1999**, *50*, 277–284.

agronomy

MDPI

Article

Use of Reflectance Indices to Assess Vine Water Status under Mild to Moderate Water Deficits

Cristina González-Flor, Lydia Serrano * and Gil Gorchs

Departament d'Enginyeria Agroalimentària i Biotecnologia, Universitat Politècnica de Catalunya, 08060 Castelldefels (Barcelona), Spain
* Correspondence: lydia.serrano@upc.edu; Tel.: +34-935-522-110

Received: 11 April 2019; Accepted: 27 June 2019; Published: 1 July 2019

Abstract: The monitoring of vine water status is of interest for irrigation management in order to improve water use while optimizing both berry yield and quality. Remote-sensing techniques might provide accurate, rapid, and non-destructive estimates of vine water status. The objective of this study was to test the capability of the reflectance-based water index (WI) and the photochemical reflectance index (PRI) to characterize *Vitis vinifera* L. cv. Xarel·lo water status under mild to moderate water deficits. The study was conducted at the leaf level in irrigated potted plants and at the plant level on five commercial rain-fed vineyards in 2009 and 2010. In potted plants, the reflectance indices PRI and WI closely tracked variation in the leaf-to-air temperature difference (ΔT) with $r^2 = 0.81$ and $r^2 = 0.83$, for WI and PRI, respectively ($p < 0.01$). In addition, in potted plants, both PRI and WI showed significant relationships with light-use efficiency (LUE)—calculated as the ratio between net CO_2 assimilation rate (A_n) and incident photosynthetic active radiation (PAR) at the leaf surface—with $r^2 = 0.92$ and $r^2 = 0.74$ for PRI and WI, respectively. At the canopy level, vine predawn water potential (Ψ_{pd}) was related to the canopy-to-air temperature difference (ΔT_m) across years ($r^2 = 0.37$, $p < 0.05$). In the years of study, the relationships between PRI and WI showed variable degrees of correlation against Ψ_{pd} and ΔT_m. Across years, PRI and WI showed significant relationships with Ψ_{pd}, with $r^2 = 0.41$ and $r^2 = 0.37$ ($p < 0.01$), for WI and PRI, respectively. Indices formulated to account for variation in canopy structure (i.e., PRI_{norm} and WI_{norm}) showed similar degrees of correlation against Ψ_{pd} to their original formulations. In addition, PRI and WI were capable of differentiating ($p < 0.01$) between mild ($\Psi_{pd} > -0.4$ MPa) and moderate ($\Psi_{pd} < -0.4$ MPa) water deficits, and a similar response was observed when PRI_{norm} and WI_{norm}—formulated to account for variation in canopy structure—were considered. Thus, at the leaf level, our result suggest that WI and PRI can be used to adequately predict the diurnal dynamics of stomatal aperture and transpiration. In addition, at the canopy level, PRI and WI effectively differentiated vines under mild water deficits from those experiencing moderate water deficits. Thus, our results show the capability of WI and PRI in characterizing vine water status under mild to moderate water deficits.

Keywords: predawn water potential; PRI; remote sensing; vineyards; water status; WI

1. Introduction

Water deficits are the major constraint for grape production in the Mediterranean region [1], and future scenarios predict further increases in the frequency and intensity of water deficits as a result of climate change [2]. As a result, irrigation is being widely adopted in order to secure more regular and predictable yields [1,3,4]. Concurrently, and due to the increasing water scarcity, as well as the rising competition between water users, deficit irrigation techniques emerged as a potential strategy to improve the productivity of water [5]. Particularly, in viticulture areas, the use of regulated deficit irrigation strategies emerged as a way of reducing water use with little or no impact on yield and a positive impact on berry quality [1,3,6]. In regulated deficit irrigation strategies, plant water status is maintained within predefined limits of deficit during certain phases of the seasonal development, normally when fruit growth is least sensitive to water reductions [5]. Thus, in order to guarantee the success of the use of this technique, an accurate control of plant water status is required for scheduling irrigation. Several physiological indicators are used to assess plant water status, with leaf water potential, stem water potential, stomatal conductance, and transpiration being the most widely used in viticulture [4,7,8]. However, measurement of these water stress indicators for practical irrigation scheduling is labor-intensive and time-consuming due to the large number of observations necessary to characterize the spatial variability. As an alternative, remote-sensing techniques might be a very useful tool to monitor vine water status because of opportunities for cost-effective generation of spatial data.

Remote-sensing methods based on thermal emission to monitor plant water status were extensively evaluated in field trials. In vineyards, infrared thermometry and thermal imaging were shown to provide reasonable estimations of whole-canopy conductance [8,9] and plant water potential in grapevines [10]. However, largely due to the effects of environmental conditions on canopy temperature [11], the practical application of thermal methods to irrigation scheduling is currently limited to regions with very constant (semi-)arid weather conditions during the growth season [11]. In addition, since grapevine cultivars present, in terms of stomatal control and water potential regulation, contrasted responses to water deficits [2], temperature-based indicators might not always properly characterize vine water status [12]. Reflectance-based indices might also provide direct or indirect estimates on vine water status. Previous studies showed the capability of reflectance indices based on water absorption features in assessing vine water status [13–15]. Particularly, the reflectance-based water index (WI) [16] was shown to track diurnal changes in stomatal conductance in irrigated vines, as well as in the canopy-to-air temperature difference in vineyards experiencing moderate to severe water stress [13]. Similarly, several studies showed the capability of the photochemical reflectance index (PRI) [17,18]—an index related to the epoxidation state of xanthophyll pigments and, thus, to photosynthetic efficiency—at detecting water stress in fruit trees grown using regulated deficit irrigation techniques [19,20] and in vineyards [12]. However, both WI and PRI are sensitive to changes in canopy structure [21,22], which might impair their capacity to assess vine water status under contrasted growing conditions (environmental), including cultural practices such as fertilization and pruning methods. Thus, because water content of vegetation depends on both leaf area and relative water content [22,23], changes in canopy structure might impair the capacity of the WI to assess vine water status. Similarly, PRI estimates of plant water status might be affected by changes in the size of constitutive pigments pools (i.e., chlorophyll and carotenoid)—which control the facultative short-term variation in PRI—as well as by changes in canopy structure [24,25]. Approaches to overcome these confounding effects consist of combining the primary index (i.e., WI or PRI) with indices of canopy structure—such as the normalized difference vegetation index (NDVI) (20,23)—or including specific bands on their formulation [12,20] that account for the effects of varying leaf area and/or pigment content.

We herein explore the capability of the reflectance indices PRI and WI as a proxy to assess vine water status under mild to moderate water deficits in *Vitis vinifera* L. cv. Xarel·lo. The specific objectives were (i) to evaluate the performance of WI and PRI in estimating physiological parameters related to water status at the leaf level on potted irrigated vines; (ii) to assess the capability of WI and PRI, as well as their normalized formulations, in estimating vine water status in field-grown vines experiencing mild to moderate water deficits; and (iii) to evaluate the capability of WI and PRI in differentiating mild from moderate water deficits levels in field-grown vines.

2. Materials and Methods

2.1. Leaf Level Study

Three-year-old *Vitis vinifera* L. cv. Xarel·lo grafted on 110R (*V. berlandieri* × *V. rupestris*) vines were grown in 17 L pots filled with a mixture of sand and peat turf (1:1, *v/v*). Vines were grown in a greenhouse and were watered daily with 0.67 L·plant^{-1} of water. In addition, once a week, 1 L·plant^{-1} of a full Hoagland solution was provided. During the growth period, all the lateral shoots, buds, and young flowers were removed in order to leave only the winter buds and leaves.

A diurnal cycle of gas exchange and reflectance measurements was carried out on 27 July 2010, approximately every two hours. Vines were placed outside the greenhouse in order to register measurements under direct sunlight. Data acquisition started at dawn (5:30 a.m. solar time) and finished late in the afternoon (7:30 p.m. solar time). Two fully expanded leaves of four vines were measured (eight leaves). Gas exchange parameters were measured with a portable gas exchange system CIRAS-2 (PP Systems Ltd., Havervill, MA, USA) under current air temperature and humidity, and leaf cuvette (Automatic Leaf Universal Cuvette, PLC6) CO_2 concentration was set to ~400 ppm using a CO_2 cartridge. The leaf cuvette had an aperture of 25 mm × 7 mm and was held to keep the leaves in their natural positions. Air vapor pressure deficit (VPD), air temperature (T_a), photosynthetic photon flux density incident on the leaf ($PPFD_i$), net CO_2 assimilation rate (A_n), transpiration rate (E), stomatal conductance (g_s), and leaf temperature (T_l) were averaged among the eight observations to represent the mean value at the measuring time. Light-use efficiency (LUE) was calculated as the ratio between A_n and $PPFD_i$.

Leaf reflectance was measured using a spectroradiometer UNISPEC (PP Systems Ltd., Havervill, MA, USA) with a 2.3-mm-diameter bifurcated fiber optic and a leaf clip (models UNI410 and UNI501, PP Systems, Havervill). The detector samples 256 bands at roughly even intervals (average band-to-band spacing of 3.3 nm) within a 400–1100-nm effective spectral range. Each leaf scan resulted from the average of three internal measurements. Apparent reflectance was obtained after standardization by a Spectralon reflectance standard measured before each cycle. Afterward, the water index (Equation (1)) [16] and the photochemical reflectance index (Equation (2)) [17,18] were formulated as follows:

$$WI = R_{900}/R_{970}, \tag{1}$$

$$PRI = (R_{531} - R_{570})/(R_{531} + R_{570}), \tag{2}$$

where R indicates spectral reflectance, and the subindices indicate the respective wavelengths in nanometers.

Vine water status was determined using a Scholander pressure chamber (Soilmoisture 3005, Soil Moisture Corp., Santa Barbara, CA, USA). At 5:30 a.m. (dawn), four leaves (i.e., one leaf per vine), which were previously wrapped in a plastic bag and covered with aluminum foil the evening before, were used to determine vine predawn water potential (Ψ_{pd}). Leaves subjected to the same coverage were used to measure stem water potential (Ψ_s) at midday (solar noon). In addition, leaf water potential was determined at midday (Ψ_m).

2.2. Field Study

The field study took place in 2009 and 2010 in five *Vitis vinifera* L. cv. Xarel·lo rain-fed vineyards (plantation years between 1989 and 1998) located in the west area of Barcelona (Alt Penedès and Anoia counties, 1°48′22″ west (W), 41°28′54″ north (N)). Vines were planted at varying density (2016 to 3086 stock·ha^{-1}) and the training system was Double Royat. In each vineyard, three plots (with three adjacent vines per plot) with contrasting vigor were studied. Nonetheless, in order to evaluate the capability of reflectance indices in assessing vine water status under mild to moderate water deficits, only plots with average values of $\Psi_{pd} > -0.6$ MPa were considered. Thus, the study comprised 14 plots in 2009 and 12 plots in 2010. Weather data were obtained from a nearby weather station located in Els Hostalets de Pierola (1°48′31″ W; 41°31′59″ N). The average temperature is around 15 °C, while the average cumulative annual precipitation is 479.2 mm. The weather water balance was computed as the difference between precipitation (P) and reference evapotranspiration (ET_0). Veraison took place between 31 July and 4 August in 2009 and from 10 August until 18 August in 2010.

Predawn water potential (Ψ_{pd}) was measured at veraison using a pressure chamber (Soilmosture 3005, Soil Moisture Corp., Santa Barbara, CA, USA). Measurements were carried out on a single mature external leaf per vine (three per plot). Additionally, the canopy-to-air temperature difference (ΔT_m) was measured at midday using a hand-held infrared thermometer (ST Pro Plus, Raytek Corp., Santa Cruz, CA, USA) at approximately 20 cm of the canopy. Measurements were taken on both the sun-exposed and the shaded sides of the canopy, and ΔT_m was computed as the average of these two measurements.

Fractional intercepted photosynthetic active radiation (fIPAR) was measured at midday using a hand-held ceptometer (Accupar, Decagon Devices Inc., Pullman, WA, USA). Seven measurements, parallel and perpendicular to the row of the vine, were collected at 10 cm of ground level. Incident light radiation was registered above the canopy. In 2010, fIPAR measurements were carried out only in eight blocks due to a failure of the instrument. In addition, exposed leaf area (ELA) was determined using the procedure proposed by Smart and Robinson [26] as follows:

$$ELA = (2\,h + e) \times (100 - T) \times d,$$

where e is the mean of three measures of canopy width, h is the mean of three measures of canopy height, T is the percent canopy gaps, and d is the distance between vines in the same row. Values of fIPAR in the remaining plots in 2010 were estimated from the regression of fIPAR against ELA as follows:

$$fIPAR = 41.37 + 12.14 \times ELA\ (r^2 = 0.86,\ p < 0.01).$$

Reflectance data measurements were conducted at midday (solar noon) on cloudless days in order to minimize variation due to differences in illumination conditions at the stage of veraison. Spectral data were collected using a narrow-band spectroradiometer (UNISPEC, PP Systems Ltd., Havervill, MA, USA), which works in a wavelength range between 310 and 1100 nm (visible and near-infrared), with a resolution of 3.3 nm. Irradiance was measured by connecting the spectroradiometer to a cosine-corrected detector lens (UNI-685 PP Systems Ltd., Havervill, MA, USA) mounted on a tripod boom and oriented to the sky above the canopy. Canopy radiance was obtained with the spectroradiometer connected to a 12° field-of-view foreoptic (UNI-710, PP Systems Ltd., Havervill, MA, USA) via a 2.3-mm-diameter fiber optic (model UNI410, PP Systems, Havervill, MA, USA). The foreoptic instrument was mounted on a tripod boom and held on a nadir orientation 0.75 m above the canopy, so that the field of view covered an area of ~15 cm in diameter. Three scans were collected and internally averaged for each vine. Apparent reflectance was calculated as the ratio between radiance and irradiance. The normalized difference vegetation index (NDVI), the photochemical reflectance index (PRI) [17], the water index (WI), the normalized WI (WI_{norm}) [16], the normalized PRI (PRI_{norm}) [12], and the structural independent pigment index (SIPI) [27] were calculated using narrow-band apparent reflectance values as follows:

$$NDVI = (R_{900} - R_{680})/(R_{900} + R_{680}),$$

$$PRI = (R_{531} - R_{570})/(R_{531} + R_{570}),$$

$$WI = R_{900}/R_{970},$$

$$WI_{norm} = WI/NDVI,$$

$$PRI_{norm} = PRI/(((R_{800} - R_{670})/(R_{800} + R_{670})^{0.5}) \times R_{700}/R_{670}),$$

$$SIPI = (R_{800} - R_{445})/(R_{800} - R_{680}),$$

where R indicates apparent reflectance, and the subindices indicate the respective wavelengths in nanometers.

2.3. Statistical Analyses

Statistical analyses were carried out using the statistical package SPSS 25.0 (SPSS Inc., Chicago, IL, USA). In the leaf level study, analyses of variance (ANOVA) were carried out to assess the changes in gas exchange parameters and reflectance indices throughout the day (i.e., time of sampling as a source of variation). In addition, Pearson correlation analyses were used to study the relationships between water status, gas exchange parameters, and reflectance indices. In the field study, differences in the variables studied were determined using ANOVA analyses while considering both year and water deficit level as sources of variation. Means were compared using the Student–Knewman–Keuls test, and the relationships among the canopy variables and reflectance data were studied by Pearson correlation analysis.

3. Results

3.1. Leaf Level Study

3.1.1. Environmental Conditions

The air temperature (Ta) and vapor pressure deficit (VPD) were typical of Mediterranean summer conditions and were characterized by a gradual increase until midday/early afternoon, followed by a gradual decrease until the end of the diurnal cycle (data not shown). During the measurement period, minimum and maximum temperatures were 24.1 °C and 32.7 °C, respectively, whereas vapor pressure deficit ranged between 1.27 kPa and 2.83 kPa. Similarly, incident photosynthetic photon flux density at the leaf surface ($PPFD_i$) showed a gradual increase from early morning until early afternoon and decreased afterward. Incident PAR on the leaf surface ranged between 60 $\mu mol \cdot m^{-2} \cdot s^{-1}$ and 1025 $\mu mol \cdot m^{-2} \cdot s^{-1}$ during the diurnal cycle (Figure 1).

3.1.2. Water Potential

Predawn water potential (Ψ_{pd}) ranged between −0.30 MPa and −0.35 MPa among plants with an average value of −0.33 ± 0.02 MPa (average ± standard error of the mean), whereas values of Ψ_s and Ψ_m were −0.49 ± 0.20 MPa and −1.00 ± 0.05 MPa, respectively.

3.1.3. Gas Exchange and Reflectance Indices

Diurnal courses of environmental conditions, gas exchange, and reflectance indices are shown in Figure 1. Net photosynthesis (A_n) showed a peak early in the morning, while it decreased in the central hours of the day, and further decreased again in the afternoon. Nonetheless, A_n did not show significant variation ($p > 0.05$) throughout the diurnal cycle with an average value of 2.37 ± 0.33 μmol $CO_2 \cdot m^{-2} \cdot s^{-1}$.

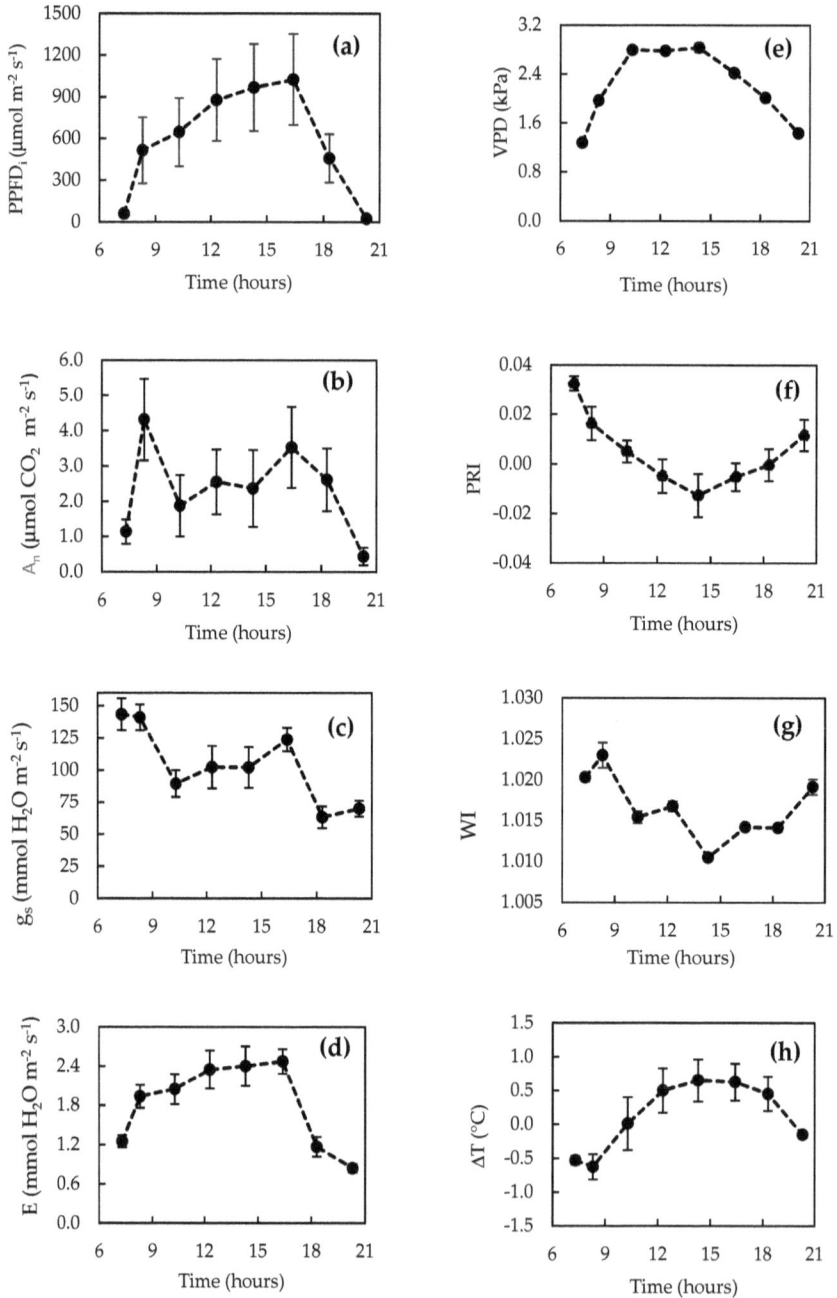

Figure 1. Diurnal time course of (**a**) photosynthetic photon flux density incident on the leaf (PPFDi), (**b**) net photosynthesis (A_n), (**c**) stomatal conductance (g_s), (**d**) transpiration rate (E), (**e**) leaf-to-air vapor pressure deficit (VPD), (**f**) photochemical reflectance index (PRI), (**g**) water index (WI), and (**h**) leaf-to-air temperature difference (ΔT). Values are means ± standard errors of the mean ($n = 8$).

In contrast, stomatal conductance (g_s) significantly decreased ($p < 0.01$) throughout the diurnal cycle from early morning (143 ± 11.4 mmol $H_2O \cdot m^{-2} \cdot s^{-1}$) to sundown ($70 \pm 6.3$ mmol $H_2O \cdot m^{-2} \cdot s^{-1}$). Transpiration rate (E) steadily increased throughout the day, reaching a peak in the early afternoon (2.47 ± 0.19 mmol $H_2O \cdot m^{-2} \cdot s^{-1}$) and significantly decreasing afterward. Similarly, ΔT varied significantly ($p < 0.01$) throughout the day, increasing from early morning until mid-afternoon (with a peak value of 0.65 ± 0.31 °C) and decreasing afterward. The narrow-band indices WI and PRI significantly varied throughout the diurnal cycle ($p < 0.01$). Both, WI and PRI showed a gradual decrease from early morning to early afternoon and increased again toward the end of the diurnal cycle (Figure 1).

Gas exchange parameters (i.e., A_n, E, and g_s) were not significantly correlated. Nonetheless, E was found to be closely related to VPD ($r = 0.85$, $p < 0.01$), whereas A_n was related to $PPFD_i$, although to a lesser extent ($r = 0.63$; $p < 0.10$). In addition, there was no significant correlation between either WI or PRI and gas exchange parameters (Table 1). Both PRI and WI showed significant correlation ($p < 0.01$) with light-use efficiency (LUE)—calculated as the ratio between A_n and incident PPFD at the leaf surface—with $r = 0.96$ and $r = 0.86$, for PRI and WI, respectively. In addition, PRI and WI were significantly correlated ($p < 0.01$) with ΔT, with $r = -0.92$ and $r = -0.90$, for PRI and WI, respectively (Figure 2).

Table 1. Correlation coefficients between leaf-to-air temperature difference (ΔT), stomatal conductance (g_s), transpiration (E), net photosynthesis (A_n), and light-use efficiency (LUE) and the reflectance indices, water index (WI) and photochemical reflectance index (PRI). Data are for average values at each sampling time ($n = 8$). Significant correlations at the 0.01 (**) level are indicated.

	PRI	WI
ΔT (°C)	−0.92 **	−0.90 **
g_s (mmol $H_2O \cdot m^{-2} \cdot s^{-1}$)	0.44	0.45
E (mmol $H_2O \cdot m^{-2} \cdot s^{-1}$)	−0.61	−0.43
A_n (μmol $CO_2 \cdot m^{-2} \cdot s^{-1}$)	−0.31	0.01
LUE (μmol $CO_2 \cdot$ μmol photon^{-1})	0.96 **	0.86 **

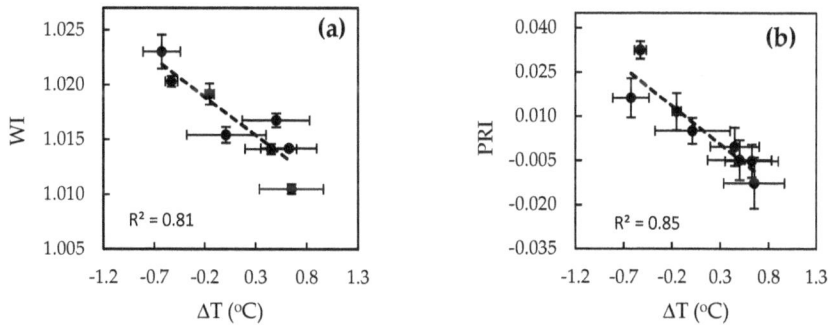

Figure 2. Relationship between the leaf-to-air temperature difference (ΔT) measured at the leaf level in Xarel·lo potted plants with (**a**) water index (WI) and (**b**) photochemical reflectance index (PRI). Values are the means ± standard errors of the mean ($n = 8$) measured at different hours throughout the day.

3.2. Field Study

3.2.1. Weather Conditions

Temperatures in the years of study were similar to the long-term average, whereas precipitation was characterized by an increase of 30% over the long-term precipitation average (479.2 mm) (Figure 3).

In addition, according to the weather water balance, there was abundant water availability over the period of study, except in summer when water deficits had a larger incidence.

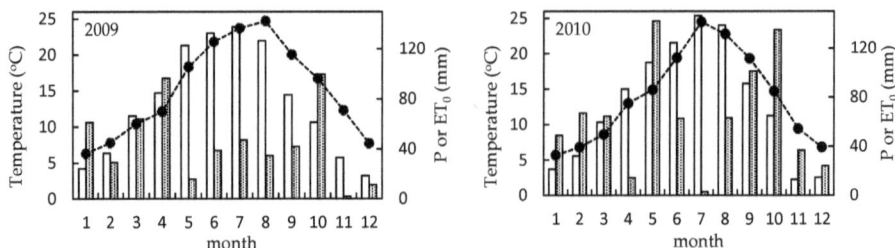

Figure 3. Monthly mean air temperature, precipitation (P) (dark columns), and reference evapotranspiration (ET_0) (white columns) at the experimental site in years 2009 and 2010. Data are from the meteorological station of Els Hostalets de Pierola (1°48′31″ west (W), 41°31′59″ north (N)).

3.2.2. Vine Water Status

Predawn water potential ranged from −0.18 MPa to −0.48 MPa among vines with an average value of −0.34 ± 0.03 MPa in 2009, whereas, in 2010, Ψ_{pd} varied from −0.27 MPa to −0.56 MPa with an average value of −0.41 ± 0.03 MPa. In addition, ΔT_m ranged from −4.85 °C to −1.08 °C among vines with an average value of −2.38 ± 0.29 °C in 2009, whereas, in 2010, ΔT_m varied from −4.10 °C to 0.33 °C with an average value of −1.75 ± 0.46 °C. In 2010, plot average values of ΔT_m and Ψ_{pd} were significantly related ($r = -0.74$, $p < 0.01$), whereas no significant correlation emerged in 2009 ($r = -0.35$, $p = 0.26$). However, in 2009, when ΔT_m temperatures acquired under partially overcast conditions were disregarded, ΔT_m and Ψ_{pd} were found to be significantly related ($r = -0.69$, $p < 0.05$). Moreover, when only data acquired under clear-sky conditions were considered, a unique relationship between ΔT_m and Ψ_{pd} emerged across years with $r^2 = 0.37$ and $p < 0.05$ (Figure 4).

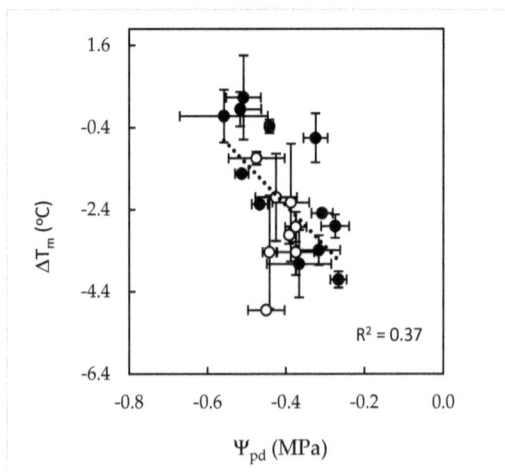

Figure 4. Relationship between the predawn water potential (Ψ_{pd}) and the canopy-to-air temperature difference (ΔT_m) in Xarel·lo field plots acquired in 2009 (open symbols) and 2010 (bold symbols) collected under clear-sky conditions. Values are means ± standard errors of the mean at each plot ($n = 8$ and $n = 12$ for 2009 and 2010, respectively). The regression line and regression coefficient are for pooled data across years.

3.2.3. Relationships between Reflectance Indices and Vine Vigor and Water Status

In the years of study, there were no significant relationships between the reflectance indices (NDVI, WI, and PRI) and fIPAR, except in 2009 when WI was negatively related with fIPAR ($r = -0.61$, $p < 0.05$), while these reflectance indices showed variable degrees of correlation with vine water status parameters (Table 2). In 2009, PRI and PRI$_{norm}$ showed significant and similar correlation with Ψ_{pd} ($r = 0.76$), whereas these correlations were not significant in 2010. Similarly, in 2009, WI and WI$_{norm}$ were significantly related to Ψ_{pd} ($r = -0.69$), whereas no significant correlations emerged in 2010.

Table 2. Correlation coefficients between predawn water potential (Ψ_{pd}), canopy-to-air temperature difference at midday (ΔT_m), and reflectance indices. Data are averaged values per plot ($n = 14$ in 2009, except for ΔT_m where $n = 8$; and $n = 12$ in 2010). Significant correlations at the 0.05 (*) and 0.01 (**) level are indicated. NDVI—normalized difference vegetation index.

	2009		2010		2009 and 2010	
	Ψ_{pd} (MPa)	ΔT_m (°C)	Ψ_{pd} (MPa)	ΔT_m (°C)	Ψ_{pd} (MPa)	ΔT_m (°C)
NDVI	0.32	0.84 **	0.19	−0.44	0.09	0.23
PRI	0.76 **	0.48	0.51	−0.60 *	0.59 **	−0.51 *
WI	−0.69 **	0.19	−0.51	0.53	−0.64 **	0.49 *
PRI$_{norm}$	0.76 **	0.58	0.57	−0.75 **	0.63 **	−0.56 *
WI$_{norm}$	−0.68 **	−0.81 **	−0.53	0.65 *	−0.60 **	0.21

In addition, in 2009, the relationships between PRI and WI and ΔT_m were either not significant or not consistent, probably due to the small sample size, and were not considered to any further extent. Contrastingly, in 2010, PRI and PRI$_{norm}$ were significantly related to ΔT_m with $r = -0.60$ ($p < 0.05$) and $r = -0.75$ ($p < 0.01$), for PRI and PRI$_{norm}$, respectively. Similarly, in 2010, WI was marginally related to ΔT_m ($r = 0.53$, $p < 0.10$), whereas the correlation between WI$_{norm}$ and ΔT_m was significant ($r = 0.65$, $p < 0.05$). When data from both years were pooled, PRI and WI were significantly ($p < 0.05$) related to ΔT_m (Table 2). Similarly, both WI and PRI were significantly related to Ψ_{pd} ($p < 0.01$) across years with $r^2 = 0.41$ and $r^2 = 0.37$ ($p < 0.01$), for WI and PRI, respectively (Figure 5). In addition, the relationship between Ψ_{pd} and PRI increased when PRI was normalized by canopy structure (i.e., PRI$_{norm}$) with $r^2 = 0.41$ ($p < 0.01$). In contrast, WI$_{norm}$ was found to be related to Ψ_{pd} to a lesser extent than WI with $r^2 = 0.36$ ($p < 0.01$).

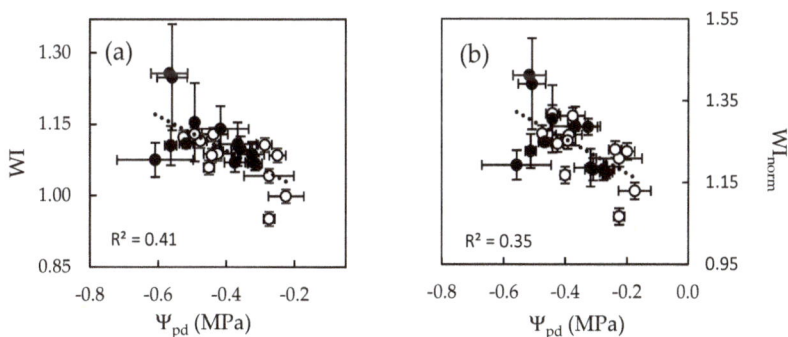

Figure 5. *Cont.*

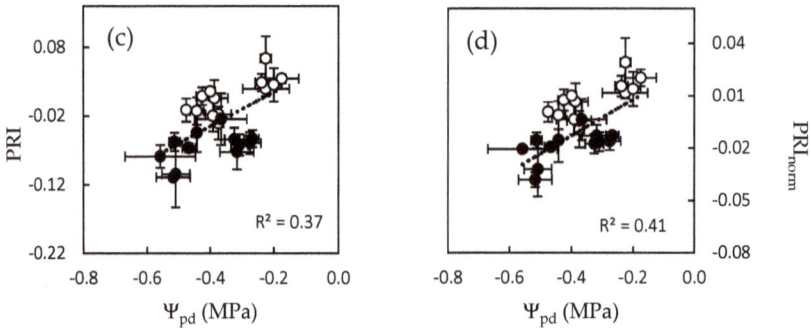

Figure 5. Relationships between predawn water potential (Ψ_{pd}) and the reflectance indices (**a**) water index (WI), (**b**) normalized water index (WI_{norm}), (**c**) photochemical reflectance index (PRI), and (**d**) normalized photochemical reflectance index (PRI_{norm}) in Xarel·lo vines. Pooled data for 2009 ($n = 14$) and 2010 ($n = 12$).

To further assess the capability of PRI and WI in detecting vine water status, data were grouped according to the water deficit level as defined by Carbonneau et al. [28]. Thus, we considered $\Psi_{pd} > -0.2$ MPa as no water deficit, -0.2 MPa $< \Psi_{pd} < -0.4$ MPa as mild water stress, and $\Psi_{pd} < -0.6$ MPa as moderate water stress. Since there were no significant differences in Ψ_p between no water deficit and mild water stress, data were pooled into a unique group (i.e., mild water deficit). Therefore, we examined two conditions, namely mild ($\Psi_{pd} < -0.4$ MPa) and moderate ($\Psi_{pd} > -0.4$ MPa) water deficits. In 2009, Ψ_{pd} values showed significant differences ($p < 0.01$) between water deficit levels with average values of -0.24 ± 0.04 MPa and -0.42 ± 0.01 MPa for mild and moderate water deficits, respectively. Similarly, in 2010, Ψ_{pd} significantly differed ($p < 0.01$) with $\Psi_{pd} = -0.31 \pm 0.02$ MPa and $\Psi_{pd} = -0.50 \pm 0.02$ MPa, for mild and moderate water deficits, respectively. Consistently, WI and PRI, as well as their respective normalized formulations (i.e., WI_{norm} and PRI_{norm}), showed significant differences between mild and moderate water deficits in the years of study (Figure 6). Differences in WI between mild and moderate water deficits were larger in 2009 ($p < 0.01$) than in 2010 ($p < 0.05$), and a similar response was observed for WI_{norm} (for more details, see Table S1, Supplementary Materials). In 2009, PRI significantly decreased ($p < 0.01$) from 0.031 ± 0.008 to -0.012 ± 0.005 between mild and moderate water deficits, and a similar response was observed in 2010, although less significant ($p < 0.05$). In addition, PRI_{norm} significantly differed ($p < 0.05$) between mild and moderate water deficit levels in the years of study.

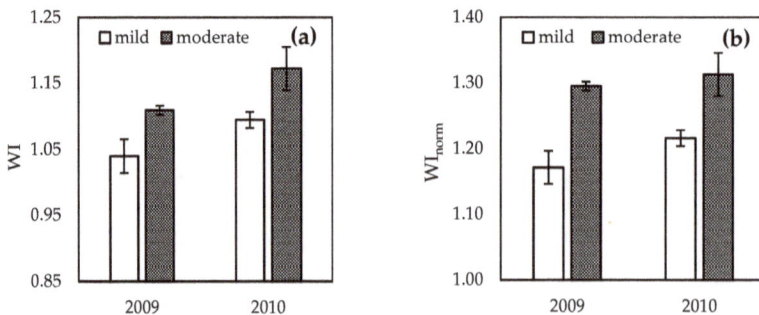

Figure 6. *Cont.*

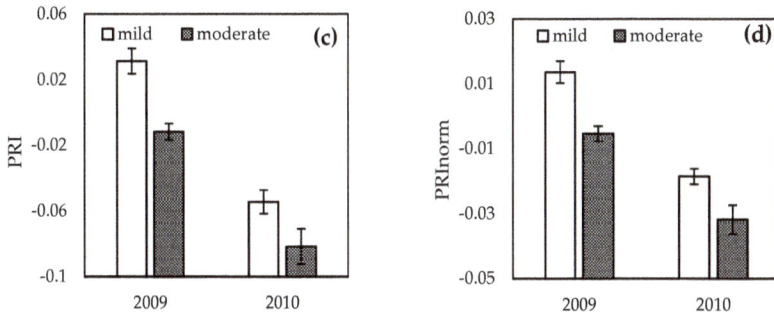

Figure 6. Effects of mild (Ψ_{pd} > −0.4 MPa; open bars) and moderate (Ψ_{pd} < −0.4 MPa; shaded bars) water deficits on (**a**) water index (WI), (**b**) normalized water index (WI$_{norm}$), (**c**) photochemical reflectance index (PRI), and (**d**) normalized photochemical reflectance index (PRI$_{norm}$) in Xarel·lo vines. Values are means ± standard errors of the mean (*n* = 6 for mild water deficits in 2009 and 2010; *n* = 8 and *n* = 7 for moderate water deficits in 2009 and 2010, respectively).

4. Discussion

4.1. Leaf Level Study

Values of Ψ_{pd} and Ψ_s in potted plants indicated that vines were subject to mild to moderate water stress [28]. In spite of ample water availability, stomatal conductance (g_s) decreased, particularly at central hours of the day under high air vapor pressure deficits (i.e., VPD > 2.0 kPa), as indicated by the close dependence of stomatal conductance on the leaf-to-air VPD (r = −0.95, p < 0.01, n = 5). Similarly, we observed that variation in E during the diurnal cycle was mainly driven by changes in VPD (r = 0.85, p < 0.01, n = 8) rather than changes in stomatal regulation. Previous studies reported a decline in stomatal conductance associated with high vapor pressure deficits even in well-watered vines [29], which results from an imbalance between water loss through evapotranspiration and water flow into the leaf [3,30]. In turn, at central hours of the day, when the photosynthetic rate was not light-limited, A$_n$ was largely determined by stomatal conductance (r = 0.99, p < 0.01, n = 5). Indeed, under high temperature and VPD deficits, such as those experienced in the summer in Mediterranean environments, a midday depression of both A$_n$ and g$_s$ was observed in agreement with previous studies [13,31,32]. Thus, stomatal conductance was particularly sensitive to developing water deficits [32]. The g$_s$ values were within the range reported for potted plants at similar Ψ_{pd} [6,13] and in field studies [33] and, on average, similar to those of vines experiencing moderate water deficits [34,35]. Therefore, it appears that, in Xarel·lo vines experiencing mild to moderate water deficits, stomatal conductance might be a good indicator of vine water status and could potentially provide a tool for irrigation scheduling [36].

Previous studies showed the sensitivity of PRI for crop water stress detection over diurnal and short time scales [12,19,37]. In our study, the PRI accounted for 92% variation in photosynthetic light-use efficiency (LUE) and 85% variation in ΔT, which agrees with the close coordination correlation observed among gas exchange parameters described above. Therefore, under the conditions of the study, PRI closely tracked diurnal changes in LUE in Xarel·lo vines. These results add to previous studies which showed the capability of PRI in estimating photosynthetic-related parameters across a wide range of water status [17,18,27,37–39], supporting the hypothesis that PRI could be a feasible indicator of plant physiological status under mild to moderate water deficits [19,20].

In addition, previous studies showed the capability of WI in tracking variation in stomatal aperture [13,16,40]. However, in the present study, WI was poorly related to g$_s$, although it was a good indicator of changes in transpiration as suggested by the close correlation with ΔT. This is consistent with the fact that, in our study, variations in transpiration rates were mainly driven by changes in VPD rather than by changes in g$_s$. Moreover, WI closely tracked changes in LUE, which might be attributed to the close dependence of net CO_2 uptake on g_s observed in our study. Thus, our results add to previous studies conducted in vines [13] by showing that WI effectively tracked changes in the leaf-to-air temperature difference in Xarel·lo vines experiencing mild to moderate water deficits.

4.2. Field Study

In the years of study, Ψ_{pd} values were similar to those reported in vines grown under deficit irrigation programs [41–43] and indicated that, at veraison, water stress was mild to moderate [28]. Under these conditions, PRI was related to Ψ_{pd} as previously reported in vineyards experiencing mild to moderate water deficits [44,45]. In addition, in our study, a negative relationship emerged between WI and Ψ_{pd}, indicating that enhanced water status (i.e., increases in Ψ_{pd}) resulted in decreased water content at midday (i.e., decreases in WI). This is consistent with the observed relationship between Ψ_{pd} and ΔT_m (Figure 4) and suggests that higher Ψ_{pd} was presumably accompanied by higher transpiration rates at midday [36]. Indeed, in grapevines experiencing mild to moderate water deficits, as those occurring in our study or under deficit irrigation programs, Ψ_{pd} is highly related to water potential and stomatal conductance measured at midday [12,43,46,47]. In addition, in our study, WI decreased as PRI increased ($r = -0.66$, $p < 0.01$), indicating that enhanced water loss (lower WI) was accompanied by increased carbon assimilation (higher PRI), which agrees with the tight dependence of leaf photosynthesis on stomatal conductance previously reported in grapevines experiencing mild to moderate water deficits [3,32,48]. Thus, both PRI and WI were feasible indicators of Ψ_{pd} under mild to moderate water deficits, whereas, in grapevines experiencing moderate to severe water stress, PRI and WI failed to estimate Ψ_{pd} [13,45]. These contrasting results might be reconciled by considering the effects of the intensity and duration of water deficits on stomatal responses [3]. In grapevines experiencing prolonged soil water deficits, Ψ_{pd} may not necessarily reflect the plant's water status later in the day since changes in cell-wall elasticity and osmotic adjustment might dictate different responses in midday water potential [48]. Increases in cell-wall elasticity lead to a larger decline in plant water potential at midday, whereas osmotic adjustment may contribute to the maintenance of open stomata at lower water potentials [3]. These responses might partly explain the lack of correlation between reflectance indices acquired at midday (i.e., PRI and WI) and Ψ_{pd} previously reported in rain-fed vineyards experiencing prolonged and severe soil water deficits [13,45]. In addition, prolonged soil water deficits during the vegetative stage might result in differences in vine leaf area, leading to different velocities in dehydration [49], which might also affect the relationship between Ψ_{pd} and both PRI and WI measured at midday. Thus, the timing of occurrence and intensity of water deficits might affect the relationships between PRI and WI and predawn water potential [13,45], whereas both PRI and WI were reliable indicators of predawn water potential and photosynthetic functioning under mild to moderate water deficits.

In spite of the variability in weather conditions (i.e., radiation, temperature, and VPD) during data acquisition, ΔT_m was found to be negatively related to Ψ_{pd} ($r^2 = 0.37$, $p < 0.05$). The correlations between PRI and WI with ΔT_m were low, although significant, when data from both years were pooled, which might be attributed to varying environmental conditions (leaf and VPD) during ΔT_m measurements [36]. In 2010, in accordance with the observed responses at both the leaf (Figure 2) and canopy (Figure 4) levels, PRI increased along with ΔT_m decreases, whereas WI decreased in parallel with ΔT_m. The relationships of both PRI and WI against ΔT_m notably improved when considering their respective normalized formulation (i.e., PRI_{norm} and WI_{norm}). Previous studies showed that PRI might be affected by changes in canopy structure and might reflect changes in pigment composition (i.e., chlorophyll to carotenoid ratios) as a result of leaf development, aging, or long-term stresses rather than the epoxidation state [39,50,51]. On the other hand, differences in vegetative growth might also affect WI values because the water content of vegetation responds to both leaf area and relative water content [22,23]. In our study, there was a close correlation between ΔT_m and SIPI ($r = 0.79$, $p < 0.01$), indicating larger carotenoid relative to chlorophyll content on more stressed vines [27]. In addition, fIPAR was significantly related to Ψ_{pd} ($r = 0.64$, $p < 0.05$) suggesting an association between vine vigor and water availability. Thus, PRI_{norm}, which accounts for variation in both canopy structure and pigment composition [52], showed higher sensitivity than PRI to changes in ΔT_m. Similarly, WI_{norm}, which accounts for variation in both leaf area and chlorophyll content, was positively related to ΔT_m, indicating that enhanced water status was associated with higher transpiration rates and, thus, lower water content, in agreement with previous studies [13,40]. Contrastingly, the relationships between PRI, WI, and Ψ_{pd} did not improve when reflectance indices were normalized, which highlights the effects of varying canopy structure on vine water status at midday (when ΔT_m was determined). Therefore, under the conditions of our study—where field data were acquired over two years on several fields and during several days within a year—correcting for variation on canopy size and pigment concentration effectively contributed to improving the performance of WI and PRI in estimating vine water status [12]. In summary, in line with recent studies, our results show the capability of the PRI [12,20,53] and WI [13,40,54], as well as their respective normalized formulations (i.e., PRI_{norm} and WI_{norm}), in monitoring vine water status.

Under regulated deficit irrigation strategies, the ability to diagnose vine water status is crucial since irrigation is normally scheduled under mild to moderate water deficits [3]. When comparing water deficit conditions, significant differences occurred between moderate ($\Psi_{pd} < -0.4$ MPa) and mild ($\Psi_{pd} > -0.4$ MPa) water deficits, in agreement with previous studies [44]. In accordance, PRI and WI, as well as their respective normalized formulations, distinguished between mild and moderate water deficits. Thus, PRI and WI were reliable as reference measures for irrigation management when assessing vine water status at veraison and proved to work in a range expanding from mild to moderate water stress, a common situation in field-grown grapevines, including those managed under regulated deficit irrigation programs [41,42]. Moreover, the results herein reported are promising, considering that these indices showed a lineal response in the range of water potentials found in this study (Figure 5). However, since regulated deficit irrigation programs require recurrent assessment of plant water status throughout the growing cycle, more studies are needed to assess the capability of PRI and WI (or their corresponding normalized formulations) in estimating vine water status at growth stages other than veraison prior to confirming their usefulness as practical tools for irrigation scheduling in vineyards.

5. Conclusions

In the present study, the capability of WI and PRI in assessing water status in vines experiencing mild to moderate water deficits was assessed. In irrigated vines, diurnal variation in stomatal conductance was particularly sensitive to developing water deficits driven by changes in the leaf-to-air vapor pressure deficit, suggesting that, in Xarel·lo vines, stomatal conductance might be a good indicator of water status. The indices PRI and WI effectively tracked diurnal changes in the leaf-to-air temperature difference. Similarly, due to the close dependence of net CO_2 uptake on stomatal conductance observed in our study, PRI and WI effectively tracked diurnal changes in light-use efficiency. Thus, PRI and WI were feasible indicators of variation in photosynthetic functioning and transpiration (i.e., the light-use efficiency and the leaf-to-air temperature difference) linked to stomatal regulation in response to mild to moderate water deficits. At the canopy level, and despite ample variability in both weather conditions between years and growing conditions among fields, differences in water availability (i.e., Ψ_{pd}) were translated into differences in transpiration rates (i.e., the canopy-to-air temperature differences, ΔT_m). Under these conditions, both WI and PRI provided consistent estimates of Ψ_{pd}. Moreover, in accordance with the observed differences in predawn water potentials, both PRI and WI effectively distinguished between mild and moderate water deficits levels. In addition, WI_{norm} and PRI_{norm}, which accounted for long-term effects of water availability on canopy structure—namely leaf area and chlorophyll content—were related to the canopy-to-air temperature difference. Therefore, PRI and WI, as well as their normalized formulations PRI_{norm} and WI_{norm}, provided estimates of key water stress indicators in vineyards within a range of mild to moderate water deficits. The capability of WI and PRI in monitoring vine water status might be of great significance in the context of increasing

Agronomy **2019**, *9*, 346

irrigated viticulture areas, particularly those under regulated deficit irrigation, as potential tools to support vineyard irrigation management. In this sense, the development of cost-effective methods for image acquisition and analysis that are commercially available to farmers is needed to make remote-sensing techniques operational for precision irrigation management.

Supplementary Materials: The following are available online at http://www.mdpi.com/2073-4395/9/7/346/s1, Table S1: Effects of mild ($\Psi_{pd} > -0.4$ MPa) and moderate (-0.4 MPa $< \Psi_{pd} < -0.6$ MPa) water deficits on the water index (WI), the normalized water index (WI_{norm}), the photochemical reflectance index (PRI), and the normalized photochemical reflectance index (PRI_{norm}).

Author Contributions: C.G.-F., field sampling, data analyses, and writing—original draft preparation; G.G., field sampling, data analyses, and writing—review and editing; L.S., conceptualization, financial support, data analyses, and writing—original draft preparation.

Funding: This research was funded by the Ministerio de Ciencia e Innovación (Spain), grant number AGL2009-13105-C03-03, and by the Universitat Politècnica de Catalunya, grant number 9-INREC-09.

Acknowledgments: We are grateful to Antoni Abad (Codorniu S.A.) for assistance in field data collection and Olga Gener for management of the greenhouse facility.

Conflicts of Interest: The authors declare no conflicts of interest.

References

1. Chaves, M.M.; Santos, T.P.; Souza, C.R.; Ortuño, M.F.; Rodrigues, M.L.; Lopes, C.M.; Maroco, J.P.; Pereira, J.S. Deficit irrigation in grapevine improves water-use efficiency while controlling vigour and production quality. *Ann. Appl. Biol.* **2007**, *150*, 237–252. [CrossRef]
2. Schultz, H.R.; Stoll, M. Some critical issues in environmental physiology of grapevines: Future challenges and current limitations. *Aust. J. Grape Wine Res.* **2010**, *16*, 4–24. [CrossRef]
3. Chaves, M.M.; Zarrouk, O.; Francisco, R.; Costa, J.M.; Santos, T.; Regalado, A.P.; Rodrigues, M.L.; Lopes, C.M. Grapevine under deficit irrigation: Hints from physiological and molecular data. *Ann. Bot.* **2010**, *105*, 661–676. [CrossRef]
4. Cifre, J.; Bota, J.; Escalona, J.M.; Medrano, H.; Flexas, J. Physiological tools for irrigation scheduling in grapevine (Vitis vinifera L.): An open gate to improve water-use efficiency? *Agric. Ecosyst. Environ.* **2005**, *106*, 159–170. [CrossRef]
5. Fereres, E.; Soriano, M.A. Deficit irrigation for reducing agricultural water use. *J. Exp. Bot.* **2007**, *58*, 147–159. [CrossRef]
6. Medrano, H.; Escalona, J.M.; Cifre, J.; Bota, J.; Flexas, J. A ten-year study on the physiology of two Spanish grapevine cultivars under field conditions: Effects of water availability from leaf photosynthesis to grape yield and quality. *Funct. Plant Biol.* **2003**, *30*, 607–619. [CrossRef]
7. Choné, X.; Van Leeuwen, C.; Dubourdieu, D.; Gaudillère, J.P. Stem water potential is a sensitive indicator of grapevine water status. *Ann. Bot.* **2001**, *87*, 477–483. [CrossRef]
8. Jones, H.G.; Stoll, M.; Santos, T.; de Sousa, C.; Chaves, M.M.; Grant, O.M. Use of infrared thermography for monitoring stomatal closure in the field: Application to grapevine. *J. Exp. Bot.* **2002**, *53*, 2249–2260. [CrossRef]
9. Grant, O.M.; Tronina, L.; Jones, H.G.; Chaves, M.M. Exploring thermal imaging variables for the detection of stress responses in grapevine under different irrigation regimes. *J. Exp. Bot.* **2006**, *58*, 815–825. [CrossRef]
10. Möller, M.; Alchanatis, V.; Cohen, Y.; Meron, M.; Tsipris, J.; Naor, A.; Ostrovsky, V.; Sprintsin, M.; Cohen, S. Use of thermal and visible imagery for estimating crop water status of irrigated grapevine. *J. Exp. Bot.* **2006**, *58*, 827–838. [CrossRef]
11. Maes, W.H.; Steppe, K. Estimating evapotranspiration and drought stress with ground-based thermal remote sensing in agriculture: A review. *J. Exp. Bot.* **2012**, *63*, 4671–4712. [CrossRef]
12. Zarco-Tejada, P.J.; González-Dugo, V.; Williams, L.E.; Suárez, L.; Berni, J.A.J.; Goldhamer, D.; Fereres, E. A PRI-based water stress index combining structural and chlorophyll effects: Assessment using diurnal narrow-band airborne imagery and the CWSI thermal index. *Remote Sens. Environ.* **2013**, *138*, 38–50. [CrossRef]
13. Serrano, L.; González-Flor, C.; Gorchs, G. Assessing vineyard water status using the reflectance based Water Index. *Agric. Ecosyst. Environ.* **2010**, *139*, 490–499. [CrossRef]
14. Rodríguez-Pérez, J.R.; Riaño, D.; Carlisle, E.C.; Ustin, S.B.; Smart, D.R. Evaluation of hyperspectral reflectance indexes to detect grapevine water status in vineyards. *Am. J. Enol. Vitic.* **2007**, *58*, 302–317.

15. De Bei, R.; Cozzolino, D.; Sullivan, W.; Cynkar, W.; Fuentes, S.; Dambergs, R.; Pech, J.; Tyerman, S.D. Non-destructive measurement of grapevine water potential using near infrared spectroscopy. *Aust. J. Grape Wine Res.* **2011**, *17*, 62–71. [CrossRef]

16. Peñuelas, J.; Filella, I.; Biel, C.; Serrano, L.; Savé, R. The reflectance at the 950–970 nm region as an indicator of plant water status. *Int. J. Remote Sens.* **1993**, *14*, 1887–1905. [CrossRef]

17. Gamon, J.A.; Serrano, L.; Surfus, J.S. The photochemical reflectance index: An optical indicator of photosynthetic radiation use efficiency across species, functional types, and nutrient levels. *Oecologia* **1997**, *112*, 492–501. [CrossRef]

18. Gamon, J.A.; Peñuelas, J.; Field, C.B. A narrow-waveband spectral index that tracks diurnal changes in photosynthetic efficiency. *Remote Sens. Environ.* **1992**, *41*, 35–44. [CrossRef]

19. Suárez, L.; Zarco-Tejada, P.J.; Sepulcre-Cantó, G.; Pérez-Priego, O.; Miller, J.R.; Jiménez-Muñoz, J.C.; Sobrino, J.A. Assessing canopy PRI for water stress detection with diurnal airborne imagery. *Remote Sens. Environ.* **2008**, *112*, 560–575. [CrossRef]

20. Stagakis, S.; González-Dugo, V.; Cid, P.; Guillén-Climent, M.L.; Zarco-Tejada, P.J. Monitoring water stress and fruit quality in an orange orchard under regulated deficit irrigation using narrow-band structural and physiological remote sensing indices. *ISPRS J. Photogramm. Remote Sens.* **2012**, *71*, 47–61. [CrossRef]

21. Barton, C.V.M.; North, P.R.J. Remote sensing of canopy light use efficiency using the photochemical reflectance index model and sensitivity analysis. *Remote Sens. Environ.* **2001**, *78*, 264–273. [CrossRef]

22. Serrano, L.; Ustin, S.L.; Roberts, D.A.; Gamon, J.A.; Peñuelas, J. Deriving water content of chaparral vegetation from AVIRIS data. *Remote Sens. Environ.* **2000**, *74*, 570–581. [CrossRef]

23. Sims, D.; Gamon, J. Estimation of vegetation water content and photosynthetic tissue area from spectral reflectance: A comparison of indices based on liquid water and chlorophyll absorption features. *Remote Sens. Environ.* **2003**, *84*, 526–537. [CrossRef]

24. Filella, I.; Porcar-Castell, A.; Munné-Bosch, S.; Bäck, J.; Garbulsky, M.F.; Peñuelas, J. PRI assessment of long-term changes in carotenoids/chlorophyll ratio and short-term changes in de-epoxidation state of the xanthophyll cycle. *Int. J. Remote Sens.* **2009**, *30*, 4443–4455. [CrossRef]

25. Wong, C.Y.S.; Gamon, J.A. Three causes of variation in the photochemical reflectance index (PRI) in evergreen conifers. *New Phytol.* **2015**, *206*, 187–195. [CrossRef]

26. Smart, R.; Robinson, M. *Sunlight into Wine: A Handbook for Winegrape Canopy Management*; Winetitles: Adelaide, Australia, 1991.

27. Peñuelas, J.; Filella, I. Visible and near-infrared reflectance techniques for diagnosing plant physiological status. *Trends Plant Sci.* **1998**, *3*, 151–156. [CrossRef]

28. Carbonneau, A. Irrigation, vignoble et produits de la vigne. In *Traité D'Irrigation*; Tiercelin, J.R., Ed.; Lavoisier Tec & Doc: Paris, France, 1998; pp. 257–298.

29. Correia, M.J.; Pereira, J.S.; Chaves, M.M.; Rodrigues, M.L.; Pacheco, C.A. ABA xylem concentrations determine maximum daily leaf conductance of field-grown Vitis vinifera L. plants. *Plant Cell Environ.* **1995**, *18*, 511–521. [CrossRef]

30. Schultz, H.R. Differences in hydraulic architecture account for near-isohydric and anisohydric behaviour of two field-grown Vitis vinifera L. cultivars during drought. *Plant Cell Environ.* **2003**, *26*, 1393–1405. [CrossRef]

31. Correia, M.J.; Chaves, M.M.C.; Pereira, J.S. Afternoon depression in photosynthesis in grapevine leaves—Evidence for a high light stress effect. *J. Exp. Bot.* **1990**, *41*, 417–426. [CrossRef]

32. Flexas, J.; Escalona, J.M.; Medrano, H. Water stress induces different levels of photosynthesis and electron transport rate regulation in grapevines. *Plant Cell Environ.* **1999**, *22*, 39–48. [CrossRef]

33. Costa, J.M.; Ortuño, M.F.; Lopes, C.M.; Chaves, M.M. Grapevine varieties exhibiting differences in stomatal response to water deficit. *Funct. Plant Biol.* **2012**, *39*, 179–189. [CrossRef]

34. Flexas, J.; Bota, J.; Escalona, J.M.; Sampol, B.; Medrano, H. Effects of drought on photosynthesis in grapevines under field conditions: An evaluation of stomatal and mesophyll limitations. *Funct. Plant Biol.* **2002**, *29*, 461–471. [CrossRef]

35. Medrano, H.; Escalona, J.M.; Bota, J.; Gulías, J.; Flexas, J. Regulation of photosynthesis of C3 plants in response to progressive drought: Stomatal conductance as a reference parameter. *Ann. Bot.* **2002**, *89*, 895–905. [CrossRef]

36. Jones, H.G. Irrigation scheduling: Advantages and pitfalls of plant-based methods. *J. Exp. Bot.* **2004**, *55*, 2427–2436. [CrossRef]

37. Sun, P.; Wahbi, S.; Tsonev, T.; Haworth, M.; Liu, S.; Centritto, M. On the use of leaf spectral indices to assess water status and photosynthetic limitations in *Olea europaea* L. during water-stress and recovery. *PLoS ONE* **2014**, *9*, e105165. [CrossRef]
38. Serrano, L.; Peñuelas, J. Assessing forest structure and function from spectral transmittance measurements: A case study in a Mediterranean holm oak forest. *Tree Physiol.* **2005**, 67–74. [CrossRef]
39. Zhang, C.; Filella, I.; Liu, D.; Ogaya, R.; Llusià, J.; Asensio, D.; Peñuelas, J. Photochemical Reflectance Index (PRI) for detecting responses of diurnal and seasonal photosynthetic activity to experimental drought and warming in a mediterranean Shrubland. *Remote Sens.* **2017**, *9*, 1189. [CrossRef]
40. Dzikiti, S.; Verreynne, J.S.; Stuckens, J.; Strever, A.; Verstraeten, W.W.; Swennen, R.; Coppin, P. Determining the water status of Satsuma mandarin trees [Citrus Unshiu Marcovitch] using spectral indices and by combining hyperspectral and physiological data. *Agric. For. Meteorol.* **2010**, *150*, 369–379. [CrossRef]
41. Patakas, A.; Noitsakis, B.; Chouzouri, A. Optimization of irrigation water use in grapevines using the relationship between transpiration and plant water status. *Agric. Ecosyst. Environ.* **2005**, *106*, 253–259. [CrossRef]
42. De Souza, C.R.; Maroco, J.P.; dos Santos, T.P.; Rodrigues, M.L.; Lopes, C.; Pereira, J.S.; Chaves, M.M. Control of stomatal aperture and carbon uptake by deficit irrigation in two grapevine cultivars. *Agric. Ecosyst. Environ.* **2005**, *106*, 261–274. [CrossRef]
43. Intrigliolo, D.S.; Castel, J.R. Vine and soil-based measures of water status in a Tempranillo vineyard. *VITIS J. Grapevine Res.* **2006**, *45*, 157–163.
44. Pôças, I.; Rodrigues, A.; Gonçalves, S.; Costa, P.M.; Gonçalves, I.; Pereira, L.S.; Cunha, M. Predicting grapevine water status based on hyperspectral reflectance vegetation indices. *Remote Sens.* **2015**, *7*, 16460–16479. [CrossRef]
45. González-Flor, C.; Serrano, L.; Gorchs, G. Predicting berry quality attributes in cv. Xarel·lo rain-fed vineyards using narrow-band reflectance-based indices. *Am. J. Enol. Vitic.* **2013**, *64*, 88–97. [CrossRef]
46. Williams, L.E.; Araujo, F.J. Correlations among predawn leaf, midday leaf, and midday stem water potential and their correlations with other measures of soil and plant water status in vitis vinifera. *J. Am. Soc. Hortic. Sci.* **2002**, *127*, 448–454. [CrossRef]
47. Winkel, T.; Rambal, S. Stomatal conductance of some grapevines growing in the field under a Mediterranean environment. *Agric. For. Meteorol.* **1990**, *51*, 107–121. [CrossRef]
48. Lovisolo, C.; Perrone, I.; Carra, A.; Ferrandino, A.; Flexas, J.; Medrano, H.; Schubert, A. Drought-induced changes in development and function of grapevine (*Vitis* spp.) organs and in their hydraulic and non-hydraulic interactions at the whole-plant level: A physiological and molecular update. *Funct. Plant Biol.* **2010**, *37*, 98–116. [CrossRef]
49. Rogiers, S.Y.; Greer, D.H.; Hutton, R.J.; Landsberg, J.J. Does night-time transpiration contribute to anisohydric behaviour in a Vitis vinifera cultivar? *J. Exp. Bot.* **2009**, *60*, 3751–3763. [CrossRef]
50. Gamon, J.A.; Surfus, J.S. Assessing leaf pigment content and activity with a reflectometer. *New Phytol.* **1999**, *143*, 105–117. [CrossRef]
51. Gamon, J.A.; Huemmrich, K.F.; Wong, C.Y.S.; Ensminger, I.; Garrity, S.; Hollinger, D.Y.; Noormets, A.; Peñuelas, J. A remotely sensed pigment index reveals photosynthetic phenology in evergreen conifers. *Proc. Natl. Acad. Sci. USA* **2016**, *113*, 13087–13092. [CrossRef]
52. Zarco-Tejada, P.J.; Guillén-Climent, M.L.; Hernández-Clemente, R.; Catalina, A.; González, M.R.; Martín, P. Estimating leaf carotenoid content in vineyards using high resolution hyperspectral imagery acquired from an unmanned aerial vehicle (UAV). *Agric. For. Meteorol.* **2013**, *171–172*, 281–294. [CrossRef]
53. Ritchie, G.L.; Sullivan, D.G.; Perry, C.D.; Hook, J.E.; Bednarz, C.W.; Sullivan, D.G.; Perry, C.D.; Member Engineer, A.; Hook, J.E.; Bednarz, C.W.; et al. Preparation of a low-cost digital camera system for remote sensing. *Appl. Eng. Agric.* **2008**, *24*, 885–896. [CrossRef]
54. Pôças, I.; Gonçalves, J.; Costa, P.M.; Gonçalves, I.; Pereira, L.S.; Cunha, M. Hyperspectral-based predictive modelling of grapevine water status in the Portuguese Douro wine region. *Int. J. Appl. Earth Obs. Geoinf.* **2017**, *58*, 177–190. [CrossRef]

Article

Semi-Minimal Pruned Hedge: A Potential Climate Change Adaptation Strategy in Viticulture

Daniel Molitor [1],*, Mareike Schultz [1,2], Robert Mannes [2], Marine Pallez-Barthel [1], Lucien Hoffmann [1] and Marco Beyer [1]

1   Luxembourg Institute of Science and Technology (LIST), Environmental Research and Innovation (ERIN) Department, 41, rue du Brill, L-4422 Belvaux, Luxembourg; mareike.schultz@ivv.etat.lu (M.S.); marine.pallez@list.lu (M.P.-B.); lucien.hoffmann@list.lu (L.H.); marco.beyer@list.lu (M.B.)

2   Institut Viti-vinicole (IVV), Section Viticulture, 8, rue Nic. Kieffer, L-5551 Remich, Luxembourg; robert.mannes@ivv.etat.lu

*   Correspondence: daniel.molitor@list.lu; Tel.: +352-275-888-5029

Received: 18 March 2019; Accepted: 29 March 2019; Published: 2 April 2019

Abstract: The low-input viticultural training system 'Semi-minimal pruned hedge' (SMPH) is progressively being more widely applied in the Central European grapegrowing regions. The present study examined the influence of (i) the training system (SMPH versus the vertical shoot position (VSP) system), (ii) the timing of shoot topping in SMPH, and (iii) the effects of mechanical thinning in SMPH on the bunch rot epidemic, grape maturity, and yield. Six-year field trials on Pinot blanc in Luxembourg demonstrated that yield levels in non-thinned SMPH treatments were 74% higher, and total soluble solids (TSS) at harvest 2.2 brix lower than in VSP. Non-thinned SMPH delayed the bunch rot epidemic and the maturity progress by 18 and 11 days compared to VSP, respectively. Different shoot-topping timings in SMPH did not affect the tested parameters. Mechanical thinning regimes reduced the yield by 28% (moderate thinning) and 53% (severe thinning) compared to non-thinned SMPH and increased TSS by 0.8 and 1.3 brix, respectively. Delayed bunch rot epidemic and maturity progress give rise to the opportunity for a longer maturity period in cooler conditions, making this system of particular interest in future, warmer climatic conditions. Providing that yield levels are managed properly, SMPH might represent an interesting climate change adaptation strategy.

Keywords: *Botrytis cinerea*; low-input; mechanical thinning; viticultural training system; yield formation

---

1. Introduction

In many traditional cool climate European grapegrowing regions, the vertical shoot positioning (VSP) system represents the standard viticultural training system [1]. However, winter pruning and canopy management in summer, in particular, are causing high production costs in VSP [2]. Minimal pruning (MP) systems developed in Australia were reported to reduce costs and susceptibility to bunch rot [2]. On the other hand, in European climate conditions, over-cropping, delayed ripening, and alternating yield levels are frequently observed in MP [3]. To overcome these limitations, Intrieri et al. [3] suggested a novel hedge-shaped training system that enables mechanized pruning and mechanized harvesting, the 'semi-minimal pruned hedge' (SMPH). Studies by Intrieri et al. [3] demonstrated reduced management costs, suitability for full mechanization, improved grape yield and quality, as well as a reduced susceptibility to bunch rot in the new system. Based on this, Intrieri et al. [3] recommended SMPH for practical applications. Meanwhile, SMPH is becoming more and more popular in several viticultural regions of Germany [1]. However, the results of multi-annual investigations into the suitability of SMPH in Central European cool climate conditions are limited in the scientific literature. Consequently, field trials were established in 2013 in Remich/Luxembourg and investigations continued over a period of six vintages.

Since crop levels in SMPH were reported to be higher than in the traditional VSP system [3,4], cluster thinning might help to avoid over-cropping and, in consequence, inadequate grape maturity. Due to the canopy characteristics of SMPH, cluster thinning is supposed to be practicable more mechanically than manually. Petrie and Clingeleffer [5] demonstrated that mechanical thinning via a modified harvest machine was able to reduce the yield, e.g., in minimally pruned vineyards, in a cost-efficient manner. To the best of our knowledge, no studies about the effect of mechanical thinning in SMPH are available in the scientific literature. For this reason, we address (i) the general suitability of mechanical thinning via a harvest machine in SMPH and (ii) the effect of different thinning strengths on bunch rot epidemic, maturity progress, and yield formation in the present work.

The timing of the first shoot topping in VSP has recently been demonstrated to have an impact on cluster compactness, bunch rot epidemics, and, in consequence, the length of the potential ripening period [6] and will therefore be considered when assessing SMPH in the present study.

Overall, these six-year studies aimed to investigate (i) the general suitability of the SMPH training system under the viticultural conditions in Luxembourg and, more specifically, (ii) the impact of differential shoot topping timings as well as different mechanical thinning regimes on the bunch rot epidemic, grape maturity progress, and yield levels. The traditional cane-pruned VSP acts as a standard (control) system to comparatively assess the results obtained in the different SMPH treatments.

2. Materials and Methods

2.1. Experimental Vineyard and Field Trial Design

Field trials were carried out in the experimental vineyard of the Institut Viti-Vinicole in Remich, Luxembourg (lat. 49.54° N; long. 6.35° E) between 2013 and 2018 on the white *Vitis vinifera* L. Pinot blanc cultivar. The vineyard in investigation was planted in 2000 and the vines, grafted onto SO4 rootstocks, were trained to a cane-pruned vertical shoot positioning system (VSP) until 2012. The cane height was 0.8 m from the ground. The training system consisted of two foliage wire pairs and one single foliage wire (horizontal distance between wire stations: 30–35 cm). The upper wire pair was installed at a height of 1.8 m from the ground.

In winter 2012/2013, the vineyard (except the plots of VSP treatment, which continued as VSP over the entire period of investigation) was transferred into the SMPH training system according to Intrieri et al. [3] in the following manner: no winter pruning took place, one additional wire was fixed at a height of approximately 1.8 m, and shoots were clamped between the upper wires with vineyard staples to avoid sliding out in the seasons to come.

The space per plant before and after transfer to SMPH was 2.4 m^2 (2 m between rows, 1.2 m between vines).

The field trial was arranged as a randomized block design (four blocks) with four replicates of eight vines per plot. The position of the experimental plots remained unchanged over the entire period of investigation (2013–2018).

The treatments tested (abbreviations in parentheses) were as follows:

- Semi-minimal pruned hedge; first shoot topping approximately one week prior to the beginning of flowering (SMPH ST 1)
- Semi-minimal pruned hedge; first shoot topping at the beginning of flowering (SMPH ST 2)
- Semi-minimal pruned hedge; first shoot topping at the end of flowering (SMPH ST 3)
- Semi-minimal pruned hedge; first shoot topping approximately one week after the end of flowering (SMPH ST 4)
- Cane-pruned vertical shoot positioning (VSP) = standard treatment
- Semi-minimal pruned hedge; moderate mechanical thinning with 320 beats/min (SMPH MT 1)
- Semi-minimal pruned hedge; severe mechanical thinning with 370 beats/min (SMPH MT 2)

In VSP, winter pruning took place every year; per plant, 12 buds remained and were bound to one horizontal cane. Here, the shoot positioning of primary shoots and lateral shoots was conducted approximately twice during the vegetation period.

No canopy management or winter pruning took place in any of the SMPH treatments. The shoot-topping dates and the developmental stages reached on these dates are given in Table S1. For the VSP, SMPH MT 1, and SMPH MT 2 treatments, shoot topping took place during all seasons on the SMPH ST 4 date.

On the shoot-topping date, shoot tips were (i) topped approximately 5 cm below the apex or (ii) if the length of a shoot already exceeded the planned final canopy height (2.0 m = upper wire + 0.2 m) or width (approximately 0.8 m), shoot lengths were limited to these dimensions. Shoots were topped with vineyard shears and this was part of all treatments only once per season; the lateral shoots that appeared were not topped.

Mechanical thinning in the SMPH MT 1 and SMPH MT 2 treatments was conducted every year in the phenophase BBCH (Biologische Bundesanstalt, Bundessortenamt und CHemische Industrie) 79 [7], representing the time period between the day of reaching BBCH 79 and the day before reaching BBCH 81. Mechanical thinning dates were 9 August 2013, 23 July 2014, 30 July 2015, 16 August 2016, 20 July 2017, and 23 July 2018. Mechanical thinning was carried out using a grape harvester with a beater amplitude of 320 beats per minute (SMPH MT 1) or 370 beats per minute (SMPH MT 2), respectively. The harvest machine (Grapeliner® 6000, ERO-Gerätebau GMBH, Simmern, Germany) was equipped with six shaker pairs at altitudes of approximately 0.6 to 2.4 m from the ground (Figure 1).

Figure 1. Mechanical thinning in the SMPH MT 1 and SMPH MT 2 treatments on 30 July 2015. SMPH, semi-minimal pruned hedge; MT, mechanical thinning.

Besides experimental treatments, all plots were managed in the same manner throughout the years. Regular background fungicide applications (at 10–12-day intervals) against *Plasmopara viticola* and *Erysiphe necator* took place in all seasons. No botryticides were applied.

2.2. Meteorological Data

Meteorological data were recorded during the period of examination by a weather station of the national agricultural administration ASTA (Administration des services techniques de l'agriculture) located in Remich/Luxembourg in direct proximity (distance <50 m) of the experimental vineyard. Air temperatures were measured at 2 m and precipitation at 1 m from the ground. The weather data can be downloaded from www.agrimeteo.lu.

2.3. Assessment of Bud Burst Percentage and Number of Inflorescences per Shoot

From 2014 to 2018, the bud-burst percentage and number of inflorescences per shoot were assessed on 50 randomly selected buds/shoots per plot. Assessments took place between the plant-growth stages BBCH 17 and 55 on 20 May 2014, 19 May 2015, 31 May 2016, 1 June 2017, and 16 May 2018, respectively.

2.4. Assessment of the Cluster Morphology

To investigate the influence of the different treatments on the cluster structure, the cluster density index, according to the protocol by Ipach et al. [8], was assessed as previously described [9]. Fifty clusters per plot were assessed in phenophase BBCH 79 (assessments took place on the date when BBCH 81 was reached in the early ripening cultivar 'Müller-Thurgau' in the same experimental vineyard to guarantee a comparable development in the different years) after the mechanical thinning of the treatments SMPH MT 1 and SMPH MT 2 (20 August 2013, 5 August 2014, 6 August 2015, 18 August 2016, 26 July 2017, 23 July 2018).

2.5. Assessment of Botrytis cinerea Disease Progress

The progress of the *B. cinerea* disease was followed at intervals of 6–14 days between veraison and harvest by examining 50 randomly selected clusters per plot. Disease severity was assessed according to the EPPO (European and Mediterranean Plant Protection Organization) guideline PP1/17 attributing visually observed disease severities to seven classes (0%; 1–5%; 6–10%; 11–25%; 26–50%; 51–75%; 76–100%). Average disease severities were calculated by summing the number of observations per class multiplied by the arithmetic mean of the class interval and dividing this sum by the total number of observations ($n = 50$) [10].

To describe the temporal progress of the disease severity, the average values per treatment were plotted against the assessment date (expressed as the day of the year (DOY)). Disease progress curves were fitted to these data according to the sigmoidal Equation (1) as described previously [6]:

$$y = \frac{100}{1 + e^{-((x-x_0)/b)}} \tag{1}$$

where y is the disease severity, x corresponds to the assessment date expressed as the day of the year (DOY), x_0 is the inflection point of the curve (disease severity of 50% reached), and b is the slope factor of the curve in the inflection point.

Solving this equation for x provides the time point at which a specific disease severity value was reached. To quantify differences in the temporal position of the annual epidemic of the different treatments, the $x_{5\%}$-values (DOY reaching a disease severity of 5%) were used following Beresford et al. [11] and Evers et al. [9].

2.6. Maturation Progress

The maturation progress was followed at intervals of 6–14 days between veraison and harvest (same dates as for bunch rot assessments) by collecting 30–40 randomly selected berries (clusters from different positions of the canopy; berries of different positions in the cluster) per plot (avoiding berries with visible bunch rot symptoms). Total soluble solids (TSS) were determined from the extracted juice (mixed sample of all berries per plot) using a digital refractometer (RHB-32ATC, Huake Instruments Co. Ltd., Lirenfuzone, Shenzhen, China).

Since berry sugar accumulation after veraison follows a sigmoidal pattern [12,13], sigmoidal maturity progress curves were fitted to observation data according to Equation (2):

$$y = \frac{a}{1 + e^{-((x-x_0)/b)}} \tag{2}$$

where y are the TSS, x corresponds to the sampling date expressed as a day of the year (DOY), x_0 is the inflection point, a is the maximum of the curve, and b is the slope factor of the curve in the inflection point.

Solving this equation for x provides the time point at which a specific TSS value was reached. In present investigations, the calculated DOYs reaching 14.17 brix (= 60 °Oechsle) were selected for a comparison of the different treatments. Sixty degree Oechsle represents the legal threshold for the production of wines with a protected designation of origin in Luxembourg.

In addition, disease progress curves were plotted against the grape maturation progress (expressed as TSS) according to Equation (1) [13]. In this case, y is the disease severity, x corresponds to the TSS, x_0 is the inflection point of the curve (disease severity of 50% reached), and b is the slope factor of the curve in the inflection point.

TSS (brix) calculated at the moment of reaching 5% disease severity was compared between the different treatments.

2.7. Yield and Total Soluble Solids at Harvest

Grapes from each plot were harvested separately (30 October 2013, 29 September 2014, 14 October 2015, 20 October 2016, 27 September 2017, 2 October 2018) and the average yield per plant was calculated. On the harvest date, defined in each year as a compromise between (i) the grape health status as well as (ii) the degree of ripeness in the different treatments, 20 grape clusters were randomly sampled per plot. After pressing, their juice (mixed sample of all clusters) was centrifuged and the TSS were measured by FT-IR (FOSS NIRSystems, Laurel, MD, USA).

2.8. Data Analyses and Statistics

Data sets consisting of average values per plot (four replicate plots per treatment) were (after testing Gaussian distribution and homogeneity of variance) analyzed for the effect of the treatment by one-way ANOVAs using SPSS Statistics 19 (IBM, Chicago, IL, USA). For the event that null-hypotheses were rejected ($p \leq 0.05$), pair-wise comparisons were performed for treatment effects according to Tukey's multiple comparison procedure.

Annual averages of (i) percentages of bud bursts, (ii) number of inflorescences per shoot, (iii) density index, (iv) yield, and (v) TSS at harvest were normalized as a ratio between the average values of the respective treatment and the standard treatment, VSP. Additionally, the annual deviations between VSP and different SMPH treatments were calculated for (i) the $x_{5\%}$-values (ii) the date reaching 14.12 brix TSS, and (iii) the TSS at the moment of reaching 5% disease severity. The yield formation of a grapevine is a two-year process [14] and the number of inflorescences per shoot is already determined in the year prior to harvest. Thus, the 2013 results might be influenced by the degree of inflorescence formation in the year before the start of the present trials (2012). Hence, average normalized values as well as average deviations were calculated for the 2014–2018 period ($n = 5$ years) without considering the 2013 results.

3. Results

3.1. Key Meteorological Data

Key meteorological data are given in Table S2. Average growing season (April–October) temperatures ranged from 14.7 °C in 2013 to 17.0 °C in 2018. The lowest average annual temperatures were measured in 2013 (9.8 °C) and the highest in 2018 (11.8). The lowest cumulative precipitation within the growing season was observed in 2018 (295 mm) and the highest cumulative precipitation in 2013 (616 mm). Annual precipitation sums ranged between 540 mm in 2015 and 813 mm in 2013 (Table S2).

3.2. Percentage of Bud Bursts and Number of Inflorescences Per Cluster

Average percentages of bud burst ranged from 69.8% (2014) and 86.9% (2018). In 2014, the percentage of bud burst in VSP was significantly higher than in SMPH ST 3. In 2017, this was the case for VSP in comparison with SMPH ST 2, SMPH ST 4, SMPH MT 1, and SMPH MT 2. No significant differences were observed in any season between any of the treatments SMPH ST 1 to SMPH ST 4 (Table S3).

The average number of inflorescences per shoot ranged between 0.5 (2014) and 1.1 (2015). In all years of observation, the number of inflorescences per shoot in VSP was significantly higher than in the SMPH treatments. No significant differences were observed in any season between any of the treatments SMPH ST 1 to SMPH ST 4 (Table S4).

3.3. Cluster Architecture

Average density indices ranged from 2.7 (2013) to 3.5 (2014). In 2014, the VSP showed significantly higher density index values than all SMPH treatments. No significant differences were observed in any season between any of the treatments SMPH ST 1 to SMPH ST 4. Average normalized density index values ranged from 0.66 in SMPH MT 2 to 1.00 in VSP (Table S5).

3.4. Bunch Rot and Maturity Progress

At the final assessment date and over all treatments, on average *B. cinerea* disease severities reached 18.5% (2013), 11.7% (2014), 13.7% (2015), 6.6% (2016), 20.4% (2017), and 2.0% (2018). On the final assessment date, the following significant differences were observed between treatments:

- In 2013, 2015, and 2018, the disease severity for VSP was significantly higher than for each of the SMPH treatments.
- In 2014, the disease severity for VSP was significantly higher than for each of the SMPH treatments with the exception of SMPH ST 2. Besides this, SMPH MT 1 showed significantly lower disease severities than SMPH ST 2.
- In 2017, the disease severity for VSP was significantly higher than for each of the SMPH treatments with the exception of SMPH ST 1. Besides this, SMPH MT 1 and SMPH MT 2 showed significantly lower disease severities than SMPH ST 1.
- In 2016, no significant differences were observed on the final assessment date.

No significant differences were observed on any assessment date between any of the treatments SMPH ST 1 to SMPH ST 4 (Table S6).

Sigmodial curves were fitted to assessment data as a function of time (Figure 2).

Coefficients of determination (R^2) of sigmoidal equations ranged from 0.72 to 1.00 with *p*-values between <0.0001 and 0.0681. Correlations between the assessment data and fitted curves observed were significant in 41 out of 42 'year x treatment-combinations' ($p \leq 0.05$) (Table S7).

Calculated dates reaching a disease severity of 5% on average for all seven treatments ranged from day of the year 252.1 in 2017 to day of the year 290 in 2013. On average, for the five years from 2014 to 2017, the four non-thinned SMPH treatments reached 5% disease severity 17.6 days later than the VSP treatments. The highest average deviation between the DOY reaching 5% disease severity compared to the standard treatment, VSP, was 31.3 days for SMPH MT 1 (Table 1).

The TSS reached for all treatments on the final assessment date was on average 16.0 brix (2013), 20.8 brix (2014), 16.5 brix (2015), 20.9 brix (2016), 19.9 brix (2017), and 20.5 brix (2018). At the final assessment date, the following significant differences were observed between treatments:

- In 2013, TSS values in VSP and SMPH MT 1 and SMPH MT 2 were significantly higher than in non-thinned SMPH treatments.
- In 2015 and 2018, TSS values in VSP and SMPH MT 2 were significantly higher than in non-thinned SMPH treatments.

- In 2016, TSS in VSP as well as in both thinned SMPH treatments were significantly lower than in SMPH ST 1 and SMPH ST 2.
- In 2014 and 2017, no significant differences were observed on the final assessment date.

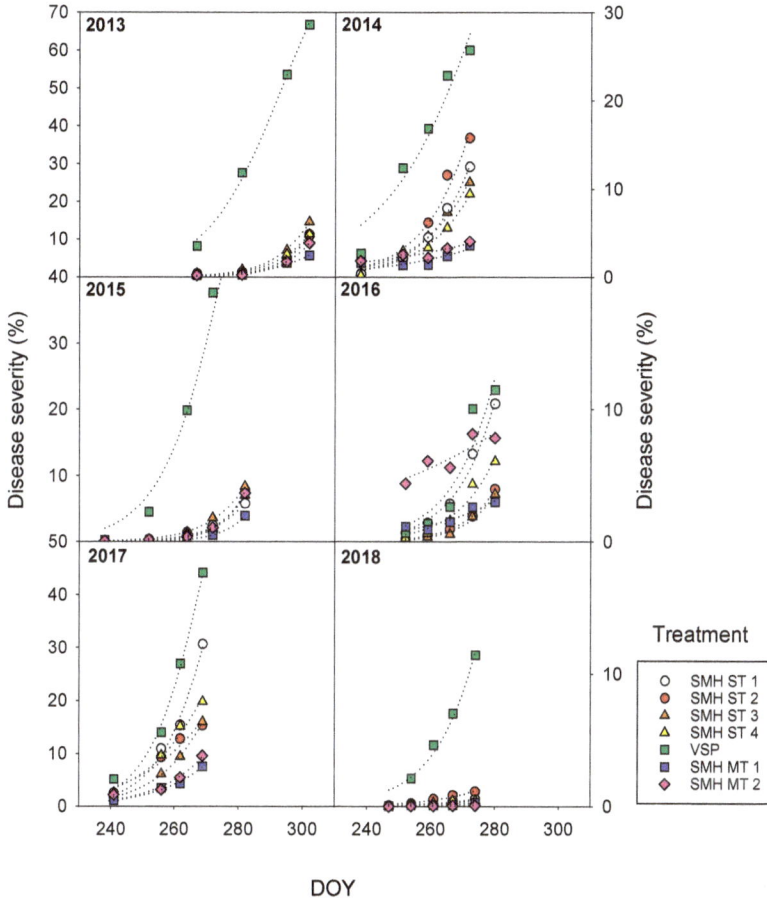

Figure 2. Progress of the disease severity of *Botrytis cinerea* in the different treatments between 2013 and 2018 as functions of the assessment date (day of the year (DOY)). Plot symbols represent the observed disease severities, with lines showing the calculated progress according to the sigmoidal equation type $y = 100/(1 + e^{-((x-x_0)/b)})$. SMPH, semi-minimal pruned hedge; ST, shoot topping; VSP, vertical shoot positioning; MT, mechanical thinning.

TSS in non-thinned SMPH treatments did not differ significantly on the final assessment date with the exception of SMPH ST 1 in 2016, which reached significantly higher TSS values than SMPH ST 3 and SMPH ST 4 (Table S8).

When fitting sigmoidal equations to data illustrating maturity progress (Figure 3), coefficients of determination (R^2) ranged from 0.96 to 1.00 with *p*-values between 0.0003 and 0.2092.

Table 1. Calculated dates (day of the year (DOY)) for reaching 5% bunch rot disease severity between 2013 and 2018 as well as deviations (Δ) between the DOY for reaching 5% disease severity in one treatment in a specific year and the DOY for reaching 5% disease severity in the standard treatment, VSP, in that year. Average (2014–2017) deviations of different treatments are depicted. Since in 2018 5% disease severity were not reached in all treatments, calculated data were not shown.

Treatment *	2013	Δ	2014	Δ	2015	Δ	2016	Δ	2017	Δ	2018	Δ	Average (2014–2017) Δ
SMPH ST 1	295.2	37.0	260.2	25.6	280.4	32.6	271.1	2.5	249.3	4.3			16.3
SMPH ST 2	295.0	36.9	256.4	21.8	278.8	31.0	282.3	13.8	246.7	1.7			17.1
SMPH ST 3	291.2	33.0	260.3	25.7	276.3	28.5	282.9	14.4	253.8	8.9			19.4
SMPH ST 4	293.4	35.3	263.2	28.6	279.0	31.2	277.1	8.5	246.7	1.7			17.5
VSP	258.1	0.0	234.6	0.0	247.8	0.0	268.5	0.0	244.9	0.0			
SMPH MT 1	300.0	41.9	283.8	49.2	283.8	36.1	290.5	22.0	262.9	18.0			31.3
SMPH MT 2	296.8	38.7	282.7	48.1	278.8	31.0	253.9	−14.7	260.1	15.2			19.9
Average	290.0	31.8	263.0	28.1	275.0	27.2	275.2	6.6	252.1	7.1			17.3

* SMPH, semi-minimal pruned hedge; ST, shoot topping; VSP, vertical shoot positioning; MT, mechanical thinning.

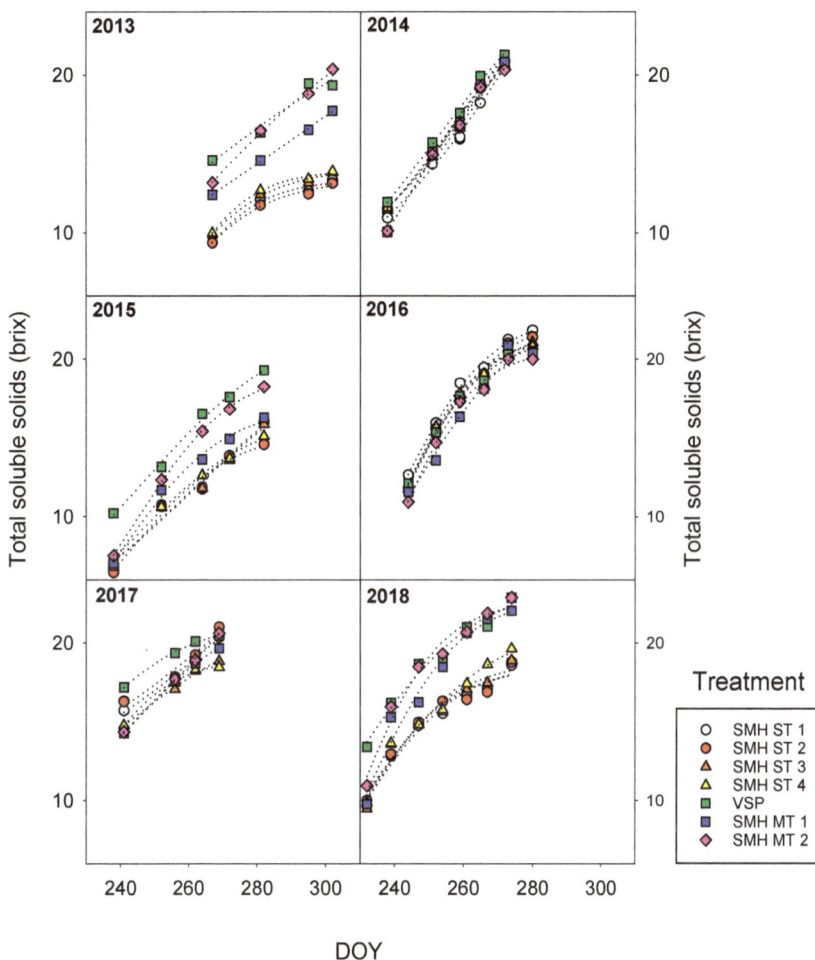

Figure 3. Progress of the total soluble solids in the different treatments between 2013 and 2018 as functions of the assessment date (DOY). Plot symbols represent the observed disease severities, with lines showing the calculated progress according to the sigmoidal equation type $y = a/(1 + e^{-((x-x_0)/b)})$. SMPH, semi-minimal pruned hedge; ST, shoot topping; VSP, vertical shoot positioning; MT, mechanical thinning.

Correlations between recorded data and fitted curves were significant in 33 of 42 'year x treatment-combinations' ($p \leq 0.05$). Non-significant correlations were limited to 2013 and 2017, for which only four assessment data sets were available (Table S9).

Calculated dates (DOY), reaching on average 14.12 brix for all seven treatments, ranged from 223.6 in 2017 to 267.8 in 2015. In 2013, 14.12 brix was not reached in the non-thinned treatments of SMPH.

In the five years from 2014 to 2018, the level of 14.12 brix was reached on average 11 days later in the four non-thinned SMPH treatments than with VSP. The highest average deviation observed between the DOY reaching 14.12 brix in an SMPH treatment and the standard VSP treatment was 12.5 days in SMPH ST 4 (Table 2).

Table 2. Calculated dates (day of the year (DOY)) for reaching 14.12 brix between 2013 and 2018 as well as deviations (Δ) between the DOY for reaching 14.12 brix in one treatment in a specific year and the DOY for reaching 14.12 in the standard treatment, VSP, in that year. Average (2014–2018) deviations of different treatments are depicted.

Treatment	2013	Δ	2014	Δ	2015	Δ	2016	Δ	2017	Δ	2018	Δ	Average (2014–2018) Δ
SMPH ST 1	NR *		251.0	5.3	272.9	18.2	246.7	−1.8	230.5	2.8	242.3	29.3	10.7
SMPH ST 2	NR		249.1	3.3	276.8	22.1	247.6	−0.9	227.6	−0.2	238.8	25.9	10.0
SMPH ST 3	NR		247.8	2.1	273.5	18.8	247.5	−1.0	237.5	9.8	241.7	28.8	11.7
SMPH ST 4	NR		248.2	2.5	273.7	19.0	247.6	−0.9	238.8	11.0	243.9	31.0	12.5
VSP	265.0	0.0	245.8	0.0	254.7	0.0	248.5	0.0	227.8	0.0	213.0	0.0	0.0
SMPH MT 1	278.4	13.3	249.1	3.3	264.9	10.2	253.4	4.9	240.3	12.5	242.5	29.5	12.1
SMPH MT 2	271.0	5.9	249.2	3.5	258.5	3.8	250.3	1.8	239.8	12.0	218.2	5.3	5.3
Average			248.6	2.8	267.8	13.1	248.8	0.3	234.6	6.8	243.3	21.4	8.9

* NR = 14.12 brix not reached.

Sigmoidal equations of the type $y = 100/(1 + e^{-((x-x_0)/b)})$, which describe the disease progress as a function of maturation progress (TSS in brix) (Figure 4), reached coefficients of determination between 0.66 and 1.00 with *p*-values between <0.0001 and 0.0949.

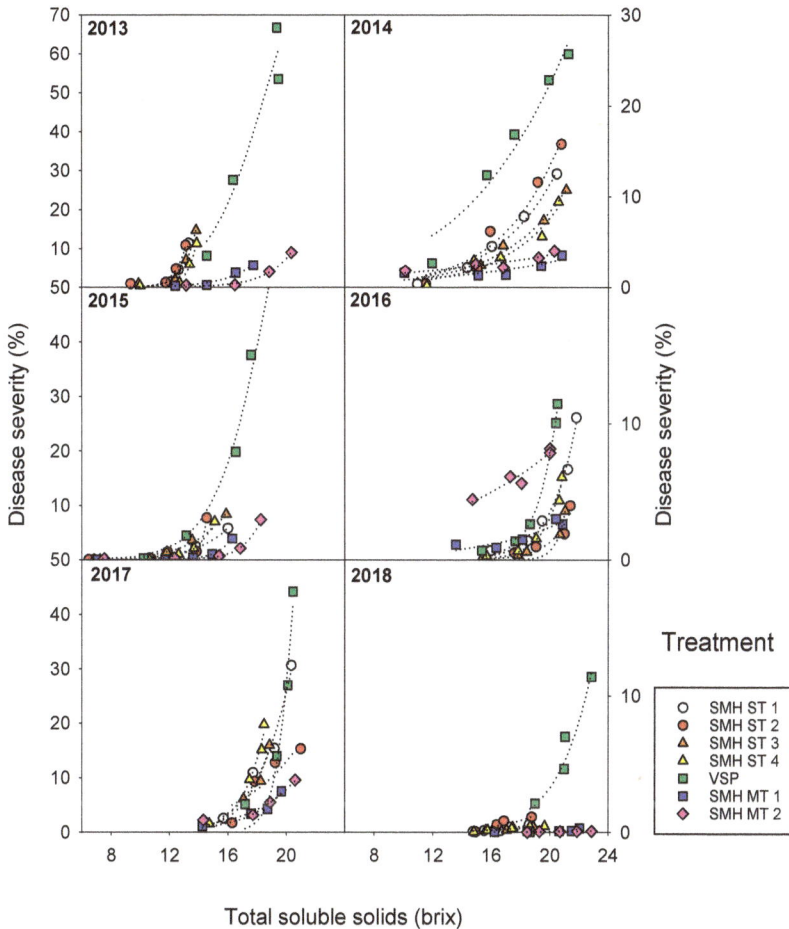

Figure 4. Progress of the disease severity of *B. cinerea* in the different treatments between 2013 and 2018 as functions of the total soluble solids. Plot symbols represent the observed disease severities, with the lines showing the calculated progress according to the sigmoidal equation type $y = 100/(1 + e^{-((x-x_0)/b)})$. SMPH, semi-minimal pruned hedge; ST, shoot topping; VSP, vertical shoot positioning; MT, mechanical thinning.

In 38 of 42 'year x treatment-combinations', disease progress was significantly correlated with the maturation progress (Table S10).

TSS calculated at the moment disease severity reached 5% ranged between 14.4 brix in 2013 and 20.5 brix in 2016 (average of all seven treatments). Over the year 2014 to 2017, 5% disease severity was reached on average at higher TSS levels in all SMPH treatments than in VSP treatments. From 2014 to 2017, the four non-thinned SMPH treatments reached on average 5% disease severity, with TSS levels being 1.8 brix higher than for VSP (Table 3).

Agronomy **2019**, *9*, 173

Table 3. Calculated total soluble solids (brix) on the day of the year (DOY) for reaching 5% disease severity between 2013 and 2018 as well as deviations (Δ) between the total soluble on the day of the year for reaching 5% disease severity in a specific year and the total soluble on the days of the year for reaching 5% disease severity in the standard treatment, VSP, in that year. Average (2014–2017) deviations of different treatments are depicted. Since in 2018 5% disease severity were not reached in all treatments, calculated data were not shown.

Treatment	2013	Δ	2014	Δ	2015	Δ	2016	Δ	2017	Δ	2018	Δ	Average (2014–2017) Δ
SMPH ST 1	12.8	0.0	16.8	5.6	15.6	2.2	20.6	1.2	16.7	−2.0			1.8
SMPH ST 2	12.6	−0.2	16.2	5.0	14.4	1.0	21.9	2.5	16.3	−2.4			1.5
SMPH ST 3	12.9	0.2	17.8	6.6	14.6	1.2	21.3	1.9	16.9	−1.7			2.0
SMPH ST 4	13.4	0.6	18.4	7.2	14.7	1.3	20.7	1.2	16.6	−2.0			1.9
VSP	12.7	0.0	11.2	0.0	13.4	0.0	19.4	0.0	18.6	0.0			0.0
SMPH MT 1	17.4	4.7	24.8	13.6	16.5	3.1	23.9	4.5	18.7	0.1			5.3
SMPH MT 2	19.3	6.6	23.9	12.7	17.8	4.4	16.0	−3.4	18.6	−0.1			3.4
Average	14.4	1.7	18.5	7.2	15.3	1.9	20.5	1.1	17.5	−1.1			2.3

Temporal day of the year (DOY) deviations, reaching 5% disease severity in the different treatments, were plotted against the temporal day of the year (DOY) deviations that reached 14.12 brix in the different treatments (Figure 5).

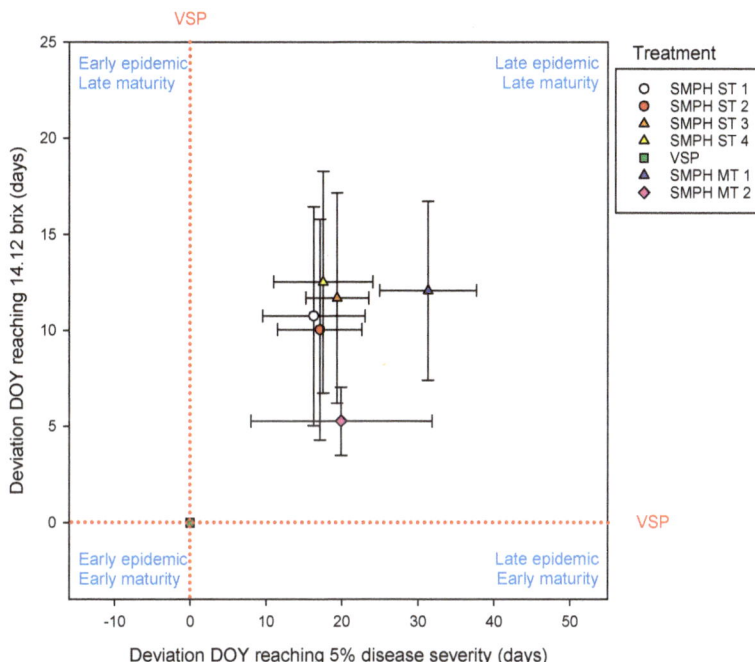

Figure 5. Temporal deviations of the day of the year (DOY) for reaching 5% disease severity in the different treatments compared to the standard treatment, VSP, plotted against the temporal deviations of the day of the year (DOY) for reaching 14.12 brix in the different treatments compared to VSP. Data from 2014 to 2017. Error bars = standard errors.

3.5. Yield and TSS at Harvest

The average yield across all treatments was 6.5 kg/plant (2013), 3.7 kg/plant (2014), 9.5 kg/plant (2015), 1.4 kg/plant (2016), 4.2 (2017) kg/plant, and 5.5 kg/plant (2018).

Significant differences between the yield levels were observed in the following cases:

- In 2013 and 2015, the yield in VSP and SMPH MT 2 was significantly lower than in the non-thinned SMPH treatments.
- In 2016, the yield in VSP was significantly higher than in SMPH ST 1 and SMPH ST 2.
- In 2018, the yield in SMPH MT 1 and SMPH MT 2 was significantly lower than in the non-thinned SMPH treatments.
- In 2014 and 2017, no significant differences were observed.

No significant differences were observed in any year between the different non-thinned treatments of SMPH ST 1.

Average normalized yields were lowest in SMPH MT 2 (0.81) and highest in SMPH 4 (1.81). Compared to the average of the non-thinned SMPH treatments, from 2014 to 2018, the yield reduction in SMPH MT 1 and SMPH MT 2 reached on average 28 and 53%, respectively (Table 4).

Table 4. Average values of the yield (kg/plant) between 2013 and 2018 as well as average normalized yield values for 2014–2018 (value of the yield of the respective treatment in a specific year/value of the standard treatment, VSP, in that year). Yield values of different treatments in the same year marked with the same letter did not differ significantly (according to Tukey's multiple comparison procedure ($p \leq 0.05$)). Average (2014–2018) normalized yield values are depicted. If no letter codes are indicated, differences were non-significant.

Treatment	2013	2014	2015	2016	2017	2018	Average Normalized Yield (2014–2018)
SMPH ST 1	9.5 c	3.8	12.5 b	0.9 a	5.3	7.6 bc	1.8
SMPH ST 2	8.8 bc	3.8	11.6 b	0.9 a	4.5	8.0 c	1.7
SMPH ST 3	9.2 c	4.6	11.4 b	1.4 ab	4.9	7.5 bc	1.7
SMPH ST 4	7.6 bc	4.4	11.3 b	1.6 ab	5.5	7.8 bc	1.8
VSP	3.3 a	3.7	4.8 a	2.7 b	1.9	3.7 ab	1.0
SMPH MT 1	4.8 ab	3.2	8.4 ab	1.2 ab	4.7	2.8 a	1.3
SMPH MT 2	2.1 a	2.6	6.4 a	1.0 a	2.5	1.2 a	0.8
Average	6.5	3.7	9.5	1.4	4.2	5.5	1.4

For all treatments, the TSS at harvest date were on average 15.5 (2013), 20.8 (2014), 15.4 (2015), 21.2 (2016), 19.9 (2017), and 21.0 brix (2018). The average normalized TSS at harvest were lowest in SMPH ST 4 (0.90) and highest in VSP (1). Compared to the average of the non-thinned SMPH treatments, from 2014 to 2018, the increase of TSS at harvest in SMPH MT 1 and SMPH MT 2 reached 0.8 and 1.3 brix on average, respectively (Table 5).

Table 5. Average values of the total soluble solids (TSS) at harvest (brix) between 2013 and 2018 as well as average normalized TSS at harvest values for 2014–2018 (value of the TSS at harvest of the respective treatment in a specific year/value of the standard treatment, VSP, in that year). TSS at harvest values of different treatments in the same year marked with the same letter did not differ significantly (according to Tukey's multiple comparison procedure ($p \leq 0.05$)). Average (2014–2018) normalized TSS at harvest values are depicted. If no letter codes are indicated, differences were non-significant.

Treatment	2013	2014	2015	2016	2017	2018	Average Normalized TSS at Harvest (2014–2018)
SMPH ST 1	12.9 a	21.1	13.3 a	22.3 b	20.4	19.7	0.9
SMPH ST 2	12.7 a	20.8	15.0 ab	21.8 ab	21.0	19.7	0.9
SMPH ST 3	13.4 a	20.5	13.6 a	21.5 ab	18.8	19.2	0.9
SMPH ST 4	13.4 a	20.6	13.7 a	21.3 ab	18.5	19.4	0.9
VSP	18.9 c	21.3	19.0 c	20.3 a	20.5	23.2	1.0
SMPH MT 1	17.3 b	20.8	15.1 ab	21.0 ab	19.7	22.9	1.0
SMPH MT 2	19.9 c	20.3	17.8 bc	20.4 a	20.6	23.2	1.0
Average	15.5	20.8	15.4	21.2	19.9	21.0	0.9

For the different treatments, average (2014–2018) normalized yields were plotted against average normalized TSS at harvest. On average, for 2014 to 2018, non-thinned treatments of SMPH as well as SMPH MT 1 obtained higher yields, as well as lower TSS than VSP, whereas SMPH MT 2 obtained lower yields and lower TSS (Figure 6).

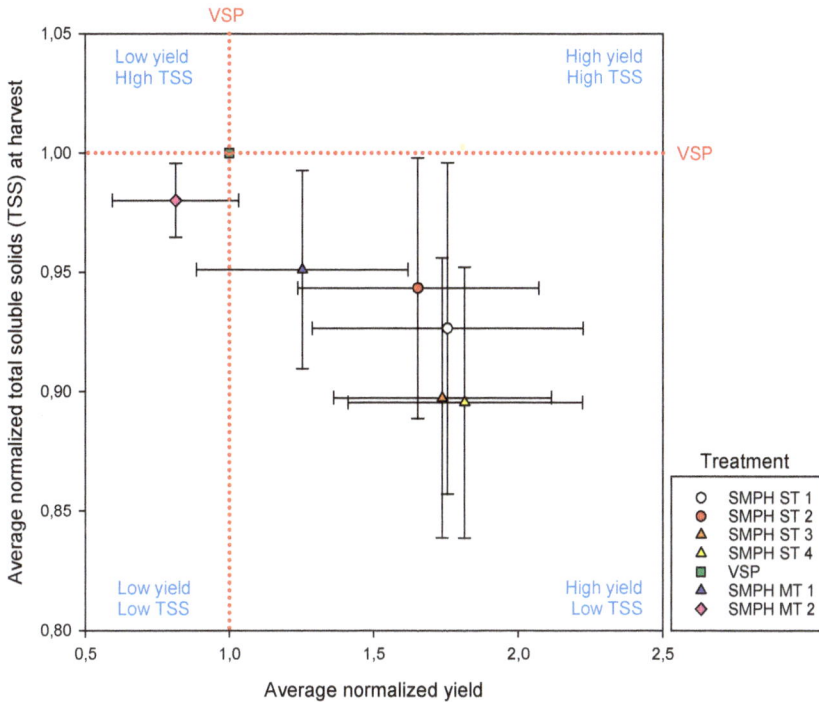

Figure 6. Average normalized yield (ratio of the yield in the different treatments and the yield in the standard treatment, VSP) plotted against the average normalized TSS (ratio of the TSS in the different treatments and the TSS in VSP) at harvest. Data from 2014 to 2018. Error bars = standard errors.

4. Discussion

4.1. General Behaviour of Non-Thinned SMPH

The number of buds per plant in SMPH was shown to be several times higher than in the traditional VSP system [3]. To avoid excessive over-cropping in SMPH, different steps of self-regulation are necessary. The present results indicate a decrease in the percentage of bud burst as well as a significantly reduced number of inflorescences per shoot in SMPH treatments compared to VSP, confirming the observations of Intrieri et al. [3].

However, the number of clusters per plant in SMPH remained much higher than in VSP [3]. As a further step in self-regulation, the size (and consequently the weight) of clusters has been observed to be smaller in SMPH than in VSP as a result of (i) the reduced size of the berries, as well as (ii) a reduced number of berries [1,4]. Furthermore, the structure of clusters in SMPH has been observed in the studies of Intrieri et al. [3] and Kraus et al. [1] to be less compact than in VSP. Present data indicate a reduced compactness of SMPH clusters even though the differences were not significant. Generally, a strong link between cluster compactness and bunch rot susceptibility was observed in several studies in different cultivars and different viticultural regions [6,15,16]. The underlying physiological and epidemiological reasons for this relationship have been discussed in detail in a previous work [17]. Consequently, on the final assessment dates, the disease severities of *B. cinerea* in SMPH were lower than in the standard treatment, VSP, in all years of the investigation, confirming the observations of Intrieri et al. [3]. Both bunch rot epidemics and TSS (as an indicator for grape maturity) progress follow a sigmoidal pattern over time [12,17] and, hence, can be adequately simulated by sigmoidal equations allowing for the calculation of precise dates when specific disease severity or maturity levels are reached. Non-thinned SMPH treatments delayed the day of the year for reaching 5% disease severity (average for 2014–2017) by 18 days on average. In the environmental conditions of the Luxembourgish Moselle region, as well as of many other cool-climate winegrowing regions, the timing of the harvest is more often determined by crop health status than by optimum grape maturity [18]. The delay of the bunch rot epidemic observed in SMPH, in fact, enabled an average temporal benefit of approximately three weeks before the grapes reached a disease severity threshold that forced the grower to harvest them before bunch rot jeopardized yield and/or wine quality.

The average yield in non-thinned SMPH treatments was considerably higher than in VSP. This was the result of the much higher number of buds per plant [3,4]. Despite the observed and aforementioned mechanisms of self-regulation, from 2014 to 2018, the average yield in non-thinned SMPH treatments was 78% higher than in the standard system. Consequently, the legal thresholds of marketable grapes/wine per ha in Luxemburg were partially exceeded. In non-thinned SMPH, the yield fluctuated greatly from year to year. Coefficients of variation (cv; standard deviation/mean) in the non-thinned SMPH treatments reached 66% between 2014 and 2018, which is twice as high as in VSP (33%). This indicates a tendency towards alternate bearing in SMPH as described by Intrieri et al. [3]. However, we observed a tendency towards more moderate yield levels (although still fluctuating) with ongoing temporal distance to the year of transfer from VSP to SMPH.

The bunch rot epidemic has to be considered in strong relationship with the maturity progress, especially in the case of marked differences in the yield levels [6]. In fact, as a consequence of the (on average) higher crop load, non-thinned SMPH treatments also delayed the maturity progress—on average by 11 days—until the legal threshold for wines marketable as wines of protected designation of origin had passed. This confirms observations by Walg et al. [4], who generally observed a maturity delay of 1–2 weeks in SMPH compared to VSP. Besides a frequently higher crop load, the canopy morphology has also been identified as a reason for this maturity delay. In fact, berry development in SMPH is more inhomogeneous than in VSP [4]. This is the case due to the lower sun and light exposure of shadowed clusters and, in consequence, lower temperatures inside the complex canopy [1]. Generally, in the present investigation, the delay of the maturity progress was observed to be lower than the delay of the bunch rot epidemic. For example, when considering a specific disease level (in our case defined as 5% following the definition of Beresford et al. [11]) as a determining factor for the harvest, non-thinned SMPH treatments would have reached, on average for 2014–2018, 1.8 more TSS at harvest than the standard system, VSP. However, generally, it must be kept in mind that (i) the length of the vegetation period is limited and (ii) the maturity progress ceases at a certain time of year. Consequently, even if the grape health status allowed for a further potential delay in the harvest date, grapes in SMPH might not ever reach full maturity in phenologically late vintages and/or in the event of high/excessive crop load (as observed in 2013 where the legal minimum TSS levels for wines of protected designation of origin were not reached by any of the non-thinned SMPH treatments). However, (i) the more moderate yield levels with ongoing temporal distance to the transfer from VSP to SMPH and (ii) the potential temporal buffer for a continuation of the maturation period after the harvest period of the traditional VSP due to climate change (earlier start of the maturation period [19]; temporal prolongation of the potential vegetation period in autumn [20]), are supposed to compensate for the delayed maturation.

4.2. Timing of Shoot Topping

While investigations by Molitor et al. [6] demonstrated the importance of the timing of the first shoot topping in VSP on the bunch rot epidemic as well as on grape maturity and yield, in the present investigations, the moment that the shoot topping was carried out showed no significant effects. This absence of significant differences may be explained by the fact that the temporal distance between the earliest and the latest moment of shoot topping was relatively low, approximately three weeks. In fact, in the trials of Molitor et al. [6] in VSP, significant effects were mainly observed when comparing early (pre-flowering or flowering) and very late (2 to 4 weeks after end of flowering) first shoot toppings.

However, (first) carrying out shoot topping in SMPH more than one week after the end of flowering might not be practically feasible (at least where the distance between rows is around 2.0 m and the plant vigor on a moderate level). Indeed, further unlimited shoot growth and, in consequence, the closure of the driving lane between rows might hinder proper plant protection treatments by tractor-driven sprayers in this period of highest susceptibility of young berries towards downy and powdery mildew, as well as black rot.

4.3. Mechanical Thinning

Mechanical thinning using a harvest machine reduced yield by between 28% (moderate thinning) and 53% (severe thinning) on average between 2014 and 2018, compared to the non-thinned SMPH treatments. Yield reduction is based on a lower berry number as well as on reduced berry size [4], both of which lead to looser cluster structures as confirmed by the tendencies in the present data. Consequently, both thinning regimes further delayed the day of the year for reaching 5% disease severity as well as the inclination of the disease progress curves. However, in some cases, an earlier start of the bunch rot epidemic was observed in the thinned treatments of SMPH. This is the result of unripe berries being damaged by the harvest machine followed by humid weather conditions in the subsequent period. Here, damaged berries were most likely infected by *Botrytis cinerea*, while, where dry weather conditions followed, damaged berries dried out completely.

Analyses of the maturity progress indicate that at early dates of maturity control, the increase of TSS caused by yield reduction due to mechanical thinning is often low or even absent compared to non-thinned treatments of SMPH, while, close to harvest, TSS levels were often comparable with the standard treatment, VSP. We assume that mechanical thinning causes a shock for the grape berries and stops ripening for several days, comparable to a ripening stop as the result of hail damage in this stage. After this period of ripening disruption, induced by the reduced remaining crop load, the further maturity progress in the thinned treatments continued at a faster pace than in the non-thinned treatments. Due to a combination of both (i) the delay of the bunch rot epidemic and (ii) the acceleration of the maturity progress of grape maturity, the level of TSS at the moment of reaching 5% disease severity in the thinned treatments is higher than in the non-thinned SMPH treatments or in VSP.

4.4. Practical Recommendations

The present study has revealed the limitations, challenges, and opportunities of the relatively novel viticultural training system SMPH in practical conditions.

Based on the present results, the impact of the timing of shoot topping appears to be limited and might be neglected in practical viticulture.

The present study, as well as that of Walg et al. [4] revealed that, especially in the year of transfer from VSP to SMPH, the yield might be too high for premium quality wine production. For example, in the year of transfer, the minimum threshold of TSS for wine of protected designation of origin was not reached in the non-thinned treatments of SMPH due to the very high crop load. This is probably the result of the two-year yield formation process [14,21], where the number of clusters per shoot is already determined in the year prior to harvest (i.e., in our case, the final year before transfer from the VSP system). Consequently, especially in years with a high yield potential (e.g., the first year after transfer from VSP), it is recommended to mechanically thin the yield to a moderate level to safeguard proper wine quality [4]. In the present trials, the percentage of yield reduction was on average 28% (moderate thinning) or 53% (severe thinning). However, the level of reduction was not stable over time and ranged between −1 (2016) and 64% (2018) for moderate thinning and between 15 (2016) and 84% (2018) for severe thinning even though the same machine and identical beat frequency were used. Observed differences in the degree of yield reduction might be caused by small differences in the phenological development (even though thinning took always place in BBCH 79) or in the annual crop load. Especially in years with high crop levels, the berry development is usually relatively inhomogeneous. Berry resistance to detachment is linked to the development of the berries. Consequently, inhomogeneous development might lead to a higher uncertainty in the degree of yield reduction caused by mechanical thinning.

In 2016, which had a generally low crop level in all SMPH treatments, the percentage of yield reduction was low in both thinning treatments. Generally, the observed fluctuation in the percentage of yield reduction indicates the challenge of exact determination of the degree of mechanical thinning. While in some years the yield reduction might be too low to avoid over-cropping, in other years the reduction might lead to economically non-profitable low yields. Consequently, further investigations are necessary to provide a more precise adjustment of the degree of yield reduction.

High disease severities were observed at early assessment dates in the thinned treatments of SMPH, especially in 2016, where the mechanical thinning took place close to veraison. Between fruit set and veraison, young, immature grape berries are highly resistant to *B. cinerea* [22], while, after veraison, the host defence progressively breaks down with ongoing maturation [23]. At the time of mechanical thinning in 2016, single berries might had already gone soft and were hence more susceptible towards *B. cinerea* than unripe, hard berries. Based on this effect observed in 2016, it might be speculated that mechanical thinning should not take place too late in the season, i.e., clearly prior to veraison.

4.5. SMPH as a Climate Change Adaptation Strategy

Generally, a ripening delay was observed in SMPH compared to VSP. This ripening delay represents a risk for full grape maturity in the case of late maturation, such as in late vintages and/or late-ripening cultivars and/or late-ripening vineyards or regions. Consequently, SMPH systems should be avoided where heat consumption usually represents a limiting factor for full maturity (non-thinned).

On the other hand, the observed delay of the maturity progress might potentially turn into a benefit in future climatic conditions. Climate change projections are afflicted by uncertainties caused by mainly three factors: natural variability, model uncertainty, and greenhouse gas (GHG) emission scenario uncertainty [24]. Even though uncertainties in climate change projections exist depending on the underlying greenhouse gas emission scenarios, an annual air temperature increase of approximately 2.6 °C is projected for the region of investigation in the far future (2061–2090) compared to the reference period 1971–2000 [19].

The observed delayed maturation in SMPH and the shift of the maturation period towards later, usually cooler parts of the year, as well as the prolongation of the maturation period (due to cooler temperatures), might contribute to conserving the freshness and lightness [25] that is especially exemplary for white wines in (by then formerly) cool climate grapegrowing regions, such as Luxembourg. Here, SMPH could represent an interesting climate change adaptation strategy.

Additionally, extreme weather events causing, e.g., hail or sunburn damage are likely to occur more frequently if climate conditions change [26]. Here, SMPH might represent an adequate low-yield risk minimization strategy. Damage caused by hail, frost, or sunburn was usually found to be lower in SMPH [4] than in VSP due to (i) the higher yield potential and (ii) the specific distribution of clusters in the canopy. While in the past, the labor and cost benefits were the main drivers of the spread of SMPH vineyards, with ongoing climate change, the morphological and physiological characteristics of SMPH [4] are likely to gain increasing importance.

The lower susceptibility towards bunch rot observed might additionally be an opportunity for the SMPH system, especially in vineyards and/or cultivars where the emerging bunch rot epidemic represents the limiting factor for full grape maturity on a routine basis. This will especially be the case in expected future climatic conditions, since temperature has been observed as the major driver of the bunch rot epidemic [17], suggesting the occurrence of more severe bunch rot epidemics in future climatic conditions.

On the one hand, the lower susceptibility of SMPH towards bunch rot might allow for a pesticide reduction in viticulture. On the other hand, the investigations of Kraus et al. [1] indicated a higher susceptibility of SMPH to downy and powdery mildew, as well as to *Drosophila suzukii* due to the micro-climatic conditions inside the extended canopy. To the best of our knowledge, the net effect concerning pesticide consumption in SMPH has not yet been quantified and thus merits further investigations.

5. Conclusions

Due to the (i) delay of the bunch rot epidemic, (ii) the delay maturity period towards later, usually cooler, times of the year, and (iii) the lower risk of inadequate yields caused by extreme weather events (such as late frost, hail, or sunburn damage), SMPH might compensate for some of the potentially negative effects of climate change and might hence represent an interesting climate change adaptation strategy for practical viticulture in (formerly) cool climate viticulture regions.

Supplementary Materials: The following are available online at http://www.mdpi.com/2073-4395/9/4/173/s1, Table S1: Dates of shoot topping and corresponding exact phenological stages (BBCH code according to Lorenz et al. (1995)) in different treatments between 2013 and 2018, Table S2: Key growing season (April–October) and annual meteorological data of Remich/Luxembourg between 2013 and 2018, Table S3: Average values of the percentage of bud burst 2014–2018 (no assessment in 2013), as well as average normalized percentage of bud burst values for 2014–2018 (value of the percentage of bud burst of the respective treatment in a specific year/value of the standard treatment, VSP, in this year), Table S4: Average values of the number of inflorescences per shoot 2014–2018 (no assessment in 2013), as well as average normalized number of inflorescences per shoot values for 2014–2018 (value of the number of inflorescences per shoot of the respective treatment in a specific year/value of the standard treatment, VSP, in this year), Table S5: Average values of the density index between 2013 and 2018 as well as average normalized density index values for 2014–2018 (value of the density index of the respective treatment in a specific year/value of the standard treatment, VSP, in this year), Table S6: *B. cinerea* disease severities in the different treatments at the different assessment dates between 2013 and 2018. Treatments in the same year at the same assessment date marked with the same letter did not differ significantly (according to Tukey's multiple comparison procedure ($p \leq 0.05$)), Table S7: Parameters describing the *Botrytis cinerea* disease progress plotted against the time according to the sigmoidal equation type $y = 100/(1 + e^{-((x-x_0)/b)})$, Table S8: Total soluble solids in the different treatments at the different assessment dates between 2013 and 2018. Treatments in the same year at the same assessment date marked with the same letter did not differ significantly (according to Tukey's multiple comparison procedure ($p \leq 0.05$)), Table S9: Parameters describing the maturation progress plotted against the time according to the sigmoidal equation type $y = a/(1 + e^{-((x-x_0)/b)})$, Table S10: Parameters describing the disease progress plotted against the maturation progress according to the sigmoidal equation type $y = 100/(1 + e^{-((x-x_0)/b)})$.

Author Contributions: All authors have made substantial contributions to the present work. The authors agree to be personally accountable for the author's own contributions and for ensuring that questions related to the accuracy or integrity of any part of the work, even ones in which the author was not personally involved, are appropriately investigated, resolved, and documented in the literature. Individual contributions: Conceptualization: D.M., R.M.; Methodology: D.M., M.B.; Investigation: D.M., M.P.-B., M.S., R.M.; Writing—Original Draft Preparation: D.M., M.P.-B., M.B.; Writing—Review & Editing: M.S., R.M., L.H.

Funding: This work was partly funded by the IVV (Remich/Luxembourg) in the framework of the "ProVino", "BioViM", and "TerroirFuture" research projects, as well as by the European Union in the framework of the "Clim4Vitis" research project (Horizon 2020 research and innovation programme; grant agreement No. 810176).

Acknowledgments: The authors would like to thank R. Rausch, L. Vesque, B. Biewers, J. Niewind, B. Untereiner, D. Dam, M. Behr (LIST—Luxembourg Institute of Science and Technology), S. Garidel, S. Cerqueira, C. Beissel, P. Zahlen, L. Gilbertz, J. Lafleur, H. Litjens, J. Koch, C. Blum, and S. Fischer (Institut Viti-vinicole, IVV) for their technical support in the experimental vineyard and in the laboratory, J. Junk (LIST) for fruitful discussions, the Weinbau-Lohnunternehmen Meierer for mechanical thinning, as well as L. Auguin (LIST) for proofreading and language-editing.

Conflicts of Interest: The authors declare no conflict of interest.

References

1. Kraus, C.; Pennington, T.; Herzog, K.; Hecht, A.; Fischer, M.; Voegele, R.T.; Hoffmann, C.; Töpfer, R.; Kicherer, A. Effects of canopy architecture and microclimate on grapevine health in two training systems. *Vitis* **2018**, *57*, 53–60.
2. Clingeleffer, P.R. Development of management systems for low cost, high quality wine production and vigour control in cool climate Australian vineyards. *Wein-Wissenschaft* **1993**, *48*, 130–134.

3.   Intrieri, C.; Filippetti, I.; Allegro, G.; Valentini, G.; Pastore, C.; Colucci, E. The semi-minimal-pruned hedge: A novel mechanized grapevine training system. *Am. J. Enol. Viticult.* **2011**, *62*, 312–318. [CrossRef]

4.   Walg, O.; Blätz, M.; Friedel, M. Minimalschnitt im Spalier–eine wirksame Möglichkeit zur Spätfrostprävention. In *Deutsches Weinbau-Jahrbuch 2019*; Schultz, H.R., Stoll, M., Eds.; Eugen Ulmer KG: Stuttgart, Germany, 2018; Volume 70, pp. 76–85.

5.   Petrie, P.R.; Clingeleffer, P. Crop thinning (hand versus mechanical), grape maturity and anthocyanin concentration: Outcomes from irrigated Cabernet Sauvignon (*Vitis vinifera* L.) in a warm climate. *Aust. J. Grape Wine Res.* **2008**, *12*, 21–29. [CrossRef]

6.   Molitor, D.; Baron, N.; Sauerwein, T.; André, C.M.; Kicherer, A.; Döring, J.; Stoll, M.; Beyer, M.; Hoffmann, L.; Evers, D. Postponing first shoot topping reduces grape cluster compactness and delays bunch rot epidemic. *Am. J. Enol. Viticult.* **2015**, *66*, 164–176. [CrossRef]

7.   Lorenz, D.H.; Eichhorn, K.W.; Bleiholder, H.; Klose, R.; Meier, U.; Weber, E. Phenological growth stages of the grapevine, *Vitis vinifera* L. ssp. vinifera. Codes and descriptions according to the extended BBCH scale. *Aust. J. Grape Wine Res.* **1995**, *1*, 100–103. [CrossRef]

8.   Ipach, R.; Huber, B.; Hofmann, H.; Baus, O. *Richtlinie zur Prüfung von Wachstumsregulatoren zur Auflockerung der Traubenstruktur und zur Vermeidung von Fäulnis an Trauben. Outline for an EPPO-Guideline*; DLR Rheinpfalz: Neustadt an der Weinstraße, Germany, 2005.

9.   Evers, D.; Molitor, D.; Rothmeier, M.; Behr, M.; Fischer, S.; Hoffmann, L. Efficiency of different strategies for the control of grey mold on grapes including gibberellic acid (Gibb3), leaf removal and/or botryticide treatments. *J. Int. Sci. Vigne Vin* **2010**, *44*, 151–159. [CrossRef]

10.   Molitor, D.; Hoffmann, L.; Beyer, M. Flower debris removal delays grape bunch rot epidemic. *Am. J. Enol. Viticult.* **2015**, *66*, 548–553. [CrossRef]

11.   Beresford, R.M.; Evans, K.J.; Wood, P.N.; Mundy, D.C. Disease assessment and epidemic monitoring methodology for bunch rot (*Botrytis cinerea*) in grapevines. *N. Z. Plant Prot.* **2006**, *59*, 1–6.

12.   Coombe, B.G. Research on development and ripening of the grape berry. *Am. J. Enol. Viticult.* **1992**, *43*, 101–110.

13.   Molitor, D.; Biewers, B.; Junglen, M.; Schultz, M.; Clementi, P.; Permesang, G.; Regnery, D.; Porten, M.; Herzog, K.; Hoffmann, L.; et al. Multi-annual comparisons demonstrate differences in the bunch rot susceptibility of nine *Vitis vinifera* L. cv. Riesling clones. *Vitis* **2018**, *57*, 17–25.

14.   Keller, M. *The Science of Grapevines. Anatomy and Physiology*, 2nd ed.; Elsevier Academic Press: London, UK, 2015.

15.   Tello, J.; Ibanez, J. What do we know about grapevine bunch compactness? A state-of-the-art review. *Aust. J. Grape Wine Res.* **2017**, *24*, 6–23. [CrossRef]

16.   Hed, B.; Ngugi, H.K.; Travis, J.W. Relationship between cluster compactness and bunch rot in Vignoles grapes. *Plant Dis.* **2009**, *93*, 1195–1201. [CrossRef] [PubMed]

17.   Molitor, D.; Baus, O.; Hoffmann, L.; Beyer, M. Meteorological conditions determine the thermal-temporal position of the annual Botrytis bunch rot epidemic on *Vitis vinifera* L. cv. Riesling grapes. *Oeno One* **2016**, *50*, 231–244. [CrossRef]

18.   Molitor, D.; Behr, M.; Hoffmann, L.; Evers, D. Impact of grape cluster division on cluster morphology and bunch rot epidemic. *Am. J. Enol. Viticult.* **2012**, *63*, 508–514. [CrossRef]

19.   Molitor, D.; Junk, J. Climate change is implicating a two-fold impact on air temperature increase in the ripening period under Central European climate conditions. *Oeno One* **2019**. submitted.

20.   Junk, J.; Görgen, K.; Rapisarda, C.; Eickermann, M. Climate change and agriculture: Impact assessment for Luxembourg based on multi-model EURO-CORDEX simulations. *Annalen der Meteorologie* **2017**, *52*, 49.

21.   Molitor, D.; Keller, M. Yield of Müller-Thurgau and Riesling grapevines is altered by meteorological conditions in the current and the previous growing seasons. *Oeno One* **2016**, *50*, 245–258.

22.   Hill, G.; Stellwaag-Kittler, F.; Huth, G.; Schloesser, E. Resistance of grapes in different developmental stages to *Botrytis cinerea*. *Phytopathologische Zeitschrift* **1981**, *102*, 328–338. [CrossRef]

23.   Kretschmer, M.; Kassemeyer, H.H.; Hahn, M. Age-dependent grey mould susceptibility and tissue-specific defence gene activation of grapevine berry skins after infection by *Botrytis cinerea*. *J. Phytopathol.* **2007**, *155*, 258–263. [CrossRef]

24.   Latif, M. Uncertainty in climate change projections. *J. Geochem. Explor.* **2011**, *110*, 1–7. [CrossRef]

25. Jackson, D.I.; Lombard, P.B. Environmental and management practices affecting grape composition and wine quality—A review. *Am. J. Enol. Viticult.* **1993**, *44*, 409–430.

26. IPCC. Managing the risks of extreme events and disasters to advance climate change adaptation. In *A Special Report of Working Groups I and II of the Intergovernmental Panel on Climate Change*; Field, C.B., Barros, V., Stocker, T.F., Qin, D., Dokken, D.J., Ebi, K.I., Mastrandrea, M.D., Mach, K.J., Plattner, G.-K., Allen, S.K., et al., Eds.; Cambridge University Press: Cambrige, UK; New York, NY, USA, 2012; p. 582.

agronomy

MDPI

Article

Influence of Foliar Kaolin Application and Irrigation on Photosynthetic Activity of Grape Berries

Andreia Garrido [1], João Serôdio [2], Ric De Vos [3], Artur Conde [1,4] and Ana Cunha [1,4,5,*]

[1] Centre for the Research and Technology of Agro-Environmental and Biological Sciences (CITAB), University of Minho, Campus de Gualtar, 4710-057 Braga, Portugal; andreiagarrido@sapo.pt (A.G.); arturconde@bio.uminho.pt (A.C.)

[2] Centre for Environmental and Marine Studies (CESAM), University of Aveiro, Campus de Santiago, 3810-193 Aveiro, Portugal; jserodio@ua.pt

[3] Wageningen Plant Research, Wageningen University and Research Centre (Wageningen-UR), PO Box 16, 6700 Wageningen, The Netherlands; ric.devos@wur.nl

[4] Centre of Molecular and Environmental Biology (CBMA), University of Minho, Campus de Gualtar, 4710-057 Braga, Portugal

[5] Centre of Biological Engineering (CEB), University of Minho, Campus de Gualtar, 4710-057 Braga, Portugal

* Correspondence: accunha@bio.uminho.pt; Tel.: +351- 253-604-046

Received: 30 September 2019; Accepted: 24 October 2019; Published: 27 October 2019

Abstract: Climate changes may cause severe impacts both on grapevine and berry development. Foliar application of kaolin has been suggested as a mitigation strategy to cope with stress caused by excessive heat/radiation absorbed by leaves and grape berry clusters. However, its effect on the light micro-environment inside the canopy and clusters, as well as on the acclimation status and physiological responses of the grape berries, is unclear. The main objective of this work was to evaluate the effect of foliar kaolin application on the photosynthetic activity of the exocarp and seeds, which are the main photosynthetically active berry tissues. For this purpose, berries from high light (HL) and low light (LL) microclimates in the canopy, from kaolin-treated and non-treated, irrigated and non-irrigated plants, were collected at three developmental stages. Photochemical and non-photochemical efficiencies of both tissues were obtained by a pulse amplitude modulated chlorophyll fluorescence imaging analysis. The maximum quantum efficiency (F_v/F_m) data for green HL-grown berries suggest that kaolin application can protect the berry exocarp from light stress. At the mature stage, exocarps of LL grapes from irrigated plants treated with kaolin presented higher F_v/F_m and relative electron transport rates ($rETR_{200}$) than those without kaolin. However, for the seeds, a negative interaction between kaolin and irrigation were observed especially in HL grapes. These results highlight the impact of foliar kaolin application on the photosynthetic performance of grape berries growing under different light microclimates and irrigation regimes, throughout the season. This provides insights for a more case-oriented application of this mitigation strategy on grapevines.

Keywords: light micro-climates; mitigation strategies; kaolin; irrigation; *Vitis vinifera* L.; grape berry tissues; pulse amplitude modulated (PAM) fluorometry; photosynthesis; photosynthetic pigments

1. Introduction

Viticulture is a historically important agronomic and socio-economic sector in Portugal. According to the last report from the International Organization of Vine and Wine (OIV), Portugal is the 11th world and 5th European wine producer [1]. With 14 winemaking regions distributed throughout the country, the Vinhos Verdes or Minho region, as well as the Douro and Alentejo, are the major contributors for national exports and growth of this sector [2].

Grapevine is influenced by a complex and interacting system commonly called *terroir*, which, according to the OIV [3], includes specific soil, topography, climate, landscape characteristics and biodiversity features, and interaction with applied vitivini-cultural practices. This complex system influences the canopy microclimate and grapevine physiology and development and, consequently, grape berry quality and the organoleptic properties of its wine, which is typical of each region.

Currently, climate change projections point to a particularly pronounced temperature variation, with an overall increase of up to 3.7 °C by the end of this century, compared to the 1985–2005 reference period [4]. These temperature changes will have great impacts in the Mediterranean wine regions [5]. According to recent investigations using very high resolution bioclimatic zoning, both temperature and dryness are predicted to increase in several economically important Portuguese viticulture regions, including the Vinhos Verdes region [6]. Therefore, Portuguese vineyards will be subject to increased stress due to the interaction of the existing high radiation levels with the foreseen elevated air temperature and drought, which, all together, can have high impact on grapevine phenology, physiology, and productivity. Several of these climate impacts have already been reported, such as: earlier phenological timings and shortenings of the grapevine growing season [7], sunburns in leaves and grape berries [8], reduction of stomatal conductance and decrease of photosynthetic rates, either by stomatal and non-stomatal limitations [9], appearance and/or intensification of grapevine-related pests and diseases [10,11], increased grape sugar concentrations that lead to higher wine alcohol levels, lower acidities, and modification of varietal aroma compounds [12], and higher inter-annual yield and wine production variability [13].

In order to mitigate these adverse climate effects, new short-term measures have recently been implemented in Portuguese vineyards, such as smart irrigation [14,15] and foliar application of kaolin [16]. Vineyards are not traditionally irrigated and there are even restrictions on this practice in some regions, such as the Douro region [15]. However, according to a recent projection model, a 10% reduction in grapevine yield is expected in the Minho region if irrigation is not applied [14]. Kaolin is a white, chemically inert, and non-toxic clay material ($Al^2Si^2O^5(OH)^4$) that can reflect radiation, including photosynthetically active (PAR), ultraviolet (UV), and infrared radiation (IR) [16]. Foliar application of this mineral has become a cost-efficient mitigation strategy to cope with water stress and excessive heat/radiation absorbed by leaves and grape berry clusters, which also proves effective in alleviating negative impacts on grapevines [17–21]. However, the amount and spectral quality of light intercepted by leaves and transmitted/ reflected into the canopy, crucial factors for leaf, and fruit physiology and development [22] are also important aspects to consider when mitigation practices are used.

Previous work done by our group, using pulse amplitude modulated (PAM) chlorophyll fluorescence imaging, has mapped grape berry photosynthesis at a histological level, and revealed both the exocarp and the seed outer integument as the main photosynthetically competent tissues [23]. More recently, we have studied the photosynthetic performance of grape berry tissues from clusters growing at three distinct light microclimates in the canopy and observed microclimate-related differences in their photosynthetic capacity and acclimation status [24]. This led to the hypothesis that, if a specific viticulture practice changes the light reaching the clusters, and alters its light microclimate, it may impact the photosynthetic activity of berry tissues and associated tissue-specific biochemical processes. In fact, foliar kaolin application may have direct implications on light distribution at the whole canopy level, and irrigation is an indirect one, through increased vegetative growth. For instance, it has already been shown that kaolin application generally reduces the photosynthetic rates of individual leaves in other agricultural crops (e.g., apple, almond, and walnut canopies) [25,26], due to a 20%–40% increase in the reflection of PAR [27]. However, the photosynthesis of the whole canopy remained unaffected or even increased (9%), because of the better light distribution within the canopy [28–30]. In another study, decreased photosynthesis was observed in the inner leaves of irrigated grapevines due to higher vegetative growth [31]. While the function of photosynthesis in fruits is still poorly understood, it can be linked with primary and secondary metabolomic pathways [32,33]. Therefore, any effect on photosynthesis may impact grape berry development and composition.

Therefore, the main objective of the present work was to evaluate the effects of foliar kaolin application on the photosynthetic activity of grape berry tissues from clusters growing at two distinct microclimates, which include high light (HL) and low light (LL) microclimates, of irrigated and non-irrigated grapevines, during the season.

2. Materials and Methods

2.1. Site Description, Applied Treatments, and Sampling

Grape berry samples were collected in 2018 from field-grown 'Alvarinho' cultivar grapevines (*Vitis vinifera* L.) in the commercial vineyard *Quinta Cova da Raposa* in the Demarcated Region of Vinho Verde, Braga, Portugal (41°34′16.4″ N, 8°23′42.0″ W). The vineyard is managed by following standard cultural practices applied in organic farming, and is arranged in terraces along a granitic hillside with high drainage. The vine training system applied for this cultivar follows the settings of Sylvoz (Simple Ascending and Recumbent Cord). The sector selected for the trial was located on a hill with NW-SE orientation and the vineyard rows with a NE-SW orientation. The treatments applied were: kaolin (K) and non-kaolin (NK) application on leaves, and irrigation (I) and non-irrigation (NI), in a complete factorial design (four treatment combinations) with two blocks, each with three to four vines per combination treatment (Figure 1B). A suspension of 5% (w/v in water) kaolin (EPAGRO®, Sunprotect, Alverca do Ribatejo, Portugal) was applied twice on leaves on both sides of the rows. On July 6 and 27, corresponding to four weeks after anthesis (WAA) or BBCH-73 (BBCH-scale used for grapes by Lorenz et al. [34]) and seven WAA or BBCH-77, respectively. Irrigation of half of the plants, started on July 26 (seven WAA, BBCH-77), (Figure 1A,B). Water was applied by drip irrigation with one dripper per vine and a drip line placed approximately 80 cm above the soil. Irrigation occurred every three days, once a day either early in the morning or late in the afternoon, for 2 h, with an average dripper capacity of 5.5 ± 1.6 L h^{-1} ($n = 12$ randomly selected drippers, $\pm$ SD). Clusters with contrasting light exposure were also selected to harvest grape berries during their development. These were called low light (LL) and high light (HL) clusters. LL clusters grew in the shaded inner zones of the canopy, which were exposed only to diffuse, reflected, and transmitted light, while HL clusters were exposed to direct or reflected sunlight most of the day. Six independent sub-clusters (three per block), each containing 15–20 grape berries, were collected randomly from clusters growing at each of the experimental conditions (four treatments × 2 microclimates) from the southeast side of rows. Berries were harvested in the morning (9–10 a.m.) at three distinct developmental stages: Green (16 July, 6 WAA, BBCH-75), *Véraison* (29 August, 12 WAA, BBCH-83), and Mature (17 September, 15 WAA, BBCH-89). The material was transported in refrigerated boxes to the Center for Environmental and Marine Studies (CESAM) laboratory and used within 2–6 h for imaging fluorometry experiments. For other assays, berries were immediately frozen in liquid nitrogen and stored at −80 °C.

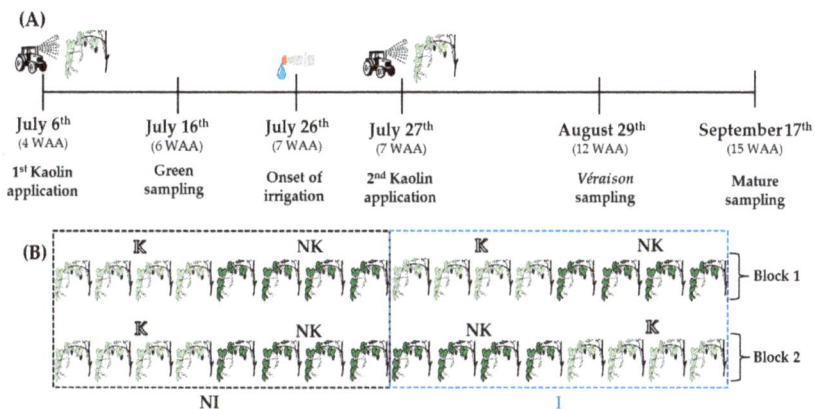

Figure 1. (**A**) Timeline of the grape growing season depicting the sampling times, foliar kaolin application dates, and the onset of irrigation. (**B**) Scheme of treatment combinations applied in the field: irrigation (I)/ non-irrigation (NI) × kaolin (K)/ non-kaolin (NK). (WAA - weeks after anthesis).

2.2. Light Intensity and Temperature Measurements for Microclimate Characterization

In order to characterize the microclimates in the vicinity of the growing clusters of all experimental conditions, light intensities and temperatures were registered on cloudless days (mean of 1500 ± 300 μmol photons m^{-2} s^{-1}), between 15 h and 17 h, at green and mature stages, as described by Garrido et al. [24]. The light intensity (μmol photons m^{-2} s^{-1}) was measured with a radiometer (LI-COR, LI-250 Light Meter, Lincoln, NE, USA) and the temperature (°C) was measured with an infrared digital thermometer (Infrared, DT8380, Beijing, China). Both parameters were determined in the frontal region of the clusters (LL and HL), at the southeast side of the row, and in full sun-exposed leaves. The devices were placed perpendicularly to the plant organ (cluster or leaf). The light sensor was placed on the organ surface facing outward, which registered the light intensity reaching at this point, and the thermometer was pointing to the organ at a distance of about 15 cm, which registered an average organ temperature. Sixteen replicate measurements per treatment were considered on randomly selected vines.

2.3. Kaolin Film Transmittance and Reflectance

Transmittance and reflectance spectra were obtained using a spectrometer (USB2000-VIS-NIR, grating #3, Ocean Optics, Duiven, The Netherlands), connected to a 400 mm-diameter fiber optic (QP400-2-VIS/NIR-BX; Ocean Optics), and recorded using the spectral acquisition software Spectra Suite (Ocean Optics, https://oceanoptics.com/). The transmittance spectrum was obtained by spreading a 5% (w/v) kaolin suspension prepared in 70% ethanol (allowing fast evaporation to prevent solvent interference), over a glass plate, which simulated particle distributions similar to those observed in the field. A halogen lamp was placed underneath to illuminate the spectrometer sensor positioned 3 cm above the glass plate. Different spectra were obtained from different areas randomly (*n* = 3). The reflectance spectrum was obtained, according to Dinis et al. [18], by pointing the fiber optics perpendicularly to the surface of collected leaves illuminated by the same halogen lamp. Three independent spectra were obtained from different leaf regions of both kaolin-treated and non-treated leaves. Transmittance and reflectance spectra were recorded for the 370–900 nm spectral range, with a spectral resolution of 0.33 nm. The transmittance spectra were expressed as a percentage of the controls (glass). Reflectance spectra were normalized to the spectrum reflected from a reference white panel (WS-1-SL Spectralon Reference Standard, Ocean Optics).

2.4. Chlorophyll Fluorescence Analysis

The photosynthetic activity of grape berry tissues was assessed as described by Garrido et al. [24]. For this, an imaging chlorophyll fluorometer was used (*Open FluorCAM 800 MF*; Photon Systems Instruments, Drásov, Czech Republic), which was comprised of four 13×13 cm LED panels emitting red light (emission peak at 621 nm, 40-nm bandwidth) and a 2/3 inch CCD camera (CCD381, Beijing, China) with a F1.2 (2.8–6 mm) objective. Two of the LED panels provided modulated measuring light (<0.1 μmol m^{-2} s^{-1}) and the other two provided saturating pulses (>7500 μmol m^{-2} s^{-1}, 0.8 s). Chlorophyll fluorescence images were captured and processed using FluorCam7 software (Photon Systems Instruments, Drásov, Czech Republic).

In a dark cabinet, exocarps and seeds were separated from dark-adapted berries and disposed in 8×8-well plates filled with water. Two independent plates were prepared for each microclimate (LL and HL), with each comprising all treatments and tissues. Exocarps and seeds were placed alternately in three rows each, using two columns per treatment, in a total of 12 biological replicates per condition and tissue ($n = 3 \times 2 \times 2 = 12$). Each plate was subjected to the experiments described below.

The maximum quantum efficiency of photosystem II [$F_v/F_m = (F_m - F_0)/F_m$], which is a chlorophyll fluorescence parameter that reflects the probability of electrons being transferred from the PSII reaction center for the transport chain of electrons by quanta absorbed [35,36], was computed following a saturation pulse (SP). The isolated tissues were then acclimated to an actinic light (AL) of 200 μmol photons m^{-2} s^{-1} for 15 min, and, after a new SP, the effective quantum yield of PSII [$\Phi_{II} = (F'_m - F_s)/F'_m$] was computed. This parameter correlated with the quantum yield of CO_2 fixation in a wide range of physiological conditions [37]. From Φ_{II} and PFR (photosynthetic photon fluence rates) (200 μmol photons m^{-2} s^{-1}), the relative electron transport rate through PSII (rETR$_{200}$ = $\Phi_{II} \times$ PFR) was calculated. Then, tissues were exposed to 1500 μmol photons m^{-2} s^{-1} for 15 min, with an SP being applied every 3 min. The last F'_m values (at 15 min) were used to calculate the non-photochemical quenching [NPQ = $(F_m - F'_m)/F'_m$].

2.5. Analysis of Chlorophylls and Carotenoids by High Performance Liquid Chromatography Coupled to A Photodiode Array Detector (HPLC-PDA)

The extraction procedure was adapted from Fraser et al. [38]. Freeze-dried material (20 mg) of grape berry tissues, which includes exocarp and seed, was extracted in 1.8 mL of chloroform/methanol (1:1) (chloroform - Emsure®, Darmstadt, Germany, methanol - Biosolve®, Dieuze, France) with both 0.1% (w/v) butylated hydroxytoluene (BHT, Sigma®, Zwijndrecht, The Netherlands) as an antioxidant and Sudan 1 (0.5 μg mL^{-1}) as the internal standard (IS). The samples were vortexed (10 s), kept on ice for 30 min (vortexed in between), and then sonicated for 15 min (Branson®, 3510 Ultrasonic Cleaner, Danbury, CT, USA). These steps were performed twice. After that, the samples were centrifuged at $16,100 \times g$ (Eppendorf®, Centrifuge 5415 R, Hamburg, Germany) and the supernatant (approx. 1200 μL) was transferred to a new Eppendorf tube with a perforated lid. The samples were dried for 1 h in a Speedvac (Savant®, SC100, Schiphol, The Netherlands) and then stored at -80 °C until the next steps. Prior to high performance liquid chromatography (HPLC) analysis, samples were dissolved in 200 μL ethylacetate solution containing 0.1% (w/v) BHT, sonicated (10 min), and then centrifuged as above. Samples were protected from light and kept on ice during all of these procedures. The supernatant (180 μL) was transferred to amber-colored 2 mL HPLC vials with a glass insert and sealed.

The HPLC-PDA procedure was adapted from Mokochinski et al. [39]. The samples (20 μL) were analyzed using an HPLC (Waters Alliance e2695 Separations Module, Milford, MA, USA) coupled to a photodiode array detector (PDA) (Waters 2996) over the 240 to 700 nm UV/Vis range. Separation was performed on a reverse-phase C$_{30}$ column (250 $\times$ 4.6 mm i.d., S-5 μm - YMC Carotenoid, Komatsu, Japan) kept at 35 °C with a flow rate of 1.0 mL min^{-1}. The compounds were identified based on comparisons of retention times and absorption spectra (240 to 700 nm) with authentic standards.

2.6. Statistical Analysis

Results were statistically analyzed using Analysis of Variance tests (two-way ANOVA), followed by post hoc multiple comparisons using the Bonferroni test whenever the factors had significant effects (GraphPad Prism version 5.00 for Windows, GraphPad Software, La Jolla, CA, USA). Significant differences ($p \leq 0.05$) between sample groups are indicated with different letters. Notation with an asterisk means that only one factor (kaolin or irrigation) was significant.

3. Results and Discussion

3.1. Climatic Conditions and Microclimate Characteristics

In order to characterize the climate during the growing season at the study site (Braga), we used the official information available from the *Instituto Português do Mar e da Atmosfera* (IPMA) [40], to determine the temperatures and total precipitation during 2018 (Figure S1). This growing season was atypical from a climatic point of view, with a relatively cold and extremely dry winter, which caused a delay of sprouting/flowering for two to three weeks [41], and a relatively cold spring with rainy periods during the vegetative growth of the grapevines.

To characterize the microclimates for the LL and HL berry clusters (two *a priori* selected distinct light microclimates within the canopy), measurements of light intensities and temperatures were performed at the cluster level, at two time points during the growing season, i.e., when the berries were still green (green stage) and, two months later, when the berries were at their mature stage of ripening (Figures 2 and 3). Figure 2 depicts the average light intensities at LL and HL clusters growing under the different experimental conditions: i.e., irrigation/non irrigation (Figure 2a,c) and with kaolin/without kaolin (Figure 2b,d). The two microclimates were clearly distinct at both ripening stages, with HL clusters receiving about three-fold more light than LL clusters. At the green stage, i.e., before the onset of irrigation (Figure 1A), no significant differences were detected between the two sets of plots assigned to the subsequent irrigation experiment (four plots for irrigated (I) plants, i.e., 2 × NK-I and 2 × K-I) and four plots for non-irrigated (NI) control plants (2 × NK-NI and 2 × K-NI), see Figure 1B) (Figure 2a), which reveals that there were no plot-related effects on microclimate light intensity. At this early ripening stage, and with the adopted measurement procedure, no differences were detected with respect to light intensities reaching the berry clusters due to foliar kaolin application (Figure 2b). However, at a mature stage, both irrigation and kaolin had a small but significant effect on the light intensity reaching the HL clusters (Figure 2c,d). Irrigation slightly reduced the light intensity, likely due to the better vegetative growth of the plants, while foliar kaolin application increased it, likely due to an increased light reflection to both the interior and lower levels of the canopy. In the LL clusters, these effects of irrigation and kaolin on light intensity were not observed, at this time of day.

HL grapes consistently experienced higher temperatures than LL ones (Figure 3), and both I and K treatments exerted significant and contrasting effects on this microclimate parameter, mainly at the mature stage. Again, and consistent with what was observed and discussed above for light intensity, no effect was detected for I treatment on the grape berry temperature at the green stage (before the onset of irrigation) (Figure 3a).

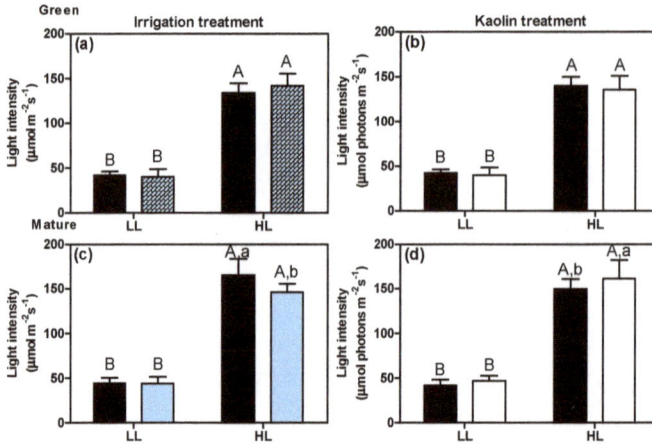

Figure 2. Light intensities received by LL and HL clusters at the green stage (**a,b**) and the mature stage (**c,d**), for plants with irrigation (blue columns, note: the textured blue columns at green stage i.e., before the onset of irrigation, represent the measurements in the plots that were later irrigated) and foliar kaolin application (white columns). Black columns correspond to the respective controls. Values represent means with a standard deviation ($n = 16$ plants). Statistical notation: per ripening stage, different capital letters refer to significant differences (two-way ANOVA, $p \leq 0.05$) between the two light microclimates within the same plant treatment, and different lowercase letters for differences between treatments within each light microclimate. If the respective factor did not have a significant effect, the lowercase letters were omitted.

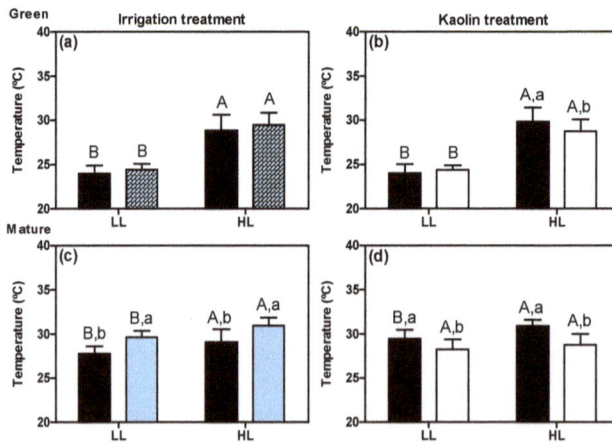

Figure 3. Temperatures of LL and HL clusters at the green stage (**a,b**) and the mature stage (**c,d**), for plants with irrigation (blue columns, note: the textured blue columns at the green stage i.e., before the onset of irrigation, represent the measurements in the plots that were later irrigated) and foliar kaolin application (white columns). Black columns correspond to the respective controls. Values represent means with standard deviation ($n = 16$ plants). Statistical notation: per ripening stage, different capital letters refer to significant differences (two-way ANOVA, $p \leq 0.05$) between the two light microclimates within the same plant treatment, and different lowercase letters for differences between treatments within each light microclimate. If the respective factor did not have a significant effect, the lowercase letters were omitted.

The Kaolin application led to a significant decrease in the temperature of the HL clusters at both green and mature stages (e.g., at this latter stage—from 31 °C to 28.7 °C), and of the LL clusters at the mature stage only (e.g., from 29.5 °C to 28 °C) (Figure 3b,d). The fact that LL clusters' temperature at the green stage were not affected by kaolin was likely related to the relatively low air temperature at this time of the growing season (early July, Figure S1). Thus, kaolin application on the leaves may increase the incident light by increasing the light reflection inside the canopy (Figure 2d), while maintaining a cooler microclimate for the growing berries, especially during the hot summer days, independently of the irrigation regime. This demonstrates one of the advantages of this mitigation strategy at the grape berry level. The kaolin solution sprayed on the leaves also resulted in leaf temperature reduction (Figure S2d). The present results are in agreement with previous studies in both grapevine leaves and berries [42,43], as well as other crops [26,27]. Therefore, it is likely that kaolin applied to leaves provides cooler temperatures throughout the grapevine by reducing the total amount of radiation transmitted into the canopy. This is also shown by thermal imaging in apple trees [30]. Furthermore, different training systems of the vineyard might influence the light intensities and temperatures inside the canopy [44,45]. For instance, the vine canopy was denser in our previous study in 2015 [24], which resulted in an LL microclimate characterized by much lower light reaching the clusters compared to that in the present study, with major impacts on grape berry photosynthetic competence.

At the mature stage, plant irrigation resulted in a significant increase in the grape temperature of both LL and HL clusters (Figure 3c). This increase was unexpected, since a previous study reported lower berry temperature as a result of irrigation, rather than a higher temperature [46]. This response is very interesting and clearly, additional studies, which are controlling/measuring the soil temperature in the rhizosphere, are required to determine the effect of irrigation and the irrigation procedure on the temperature of grape clusters.

3.2. Kaolin Film Transmittance and Reflectance Spectral Properties

To characterize the potential effects of kaolin applied to the leaves with regard to light intensity and quality, we performed transmittance and reflectance studies. For this purpose, the transmittance spectrum of a film of kaolin solution (5% w/v) on a glass plate was determined (Figure 4a). Although a high percentage of most photosynthetically active radiation (PAR) wavelengths is transmitted by the kaolin film, the blue light range is the least transmitted. The reflectance spectra obtained for leaves of grape plants sprayed with and without kaolin are represented in Figure 4b. Our results showed a relevant percentage of PAR is reflected by this white mineral, as compared to non-spayed (NK) vine leaves, rather than exclusively or mainly reflecting in the ultraviolet (UV) and infrared radiation (IR) ranges [47]. Additionally, kaolin was more efficient in reflecting UV light than IR light (in the measured ranges). These results are in accordance with previous studies using grapevine leaves [18] and other crops [27,28,48,49]. Thus, the beneficial effect of kaolin application is related to the reflection of excess radiation outwards, which reduces the risk of light stress-induced damage to leaves and fruit [47], while transmitting a very significant proportion of PAR.

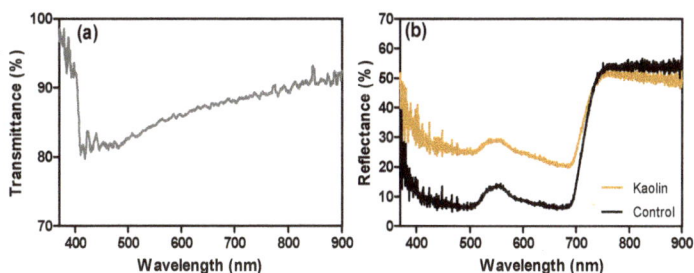

Figure 4. Transmittance (%) spectrum (**a**) of a kaolin suspension (5% w/v) and reflectance (%) spectra (**b**) of leaves with and without kaolin (control) (*n* = 3).

Together, these results call attention to the fact that foliar kaolin application may directly impact the photosynthesis of the sprayed leaves but also have an indirect effect on the non-sprayed leaves and grape berry clusters inside the canopy.

3.3. Effects on Berry Photosynthesis and Photosynthetic Pigments

3.3.1. Maximum Quantum Efficiency of PSII

The maximum quantum efficiency of PSII (F_v/F_m) was determined *ex planta* under controlled conditions, using both exocarps and seeds from grape berries grown under the different treatments, microclimates, and three ripening stages (Figure 5). At the green stage, F_v/F_m was similar in both berry tissues (Figure 5a,b). Upon further ripening, the exocarp kept its F_v/F_m values (~0.7), while the seeds showed a significant decrease in this parameter, which reached values around 0.4–0.5 at the mature stage, which was in accordance with what was reported by Garrido et al. [24].

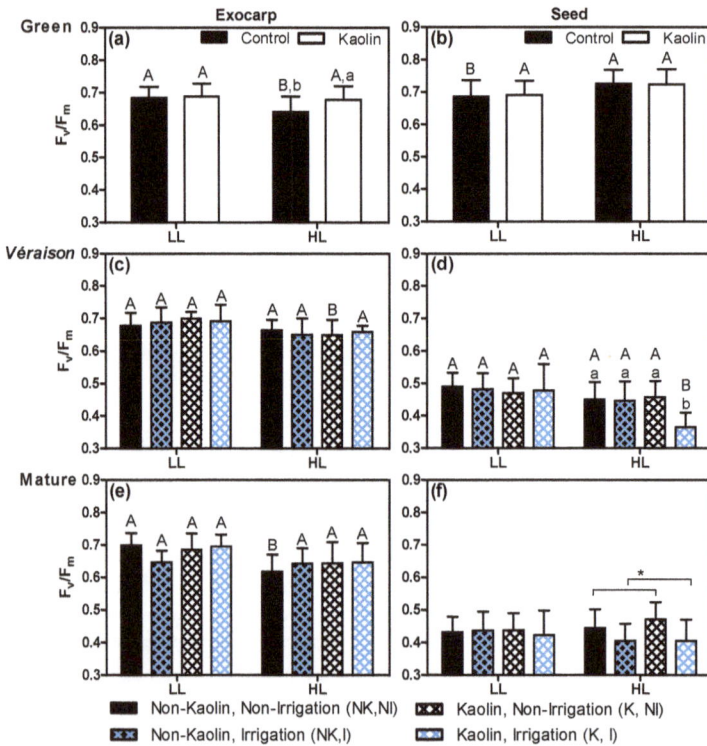

Figure 5. Maximum quantum efficiency of PSII (F_v/F_m) mean values (*n* = 12–24 berries, +SD) of exocarps and seed integuments obtained from dark-adapted LL and HL grape berries grown under the four combinations of the two treatments applied: irrigation (I)/ non-irrigation (NI) × kaolin (K)/ non-kaolin (NK). Samples were collected at three development stages (green, véraison, and mature). Statistical notation: for each developmental stage, capital letters refer to differences between light microclimates within the same treatment combination, and lowercase letters refer to differences between treatment combinations within each light microclimate (mean values with a common letter were not significantly different). When capital and lowercase letters are omitted, the respective factor did not have a significant effect (two-way ANOVA *p* > 0.05). Notation with an asterisk means that only one factor (kaolin or irrigation) was significant.

The cluster microclimate (LL vs. HL) had a significant effect on the F_v/F_m values in both tissues and all developmental stages, with the exception of seeds at the mature stage, but exocarps and seed integuments responded differently to a light microclimate [24]. At the green stage, the exocarps from berries under control conditions, showed lower F_v/F_m values in HL clusters, while their seeds showed significantly higher values than those from LL berries (Figure 5a,b). This was likely due to their inner location where the light transmitted through the skin and flesh tissues reaches values as low as 2% of the incident photon flux density (PFD) [50], which eventually translates a light limitation effect in LL clusters. These microclimate effects were more or less maintained in exocarps upon subsequent ripening (see NK + NI Figure 5a,c,e, while not being significant in véraison), while, in seeds, the difference between LL and HL clusters disappeared (Figure 5b,d,f), likely related to the (large) intrinsic ripening-dependent decrease in photosynthetic competence of this tissue.

Both kaolin and irrigation treatments of plants differentially influenced the F_v/F_m values of the two berry tissues, with the seeds globally more responsive than exocarps, particularly to the irrigation treatment, which induced a significant effect at véraison and mature stages (Figure 5d,f). On the other hand, at these latter stages, no effects from treatments were detected on the F_v/F_m of exocarps. This tissue only responded to kaolin treatment at the green stage, where HL berries showed an increment (6%) on F_v/F_m by kaolin application (Figure 5a). In fact, the decrease in F_v/F_m values of exocarps when comparing LL with HL berries at a green stage in control conditions (Figure 5a) had already been observed [24], which revealed that microclimates with higher luminosity can decrease F_v/F_m values of exocarps at this stage. Together, these results suggest that foliar kaolin may protect the berry exocarp from excess light at a stage when the grape berry photosynthetic phenotype is still developing [24]. In HL seeds, the most prominent effect was a decrease in F_v/F_m in irrigated-treated plants (Figure 5d,f). However, at véraison, this effect was observed only in kaolin-treated plants (Figure 5d). This apparent paradoxical effect may be related with the increased temperatures observed in irrigated grape berries in the hottest months (Figure 3c and Figure S1). In fact, the inhibitory effect of higher temperatures on F_v/F_m was already reported for grapevine leaves [51].

3.3.2. Relative Rate of Electron Transport Through PSII (rETR$_{200}$)

The relative electron transport rate through photosystem II (rETR) was determined after acclimation of the berry tissues to an AL intensity of 200 μmol photons m^{-2} s^{-1} (rETR$_{200}$) in order to simulate average field light conditions.

At the green stage, the rETR$_{200}$ was higher in HL grape berries than in LL berries (both in control and kaolin-treated plants), especially in seeds (Figure 6a,b). Interestingly, a positive influence of kaolin application was observed in LL exocarps at this stage (Figure 6a), which suggests that more PAR reflected by kaolin is reaching the inside of the canopy, which may improve the exocarp photosynthesis of berries in shaded microclimates. At the véraison stage, this positive effect of kaolin was not detected, while it was evident at the mature stage (Figure 6e). In addition, a clear decrease in rETR$_{200}$ was observed in seeds, especially in HL clusters ($p < 0.0001$) during ripening (Figure 6b,d,f), which is similar to the results obtained for F_v/F_m. In addition, in seeds from HL-grown berries at the véraison stage of irrigated plants, the kaolin treatment led to a significant decrease in rETR$_{200}$ (Figure 6d: NK, I compared to K, I), which was observed for the parameter F_v/F_m (Figure 5d). This pointed to a possible effect of berry temperature on seed photosynthesis. No reports were found for the effect of temperature on grape berry photosynthesis, but a continuous four-day exposure to high temperatures (38–40 °C) led to a decrease in photosynthetic activity of grapevine leaves [52–54].

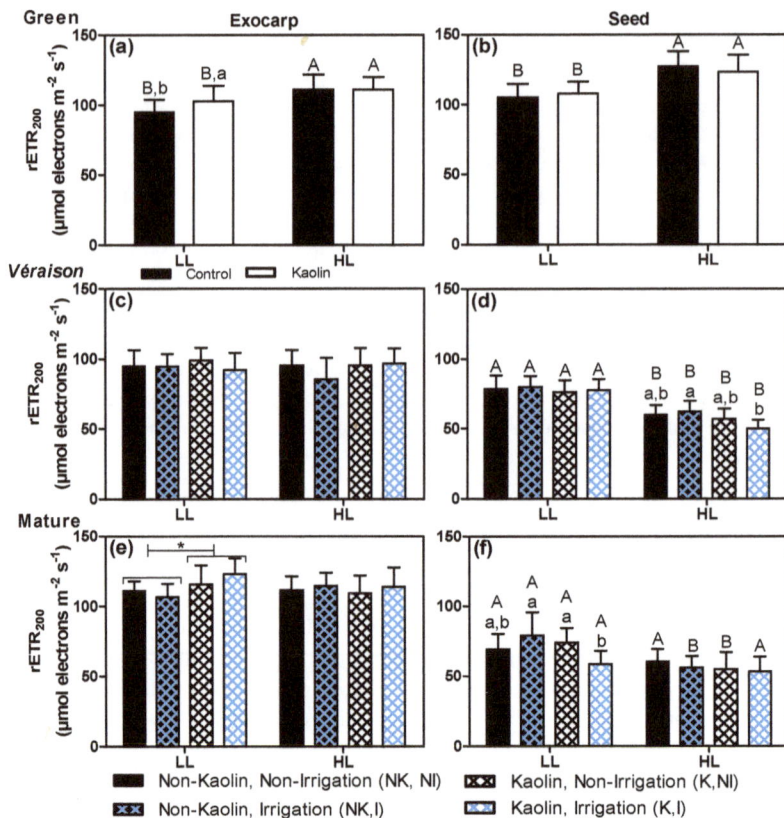

Figure 6. Mean values (n = 12–24 berries, +SD) of relative rate of electron transport through PSII at 200 µmol photons m^{-2} s^{-1} (rETR$_{200}$). All the microclimate conditions, treatment combinations and statistical information are the same as in Figure 5.

3.3.3. Non-Photochemical Quenching

A major component of non-photochemical quenching (NPQ) is the primary protective mechanism against light-induced photoinhibition, which involves various processes dissipating excessive non-radiative energy [55], including the xanthophylls cycle (e.g., as shown in grapevines, [56]) and phosphorylation/dephosphorylation of light harvesting complexes [57].

The NPQ results are represented in Figure 7. When comparing the berry tissues, it can be concluded that the exocarp tissue consistently exhibits roughly two-fold higher NPQ values than seeds. This result suggests that the exocarp exhibits more developed mechanisms of photoprotection, which is consistent with the fact that it is an external, more exposed tissue. During berry ripening, NPQ values decreased in both tissues, especially in seeds. Seeds attained very low NPQ values at later stages, which is in line with their F_v/F_m (Figure 5b,d,f) and rETR$_{200}$ (Figure 6b,d,f) profiles, which likely reflect the normal ripening-related loss of photosynthetic functioning of seeds. For the exocarps, dissipation or quenching mechanisms other than NPQ may explain the result, since this tissue maintains high photosynthetic activity until the mature stage (Figure 6a,c,e). Accumulation of carotenoids in white berries ('Sauvignon Blanc') was also increased in response to increasing levels of solar light in the canopy, which shows that the berries utilize these photosynthesis-related pigments in photo-acclimation responses and/or as "sunscreens" [58].

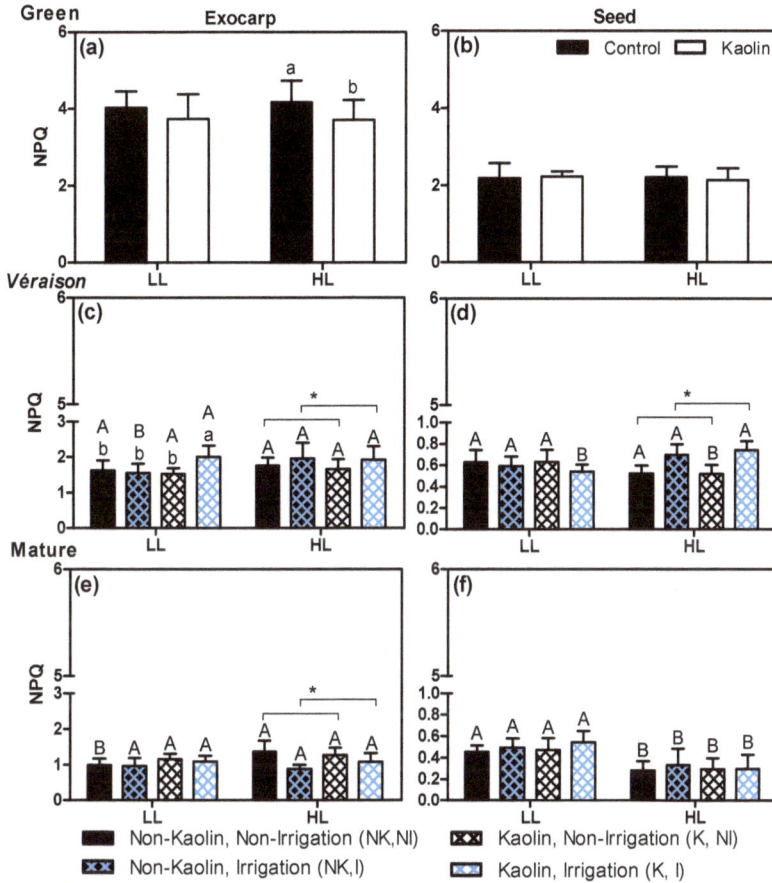

Figure 7. Non-photochemical quenching (NPQ) mean values (n = 12–24 berries, +SD). All the microclimate conditions, treatment combinations, and statistical information are the same as in Figure 5.

Regarding the effect of treatments, we found that the foliar kaolin application promoted lower NPQ in HL exocarps at the green stage, when compared with their NK controls (Figure 7a), which suggests that kaolin helps to protect these HL berries from excessive radiation absorption. This is similar to the F_v/F_m results (Figure 5a).

At the véraison stage, both HL exocarps and HL seeds showed increased NPQ in irrigated-treated grape berries (Figure 7c,d). In LL exocarps, NPQ was also increased in K,I berries, when compared to the remaining treatment combinations (Figure 7c). This increase in NPQ values of grape berries in irrigated plants, was lost at the mature stage, even though it is important to note that NPQ values were already very low at this stage (Figure 7e,f). This irrigation-related feature had already been observed with other parameters and is discussed above. The increased temperatures registered at later developmental stages of grape berries (Figure 3c), can impose other limitations or impairments, and, thus, recruit more energy-dissipation by NPQ, and eventually by other dissipative mechanisms.

3.3.4. Photosynthetic Pigments

To better evaluate the impact of foliar kaolin application on the light microclimate of grape berry clusters and its relationship with berry photosynthesis, and non-photochemical mechanisms, photosynthetic pigments were quantified in exocarps and seeds of both LL and HL-exposed grapes.

Results obtained for the green stage are depicted in Figure 8 (for later stages, see Supplementary Materials). At control conditions, the HL berries had higher levels of both chlorophylls and carotenoids than LL berries, in both tissues. Additionally, and in line with the $rETR_{200}$ results (Figure 6a), kaolin application resulted in a marked increase by 26% in chlorophylls and 82% in carotenoids content in exocarps from LL berries (Figure 8a,c), which support the idea that more light reached the inner parts of the kaolin-sprayed canopy. This is fundamental to build the photosynthetic machinery [59]. During ripening, the photosynthetic pigments decrease in both tissues and especially in the seed integuments (Figures S3 and S4) and no consistent and conspicuous effects by combined mitigation treatments were observed (Figure S3).

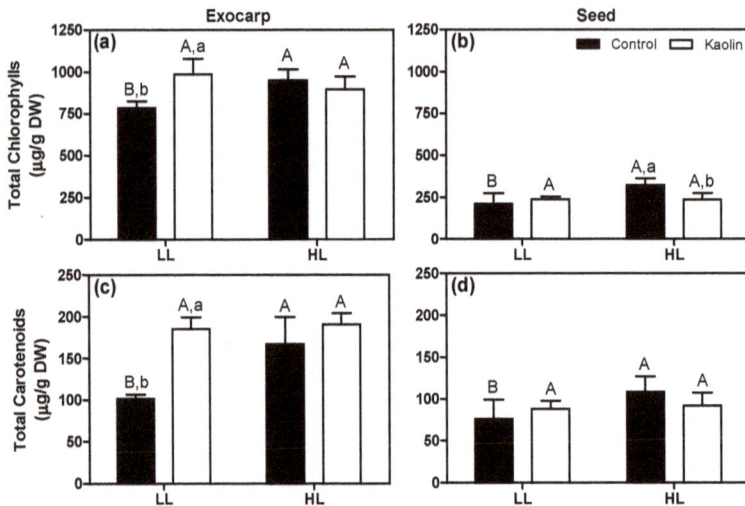

Figure 8. Chlorophylls (**a,b**) and carotenoids (**c,d**) concentration mean values ($n = 3$, +SD) of exocarps and seeds obtained from LL and HL grape berries grown under non-kaolin (black columns) and kaolin (white columns) application, and collected at the green stage. Statistical notation: different capital letters refer to significant differences (two-way ANOVA, $p \leq 0.05$) between the two light microclimates within the same plant treatment, and different lowercase letters to differences between treatments within each light microclimate. If the respective factor did not have a significant effect, the letters were omitted.

In addition, and supporting the view discussed above, the higher grape berry temperature was registered in irrigated treatments at later developmental stages (Figure 3c), by imposing physiological impairments. The temperature recruits more energy-dissipation by NPQ (Figure 7), which is the fact that carotenoids contents (Figure S4), but not chlorophylls (Figure S3), were also increased by irrigation treatment, for both tissues at the véraison stage.

Overall, the results obtained by pulse amplitude modulated fluorometry showed that, for the external tissue, exocarp, and foliar kaolin application led to an increase of F_v/F_m (Figure 5a, HL), $rETR_{200}$ (Figure 6a,e LL), and a reduction in non-photochemical quenching (Figure 7a, HL). To our best knowledge, this is the first work assessing the impact of foliar kaolin application on photochemical and non-photochemical functions in grape berries. Recently, it was verified that grapevine leaves with kaolin display the same response, i.e., an increase in F_v/F_m, Φ_{II}, and ETR, and a decrease in NPQ [18,60]. Similar results were also reported for olive leaves [61]. In this way, and in terms of photochemical processes, those kaolin-treated leaves have lower photo-inhibitory damage [17,62], and the open PSII reaction centers captured the light absorbed by PSII antenna more efficiently [17,35]. This response was likely due to a reduced loss of excitation energy by thermal dissipation, which could compete with its transference to PSII reaction centers, as shown by the lower NPQ values [17,35].

For exocarps of grape berries growing in inner parts of the canopy (LL microclimate), the photosynthetic results revealed that foliar kaolin application, may cause an extra "sunscreen" effect, and did not have a negative effect on those parameters, which we conjectured in our previous work [24]. The increased reflection provided by this mineral to inner parts of the canopy allowed good photochemical performance of LL exocarps, which is contrary to what we hypothesized in our previous work [24]. This contributes to higher carbon gains at the whole canopy level and also at the fruit level.

Regarding the results for the seed integument (internal organ), the positive effects of kaolin were observed mainly in non-irrigated plants such as an increase in F_v/F_m (Figure 5f, HL) and a decrease in NPQ (Figure 7d, HL). In more temperate or Mediterranean regions, this seems like a positive effect, but these results also show the importance of the irrigation system. The interaction between kaolin application and irrigation treatments on grapevine leaves have been studied before [42,63–66]. However, based on our knowledge, no study has approached the impacts on photosynthetic activity at the grape berry level, using chlorophyll fluorescence analysis.

4. Conclusions

The purpose of the current study was to assess the effects of foliar application of kaolin and irrigation, as abiotic stress mitigation strategies, on the photosynthetic activity of exocarps (skins) and seeds of grape berries growing under different light microclimates in the canopy. One of the most relevant findings was that the kaolin applied to leaves increased the photosynthetic activity of both exocarps and seed integuments of berries growing under low light conditions in the canopy. This is likely due to higher reflection of PAR to the inner zones. We believe, though, that the beneficial effects will depend on the canopy structure and on the incident radiation, with denser canopies and higher radiations conferring higher overall photosynthetic gains. Somewhat puzzling was the observation that seeds of irrigated plants showed lower photosynthetic activities, in the véraison and mature stages, especially under kaolin treatment. Several causes may explain this unexpected phenomenon, so more detailed and ad-hoc design studies should be conducted to address this relevant finding.

This comprehensive study provides the first evidence of foliar kaolin application as a procedure allowing the modulation of photosynthesis in the grape berry, but also calls attention to the importance of the irrigation system. In this way, this knowledge can be used by farmers to support their decisions concerning sustainable adaptation strategies applied on vineyards. Research to unveil the function of berry tissues' photosynthesis on the metabolome of the grapes is already underway, which ultimately contributes to the final quality of the fruit and wine.

Supplementary Materials: The following are available online at http://www.mdpi.com/2073-4395/9/11/685/s1. Figure S1. Meteorological elements from IPMA Institute from Braga city. (a) Temperature (°C) maximal, average, and minimal. (b) Total precipitation (mm). Figure S2. Temperatures of full exposed leaves at the green stage (a,b) and the mature stage (c,d), for plants with irrigation (blue columns, note: the textured blue columns at green stage i.e., before the onset of irrigation, represent the measurements in the plots that were later irrigated) and foliar kaolin application (white columns). Black columns correspond to the respective controls. Values represent means with standard deviation ($n = 16$ plants). Statistical notation: per ripening stage, different lowercase letters refer to significant differences ($p \leq 0.05$) between treatments. Whenever letters are omitted, it means that the respective factor did not have a significant effect. Figure S3. Chlorophylls concentration mean values ($n = 3$, +SD) of exocarp and seed obtained from LL and HL grape berries grown under the four combinations of the two treatments applied: irrigation (I)/ non-irrigation (NI) × kaolin (K)/ non-kaolin (NK). Samples were collected at three development stages (green, véraison, and mature). Statistical notation: for each developmental stage, capital letters refer to differences between light microclimates within the same treatment combination, and lowercase letters refer to differences between treatment combinations within each light microclimate (mean values with a common letter were not significantly different). When capital and lowercase letters are omitted, the respective factor did not have a significant effect (two-way ANOVA $p > 0.05$). Figure S4. Carotenoids concentration mean values ($n = 3$, +SD) of exocarp and seeds. All the microclimate conditions, treatment combinations, and statistical information are the same as in Figure S3.

Author Contributions: Conceptualization, A.G., A.C. (Artur Conde), R.V., and A.C. (Ana Cunha) Methodology, A.G., J.S., and A.C. (Ana Cunha) Formal analysis, A.G. and A.C. (Ana Cunha). Investigation, A.G. and A.C. (Ana Cunha). Resources, J.S. and R.V. Writing—original draft preparation, A.G. Writing—review and editing, A.C. (Ana Cunha), J.S., R.V., and A.C. (Artur Conde). Supervision, A.C. (Artur Conde), R.V., and A.C. (Ana Cunha) Project administration, A.C. (Ana Cunha).

Funding: The FCT-Portuguese Foundation for Science and Technology by the grant provided to Andreia Garrido (PD/BD/128275/2017), under the Doctoral Program "Agricultural Production Chains – from fork to farm" (PD/00122/2012), funded this research and APC.

Acknowledgments: The National Funds by FCT - Portuguese Foundation for Science and Technology, under the strategic programmes UID/AGR/04033/2019 and UID/BIA/04050/2019, and the project "INTERACT - VitalityWine - NORTE-01-0145-FEDER-000017 – funded by Norte2020 supported the work. The FCT and FEDER/COMPETE/POCI - Operational Competitiveness and Internationalization Program, under Project the projects MitiVineDrought – PTDC/BIA-FBT/30341/2017 (POCI-01-0145-FEDER-030341), and POCI-01-0145-FEDER-006958 also supported this work. Artur Conde was supported with a post-doctoral fellow of the mentioned INTERACT/VitalityWine project with the Reference BPD/UTAD/INTERACT/VW/218/2016, and also supported by a post-doctoral researcher contract/position within the project "MitiVineDrought" (PTDC/BIA-FBT/30341/2017 and POCI-01-0145-FEDER-030341). This work also benefited from the networking activities within the European Union-funded COST Action CA17111 – "INTEGRAPE - Data Integration to maximize the power of omics for grapevine improvement". Authors acknowledge the owner from *Quinta Cova da Raposa*, Manuel Taxa, who provided the samples, Susana Chaves (from CBMA) for her English grammar revision, and also all support given by the Biology Department of the School of Sciences from the University of Minho.

Conflicts of Interest: The authors declare no conflict of interest.

References

1.   *International Organisation of Vine and Wine Statistical Report on World Vitiviniculture*; OIV: Paris, France, 2018.
2.   Lavrador da Silva, A.; João Fernão-Pires, M.; Bianchi-de-Aguiar, F. Portuguese vines and wines: Heritage, quality symbol, tourism asset. *Ciência Técnica Vitivinícola* **2018**, *33*, 31–46. [CrossRef]
3.   OIV. *Definition of Vitivinicultural "Terroir"*; OIV: Paris, France, 2010.
4.   IPCC (International Panel on Climate Change). *Special Report on Climate Change, Desertification, Land Degradation, Sustainable Land Management, Food Security, and Greenhouse gas fluxes in Terrestrial Ecosystems. Summary for Policymakers*; IPCC: London, UK, 2019.
5.   Ferrise, R.; Moriondo, M.; Trombi, G.; Miglietta, F.; Bindi, M. Climate Change Impacts on Typical Mediterranean Crops and Evaluation of Adaptation Strategies to Cope With. In *Advances in Global Change Research*; Springer: Dordrecht, The Netherlands, 2013.
6.   Fraga, H.; Malheiro, A.C.; Moutinho-Pereira, J.; Jones, G.V.; Alves, F.; Pinto, J.G.; Santos, J.A. Very high resolution bioclimatic zoning of Portuguese wine regions: Present and future scenarios. *Reg. Environ. Chang.* **2014**, *14*, 295–306. [CrossRef]
7.   Fraga, H.; Santos, J.A.; Moutinho-Pereira, J.; Carlos, C.; Silvestre, J.; Eiras-Dias, J.; Mota, T.; Malheiro, A.C. Statistical modelling of grapevine phenology in Portuguese wine regions: Observed trends and climate change projections. *J. Agric. Sci.* **2016**, *154*, 795–811. [CrossRef]
8.   Van Leeuwen, C.; Darriet, P. The Impact of Climate Change on Viticulture and Wine Quality. *J. Wine Econ.* **2016**, *11*, 150–167. [CrossRef]
9.   Moutinho-Pereira, J.; Magalhães, N.; Gonçalves, B.; Bacelar, E.; Brito, M.; Correia, C. Gas exchange and water relations of three *Vitis vinifera* L. cultivars growing under Mediterranean climate. *Photosynthetica* **2007**, *45*, 202–207. [CrossRef]
10. Caffarra, A.; Rinaldi, M.; Eccel, E.; Rossi, V.; Pertot, I. Modelling the impact of climate change on the interaction between grapevine and its pests and pathogens: European grapevine moth and powdery mildew. *Agric. Ecosyst. Environ.* **2012**, *148*, 89–101. [CrossRef]
11. Bois, B.; Zito, S.; Calonnec, A.; Ollat, N. Climate vs. grapevine pests and diseases worldwide: The first results of a global survey. *J. Int. Des Sci. La Vigne Du Vin* **2017**, *51*, 133–139. [CrossRef]
12. Mira de Orduña, R. Climate change associated effects on grape and wine quality and production. *Food Res. Int.* **2010**, *43*, 1844–1855. [CrossRef]
13. Cunha, M.; Richter, C. The impact of climate change on the winegrape vineyards of the Portuguese Douro region. *Clim. Chang.* **2016**, *138*, 239–251. [CrossRef]

14. Fraga, H.; García de Cortázar Atauri, I.; Santos, J.A. Viticultural irrigation demands under climate change scenarios in Portugal. *Agric. Water Manag.* **2018**, *196*, 66–74. [CrossRef]

15. Costa, J.M.; Vaz, M.; Escalona, J.; Egipto, R.; Lopes, C.; Medrano, H.; Chaves, M.M. Modern viticulture in southern Europe: Vulnerabilities and strategies for adaptation to water scarcity. *Agric. Water Manag.* **2016**, *164*, 5–18. [CrossRef]

16. Brito, C.; Dinis, L.T.; Moutinho-Pereira, J.; Correia, C. Kaolin, an emerging tool to alleviate the effects of abiotic stresses on crop performance. *Sci. Hortic. (Amst.)* **2019**, *250*, 310–316. [CrossRef]

17. Dinis, L.T.; Ferreira, H.; Pinto, G.; Bernardo, S.; Correia, C.M.; Moutinho-Pereira, J. Kaolin-based, foliar reflective film protects photosystem II structure and function in grapevine leaves exposed to heat and high solar radiation. *Photosynthetica* **2016**, *54*, 47–55. [CrossRef]

18. Dinis, L.T.; Malheiro, A.C.; Luzio, A.; Fraga, H.; Ferreira, H.; Gonçalves, I.; Pinto, G.; Correia, C.M.; Moutinho-Pereira, J. Improvement of grapevine physiology and yield under summer stress by kaolin-foliar application: Water relations, photosynthesis and oxidative damage. *Photosynthetica* **2018**, *56*, 641–651. [CrossRef]

19. Conde, A.; Neves, A.; Breia, R.; Pimentel, D.; Dinis, L.T.; Bernardo, S.; Correia, C.M.; Cunha, A.; Gerós, H.; Moutinho-Pereira, J. Kaolin particle film application stimulates photoassimilate synthesis and modifies the primary metabolome of grape leaves. *J. Plant Physiol.* **2018**, *223*, 47–56. [CrossRef] [PubMed]

20. Dinis, L.T.; Bernardo, S.; Conde, A.; Pimentel, D.; Ferreira, H.; Félix, L.; Gerós, H.; Correia, C.M.; Moutinho-Pereira, J. Kaolin exogenous application boosts antioxidant capacity and phenolic content in berries and leaves of grapevine under summer stress. *J. Plant Physiol.* **2016**, *191*, 45–53. [CrossRef] [PubMed]

21. Conde, A.; Pimentel, D.; Neves, A.; Dinis, L.-T.; Bernardo, S.; Correia, C.M.; Gerós, H.; Moutinho-Pereira, J. Kaolin Foliar Application Has a Stimulatory Effect on Phenylpropanoid and Flavonoid Pathways in Grape Berries. *Front. Plant Sci.* **2016**, *7*, 1–14. [CrossRef]

22. Poni, S.; Gatti, M.; Palliotti, A.; Dai, Z.; Duchêne, E.; Truong, T.T.; Ferrara, G.; Matarrese, A.M.S.; Gallotta, A.; Bellincontro, A.; et al. Grapevine quality: A multiple choice issue. *Sci. Hortic. (Amst.)* **2018**, *234*, 445–462. [CrossRef]

23. Breia, R.; Vieira, S.; Da Silva, J.M.; Gerós, H.; Cunha, A. Mapping grape berry photosynthesis by chlorophyll fluorescence imaging: The effect of saturating pulse intensity in different tissues. *Photochem. Photobiol.* **2013**, *89*, 579–585. [CrossRef]

24. Garrido, A.; Breia, R.; Serôdio, J.; Cunha, A. Impact of the Light Microclimate on Photosynthetic Activity of Grape Berry (*Vitis vinifera*): Insights for Radiation Absorption Mitigations' Measures. In *Theory and Practice of Climate Adaptation*; Springer: Cham, Switzerland, 2018; pp. 419–441.

25. Le Grange, M.; Wand, S.J.E.; Theron, K.I. Effect of kaolin applications on apple fruit quality and gas exchange of apple leaves. *Acta Hortic.* **2004**, *636*, 545–550. [CrossRef]

26. Wünsche, J.-N.; Lombardini, L.; Greer, D.H.; Palmer, J.W. "Surround" particle film applications—The effect on whole canopy physiology of apple. *Acta Hortic.* **2004**, *636*, 565–571. [CrossRef]

27. Rosati, A.; Metcalf, S.G.; Buchner, R.P.; Fulton, A.E.; Lampinen, B.D. Physiological effects of kaolin applications in well-irrigated and water-stressed walnut and almond trees. *Ann. Bot.* **2006**, *98*, 267–275. [CrossRef] [PubMed]

28. Rosati, A.; Metcalf, S.G.; Buchner, R.P.; Fulton, A.E.; Lampinen, B.D. Effects of Kaolin application on light absorption and distribution, radiation use efficiency and photosynthesis of almond and walnut canopies. *Ann. Bot.* **2007**, *99*, 255–263. [CrossRef] [PubMed]

29. Glenn, D.M.; Puterka, G.J. The use of plastic films and sprayable reflective particle films to increase light penetration in apple canopies and improve apple color and weight. *HortScience* **2007**, *42*, 91–96. [CrossRef]

30. Glenn, D.M. Particle Film Mechanisms of Action That Reduce the Effect of Environmental Stress in 'Empire' Apple. *J. Am. Soc. Hortic. Sci.* **2009**, *134*, 314–321. [CrossRef]

31. Escalona, J.M.; Flexas, J.; Bota, J.; Medrano, H. Distribution of leaf photosynthesis and transpiration within grapevine canopies under different drought conditions. *Vitis* **2003**, *42*, 57–64.

32. Obiadalla-Ali, H.; Fernie, A.R.; Lytovchenko, A.; Kossmann, J.; Lloyd, J.R. Inhibition of chloroplastic fructose 1,6-bisphosphatase in tomato fruits leads to decreased fruit size, but only small changes in carbohydrate metabolism. *Planta* **2004**, *219*, 533–540. [CrossRef]

33. Cocaliadis, M.F.; Fernández-Muñoz, R.; Pons, C.; Orzaez, D.; Granell, A. Increasing tomato fruit quality by enhancing fruit chloroplast function. A double-edged sword? *J. Exp. Bot.* **2014**, *65*, 4589–4598. [CrossRef]

34. Lorenz, D.H.; Eichhorn, K.W.; Bleiholder, H.; Klose, R.; Meier, U.; Weber, E. Growth Stages of the Grapevine: Phenological growth stages of the grapevine (*Vitis vinifera* L. ssp. *vinifera*)—Codes and descriptions according to the extended BBCH scale. *Austr. J. Grape Wine Res.* **1995**, *1*, 100–110. [CrossRef]

35. Baker, N.R. Chlorophyll fluorescence: A probe of photosynthesis in vivo. *Annu. Rev. Plant Biol.* **2008**, *59*, 89–113. [CrossRef]

36. Schreiber, U. Pulse-Amplitude-Modulation (PAM) Fluorometry and Saturation Pulse Method: An Overview. In *Chlorophyll a Fluorescence: A Signature of Photosynthesis*; Kluwer Academic: Dordrecht, The Netherlands, 2004; pp. 279–319.

37. Genty, B.; Briantais, J.-M.; Baker, N.R. The relationship between the quantum yield of photosynthetic electron transport and quenching of chlorophyll fluorescence. *Biochim. Biophys. Acta Gen. Subj.* **1989**, *990*, 87–92. [CrossRef]

38. Fraser, P.D.; Pinto, M.E.S.; Holloway, D.E.; Bramley, P.M. Application of high-performance liquid chromatography with photodiode array detection to the metabolic profiling of plant isoprenoids. *Plant J.* **2000**, *24*, 551–558. [CrossRef] [PubMed]

39. Mokochinski, J.B.; Mazzafera, P.; Sawaya, A.C.H.F.; Mumm, R.; de Vos, R.C.H.; Hall, R.D. Metabolic responses of *Eucalyptus* species to different temperature regimes. *J. Integr. Plant Biol.* **2018**, *60*, 397–411. [CrossRef]

40. IPMA Instituto Português do Mar e da Atmosfera. Available online: http://www.ipma.pt/pt/otempo/obs.superficie/index-map-dia-chart.jsp#Braga,Merelim (accessed on 2 June 2019).

41. IVV, Instituto da Vinha e do Vinho, I.P.I. de Mercado. Nota Informativa, No3/2018, 30/07/2018, Previsão de Colheita—Campanha 2018/2019. Available online: https://www.ivv.gov.pt/np4/8955.html (accessed on 1 July 2019).

42. Shellie, K.C.; King, B.A. Kaolin particle film and water deficit influence malbec leaf and berry temperature, pigments, and photosynthesis. *Am. J. Enol. Vitic.* **2013**, *64*, 223–230. [CrossRef]

43. Oliveira, M. Viticulture in Warmer Climates: Mitigating Environmental Stress in Douro Region, Portugal. In *Grapes and Wines—Advances in Production, Processing, Analysis and Valorization*; InTech: Rijeka, Croatia, 2018.

44. Reynolds, A.G.; Heuvel, J.E.V. Influence of grapevine training systems on vine growth and fruit composition: A review. *Am. J. Enol. Vitic.* **2009**, *60*, 251–268.

45. Kraus, C.; Pennington, T.; Herzog, K.; Hecht, A.; Fischer, M.; Voegele, R.T.; Hoffmann, C.; Töpfer, R.; Kicherer, A. Effects of canopy architecture and microclimate on grapevine health in two training systems. *Vitis J. Grapevine Res.* **2018**, *57*, 53–60. [CrossRef]

46. Dos Santos, T.P.; Lopes, C.M.; Lucília Rodrigues, M.; de Souza, C.R.; Ricardo-da-Silva, J.M.; Maroco, J.P.; Pereira, J.S.; Manuela Chaves, M. Effects of deficit irrigation strategies on cluster microclimate for improving fruit composition of Moscatel field-grown grapevines. *Sci. Hortic. (Amst.)* **2007**, *112*, 321–330. [CrossRef]

47. Sharma, R.R.; Vijay Rakesh Reddy, S.; Datta, S.C. Particle films and their applications in horticultural crops. *Appl. Clay Sci.* **2015**, *116–117*, 54–68. [CrossRef]

48. Glenn, D.M.; Prado, E.; Erez, A.; McFerson, J.; Puterka, G.J. A Reflective, Processed-Kaolin Particle Film Affects Fruit Temperature, Radiation Reflection, and Solar Injury in Apple. *J. Am. Soc. Hortic. Sci.* **2002**, *127*, 188–193. [CrossRef]

49. Glenn, M.D. The mechanisms of plant stress mitigation by kaolin-based particle films and applications in horticultural and agricultural crops. *HortScience* **2012**, *47*, 710–711. [CrossRef]

50. Aschan, G.; Pfanz, H. Non-foliar photosynthesis—A strategy of additional carbon acquisition. *Flora* **2003**, *198*, 81–97. [CrossRef]

51. Kadir, S.; Von Weihe, M.; Al-Khatib, K. Photochemical Efficiency and Recovery of Photosystem II in Grapes After Exposure to Sudden and Gradual Heat Stress. *J. Am. Soc. Hortic. Sci.* **2007**, *132*, 764–769. [CrossRef]

52. Greer, D.H.; Weston, C. Heat stress affects flowering, berry growth, sugar accumulation and photosynthesis of *Vitis vinifera* cv. Semillon grapevines grown in a controlled environment. *Funct. Plant Biol.* **2010**, *37*, 206–214. [CrossRef]

53. Luo, H.B.; Ma, L.; Xi, H.F.; Duan, W.; Li, S.H.; Loescher, W.; Wang, J.F.; Wang, L.J. Photosynthetic responses to heat treatments at different temperatures and following recovery in grapevine (*Vitis amurensis* L.) leaves. *PLoS ONE* **2011**, *6*. [CrossRef]

54. Xiao, F.; Yang, Z.Q.; Lee, K.W. Photosynthetic and physiological responses to high temperature in grapevine (*Vitis vinifera* L.) leaves during the seedling stage. *J. Hortic. Sci. Biotechnol.* **2017**, *92*, 2–10. [CrossRef]

55. Krause, G.H.; Jahns, P. Non-photochemical Energy Dissipation Determined by Chlorophyll Fluorescence Quenching: Characterization and Function. In *Chlorophyll a Fluorescence*; Springer: Dordrecht, The Netherlands, 2007.
56. Medrano, H.; Bota, J.; Abadía, A.; Sampol, B.; Escalona, J.M.; Flexas, J. Effects of drought on light-energy dissipation mechanisms in high-light-acclimated, field-grown grapevines. *Funct. Plant Biol.* **2002**, *29*, 1197–1207. [CrossRef]
57. Szabó, I.; Bergantino, E.; Giacometti, G.M. Light and oxygenic photosynthesis: Energy dissipation as a protection mechanism against photo-oxidation. *EMBO Rep.* **2005**, *6*, 629–634. [CrossRef]
58. Joubert, C.; Young, P.R.; Eyéghé-Bickong, H.A.; Vivier, M.A. Field-Grown Grapevine Berries Use Carotenoids and the Associated Xanthophyll Cycles to Acclimate to UV Exposure Differentially in High and Low Light (Shade) Conditions. *Front. Plant Sci.* **2016**, *7*, 1–17. [CrossRef]
59. Tikkanen, M.; Grieco, M.; Nurmi, M.; Rantala, M.; Suorsa, M.; Aro, E.M. Regulation of the photosynthetic apparatus under fluctuating growth light. *Philos. Trans. R. Soc. B Biol. Sci.* **2012**, *367*, 3486–3493. [CrossRef]
60. Frioni, T.; Tombesi, S.; Luciani, E.; Sabbatini, P.; Berrios, J.G.; Palliotti, A. Kaolin treatments on Pinot noir grapevines for the control of heat stress damages. *Bio Web Conf.* **2019**, *13*, 04004. [CrossRef]
61. Brito, C.; Dinis, L.T.; Luzio, A.; Silva, E.; Gonçalves, A.; Meijón, M.; Escandón, M.; Arrobas, M.; Rodrigues, M.Â.; Moutinho-Pereira, J.; et al. Kaolin and salicylic acid alleviate summer stress in rainfed olive orchards by modulation of distinct physiological and biochemical responses. *Sci. Hortic. (Amst.)* **2019**, *246*, 201–211. [CrossRef]
62. Maxwell, K.; Johnson, G.N. Chlorophyll fluorescence—A practical guide. *J. Exp. Bot.* **2000**, *51*, 659–668. [CrossRef] [PubMed]
63. Cooley, N.M.; Glenn, D.M.; Clingeleffer, P.R.; Walker, R.R. The Effects of Water Deficit and Particle Film Technology Interactions on Cabernet Sauvignon Grape Composition. *Acta Hortic.* **2008**, 193–200. [CrossRef]
64. Shellie, K.; Glenn, D.M. Wine grape response to foliar kaolin particle film under differing levels of preveraison water stress. *HortScience* **2008**, *43*, 1392–1397. [CrossRef]
65. Glenn, M.; Cooley, N.; Walker, R.; Clingeleffer, P.; Shellie, K. Impact of kaolin particle film and water deficit on wine grape water use efficiency and plant water relations. *HortScience* **2010**, *45*, 1178–1187. [CrossRef]
66. Shellie, K.C.; King, B.A. Kaolin particle film and water deficit influence red winegrape color under high solar radiation in an arid climate. *Am. J. Enol. Vitic.* **2013**, *64*, 214–222. [CrossRef]

![agronomy logo] *agronomy*

MDPI

Article

Vineyard Variability Analysis through UAV-Based Vigour Maps to Assess Climate Change Impacts

Luís Pádua [1,2,*], Pedro Marques [1,3], Telmo Adão [1,2], Nathalie Guimarães [1], António Sousa [1,2], Emanuel Peres [1,2] and Joaquim João Sousa [1,2]

1   Engineering Department, School of Science and Technology, University of Trás-os-Montes e Alto Douro, 5000-801 Vila Real, Portugal; pedro.marques@utad.pt (P.M.); telmoadao@utad.pt (T.A.); nsguimaraes@utad.pt (N.G.); amrs@utad.pt (A.S.); eperes@utad.pt (E.P.); jjsousa@utad.pt (J.J.S.)
2   Centre for Robotics in Industry and Intelligent Systems (CRIIS), INESC Technology and Science (INESC-TEC), 4200-465 Porto, Portugal
3   Centre for the Research and Technology of Agro-Environmental and Biological Sciences, University of Trás-os-Montes e Alto Douro, 5000-801 Vila Real, Portugal
*   Correspondence: luispadua@utad.pt; Tel.: +351-259-350-762 (ext. 4762)

Received: 29 August 2019; Accepted: 20 September 2019; Published: 25 September 2019

Abstract: Climate change is projected to be a key influence on crop yields across the globe. Regarding viticulture, primary climate vectors with a significant impact include temperature, moisture stress, and radiation. Within this context, it is of foremost importance to monitor soils' moisture levels, as well as to detect pests, diseases, and possible problems with irrigation equipment. Regular monitoring activities will enable timely measures that may trigger field interventions that are used to preserve grapevines' phytosanitary state, saving both time and money, while assuring a more sustainable activity. This study employs unmanned aerial vehicles (UAVs) to acquire aerial imagery, using RGB, multispectral and thermal infrared sensors in a vineyard located in the Portuguese Douro wine region. Data acquired enabled the multi-temporal characterization of the vineyard development throughout a season through the computation of the normalized difference vegetation index, crop surface models, and the crop water stress index. Moreover, vigour maps were computed in three classes (high, medium, and low) with different approaches: (1) considering the whole vineyard, including inter-row vegetation and bare soil; (2) considering only automatically detected grapevine vegetation; and (3) also considering grapevine vegetation by only applying a normalization process before creating the vigour maps. Results showed that vigour maps considering only grapevine vegetation provided an accurate representation of the vineyard variability. Furthermore, significant spatial associations can be gathered through (i) a multi-temporal analysis of vigour maps, and (ii) by comparing vigour maps with both height and water stress estimation. This type of analysis can assist, in a significant way, the decision-making processes in viticulture.

Keywords: unmanned aerial vehicles; vigour maps; spatial variability; normalized difference vegetation index; crop water stress index; crop surface model; precision viticulture; climate change; multi-temporal analysis

1. Introduction

About 70% of the available worldwide clean water is used in agriculture [1]. Moreover, by the year 2050, there will have to be an estimated 70% increase in food production [1] to sustain Earth's population. Therefore, to attain a sustainable agriculture, it is essential to ensure proper water management. Global warming evolution throughout the years means this phenomena is one of the major threats to agricultural production, also with effects on society [2–5]. Less precipitation, associated with more frequent and longer drought periods [6], ultimately leads to an increase in the use of water in agricultural activity. To improve water usage efficiency, the United Nations (UN) set sustainable development goals with the aim to create an expected increase in efficiency in all sectors by the year 2030. This will ensure sustainable extractions and the implementation of integrated water resources management [7]. It is crucial that the agricultural sector contributes to this effort by developing and implementing controlled irrigation management systems [8,9]. As such, it is necessary to have an efficient analysis of crops' water status.

The enduring search for resource use optimization, risks reduction, and minimizing environmental impacts led to the emergence of precision agriculture (PA) [10]. To understand both spatial and temporal variabilities of a production unit, PA's tools and technologies enable the acquisition and processing of large data volumes (e.g., image processing techniques, geo-statistical methods) [10,11]. The precision viticulture (PV) concept derived from PA involves applying different technologies to vineyard management and grape production [12,13]. However, grapevine (*Vitis vinifera* L.) development is strongly related to spatial heterogeneity, which depends on several factors to determine both its production and quality [14]. Some of the more relevant factors are soil quality and type, vegetation management operations, irrigation systems, nutritional status, pest and disease control, air temperature, and precipitation levels [13,15]. Changes in one of these factors may result in the occurrence of biotic and abiotic problems. Depending on its severity, it may result in a significant decrease in production or quality, and therefore, considerable economic losses [16]. The Douro Demarcated Region (DDR, north-eastern Portugal) spatial variability is high due mainly to the terrain's topographic profile, climatic variations, and soil characteristics, which causes vineyards to be unique throughout the DDR [12].

In the last few years, due to their flexibility and efficiency in diverse environments, the use of unmanned aerial vehicles (UAVs) emerged in agriculture applications [17]. UAVs can acquire georeferenced data with a high spatial resolution while using different types of sensors (RGB, near infrared, multi and hyper-spectral, thermal infrared (TIR) and LiDAR) [18], which allow for the output of several digital products, such as ortho-rectified mosaics, digital elevation models (DEMs), land surface temperature, and vegetation indices (VIs) [18]. Indeed, their ability to carry different types of sensors make UAVs a suitable solution for agricultural applications. While multispectral sensors acquire data from the electromagnetic spectrum in the near and visible infrared region (400 to 1000 nm), thermal sensors can acquire data in the far infrared zone (5000 to 18,000 nm), where the reflection value of each pixel can be transformed into a temperature value [18]. Among the different VIs, which can be considered as a set of arithmetic operations applied in different bands used to extract different vegetation characteristics [18], the normalized difference vegetation index (NDVI) [19] must be highlighted as it is frequently used in agricultural applications to estimate different crop-related parameters: biomass [20]; canopy structure, leaf area index (LAI), crop management [21]; and mapping vigour zones [22]. Moreover, it was found to correlate well with grape quality properties [23]. As for temperature-based indices, they constitute a quick and practical way to estimate crop water status, therefore indicating the plants' water content. The crop water stress index (CWSI) [24] is widely used in remote sensing to monitor plants' water status and consequent irrigation management [25]. TIR-based indices were employed to different crops, such as olives [26], grapevines [27], cotton [28], wheat [29], rice [30], sugar-beet [31], and maize [32]. Remote sensing platforms can also be a helpful tool for a better understanding of spatial variability, which has a significant meaning in vineyard management activities. Actually, UAVs have already been used to, e.g., estimate the leaf area index [33,34], irrigation management and water stress mapping [27,35,36], disease detection and mapping [37,38], and detection of nutritional deficiencies [39].

UAVs have already proved to be a cost-effective and flexible alternative for remote sensing, within a PA context. They present an improved decision-making process to the farmer and provide greater flexibility, when compared to other remote sensing platforms [13].

As for PV, vineyards have significant areas occupied by elements other than grapevines (e.g., inter-row vegetation, man-made structures, vegetation that usually surrounds the plot, and grapevines' shadows) [13,40]. These elements can be automatically identified by means of digital image processing methods. Indeed, several methods have been proposed to deal with UAV-based aerial imagery or with the resulting digital products from the photogrammetric processing. For example, grapevine segmentation [41,42], supervised and unsupervised machine learning [43], point clouds derived from photogrammetric processing [44,45], and DEMs [16,33,40]. Regarding VIs, they are one of the most common segmentation techniques applied in a remote sensing [46], mainly to segment a given image into two classes: vegetation or non-vegetation [47]. However, when considering vineyard vegetation, VIs acknowledges all types of vegetation without distinguishing grapevines from non-grapevines (e.g., inter-row vegetation). By using the DEM—or more specifically, the canopy surface model (CSM), which can be obtained by subtracting the digital terrain model (DTM) from the digital surface model (DSM)—quantifying and removing non-grapevine vegetation in a vineyard's segmentation process can be done as plant height is provided [48].

While different digital outputs can be generated from UAV-based imagery, the amount of data and its complexity can be overwhelming for the common farmer to interpret. Straightforward useful crop-related information is needed. Vigour maps are an example where by using the NDVI, vegetation is classified into different classes according to its characteristics. By applying it to PV, grapevines' vigour can be defined as the measure of the growth rate during a given time period (e.g., the growing season). This not only enables the classification of vineyard homogeneity zones [49], which is a way to represent the impact of both environmental conditions and soil fertility [50]. There have been some related works done in this area. Khaliq et al. [51] compared satellite imagery with UAV-based multispectral data in four different epochs of the grapevines' vegetative cycle. Different comparisons were made by considering: (i) the whole vineyard, (ii) only the grapevines' vegetation, and (iii) only inter-row areas. The authors reported that satellite multispectral imagery presented limitations due to the ground sampling distance (GSD, 10 m) and to the influence of inter-row information. Primicerio et al. [22] evaluated vigour maps produced for the whole vineyard and only encompassing grapevines' vegetation by applying an automatic segmentation method [41]. Campos et al. [52] used UAV-based vigour maps to create prescription maps for vineyard spraying operations.

Studies supported by imagery acquired in one flight mission alone mainly focused on assessing non-grapevine vegetation removal when considering the whole vineyard, and in creating task-oriented vigour maps [22,52–54]. With reference to multi-temporal studies, there are those whose aim is to compare different growing seasons by evaluating biophysical grapevines parameters [54–56]. Furthermore, studies utilizing intra-season multi-temporal data, considered the whole vineyard information [57] or vineyard changes were not the main focus [51]. As found in Primicerio et al. [22], vigour maps using only grapevines' vegetation showed a better representation of the variability within the vineyard. The spatial variability in grapevines' water status can be assessed thought both multispectral and TIR imagery, where TIR imagery serves as an immediate way to estimate crops' water status, while multispectral data can show cumulative water deficits [35]. As such, the TIR data has the potential to help understand water stress for near-real-time decision-making support [58]. By integrating TIR and multispectral data, datasets to study grapevines' response to climate change [59] can be created.

This study aimed to evaluate vineyard vigour maps (NDVI) created using UAV-based multispectral imagery within a multi-temporal context and in different grapevines' phenological stages. The main goal was to study grapevines' vegetation dynamics during the growing season up until harvesting. Two approaches were used: (i) considering the whole vineyard area, and (ii) considering only automatically detected grapevines' vegetation. Spatial assessment between the generated vigour maps, and grapevines' canopy temperature and height data—obtained from UAV-based TIR and RGB imagery, respectively—were conducted with the objective to correlate vigour maps with potential grapevines' water stress and canopy height. This allowed for the assessment of non-grapevine features when analyzing vigour maps.

The next section presents the study area and the methods used both for data acquisition and processing. Results are presented in Section 3 and discussed in Section 4. Lastly, the most significant conclusions are shown in Section 5.

2. Materials and Methods

2.1. Study Area and Environmental Context

This study was conducted in a 0.30 ha vineyard located in the University of Trás-os-Montes e Alto Douro campus, Vila Real, Portugal (41°17′13.2″ N 7°44′08.7″ W WGS84, altitude: 462 m), in the DDR (Figure 1). The vineyard (cv. *Malvasia Fina*) is trained in a double Guyot system, where each row has grapevines 1.20 m apart and there is 1.80 m distance in between rows. There is a total of 22 rows with a NE–SW orientation. Furthermore, it is a rainfed vineyard, with fertilization applied using foliar spraying and with phytosanitary management operations taking place throughout the entire season. Inter-row areas are composed of spontaneous vegetation, which is managed using mechanical interventions at least twice per season.

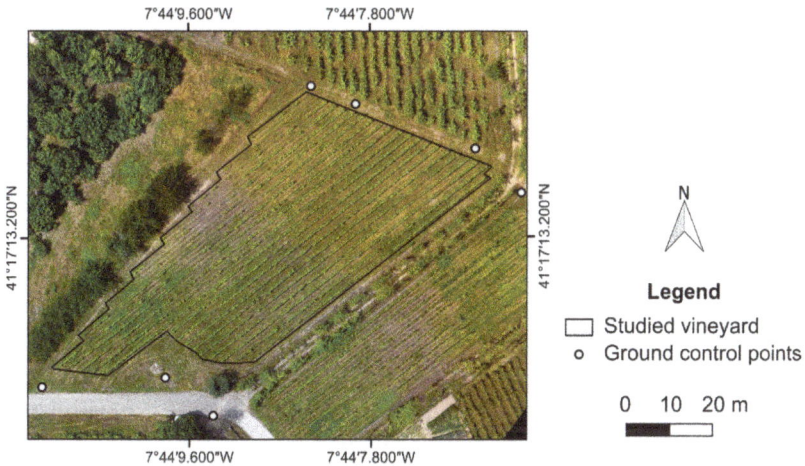

Figure 1. General overview of the studied area delimited by a polygon. Coordinates in WGS84 (EPSG:4326).

During the studied period (May to September 2018), a total of 170 mm of precipitation was registered, along with 590 mm of potential evapotranspiration. Mean values for maximum, mean, and minimum air temperatures were 29 °C, 20 °C, and 13 °C, respectively. Monthly values are presented in Figure 2. Higher air temperature values were observed in July, August, and September, while May and June presented higher precipitation values. In contrast, there was almost no precipitation in August. This environmental data was acquired using a weather station located some 400 m away from the study area.

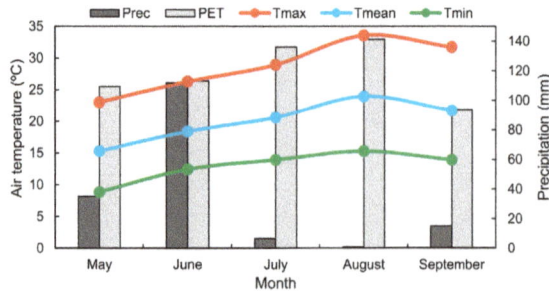

Figure 2. Monthly mean values for maximum (Tmax), mean (Tmean), and minimum (Tmin) air temperatures; precipitation (Prec); and potential evapotranspiration (PET) for the studied area in the period ranging from May to September 2018.

2.2. UAV-Based Data Acquisition

RGB, multispectral and TIR imagery were acquired using both a DJI Phantom 4 (DJI, Shenzhen, China) and a Sensefly eBee (senseFly SA, Lausanne, Switzerland). The former is a low-cost UAV equipped with an RGB sensor (12.4 MP resolution) attached to a three-axis electronic gimbal. For the purpose of this study, it was modified to support a multispectral sensor: the Parrot SEQUOIA (Parrot SA, Paris, France). This sensor consisted of a four-camera array, which was able to acquire data in the green (550 nm), red (660 nm), red-edge (735 nm), and near infrared (790 nm) parts of the electromagnetic spectrum, with a 1 MP resolution. Moreover, a Sunshine sensor (Parrot SA, Paris, France) was also added to the UAV's top. It is responsible for acquiring the irradiance conditions during the flight mission in the same spectral bands as the multispectral sensor and to geolocate the acquired imagery.

As for the Sensefly eBee, it is a fixed-wing UAV used to acquire TIR imagery with the thermoMAP (senseFly SA, Lausanne, Switzerland) sensor (between 7500 nm to 13,500 nm, with 640×512 pixels and a temperature resolution of 0.1 °C), with automatic in-flight thermal image-based calibration. Ground control points (GCPs), used for aligning the acquired imagery during the photogrammetric processing, were measured using a Global Navigation Satellite System (GNSS) receiver in real-time kinematic (RTK) mode based on the TM06/ETRS89 coordinate system (GCP's location in Figure 1). While the multi-rotor UAV was used mainly due to its capability to survey areas at lower flight heights, which provides higher spatial resolution [18], the fixed-wing UAV surveyed a larger area, which included the studied area. Furthermore, the TIR sensor only operated as a fixed-wing UAV.

Data acquisition was conducted in five flight campaigns, from 17 May 2018 to 21 September 2018. Each flight campaign corresponded to distinct grapevine phenological stages: flowering (May and June), fruit set (July), veraison (August), and harvest (September). Details are presented in Figure 3. All flight campaigns were conducted near solar noon to minimize sun and shadow influences. Flights for both the RGB and multispectral sensors were done at a 40 m height, with a forward overlap of 80% and 70% side overlap between images. The GSD was approximately 1.8 cm for the RGB and of 4.4 cm for the multispectral imagery. Regarding flights for TIR imagery acquisition, they were carried out at a 75 m flight height, with a 90% forward overlap and 75% side overlap between images, resulting in an approximate 17.5 cm GSD. All flight campaigns utilized RGB and multispectral imagery, while TIR imagery was only acquired from F3 onward (see Figure 3), due to both in-field observations and the environmental context, since rainfall can induce an error in the remotely sensed grapevine water status in the subsequent days [60]. Moreover, a radiometric calibration was performed prior to each flight for the multispectral imagery using a reflectance panel provided by the manufacturer, along with the irradiance data from the sunshine sensor. Irradiance and reflectance data enabled a reliable radiometric workflow for the collection of repeatable reflectance data over different flights, dates, and weather conditions.

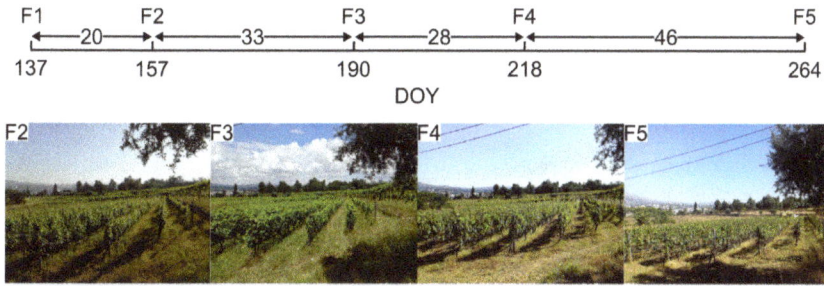

Figure 3. Flight campaigns' details: flight number (F#), Day of Year (DOY), and temporal difference (in days) between flights. Vineyard images in different flight campaigns are also shown.

2.3. Data Processing and Parameters Extraction

Imagery acquired in each flight campaign was processed using the Pix4Dmapper Pro (Pix4D SA, Lausanne, Switzerland). This software makes use of structure from motion (SfM) algorithms to identify common points in the images. It can create point clouds, and by interpolating them, generate different orthorectified outcomes depending on the sensor used. Imagery from each sensor was processed in different projects. The default processing options for each sensor were applied, but point clouds were generated with a high-point density. Point cloud interpolation was achieved using inverse distance weighting (IDW) and by applying noise filters. The generated digital outcomes were: (i) RGB–orthophoto mosaic, DSM, and DTM; (ii) multispectral–VIs; and (iii) TIR–land surface temperature. By subtracting the DTM from the DSM, the CSM was obtained. From the multispectral imagery, the NDVI [19] was obtained using a normalization between the near-infrared (NIR) and red bands, as given in Equation (1).

$$NDVI = \frac{NIR - RED}{NIR + RED} \tag{1}$$

The land surface temperature was used to compute the CWSI through the empirical model presented in Equation (2). It was based in the usage of canopy temperature, T_c, and the lower and upper canopy temperature limits (T_{dry} and T_{wet}), corresponding, respectively, to well-watered and non-transpiring leaves. These values can be directly obtained in the field or by using UAV-based thermal infrared imagery [61]. CWSI values can vary between 0 (no stress signs) and 1 (high levels of stress). In this study, T_{wet} and T_{dry} values were obtained as described in the work of Matese and Di Gennaro [23]: T_{wet} was obtained by wetting some leaves and immediately measuring their temperature, while T_{dry} values were obtained by applying petroleum jelly in the leaves and registering their temperatures after some minutes had gone by. Temperature values were measured using a handheld infrared thermometer (Shenzhen Jumaoyuan Science and Technology Co., Ltd., Shenzhen, China), with a ±1.5 °C precision and operating between 8000 nm to 14,000 nm.

$$CWSI = \frac{T_c - T_{wet}}{T_{dry} - T_{wet}} \tag{2}$$

To remove non-grapevine elements from the acquired imagery, segmentation was performed by using the method proposed in Pádua et al. [62]. Both the CSM and the G% index [63], computed from the orthophoto mosaic, were used as inputs, and through thresholding, considering both vegetation and height thresholds, it identified all vegetation within a given height range. While G% was automatically obtained using Otsu's method for thresholding, CSM used a defined height range. An accurate grapevine segmentation was obtained, filtering out non-grapevine objects such as soil and inter-row vegetation.

This method has already been used in a multi-temporal analysis of grapevines' vegetation evolution throughout a season in two vineyard plots in Pádua et al. [64]. Similarly, in this study, the method to segment grapevines' vegetation [62] was applied to evaluate the multi-temporal vineyard evolution when regarding grapevine area and canopy volume, as well as the inter-row vegetation area. The grapevine canopy volume was computed according to Pádua et al. [64], using the mean height of each cluster of pixels obtained during the segmentation process multiplied by its area, where the sum of the volume of each cluster represents the total vineyard volume.

As such, the orthorectified outputs from each flight campaign were used for different purposes. The grapevines' vegetation was detected and then CSM, NDVI, and CWSI values from the detected parts were considered, while non-grapevine pixels were discarded. Within the scope of this study, three different approaches were tested to create vigour maps. Figure 4 describes the main steps in each approach. Moreover, vigour classes were set to low, medium, and high. The workflow consisted in loading the orthorectified outcomes, followed by the vineyard segmentation method, depending on the used approach. Then, vigour maps were created by a applying a mean filter to the image, using a 2×2 m sliding window. Data could then be normalized according to Equation (3) before the vigour map was created. Again, this last step depended on the approach being used.

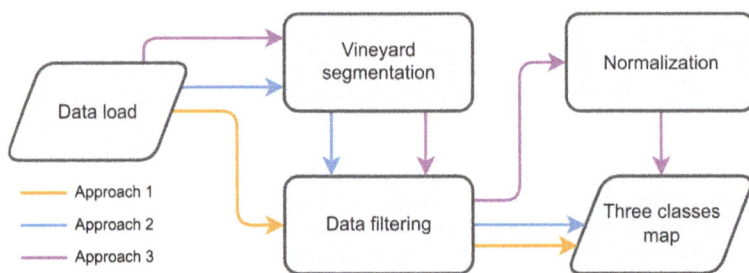

Figure 4. Approaches tested to produce vigour maps using three vigour classes.

The first approach relied on the usage of data from the whole vineyard. The outcome was directly smoothed and divided into three classes, using terciles. As for the second approach, it was similar to the first, but it only considered the grapevines' vegetation. Lastly, the third approach, similar to the second approach, considered only normalized grapevines' vegetation. Normalization was done based on the mean value of the 10% higher and lower values of the smoothed grapevines' vegetation values. Then, three vigour classes are created by dividing the values in the normalized raster according to fixed thresholds: (i) values lower or equal to 0.4 were considered low vigour; (ii) between 0.4 and 0.7 were considered medium vigour; (iii) and values above 0.7 were considered high vigour.

2.4. Vigour Maps versus Spatial Statistics

Vigour maps obtained from each flight campaign were compared with the CSM and the CWSI using statistical techniques that consider geospatial variability. This comparison was done by converting the three vigour classes maps to a 4×4 m grid. The grid size was selected by considering the studied vineyard's characteristics: each grid square was confined to two vine rows. This pipeline was proposed by Matese et al. [56]. Regarding the methods used in this comparison process, they were the local bivariate Moran's index (MI) and the bivariate local indicators of spatial association (LISA) [65]. Local MI (LMI) is based in the Moran's index [66], which measures the global data correlation. While a positive correlation represents similar values in the area's neighbourhood, a negative value represents the opposite, and zero represents a random spatial agreement. Regarding the LMI, a value is provided for each observation through permutation. The local bivariate MI was used in this study to assess the correlation between a defined variable and a different variable in the nearby areas. In turn, LISA measures the local spatial correlation, providing maps of local clusters with a similar behaviour, which is based on MI. This way, spatial clusters and its dispersion can be assessed. Bivariate LISA (BILISA) [65] was used as in Anselin [67] to examine the spatial relationship between the CSM and

CWSI and the vigour maps. This comparison was made using GeoDa software [68]. Spatial weights were necessary to perform these analyses: a eight-connectivity approach (3×3 matrix) was used to create the weights map and BILISA was executed with 999 random permutations. The computed cluster maps and its significance were used. Cluster maps specify positive and negative spatial associations and are divided into four classes, based on the correlation of the value with its neighbourhood. The obtained associations are: (i) high–high (HH), where high values correlated with high values in the neighbourhood; (ii) low–low (LL), in which low values correlated with low values in the neighbourhood; (iii) high–low (HL); (iv) and low–high (LH). The three classes of vigour maps computed through the different approaches were compared with their correspondent vigour map in the following flight campaign, as well as with the CSM and CWSI three classes maps.

3. Results

This study yielded different digital products through the methods employed, from which it is important to highlight the vineyard status, vigour areas, potential water stress areas, and a multi-temporal vineyard characterization.

3.1. Multi-Temporal Vineyard Characterization

Figure 5 presents the orthorectified outcomes from the photogrammetric processing. There was a noticeable overall NDVI decline throughout the season (Figure 5a). However, grapevines' canopy height (Figure 5b) presented a growth from the first to the third flight campaign, while remaining constant from then on. As for the temperature (Figure 5c), a high temporal variability was observed due to both the day temperature and the inter-row vegetation. For example, in the third flight campaign, temperature differences between areas with or without grapevines' vegetation were smaller, about 1.0 °C, than in the other flight campaigns: approximately 2.2 °C for F4 and 1.4 °C for F5. Moreover, registered land surface temperatures presented the same behaviour as the maximum air temperature (Figure 2) registered in each month. Indeed, they were lower in July (followed by September), and higher in August.

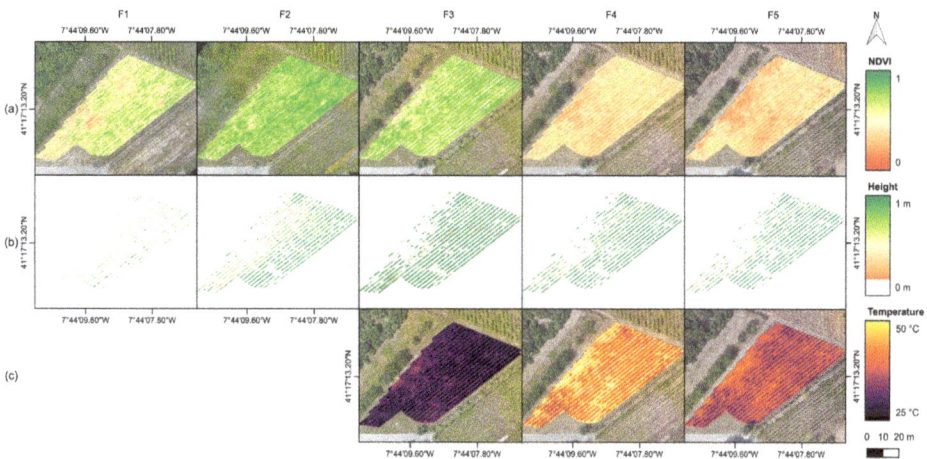

Figure 5. Orthorectified outcomes generated with data acquired in each flight campaign using a colour-code representation: (**a**) normalized difference vegetation index, (**b**) crop surface model, and (**c**) land surface temperature. Orthophoto mosaics are presented as the background of (a) and (c).

Due to early vegetation development in grapevines by the time the first flight campaign took place, the minimum height to consider as grapevines' vegetation was 0.2 m. As for the remainder of the flight campaigns, minimum and maximum heights were set to 0.5 and 1.9 m, respectively.

Table 1 presents the differences in NDVI, CSM, land surface temperature, and CWSI values when considering the whole vineyard plot and when analyzing only detected grapevines' vegetation. Generally, mean and minimum NDVI values were higher when considering only grapevines' vegetation. As for maximum values, some high values were accounted for in areas other than with grapevines' vegetation. The same tendency was verified in the mean and minimum height values, obtained through the CSM. However, maximum values were practically similar, except for the first flight campaign. An inverse tendency was verified when analyzing the land surface temperature and CWSI, i.e., higher values were found when analyzing the whole vineyard plot.

Table 1. Maximum, mean, and minimum values of the normalized difference vegetation index (NDVI), crop surface model (CSM), surface temperature, and crop water stress index (CWSI) when considering the whole vineyard plot and only grapevines' vegetation in the five flight campaigns.

Type	Outcome	Parameter	F1	F2	F3	F4	F5
Whole area	NDVI	Max	0.88	0.91	0.89	0.78	0.78
		Mean	0.57	0.74	0.68	0.42	0.38
		Min	0.13	0.26	0.27	0.17	0.01
	CSM (m)	Max	1.17	1.48	1.59	1.51	1.53
		Mean	0.06	0.19	0.35	0.22	0.19
		Min	0.00	0.00	0.00	0.00	0.00
	Temp (°C)	Max	–	–	38.74	59.90	45.84
		Mean	–	–	29.89	44.35	37.20
		Min	–	–	27.12	37.26	32.49
	CWSI	Max	–	–	1.00	1.00	1.00
		Mean	–	–	0.60	0.83	0.78
		Min	–	–	0.04	0.23	0.07
Grapevines' vegetation only	NDVI	Max	0.87	0.89	0.89	0.75	0.78
		Mean	0.70	0.82	0.80	0.62	0.59
		Min	0.41	0.59	0.64	0.37	0.25
	CSM (m)	Max	1.07	1.48	1.59	1.51	1.53
		Mean	0.40	0.89	1.16	1.01	0.99
		Min	0.20	0.47	0.52	0.27	0.20
	Temp (°C)	Max	–	–	31.20	47.81	39.36
		Mean	–	–	28.92	42.17	35.84
		Min	–	–	27.12	37.26	32.49
	CWSI	Max	–	–	0.82	1.00	0.91
		Mean	–	–	0.38	0.68	0.48
		Min	–	–	0.04	0.23	0.07

Extracted vineyard parameters allowed for a multi-temporal analysis of both grapevines' vegetation area and volume, as well as for other vegetation present in the studied area. Figure 6 contains these results. The first flight campaign presented the lower values for the grapevines' vegetation area: 82 m^2, representing 3% of the vineyard plot. The grapevines' vegetation area increased until the third flight campaign, from which a significant decline was verified in the following flight campaigns. The grapevines' canopy volume presented the same behaviour. As for inter-row vegetation, a growth happened between the first and the second flight campaigns, from 6% to 20% of the vineyard plot. After the fourth flight campaign, inter-row vegetation area decreased to 26 m^2 (1% of the vineyard plot), whilst a small increase was verified in the last flight campaign.

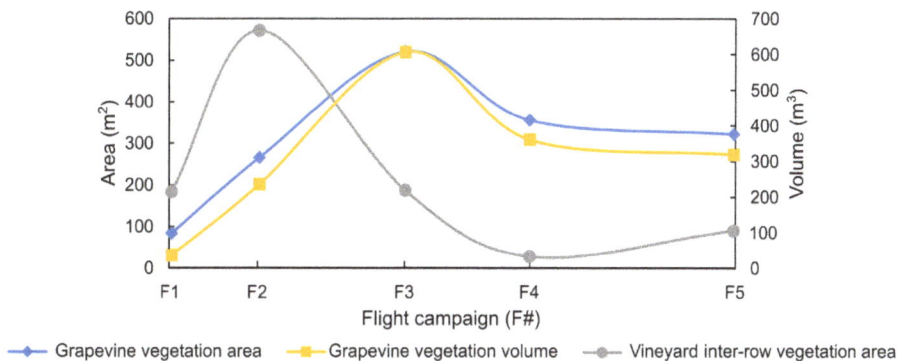

Figure 6. Estimated grapevines' vegetation area and volume, and inter-row vineyard vegetation, in each flight campaign.

3.2. Generated Vigour Maps

Vigour maps were generated as described in Section 2.3 and assessment values are presented in this section. Each map was classified as one of three classes, namely as a low, medium, or high vigour area.

3.2.1. Visual Assessment

Figure 7 presents the vigour maps generated using three approaches. When encompassing the whole vineyard (i.e., considering bare soil and all existing vegetation), as presented in Figure 7a, a perspective of the plot's homogeneity throughout the season was obtained. Approaches considering only detected vineyard vegetation presented a higher diversity, providing a deeper perspective on the grapevines' vegetation spatial variability (Figure 7b,c). Still, a tendency for a lower vigour classification in the left part of the studied area was noticeable in all approaches. The same situation was verified in the southern central part of the vineyard plot. This assessment was more pronounced in the first approach but had more detail in both the second and third approaches.

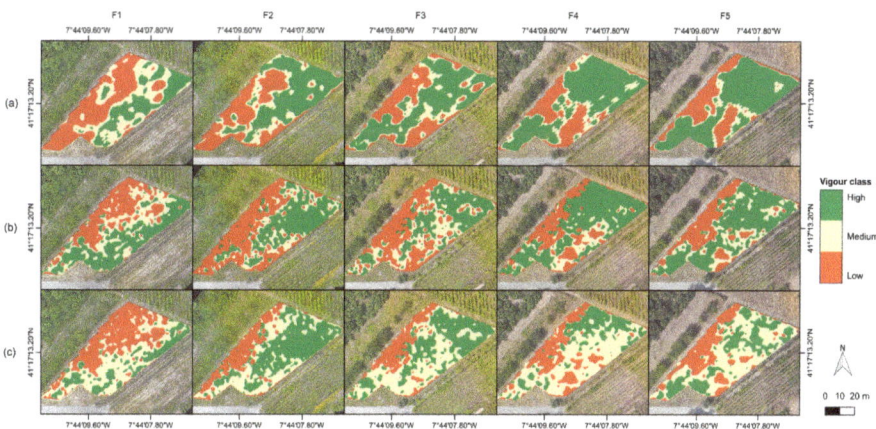

Figure 7. Generated vigour maps, based on the normalized difference vegetation index (NDVI), with three vigour classes (high, medium, and low) for each flight campaign, with the three evaluated approaches: (**a**) considering all vegetation present, (**b**) regarding only the grapevines' vegetation, and (**c**) considering only normalized grapevines' vegetation.

Vineyard areas classified with high, medium, or low vigour were evaluated in all flight campaigns. Their percentages are presented in Figure 8a. As for the first approach, the vineyard plot showed a higher percentage of vegetation in the high vigour class (mean overall percentage of 48%). However, in the first flight campaign, there was a higher area classified in the low vigour class (mean overall percentage of 31%). The medium vigour class presented the lower mean overall percentage (21%). As for the second approach, the overall mean area percentage was similar: 43% in the high vigour class, followed by 33% in the low vigour class and 24% in the medium vigour class. Regarding the third approach, the medium vigour class presented the higher mean overall occupation area (42%), followed by the high vigour class (31%) and the low vigour class (27%).

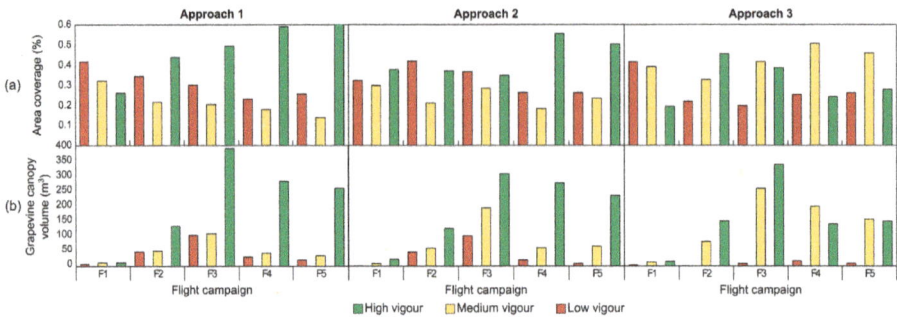

Figure 8. Vineyard area (**a**) and grapevines' canopy volume (**b**) per vigour class and approach in all flight campaigns (F#).

The vineyard vigour area behaviour may not correspond to the grapevines' vegetation. As such, Figure 8b shows the grapevines' canopy volume present in each class throughout all the flight campaigns. This was achieved by intercepting vigour classes with the detected grapevines' vegetation canopy volume. There were variations when comparing the applied approach and when analyzing the flight campaigns in the same approach: the overall value corresponded to the grapevines' canopy volume presented in Figure 6. When considering the NDVI values for the whole vineyard to generate a canopy map, the grapevines' canopy volume presented a higher predominance in the high vigour class. However, when comparing this with the approaches that consider only the grapevines' vegetation, the grapevines' canopy volume was significantly lower in the low vigour class for the latter approach. Regarding the approach where only grapevines' vegetation was considered, a clear distinction among the grapevines' vegetation volume was clear: the high vigour class had a greater grapevines' canopy volume, followed by the medium and low vigour classes. As for the third approach (normalized grapevines' vegetation), in the last two flight campaigns (F4 and F5), there was a higher volume in the medium vigour class, corresponding to the detected vineyard area (Figure 8a).

3.2.2. Spatial Correlations

To undergo a spatial assessment, the three approaches to generate vigour maps were applied to the CSM and CWSI outcomes of each flight campaign, when available. Ergo, maps with height values sorted in classes—low, medium and high height—could be obtained from the CSM. These results are presented in Figure 9. Height maps presented a high homogeneity among all approaches, especially from the third flight campaign onward.

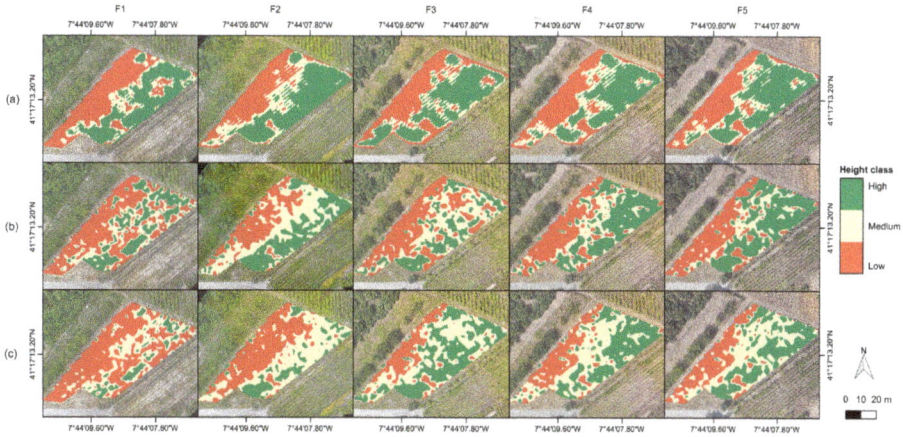

Figure 9. Generated height maps obtained from the crop surface models (CSM) for each flight campaign. Each height value was sorted into one of three height classes (low, medium, or high). The whole vineyard (**a**), grapevines' vegetation only (**b**), and normalized grapevines' vegetation (**c**) was considered.

From the CWSI, maps that could potentially point out grapevines' water stress were obtained. They are presented in Figure 10. Again, three classes were considered to sort out each value on every map: low, medium, and high water stress. Results from considering all vegetation present in the vineyard (Figure 10a) showed a high homogeneity across the plot for all flight campaigns. However, when considering only grapevines' vegetation (approaches two and three) the behaviour was different (Figure 10b,c).

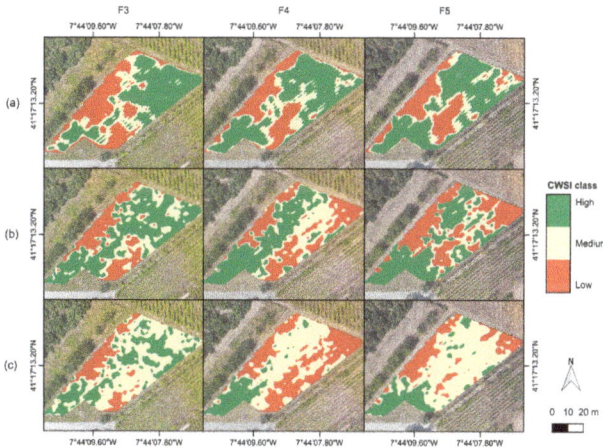

Figure 10. Generated crop water stress index (CWSI) maps for each flight campaign. Each CSWI value was sorted into one of three classes (low, medium, or high). The whole vineyard (**a**), grapevines' vegetation only (**b**), and normalized grapevines' vegetation (**c**) was considered.

Maps presented in Figures 9 and 10 were compared with the vigour maps presented in Figure 7 in a 4 × 4 m grid using the LMI to measure their spatial correlation. Table 2 presents these results. Considering all the vineyards' vegetation (first approach), stronger correlations were observed for the CSM. In turn, the other two approaches presented a more balanced trend for the CSM and CWSI. Stronger correlation values were found among vigour maps using data from the fourth flight campaign with the third approach (LMI = 0.70 for the CSM and LMI = 0.66 for the CWSI). Lower correlation values were observed in the height maps when considering all the vineyard's vegetation with data from the first flight campaign. The same was verified in the fourth flight campaign for the CWSI.

Table 2. Quantitative comparison using the local Moran's index of the normalized difference vegetation index (NDVI) vigour classes in the three different approaches considered to the crop surface model (CSM) and crop water stress index (CWSI) classes with a *p*-value < 0.001, for each flight campaign (F#).

Vigour map	Approach 1		Approach 2		Approach 3	
F#	CSM	CWSI	CSM	CWSI	CSM	CWSI
1	0.32	–	0.39	–	0.35	–
2	0.53	–	0.50	–	0.50	–
3	0.41	0.44	0.37	0.43	0.36	0.41
4	0.65	0.40	0.67	0.63	0.70	0.66
5	0.59	0.39	0.66	0.59	0.67	0.57

The local spatial autocorrelation enabled the creation of clusters maps using BILISA to evaluate HH, LL, LH, and HL patterns between vigour maps of the different flight campaigns and between vigour maps and their correspondent height and water stress maps.

BILISA cluster map for the three evaluated vigour map approaches and its association with height maps is presented in Figure 11. As for the first approach (Figure 11a), there was a clear spatial correlation with a higher significance in the left and right sides of the vineyard plot, corresponding, respectively, to LL and HH associations. However, a smaller number of significant LH and HL clusters were detected. Regarding the other two approaches (Figure 11b,c) that considered only the grapevines' vegetation, similar spatial patters were found for HH and LL. Furthermore, a significant HL cluster could be found in the southwestern part of the vineyard plot in the fourth and fifth flight campaigns. Significant LH clusters were found in the southeastern part of the vineyard in the second, third, and fourth flight campaigns for the second approach (Figure 11b).

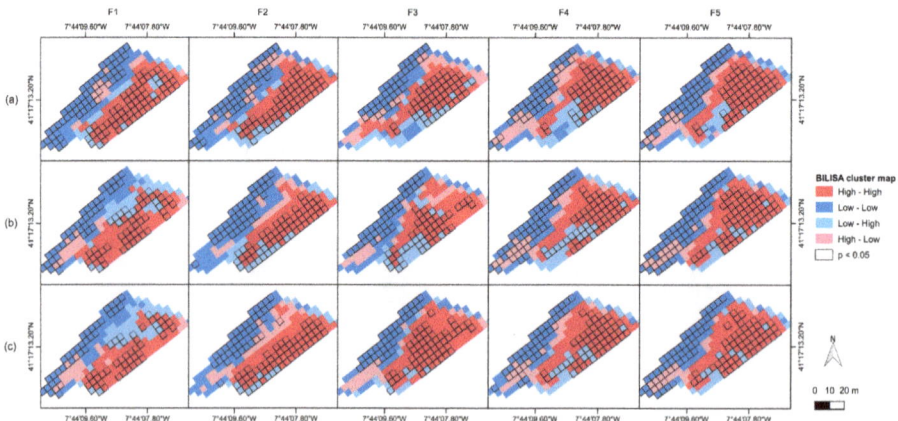

Figure 11. BILISA cluster maps between NDVI vigour maps and CSM height maps for the three evaluated approaches: (**a**) first approach, (**b**) second approach, and (**c**) third approach. Associations with a *p*-value < 0.05 are highlighted with a black border.

Figure 12 presents the BILISA cluster maps generated from the spatial associations among vigour maps and water stress maps (Figure 10). Significant associations were found when using the first approach, with a representative HH cluster present in the northeastern region of the vineyard plot and a LL cluster in the vineyard's left side. When considering only the grapevines' vegetation, a similar behaviour was observed in the third flight campaign. In the remaining flight campaigns, a significant

LL cluster existed in the left part of the vineyard, but a lower significance was found for HH in the northeastern part. A high significance among the values was detected in the southern region, which presented HH and LH associations.

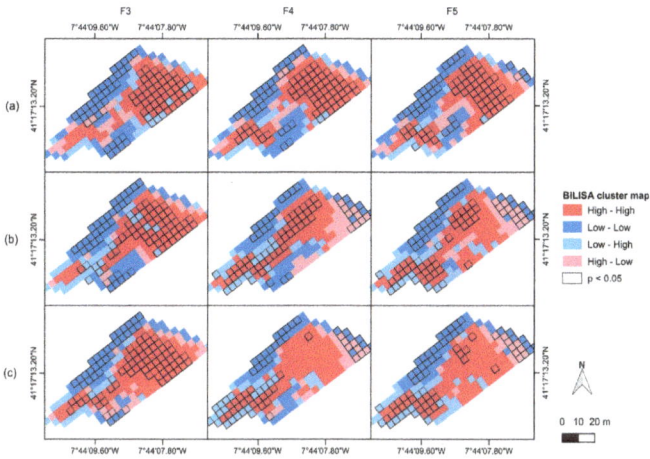

Figure 12. BILISA cluster maps between NDVI vigour maps and CWSI maps for the three evaluated approaches: (**a**) first approach, (**b**) second approach, and (**c**) third approach. Associations with a *p*-value < 0.05 are highlighted with a black border.

Considering the BILISA clusters maps from the vigour maps for each evaluated approach when comparing consecutive flight campaigns (Figure 13), similar patterns were observed in all approaches and significant LH clusters were identified when comparing the first and second flight campaigns considering only the grapevines' vegetation (Figure 13b,c).

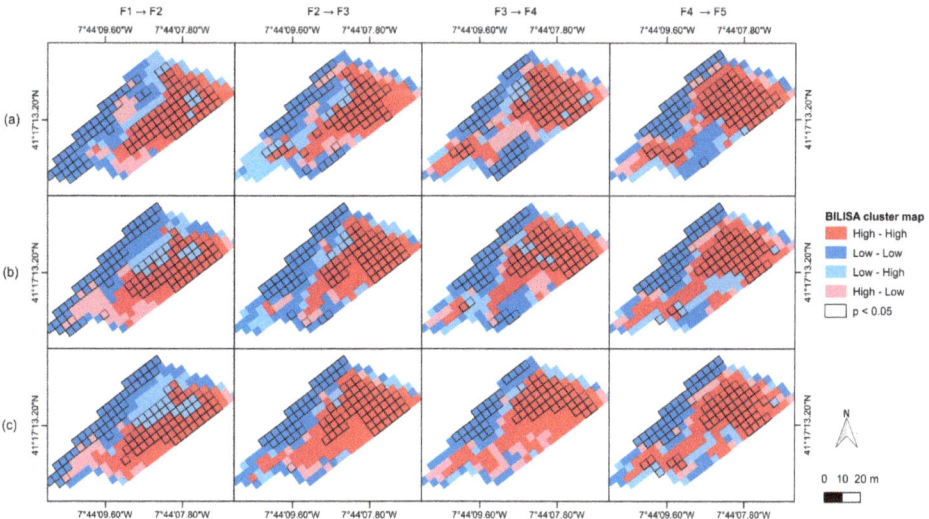

Figure 13. BILISA cluster maps between NDVI vigour maps of two consecutive flight campaigns for the three evaluated approaches: (**a**) first approach, (**b**) second approach, and (**c**) third approach. Associations with a *p*-value < 0.05 are highlighted with a black border.

4. Discussion

In this section the most meaningful results achieved in this study are discussed: (i) the multi-temporal analysis of the studied vineyard plot; (ii) the generated vigour maps; and (iii) spatial correlations between vigour maps, grapevines' height, and potential water stress.

4.1. Multi-Temporal Analysis

The vineyard multi-temporal dynamics can be better understood using the orthorectified results obtained via photogrammetric processing of the UAV-based imagery (Figure 5) though their visual inspection in a geographical information system (GIS) [69].

Orthophoto mosaics can be used to detect missing grapevines and to manage vineyard in-field operations [64]. Vegetation indices (e.g., NDVI) can provide an overall assessment of vegetation vigour and potentially detect phytosanitary problems, such as *flavescence dorée* [37] and esca [70]. Leaf canopy temperature maps and CWSI can suppress the need to manually measure leaf water potential in the field [35]—a time-consuming approach, usually not performed in the whole vineyard—as well as be used for irrigation management [71].

In this study, an overall NDVI decline was noticeable from the third flight campaign onward (Figure 5a F4, F5). This was related to the grapevines' vegetative cycle and to the decline of inter-row vegetation. Regarding height values obtained from each flight campaign's CSM (Figure 5b), a clear distinction existed between grapevine and non-grapevine vegetation (e.g., soil and inter-row vegetation), except for in the first flight campaign (Figure 5b F1). Land surface temperature (Figure 5c) was clear-cut between flight campaigns. In fact, in the fourth and fifth flight campaigns (Figure 5c F4 and F5), there were some signs of the grapevines' water stress.

Removing non-grapevine elements from the vineyard imagery provided a different perspective on the results, as confirmed in Table 1. Indeed, this enabled the production of estimate parameters such as the overall inter-row vegetation and the grapevines' area and volume (Figure 6). The estimated grapevines' vegetation area in the first flight campaign was 81 m^2 (3% of the vineyard plot) and in the second flight campaign, a 181 m^2 growth took place (262 m^2, 9% of the vineyard plot). As for the third flight campaign, there was a growth of 255 m^2 to 518 m^2 (18% of the vineyard plot). In the following flight campaigns, the grapevines' vegetation area was reduced by 199 m^2 (−38%) to 319 m^2. Regarding the grapevines' canopy volume, it was modified by +634%, +160%, −41%, and −12% in between each successive flight campaign, respectively. Concerning the inter-row vegetation area, it presented a behaviour consistent with the available precipitation data (Figure 2). Indeed, it had a 214% growth in between the first two flight campaigns (from 181 m^2 to 569 m^2), representing 20% of the vineyard plot area. A steep decline was noticeable in the third and fourth flight campaigns (a decline of 95% to 26 m^2), followed by a growth in the last flight campaign (88 m^2). As such, the vineyard inter-row vegetation was a good indicator of soil water status. The same tendency had already been verified in Pádua et al. [64].

By comparing the mean, maximum, and minimum values observed in the different outcomes, either when considering the whole vineyard or only grapevines' vegetation (Table 1), there was a clear difference among the flight campaigns. Mean NDVI values were superior in all flight campaigns when considering only the grapevines' vegetation. The same tendency was verified in the CSM. This can be explained by the presence of a significant amount of lower values in the non-grapevine vegetation areas. However, the maximum NDVI values in the first, second, and fourth flight campaigns were registered in non-grapevine vegetation areas. Inter-row vegetation can account for this. Regarding maximum CSM values, they were similar in all flight campaigns, except for the first one, where the maximum height was detected in a non-grapevine area (probably a vineyard post). Minimum CSM and NDVI values were lower in non-grapevine areas. As for temperature-based outcomes (land surface temperature and CWSI), the opposite behaviour was found for the maximum values: they were located in non-grapevine vegetation areas. Mean temperature and CWSI values were lower in the grapevines' vegetation areas, as it was expected due to the existence of bare soil areas in the vineyard. Minimum temperature and CWSI values were similar in both approaches since they were found in the grapevines' vegetation areas. These results showed the importance of grapevines' vegetation segmentation when analyzing a whole vineyard plot. The grapevines' vegetation segmentation could improve the results in studies where this operation was not automatically performed, which is beneficial for removing non-grapevine elements from the analysis. Such an automatic procedure could help in the evaluation of vegetation indices [21], to detect *flavescence dorée* and grapevine trunk diseases [72], and to estimate grapevines' biophysical and geometrical parameters [73].

4.2. Vigour Maps

Vigour maps generated when considering the whole vineyard provided an overall perspective (Figure 7a) about the studied area. Indeed, influences from bare soil and especially inter-row vegetation were clearly noticeable. Generally, the medium vigour class had the smaller area (Figure 8a) and the high vigour class encompassed the majority of the grapevines' canopy volume (Figure 8b). The latter was, on average, 150% higher than the other vigour classes. A high homogeneity was verified for the last two flight campaigns. The same happened from the second to the last flight campaigns, when computing height maps from the CSM and all campaigns with CWSI. The whole vineyard was considered in both. Positive correlation values were found for the LMI (Table 2). Moreover, the verified homogeneity resulted in meaningful HH and LL areas when comparing vigour maps with CSM and CWSI in the same flight campaign.

Different results were obtained in the other two approaches, where only the grapevines' vegetation was considered to create vigour maps. The higher incidence of missing grapevine plants in the left area of the vineyard remained almost the same throughout all flight campaigns. This was not noticeable in the first two flight campaigns' vigour maps when considering the whole vineyard, probably due to an effect caused by inter-row vegetation. Other studies reported similar trends using vigour maps produced from the UAV-based NDVI [22,51] when excluding inter-row vegetation. Moreover, Vanegas et al. [74] found positive correlations when comparing vigour maps created from UAV-based data and a vineyard expert assessment.

As for vineyard area, when considering only grapevine vegetation, it presented a more balanced behaviour. The third approach, normalized grapevines' vegetation, showed a considerable area of medium vigour class, particularly in the last three flight campaigns due to the fixed cut-off values to create vigour classes. Both approaches, grapevines' vegetation and normalized grapevines' vegetation, presented insignificant grapevines' canopy volume values in the lower classes. Moreover, when considering normalized grapevines' vegetation, canopy volume values were predominant in the medium vigour class, in agreement with the vineyard's overall vegetative growth and decline (growth from first to the third flight campaigns and decline onward). Similar relations between vigour and the grapevines' canopy volume were reported in other studies [73,75]. A higher heterogeneity was verified when observing both the CSM and CWSI maps generated with the approach that considered the normalized grapevines' vegetation. In fact, when analyzing the CWSI maps from the last two flight campaigns (Figure 10), a period of the grapevines' water stress was observed. However, this period was not clearly distinguishable in a visual map inspection based on data from the first approach (when the whole vineyard was considered). These correlations were observed in the BILISA cluster maps (Figure 12b,c), where areas with a high vigour showed a HL relationship with the CWSI maps, and significant agreements could be observed in the third flight campaign. A similar trend was reported in Matese and Di Gennaro [23]. Significant spatial associations were found in all approaches—whole vineyard, grapevines' vegetation, and normalized grapevines' vegetation—when analyzing the height class maps (Figure 11). Although lesser associations were found in the first flight campaign, this can be explained with the grapevines' growth cycle. In this case, significant HL areas were found in the approaches considering only the grapevine vegetation. Similarly, Matese et al. [75] observed that some areas with a higher vigour were linked to areas with higher heights.

This study analyzed a vineyard's behaviour throughout a season with a multi-temporal approach based on multispectral data acquired using a UAV. Furthermore, correlations between the different digital outcomes were found. This presents a potential tool for multi-temporal vineyard assessment and can serve as a base to provide prescription maps, similar to Campos et al. [52], since they can be correlated with agronomical variables (e.g., yield, berry weight, and total soluble solids), as shown in Matese et al. [56]. Indeed, patterns detected when comparing vigour maps from consecutive flight campaigns (Figure 13) highlighted differences in the multi-temporal data, which helps to understand local and spatial grapevines' vegetative development dynamics throughout the season. However, filtered data considering only values representing grapevines' vegetation, therefore representing the plants' physiological status, was proven to be more reliable when comparing the evaluated approaches (Table 2); that is to say, it had a higher overall correlation. As such, it stands to be an excellent tool for decision support systems within vineyard management processes.

5. Conclusions

Climate change can heighten key environmental vectors that negatively impact vineyards. Grapevines can be weakened by both water stress and exposure to higher temperatures, which will increase their vulnerability to phytosanitary issues. UAVs equipped with different sensors can be used to regularly monitor grapevines, documenting changes in the vegetation or signs of diseases/infestation, as well as any stress caused by environmental constraints.

In this context, the need to evaluate current vineyard behaviour is crucial to proceed toward PV. Vigour maps can help to provide relevant insights, helping farmers and/or winemakers to understand their vineyards status and enabling timely actions to tackle problematic areas or observing response to treatments. Furthermore, the methods employed in this study to filter out non-grapevine vegetation presented a better vineyard representation, which can be used to assess a vineyard's variability, but also to help in managing field-operations, such as those to inspect grapevines or to improve grapevines' physiological status.

The use of methods to compare spatial correlations allowed us to obtain a spatial distribution of significant clusters among the different approaches evaluated for creating vigour maps. The importance of using different UAV-based outcomes to estimate biophysical and geometrical parameters shows the suitability of UAVs as a remote sensing platform for vineyard multi-temporal monitoring operations. This study allowed us to conclude that the need for UAV-based data can be tracked according to a vineyard's phenology. Moreover, TIR data should be acquired in periods of higher temperatures to assess areas potentially affected by water stress. Nevertheless, the analysis presented in this study should be assessed in other vineyard types, such as those with irrigation systems, with a lower rate of missing grapevines, and in other wine producing regions with different grapevine training parameters.

Author Contributions: Conceptualization, L.P. and J.J.S.; data curation, L.P. and P.M.; formal analysis, L.P.; funding acquisition, A.S., E.P., and J.J.S.; investigation, L.P., P.M., T.A., and N.G.; methodology, L.P. and J.J.S.; project administration, L.P., E.P., and J.J.S.; resources, L.P., T.A., E.P., and J.J.S.; software, L.P., P.M., and T.A.; supervision, A.S., E.P., and J.J.S.; validation, L.P. and J.J.S.; visualization, L.P. and N.G.; writing—original draft, L.P. and P.M.; writing—review and editing, A.S., E.P., and J.J.S.

Funding: Financial support provided by the FCT-Portuguese Foundation for Science and Technology (PD/BD/150260/2019) to Pedro Marques, under the Doctoral Programme "Agricultural Production Chains – from fork to farm" (PD/00122/2012) and to Luís Pádua (SFRH/BD/139702/2018).

Acknowledgments: The authors would like to thank Miguel Fonseca for providing the photographs of the vineyard.

Conflicts of Interest: The authors declare no conflict of interest.

References

1. Gilbert, N. Water under pressure. *Nature* **2012**, *483*, 256–257. [CrossRef]
2. Wang, J.; Mendelsohn, R.; Dinar, A.; Huang, J.; Rozelle, S.; Zhang, L. The impact of climate change on China's agriculture. *Agric. Econ.* **2009**, *40*, 323–337. [CrossRef]
3. Asseng, S.; Ewert, F.; Martre, P.; Rötter, R.P.; Lobell, D.B.; Cammarano, D.; Kimball, B.A.; Ottman, M.J.; Wall, G.W.; White, J.W.; et al. Rising temperatures reduce global wheat production. *Nat. Clim. Chang.* **2015**, *5*, 143–147. [CrossRef]
4. Gerald, C.N.; Mark, W.R.; Jawoo, K.; Richard, R.; Timothy, S.; Tingju, Z.; Claudia, R.; Siwa, M.; Amanda, P.; Miroslav, B.; et al. *Climate Change: Impact on Agriculture and Costs of Adaptation*; International Food Policy Research Institute: Washington, DC, USA, 2009; ISBN 978-0-89629-535-3.
5. Lobell, D.B.; Gourdji, S.M. The Influence of Climate Change on Global Crop Productivity. *Plant Physiol.* **2012**, *160*, 1686–1697. [CrossRef] [PubMed]
6. Schmidhuber, J.; Tubiello, F.N. Global food security under climate change. *Proc. Natl. Acad. Sci. USA* **2007**, *104*, 19703–19708. [CrossRef] [PubMed]
7. United Nations Transforming Our World: the 2030 Agenda for Sustainable Development. Resolution Adopted by the General Assembly. United Nations General Assembly, New York, NY, USA. 25 September 2015. Available online: https://sustainabledevelopment.un.org/post2015/transformingourworld (accessed on 29 July 2018).
8. Gago, J.; Douthe, C.; Coopman, R.E.; Gallego, P.P.; Ribas-Carbo, M.; Flexas, J.; Escalona, J.; Medrano, H. UAVs challenge to assess water stress for sustainable agriculture. *Agric. Water Manag.* **2015**, *153*, 9–19. [CrossRef]
9. Cancela, J.; Fandiño, M.; Rey, B.; Dafonte, J.; González, X. Discrimination of irrigation water management effects in pergola trellis system vineyards using a vegetation and soil index. *Agric. Water Manag.* **2017**, *183*, 70–77. [CrossRef]
10. Gebbers, R.; Adamchuk, V.I. Precision Agriculture and Food Security. *Science* **2010**, *327*, 828–831. [CrossRef]
11. Zarco-Tejada, P.J.; Hubbard, N.; Loudjani, P. *Precision Agriculture: An Opportunity for EU Farmers—Potential Support with the CAP 2014–2020*; Joint Research Centre (JRC) of the European Commission Monitoring Agriculture ResourceS (MARS): Brussels, Belgium, 2014.
12. Morais, R.; Fernandes, M.A.; Matos, S.G.; Serôdio, C.; Ferreira, P.J.S.G.; Reis, M.J.C.S. A ZigBee multi-powered wireless acquisition device for remote sensing applications in precision viticulture. *Comput. Electron. Agric.* **2008**, *62*, 94–106. [CrossRef]
13. Matese, A.; Toscano, P.; Di Gennaro, S.F.; Genesio, L.; Vaccari, F.P.; Primicerio, J.; Belli, C.; Zaldei, A.; Bianconi, R.; Gioli, B. Intercomparison of UAV, Aircraft and Satellite Remote Sensing Platforms for Precision Viticulture. *Remote Sens.* **2015**, *7*, 2971–2990. [CrossRef]
14. Proffitt, A.P.B.; Bramley, R.; Lamb, D.; Winter, E. *Precision Viticulture: A New Era in Vineyard Management and Wine Production*; Winetitles: Ashford, Australia, 2006; ISBN 978-0-9756850-4-4.
15. Steyn, J.; Tudó, J.L.A.; Benavent, J.L.A. Grapevine vigour and within vineyard variability: A review. *Int. J. Sci. Eng. Res.* **2016**, *7*, 1056–1065.
16. Baofeng, S.; Jinru, X.; Chunyu, X.; Yulin, F.; Yuyang, S.; Fuentes, S. Digital surface model applied to unmanned aerial vehicle based photogrammetry to assess potential biotic or abiotic effects on grapevine canopies. *Int. J. Agric. Biol. Eng.* **2016**, *9*, 119.
17. Adão, T.; Hruška, J.; Pádua, L.; Bessa, J.; Peres, E.; Morais, R.; Sousa, J.J. Hyperspectral Imaging: A Review on UAV-Based Sensors, Data Processing and Applications for Agriculture and Forestry. *Remote Sens.* **2017**, *9*, 1110. [CrossRef]
18. Pádua, L.; Vanko, J.; Hruška, J.; Adão, T.; Sousa, J.J.; Peres, E.; Morais, R. UAS, sensors, and data processing in agroforestry: A review towards practical applications. *Int. J. Remote Sens.* **2017**, *38*, 2349–2391. [CrossRef]
19. Rouse, J.W., Jr.; Haas, R.H.; Schell, J.A.; Deering, D.W. Monitoring Vegetation Systems in the Great Plains with Erts. *NASA Spec. Publ.* **1974**, *351*, 309.
20. Bendig, J.; Yu, K.; Aasen, H.; Bolten, A.; Bennertz, S.; Broscheit, J.; Gnyp, M.L.; Bareth, G. Combining UAV-based plant height from crop surface models, visible, and near infrared vegetation indices for biomass monitoring in barley. *Int. J. Appl. Earth Obs. Geoinf.* **2015**, *39*, 79–87. [CrossRef]

21. Candiago, S.; Remondino, F.; De Giglio, M.; Dubbini, M.; Gattelli, M. Evaluating Multispectral Images and Vegetation Indices for Precision Farming Applications from UAV Images. *Remote Sens.* **2015**, *7*, 4026–4047. [CrossRef]

22. Primicerio, J.; Gay, P.; Ricauda Aimonino, D.; Comba, L.; Matese, A.; di Gennaro, S.F. NDVI-based vigour maps production using automatic detection of vine rows in ultra-high resolution aerial images. In *Precision agriculture'15*; Wageningen Academic Publishers: Wageningen, The Netherlands, 2015; pp. 465–470, ISBN 978-90-8686-267-2.

23. Matese, A.; Di Gennaro, S. Practical Applications of a Multisensor UAV Platform Based on Multispectral, Thermal and RGB High Resolution Images in Precision Viticulture. *Agriculture* **2018**, *8*, 116. [CrossRef]

24. Idso, S.B.; Jackson, R.D.; Pinter, P.J.; Reginato, R.J.; Hatfield, J.L. Normalizing the stress-degree-day parameter for environmental variability. *Agric. Meteorol.* **1981**, *24*, 45–55. [CrossRef]

25. Alderfasi, A.A.; Nielsen, D.C. Use of crop water stress index for monitoring water status and scheduling irrigation in wheat. *Agric. Water Manag.* **2001**, *47*, 69–75. [CrossRef]

26. Berni, J.A.J.; Zarco-Tejada, P.J.; Sepulcre-Cantó, G.; Fereres, E.; Villalobos, F. Mapping canopy conductance and CWSI in olive orchards using high resolution thermal remote sensing imagery. *Remote Sens. Environ.* **2009**, *113*, 2380–2388. [CrossRef]

27. Bellvert, J.; Zarco-Tejada, P.J.; Girona, J.; Fereres, E. Mapping crop water stress index in a 'Pinot-noir' vineyard: Comparing ground measurements with thermal remote sensing imagery from an unmanned aerial vehicle. *Precis. Agric* **2013**, *15*, 361–376. [CrossRef]

28. Sullivan, D.G.; Fulton, J.P.; Shaw, J.N.; Bland, G. Evaluating the sensitivity of an unmanned thermal infrared aerial system to detect water stress in a cotton canopy. *Trans. ASABE* **2007**, *50*, 1963–1969. [CrossRef]

29. Banerjee, K.; Krishnan, P.; Mridha, N. Application of thermal imaging of wheat crop canopy to estimate leaf area index under different moisture stress conditions. *Biosyst. Eng.* **2018**, *166*, 13–27. [CrossRef]

30. Liu, T.; Li, R.; Zhong, X.; Jiang, M.; Jin, X.; Zhou, P.; Liu, S.; Sun, C.; Guo, W. Estimates of rice lodging using indices derived from UAV visible and thermal infrared images. *Agric. For. Meteorol.* **2018**, *252*, 144–154. [CrossRef]

31. Quebrajo, L.; Perez-Ruiz, M.; Pérez-Urrestarazu, L.; Martínez, G.; Egea, G. Linking thermal imaging and soil remote sensing to enhance irrigation management of sugar beet. *Biosyst. Eng.* **2018**, *165*, 77–87. [CrossRef]

32. Romano, G.; Zia, S.; Spreer, W.; Sanchez, C.; Cairns, J.; Araus, J.L.; Müller, J. Use of thermography for high throughput phenotyping of tropical maize adaptation in water stress. *Comput. Electron. Agric.* **2011**, *79*, 67–74. [CrossRef]

33. Kalisperakis, I.; Stentoumis, C.; Grammatikopoulos, L.; Karantzalos, K. Leaf area index estimation in vineyards from UAV hyperspectral data, 2D image mosaics and 3D canopy surface models. *Int. Arch. Photogramm. Remote Sens. Spat. Inf. Sci.* **2015**, *40*, 299. [CrossRef]

34. Mathews, A.J.; Jensen, J.L.R. Visualizing and Quantifying Vineyard Canopy LAI Using an Unmanned Aerial Vehicle (UAV) Collected High Density Structure from Motion Point Cloud. *Remote Sens.* **2013**, *5*, 2164–2183. [CrossRef]

35. Baluja, J.; Diago, M.P.; Balda, P.; Zorer, R.; Meggio, F.; Morales, F.; Tardaguila, J. Assessment of vineyard water status variability by thermal and multispectral imagery using an unmanned aerial vehicle (UAV). *Irrig. Sci.* **2012**, *30*, 511–522. [CrossRef]

36. Romero, M.; Luo, Y.; Su, B.; Fuentes, S. Vineyard water status estimation using multispectral imagery from an UAV platform and machine learning algorithms for irrigation scheduling management. *Comput. Electron. Agric.* **2018**, *147*, 109–117. [CrossRef]

37. Albetis, J.; Duthoit, S.; Guttler, F.; Jacquin, A.; Goulard, M.; Poilvé, H.; Féret, J.B.; Dedieu, G. Detection of Flavescence dorée Grapevine Disease Using Unmanned Aerial Vehicle (UAV) Multispectral Imagery. *Remote Sens.* **2017**, *9*, 308. [CrossRef]

38. Matese, A.; Primicerio, J.; Di Gennaro, F.; Fiorillo, E.; Vaccari, F.P.; Genesio, L. Development and Application of an Autonomous and Flexible Unmanned Aerial Vehicle for Precision Viticulture. *Acta Hortic.* **2013**, 63–69. [CrossRef]

39. Martín, P.; Zarco-Tejada, P.J.; González, M.R.; Berjón, A. Using hyperspectral remote sensing to map grape quality in "Tempranillo" vineyards affected by iron deficiency chlorosis. *VITIS J. Grapevine Res.* **2015**, *46*, 7.

40. Burgos, S.; Mota, M.; Noll, D.; Cannelle, B. Use of very high-resolution airborne images to analyse 3D canopy architecture of a vineyard. *Int. Arch. Photogramm. Remote Sens. Spat. Inf. Sci.* **2015**, *40*, 399. [CrossRef]

41. Comba, L.; Gay, P.; Primicerio, J.; Ricauda Aimonino, D. Vineyard detection from unmanned aerial systems images. *Comput. Electron. Agric.* **2015**, *114*, 78–87. [CrossRef]
42. Nolan, A.; Park, S.; Fuentes, S.; Ryu, D.; Chung, H. Automated detection and segmentation of vine rows using high resolution UAS imagery in a commercial vineyard. In Proceedings of the 21st International Congress on Modelling and Simulation, Gold Coast, Australia, 1 December 2015; Volume 29, pp. 1406–1412.
43. Poblete-Echeverría, C.; Olmedo, G.F.; Ingram, B.; Bardeen, M. Detection and Segmentation of Vine Canopy in Ultra-High Spatial Resolution RGB Imagery Obtained from Unmanned Aerial Vehicle (UAV): A Case Study in a Commercial Vineyard. *Remote Sens.* **2017**, *9*, 268. [CrossRef]
44. Weiss, M.; Baret, F. Using 3D Point Clouds Derived from UAV RGB Imagery to Describe Vineyard 3D Macro-Structure. *Remote Sens.* **2017**, *9*, 111. [CrossRef]
45. Comba, L.; Biglia, A.; Ricauda Aimonino, D.; Gay, P. Unsupervised detection of vineyards by 3D point-cloud UAV photogrammetry for precision agriculture. *Comput. Electron. Agric.* **2018**, *155*, 84–95. [CrossRef]
46. Ponti, M.P. Segmentation of Low-Cost Remote Sensing Images Combining Vegetation Indices and Mean Shift. *IEEE Geosci. Remote Sens. Lett.* **2013**, *10*, 67–70. [CrossRef]
47. Peña-Barragán, J.M.; Ngugi, M.K.; Plant, R.E.; Six, J. Object-based crop identification using multiple vegetation indices, textural features and crop phenology. *Remote Sens. Environ.* **2011**, *115*, 1301–1316. [CrossRef]
48. Jiménez-Brenes, F.M.; López-Granados, F.; Torres-Sánchez, J.; Peña, J.M.; Ramírez, P.; Castillejo-González, I.L.; Castro, A.I. de Automatic UAV-based detection of Cynodon dactylon for site-specific vineyard management. *PLoS ONE* **2019**, *14*, e0218132. [CrossRef] [PubMed]
49. Proffitt, T.; Turner, N. Reducing vineyard management overheads from above. *Aust. N. Z. Grapegrow. Winemak.* **2017**, *647*, 46.
50. van Leeuwen, C. Terroir: The effect of the physical environment on vine growth, grape ripening and wine sensory attributes. In *Managing Wine Quality*; Reynolds, A.G., Ed.; Woodhead Publishing Series in Food Science, Technology and Nutrition; Woodhead Publishing: Cambridge, UK, 2010; pp. 273–315, ISBN 978-1-84569-484-5.
51. Khaliq, A.; Comba, L.; Biglia, A.; Ricauda Aimonino, D.; Chiaberge, M.; Gay, P. Comparison of Satellite and UAV-Based Multispectral Imagery for Vineyard Variability Assessment. *Remote Sens.* **2019**, *11*, 436. [CrossRef]
52. Campos, J.; Llop, J.; Gallart, M.; García-Ruiz, F.; Gras, A.; Salcedo, R.; Gil, E. Development of canopy vigour maps using UAV for site-specific management during vineyard spraying process. *Precis. Agric.* **2019**, 1–21. [CrossRef]
53. Costa Ferreira, A.-M.; Germain, C.; Homayouni, S.; Da Costa, J.P.; Grenier, G.; Marguerit, E.; Roby, J.P.; Van Leeuwen, C. Transformation of high resolution aerial images in vine vigour maps at intra-block scale by semi-automatic image processing. In Proceedings of the International Symposium of the GESCO, Porec, Croatia, 20–23 June 2007; pp. 1372–1381.
54. Rey-Caramés, C.; Diago, M.P.; Martín, M.P.; Lobo, A.; Tardaguila, J. Using RPAS Multi-Spectral Imagery to Characterise Vigour, Leaf Development, Yield Components and Berry Composition Variability within a Vineyard. *Remote Sens.* **2015**, *7*, 14458–14481. [CrossRef]
55. Bonilla, I.; de Toda, F.M.; Martínez-Casasnovas, J.A. Vine vigor, yield and grape quality assessment by airborne remote sensing over three years: Analysis of unexpected relationships in cv. Tempranillo. *Span. J. Agric. Res.* **2015**, *13*, 0903. [CrossRef]
56. Matese, A.; Di Gennaro, S.F.; Santesteban, L.G. Methods to compare the spatial variability of UAV-based spectral and geometric information with ground autocorrelated data. A case of study for precision viticulture. *Comput. Electron. Agric.* **2019**, *162*, 931–940. [CrossRef]
57. Marcal, A.R.S.; Cunha, M. Vineyard monitoring in Portugal using multi-sensor satellite images. In *GeoInformation in Europe*; Millpress: Rotterdam, The Netherlands, 2007; p. 704, ISBN 978-90-5966-061-8.
58. Espinoza, C.Z.; Khot, L.R.; Sankaran, S.; Jacoby, P.W. High Resolution Multispectral and Thermal Remote Sensing-Based Water Stress Assessment in Subsurface Irrigated Grapevines. *Remote Sens.* **2017**, *9*, 961. [CrossRef]
59. Di Gennaro, S.F.; Matese, A.; Gioli, B.; Toscano, P.; Zaldei, A.; Palliotti, A.; Genesio, L. Multisensor approach to assess vineyard thermal dynamics combining high-resolution unmanned aerial vehicle (UAV) remote sensing and wireless sensor network (WSN) proximal sensing. *Sci. Hortic.* **2017**, *221*, 83–87. [CrossRef]

60. Bellvert, J.; Zarco-Tejada, P.J.; Marsal, J.; Girona, J.; González-Dugo, V.; Fereres, E. Vineyard irrigation scheduling based on airborne thermal imagery and water potential thresholds. *Aust. J. Grape Wine Res.* **2016**, *22*, 307–315. [CrossRef]

61. Matese, A.; Baraldi, R.; Berton, A.; Cesaraccio, C.; Di Gennaro, S.F.; Duce, P.; Facini, O.; Mameli, M.G.; Piga, A.; Zaldei, A. Estimation of Water Stress in Grapevines Using Proximal and Remote Sensing Methods. *Remote Sens.* **2018**, *10*, 114. [CrossRef]

62. Pádua, L.; Marques, P.; Hruška, J.; Adão, T.; Bessa, J.; Sousa, A.; Peres, E.; Morais, R.; Sousa, J.J. Vineyard properties extraction combining UAS-based RGB imagery with elevation data. *Int. J. Remote Sens.* **2018**, *39*, 5377–5401. [CrossRef]

63. Richardson, A.D.; Jenkins, J.P.; Braswell, B.H.; Hollinger, D.Y.; Ollinger, S.V.; Smith, M.-L. Use of digital webcam images to track spring green-up in a deciduous broadleaf forest. *Oecologia* **2007**, *152*, 323–334. [CrossRef] [PubMed]

64. Pádua, L.; Marques, P.; Hruška, J.; Adão, T.; Peres, E.; Morais, R.; Sousa, J.J. Multi-Temporal Vineyard Monitoring through UAV-Based RGB Imagery. *Remote Sens.* **2018**, *10*, 1907. [CrossRef]

65. Anselin, L. Local indicators of spatial association—LISA. *Geogr. Anal.* **1995**, *27*, 93–115. [CrossRef]

66. Moran, P.A. Notes on continuous stochastic phenomena. *Biometrika* **1950**, *37*, 17–23. [CrossRef]

67. Anselin, L. *Modern Spatial Econometrics in Practice: A Guide to GeoDa, GeoDaSpace and PySAL*; GeoDa Press: Chicago, IL, USA, 2014; ISBN 0-9863421-0-6.

68. Anselin, L.; Syabri, I.; Kho, Y. GeoDa: An Introduction to Spatial Data Analysis. *Geogr. Anal.* **2006**, *38*, 5–22. [CrossRef]

69. Ozdemir, G.; Sessiz, A.; Pekitkan, F.G. Precision viticulture tools to production of high quality grapes. *Sci. Pap. Ser. B Hortic.* **2017**, *61*, 209–218.

70. Gennaro, S.F.D.; Battiston, E.; Marco, S.D.; Facini, O.; Matese, A.; Nocentini, M.; Palliotti, A.; Mugnai, L. Unmanned Aerial Vehicle (UAV)-based remote sensing to monitor grapevine leaf stripe disease within a vineyard affected by esca complex. *Phytopathol. Mediterr.* **2016**, *55*, 262–275.

71. Bellvert, J.; Girona, J. The use of multispectral and thermal images as a tool for irrigation scheduling in vineyards. In *The Use of Remote Sensing and Geographic Information Systems for Irrigation Management in Southwest Europe*; Erena, M., López-Francos, A., Montesinos, S.Y., Berthoumieu, J.F., Eds.; Options Méditerranéennes. Serie B: Studies and Researchs; CIHEAM-IMIDA-SUDOE Interreg IVB (EU-ERDF): Zaragoza, Spain, 2012; pp. 131–138.

72. Albetis, J.; Jacquin, A.; Goulard, M.; Poilvé, H.; Rousseau, J.; Clenet, H.; Dedieu, G.; Duthoit, S. On the Potentiality of UAV Multispectral Imagery to Detect Flavescence dorée and Grapevine Trunk Diseases. *Remote Sens.* **2019**, *11*, 23. [CrossRef]

73. Caruso, G.; Tozzini, L.; Rallo, G.; Primicerio, J.; Moriondo, M.; Palai, G.; Gucci, R. Estimating biophysical and geometrical parameters of grapevine canopies ('Sangiovese') by an unmanned aerial vehicle (UAV) and VIS-NIR cameras. *VITIS J. Grapevine Res. Vitis* **2017**, *56*, 63–70.

74. Vanegas, F.; Bratanov, D.; Powell, K.; Weiss, J.; Gonzalez, F. A Novel Methodology for Improving Plant Pest Surveillance in Vineyards and Crops Using UAV-Based Hyperspectral and Spatial Data. *Sensors* **2018**, *18*, 260. [CrossRef] [PubMed]

75. Matese, A.; Di Gennaro, S.F.; Berton, A. Assessment of a canopy height model (CHM) in a vineyard using UAV-based multispectral imaging. *Int. J. Remote Sens.* **2016**, *38*, 2150–2160. [CrossRef]

agronomy

MDPI

Article

Genetical, Morphological and Physicochemical Characterization of the Autochthonous Cultivar 'Uva Rey' (*Vitis vinifera* L.)

Pau Sancho-Galán, Antonio Amores-Arrocha *, Víctor Palacios and Ana Jiménez-Cantizano

Department of Chemical Engineering and Food Technology, Vegetal Production Area, University of Cadiz, Agrifood Campus of International Excellence (ceiA3), IVAGRO, P.O. Box 40, 11510 Puerto Real, Spain; pau.sancho@uca.es (P.S.-G.); victor.palacios@uca.es (V.P.); ana.jimenezcantizano@uca.es (A.J.-C.)
* Correspondence: antonio.amores@uca.es; Tel.: +34-9-5601-6398

Received: 31 July 2019; Accepted: 12 September 2019; Published: 18 September 2019

Abstract: 'Uva Rey' is considered an Andalusian (Spain) ancient autochthonous cultivar with hard white grapes used for the production of wine and raisins and also for raw consumption. Currently, this cultivar is not included in the official register of Spanish grapevine varieties and there is neither a description nor a characterization that could facilitate its insertion in this register. In order to study this genetic resource, a genetic and morphological characterization of 'Uva Rey' has been carried out in comparison with 'Palomino Fino', the main cultivar in Andalusia (Spain). Additionally, grape must physicochemical characterization and grape berry texture profile analyses were performed. Genetically, 'Uva Rey' was synonymous with the cultivar 'De Rey'. 'Uva Rey' grape must physicochemical results showed a lower sugar concentration and a higher malic acid content compared to 'Palomino Fino' must, while the analysis of the grape berry texture profile proved to be more consistent and cohesive. These results can be attributed to the longer phenological cycle presented by 'Uva Rey'. All these facts could lead to consideration of 'Uva Rey' as a cultivar for the production of white wines in warm climate regions.

Keywords: *Vitis vinifera*; autochthonous cultivar; 'Uva Rey'

---

1. Introduction

Grapevine (*Vitis vinifera* L.) is one of the most ancient and important fruit crops worldwide [1]. Around 12,500 cultivars have been registered in the *Vitis* International Variety Catalogue [2]. However, based on their DNA profiles, the number of grapevine varieties is estimated at around 5000, many of them closely related [3,4].

Nowadays, 7.4 mHa of the Earth area is covered by grapevines, with Spain being the first country in terms of cultivated land extension. Spanish vineyards cover thousands of hectares and produce approximately 44.4 mHL of wine per year [5]. For that reason, viticulture could be considered as one of the most important socioeconomic sectors in the Spanish agro-industrial network. Grapevine cultivation throughout the country, and the significance over time, have led to a grapevine heritage of great magnitude. Spain's varietal heritage had continuously increased from its origin until the arrival of diseases and pathogens from America (mildews and *Phylloxera*) [6]. According to García de los Salmones [7], the first *Phylloxera* outbreak in Spain was detected in Malaga (Andalusia) in 1876. From that moment on, this pathogen spread throughout the whole country and destroyed more than 1,000,000 ha, which caused serious damages to the Spanish native germplasm [8]. In order to preserve the maximum number of *Vitis vinifera* genetic diversity, a number of germplasm banks were created. 'El Encín', the most important germplasm bank in Spain, was established in 1914 in Alcalá de Henares

(Madrid, Spain) [9]. Later on, the currently germplasm bank known as 'Rancho de la Merced', was created in 1940, with the first collection of grapevines in Jerez de la Frontera (Andalusia, Spain) [10].

From then on, the prospection, collection and conservation of different grapevine cultivars as a genetic resource have been the subject of numerous studies that intend to preserve those cultivars considered as autochthonous [7–11]. For the identification of that genetic material, molecular characterization using Simple Sequence Repeats (SSR) markers [12], ampelographic [4] and physicochemical [13] techniques have been used. Grapevine genotypes are highly heterozygous and the relevance of near-homozygous lines was not considered until recently due to the need to generate high quality reference sequences [14], and has been maintained in cultivated plants through vegetative propagation [15].

Modern wine industries only use a limited number of *Vitis vinifera* cultivars [16]. In Spain, by virtue of the Spanish Royal-Decree-Law (RD) 1338/2018, only those varieties that have been properly registered can be planted [17]. However, there is a current trend towards the production of genuine and characteristic wines [18]. Currently, the changing climate is expected to impose new challenges to varietal selection. Since grapevine varietal suitability is strongly linked to regional environmental conditions, growers are prone to select varieties that are best suited to these changing agroclimatic factors [19].

As a result, autochthonous cultivars, such as 'Uva Rey' would require to be identified and characterized, since they were already used for wine making in the 19th century in southwestern regions in Andalusia [20]. Roxas Clemente [21] included this variety in Tribe III of the First Section and indicated that it was cultivated under different denominations in different districts within Cadiz and Seville provinces in Andalusia. Regarding its grapes, this author described them as very large, round, somewhat golden and with a long cycle. With regards to its winemaking potential, Abela [22] confirmed that this grape variety was able to produce fine wines with plenty of mouth-feel and acidity.

The main objective of this research work is to complete the characterization of the cultivar 'Uva Rey' as currently kept in a specific vineyard located in Andalusia (Spain). For this purpose, the genetic identification, the ampelographic characterization, the grape berry texture profile analysis and the physicochemical characterization of the grape musts have been carried out.

2. Materials and Methods

2.1. Plant Material and Experimental Design

A total of 10 plants of 'Uva Rey' from a vineyard in the town of Chiclana de la Frontera municipal district (Andalusia, Spain) were selected (lat. 36°27′30.6″ N; long. 6°05′46.2″ W; 69 m above sea level). In addition, 'Palomino Fino' was used as a reference cultivar for all the studies, as it is the most widespread variety in the southwest of Andalusia [23]. Both cultivars were 15 years old and had been grown with the same vine spacing (2.30 × 1.15 m) as well as trained according to the 'Vara y Pulgar' (stick and thumb) system. Additional Figures S1a–c and S2a–c show the temperature, humidity, radiation and rainfall during the period from July (veraison) to September (harvest) for 2016 and 2017 respectively. For the genetic characterization of the cultivar, four varieties: 'Cabernet Sauvignon', 'Chardonnay', 'Muscat a Petits Grains Blancs' and 'Pinot Noir' were included as reference to compare their genotype databases and confirm the new cultivar accession identity (Table 1).

The morphological description and the texture profile analysis (TPA) of the berries as well as the grape must characterization were carried out for 'Uva Rey' and 'Palomino Fino' cultivars from the same vineyard and in two consecutive years (2016 and 2017) in order to study the vintage effect on the different cultivars. Both cultivars were grown at the same vineyard and under the same agroclimatic conditions, the cultural practices and were harvested in the same period (first week in September). In order to minimize variability due to grapevine sampling, Santesteban et al. [24] criterion was applied. For this purpose, the trunk cross sectional area (TCSA) of a total of 50 vines were measured at 30 cm height using a digital Verner calliper Maurer 93,110 (Padova, Italy). Of all the vines measured, 10 were selected and marked as their TCSA value was the closest to the TCSA average ± 10%.

2.2. Microsatellite Analysis

Two young fresh leaves from each accession were collected at the vineyard and kept at –80 °C until analysis. DNA extraction was carried out using a DNeasy Plant Mini Kit (Qiagen, Hilden, Germany). Varietal identification was performed using 22 nuclear microsatellite loci. The first set of 20 microsatellite loci located in the 19 linkage groups of grapevine genome (VMC1B11 (GeneBank, Accession Number BV681754), VMC4F3-1 [25]; VVMD5, VVMD7, VVMD21, VVMD24, VVMD25, VVMD27, VVMD28, VVMD28, VVMD32 [26,27]; VVS2 [28]; VV1B01, VVIH54, VVIN16, VVIN73, VVIP31, VVIP 60, VVIQ52, VVIV37, VVIV67 [29]) were analysed as described by Vargas et al. [30], using two multiplex Polimerase Chain Reactions (PCR). An additional set of two microsatellite loci (VrZAG62 and VRZAG79) [31] were analyed following the conditions described in detail by Jiménez-Cantizano et al. [32], in order to complete the list of loci authorized by the International Organisation of Vine and Wine (OIV). PCR amplifications were performed using a 9700 thermocycler and the amplified products were separated by capillary electrophoresis using an automated sequencer ABI Prism 3130 (Applied Biosystems, Foster City, CA, USA). Fluorescent labelled fragments (6-FAM, VIC, PET and NED) were detected and sized using GeneMapper v. 3.7 and fragment lengths were assessed with the help of internal standards GeneScan-500 LIZTM (Applied Biosystems, Foster City, CA, USA). The microsatellite genotypes obtained after the analysis were compared with the genetic profiles provided by Lacombe et al. [33] and the data contained in the microsatellite databases *Vitis* International Variety Catalogue [34], Rancho de la Merced Germplasm Bank genotype database [35] and the *Vitis* Germplasm Bank at Finca el Encín [25,36,37]. The SSR profiles obtained were compared using the microsatellite toolkit v. 9.0 software [38].

2.3. Morphological Characterization

For the morphological analysis, Benito et al. [39] criterion was followed. A total of 10 young shoots, young and mature leaves, flowers, bunches and berries from each accession were analysed using 34 descriptors from the Organisation Internationale de la Vigne et du Vin descriptor list [40]. Each accession from two different vintages was described by five ampelographers and the modal value was selected as the final description.

2.4. Physicochemical Characterization of Grape Berries and Musts

Grapevine berries ($n = 50$) were evaluated using a texture-meter (Lloyd Material Testing Machine, West Sussex, UK) fitted with a 2 mm cylindrical flat probe at 1 mm/s. The results regarding consistency, firmness, work of penetration (WoP) and cohesiveness were calculated as the average values for 50 berries.

Once harvested, 5 kg of berries of each cultivar (500 g from each vine) were destemmed, grounded and pressed. pH determinations were carried out using a Crisson-2001 digital pH-meter (Loveland, CO, USA). Sugar concentration (°Bé) was determined using a calibrated Dujardin-Salleron hydrometer (Laboratories Dujardin-Salleron, Arcueil Cedex, France). Total acidity (TA) was calculated according to the official methods of analysis [41]. Ripening index (RI) was calculated following the equation proposed by Hidalgo [42]. Yeast assimilable nitrogen (YAN) was determined according to Aerny [43]. Citric, tartaric and malic acids were assessed following the methodology proposed by Sancho-Galán et al [44]. Organic acids concentrations were obtained by ionic chromatography using a Metrohm 930 compact IC Flex ionic chromatographer equipped with a conductimetric detector on a Metrosep Organic Acids column-250/7.8 (Herisau, Switzerland). Organic acids separation was performed using as eluent H_2SO_4 0.4 mM in a 12% acetone solution with an isocratic 0.4 mL/min flow. All the physicochemical measurements were destructive analysis and were conducted in triplicate to ensure statistical significance.

2.5. Statistical Analysis

Means and standard deviations were calculated using Microsoft Office Excel 2016 for Windows 10. Significant differences were evaluated by two-way ANOVA and Bonferroni's multiple range (BSD) test; $p < 0.05$ was considered significant (GraphPad Prism version 6.01 for Windows, GraphPad Software, San Diego, CA, USA).

3. Results

3.1. Microsatellite Analysis

The allele profiles obtained for 'Uva Rey' and the five reference cultivars at 22 microsatellite loci are shown in Table 1. The genotype obtained for 'Uva Rey' was compared with the Rancho de la Merced Germplasm Bank genotype database [14,35], the *Vitis* Germplasm Bank at the Finca El Encín [30,36,37] and European databases [33,34]. 'Uva Rey' showed the same genotype as 'Mantuo de Pilas' kept in Rancho de la Merced Germplasm Bank at 22 SSR loci and 'De Rey' at Finca El Encín at 20 SSR loci.

Table 1. Genetic profiles of 'Uva Rey' and reference cultivars at 22 microsatellite loci. Alleles sizes are given in base pairs.

Locus	'Uva Rey'		'Palomino Fino' [a]		'Cabernet Sauvignon' [a]		'Chardonnay' [a]		'Muscat a Petits Grains Blancs' [a]		'Pinot Noir' [a]	
VVIB01	307	307	291	307	291	291	289	295	291	295	289	295
VMC1b11	184	188	184	188	184	184	166	184	184	188	166	172
VMC4F31	184	190	176	206	174	178	174	180	168	206	174	180
VVMD5	224	232	226	238	228	238	232	236	226	324	226	236
VVMD7	244	246	236	246	236	236	236	240	323	246	236	240
VVMD21	243	249	243	249	249	257	249	249	249	265	249	249
VVMD24	209	209	209	209	209	217	209	217	213	217	215	217
VVMD25	238	252	240	240	238	246	238	252	240	246	238	246
VVMD27	180	182	186	194	176	190	182	190	180	194	186	190
VVMD28	246	248	238	250	236	238	220	230	248	270	220	238
VVMD32	270	270	254	256	238	238	238	270	262	270	238	270
VVIH54	166	168	166	166	166	182	164	168	166	166	164	168
VVIN16	151	153	151	151	153	153	151	151	149	149	151	159
VVIN73	264	264	256	264	264	268	264	266	264	264	264	266
VVIP31	176	190	188	190	188	188	180	184	184	188	180	180
VVIP60	318	326	318	322	306	314	318	322	318	318	318	320
VVIQ52	85	89	85	85	83	89	83	89	83	83	89	89
VVS2	131	142	131	144	137	151	135	142	131	131	135	151
VVIV37	161	161	163	167	163	163	153	163	163	165	153	163
VVIV67	372	375	364	366	364	372	364	372	364	375	364	372
VrZAG62	187	193	187	193	187	193	187	195	185	195	187	193
VrZAG79	242	248	250	260	246	246	242	244	250	254	238	244
Variety [b]	'De Rey'											

[a] Reference cultivars. [b] Prime names according to *Vitis* International Variety Catalogue (VIVC).

3.2. Morphological Characterization

Modal values for the ampelographic descriptions of 'Uva Rey' cultivar corresponding to years 2016 and 2017 are shown in Table 2 compared to the reference cultivar 'Palomino Fino'.

Table 2. Ampelographic description of 'Uva Rey' and 'Palomino Fino' cultivars using the International Organisation of Vine and Wine (OIV) descriptors.

Code	Descriptor	'Uva Rey'	'Palomino Fino'
OIV 001	Young shoot: opening of the shoot tip. 1 closed, 3 half open, 5 fully open.	5	5
OIV 003	Young shoot: intensity of anthocyanin coloration on prostrate hairs of the shoot tip. 1 none or very low, 3 low, 5 medium, 7 high, 9 very high.	3	5
OIV 004	Young shoot: density of prostrate hairs on the shoot tip. 1 none or very low, 3 low, 5 medium, 7 high, 9 very high.	7	5
OIV 006	Shoot: attitude (before tying). 1 erect, 3 semi-erect, 5 horizontal, 7 semi-drooping, 9 drooping.	3	3
OIV 007	Shoot: colour of the dorsal side of internodes. 1 green, 2 green and red, 3 red.	1	2
OIV 008	Shoot: colour of the ventral side of internodes. 1 green, 2 green and red, 3 red.	1	2
OIV 015-1	Shoot: distribution of anthocyanin coloration on the bud scales. 1 absent, 2 basal, 3 up to 3/4 of bud scale, 4 almost on the whole bud scale.	1	3
OIV 016	Shoot: number of consecutive tendrils. 1 two or less, 2 three or more.	1	1
OIV 051	Young leaf: colour of upper side of blade (4th leaf). 1 green, 2 yellow, 3 bronze, 4 copper-reddish.	3	3
OIV 053	Young leaf: density of prostrate hairs between main veins on lower side of blade (4th leaf). 1 none or very low, 3 low, 5, medium, 7 high, 9 very high.	9	5
OIV 065	Mature leaf: size of blade. 1 very small, 3, small, 5 medium, 7 large, 9 very large.	7	7
OIV 067	Mature leaf: shape of blade. 1 cordate, 3 wedge-shaped, 3 pentagonal, 4 circular, 5 kidney-shaped.	3	3
OIV 068	Mature leaf: number of lobes. 1 one, 2 three, 3 five, 4 seven, 5 more than seven.	3	3
OIV 070	Mature leaf: area of anthocyanin coloration of main veins on upper side of blade. 1 absent, 2 only at the petiolar point, 3 up to the 1st bifurcation, 4 up to the 2nd bifurcation, 5 beyond the 2nd bifurcation.	1	3
OIV 072	Mature leaf: goffering of blade. 1 absent or very weak, 3 weak, 5 medium, 7 strong, 9 very strong.	7	5
OIV 074	Mature leaf: profile of blade in cross section. 1 flat, 2 V-shaped, 3 involute, 4 revolute, 5 twisted.	5	4
OIV 075	Mature leaf: blistering of upper side of blade. 1 absent or very weak, 2 weak, 3 medium, 4 strong, 9 very strong.	5	3
OIV 076	Mature leaf: shape of teeth. 1 both sides concave, 2 both sides straight, 3 both sides convex, 4 one side concave on side convex, 5 mixture between both sides straight and both sides convex.	2	3
OIV 079	Mature leaf: degree of opening/overlapping of petiole sinus. 1 very wide open, 3 open, 5 closed, 7 overlapped, 9 strongly overlapped.	3	5
OIV 080	Mature leaf: shape of base petiole sinus. 1 U-shaped, 2 brace-shaped, 3 V-shaped.	3	3

Table 2. *Cont.*

Code	Descriptor	'Uva Rey'	'Palomino Fino'
OIV 081-1	Mature leaf: teeth in the petiole sinus. 1 none, 9 present.	1	1
OIV 081-2	Mature leaf: petiole sinus base limited by vein. 1 not limited, 3 on one side, 3 on both sides.	1	1
OIV 083-2	Mature leaf: teeth in the upper lateral sinuses. 1 none, 9 present.	1	1
OIV 084	Mature leaf: density of prostrate hairs between main veins on lower side of blade. 1 none or very low, 3 low, 5 medium, 7 high, 9 very high.	7	7
OIV 087	Mature leaf: density of erect hairs on main veins on lower side of blade. 1 none or very low, 3 low, 5 medium, 7 high, 9 very high.	9	1
OIV 151	Flower: sexual organs. 1 fully developed stamens and no gynoecium, 2 fully developed stamens and reduced gynoecium, 3 fully developed stamens and fully developed gynoecium, 4 reflexed stamens and fully developed gynoecium.	3	3
OIV 202	Bunch: length (peduncle excluded). 1 very short, 3 short, 5 medium, 7 long, 9 very long.	5	7
OIV 203	Bunch: width. 1 very narrow, 3 narrow, 5 medium, 7 wide, 9 very wide.	5	5
OIV 204	Bunch: density. 1 very loose, 3 loose, 5 medium, 7 dense, 9 very dense.	5	5
OIV 206	Bunch: length of peduncle of primary bunch. 1 very short, 3 short, 5 medium, 7 long, 9 very long.	3	1
OIV 220	Berry: length. 1 very short, 3 short, 5 medium, 7 long, 9 very long.	5	3
OIV 221	Berry: width. 1very narrow, 3 narrow, 5 medium, 7 wide, 9 very wide.	5	3
OIV 223	Berry: shape. 1 obloid, 2 globose, 3 broad ellipsoid, 4 narrow ellipsoid, 5 cylindrical, 6 obtuse ovoid, 7 ovoid, 8 obovoid, 9 horn shaped, 10 finger shaped.	7	2
OIV 225	Berry: colour of skin. 1 green yellow, 2 rose, 3 red, 4, grey, 5 dark red violet, 6 blue black.	1	1

A total of 34 descriptors were studied, eight of which correspond to shoots, 17 to leaves, one to inflorescence, four to bunches and four to berries. In regard to the density of prostate hairs between the main veins on lower side of blade (OIV 053), 'Uva Rey' showed very high density while 'Palomino Fino' prostate hair density was medium. Also, the density of erect hairs on the main veins on the lower side of tthe blade (OIV 087) was high for 'Uva Rey' and non-existent or low for 'Palomino Fino' cultivar. Finally, grape berries were green yellow in both cases (OIV 225), but their shapes differed (OIV 223), being ovoid for 'Uva Rey' and globose for 'Palomino Fino'.

3.3. Physicochemical Characterization of Grapes and Musts

'Uva Rey' and 'Palomino Fino' grape must physicochemical characterizations and berry texture profile analyses (TPA) from two vintages (2016 and 2017) are displayed in Table 3.

Table 3. 'Uva Rey' and 'Palomino Fino' grape berry texture profile analysis (TPA) and must characterization.

	2016		2017	
	'Palomino Fino'	**'Uva Rey'**	**'Palomino Fino'**	**'Uva Rey**
Physicochemical Parameters				
pH	3.93 ± 0.01 [a]	3.87 ± 0.07 [a]	4.02 ± 0.03 [a]	3.97 ± 0.02 [a]
Total Acidity (g/L TH$_2$)	3.74 ± 0.05 [a]	3.51 ± 0.07 [a]	3.15 ± 0.08 [b]	3.25 ± 0.21 [b]
Sugar (°Bé)	12.85 ± 0.01 [a]	8.45 ± 0.02 [b]	11.70 ± 0.02 [c]	7.40 ± 0.06 [d]
Ripening Index (RI)	3.44 ± 0.02 [a]	2.41 ± 0.01 [b]	3.71 ± 0.02 [a]	2.28 ± 0.01 [b]
YAN (mg/L)	200.00 ± 2.00 [a]	140.00 ± 2.00 [b]	161.00 ± 6.00 [c]	140.00 ± 3.00 [b]
Tartaric Acid (g/L)	3.140 ± 0.050 [a]	2.720 ± 0.008 [b]	2. 470 ± 0.100 [b]	2.600 ± 0.200 [b]
Citric Acid (g/L)	0.030 ± 0.005 [a]	0.100 ± 0.001 [b]	0.030 ± 0.010 [a]	0.150 ± 0.002 [c]
Malic acid (g/L)	0.420 ± 0.020 [a]	0.650 ± 0.003 [b]	0.100 ± 0.020 [c]	0.600 ± 0.010 [d]
TPA				
Consistency (Nmm)	89.58 ± 1.59 [a]	138.24 ± 8.47 [b]	93.66 ± 2.27 [a]	152.42 ± 11.18 [c]
Hardness (Nmm)	237.57 ± 4.58 [a]	239.20 ± 7.56 [a]	237.29 ± 5.18 [a]	245.05 ± 12.08 [a]
WoP (Nmm)	260.47 ± 12.87 [a]	351.35 ± 14.98 [b]	280.13 ± 16.70 [a]	409.93 ± 23.70 [c]
Cohesiveness	0.21 ± 0.02 [a]	0.41 ± 0.02 [b]	0.23 ± 0.02 [a]	0.40 ± 0.03 [b]

Different superscript letters mean statistically significant differences between samples at *p*-adjust < 0.05 obtained by two-way ANOVA and Bonferroni's multiple range (BSD) test. Results are the means ± SD of three repetitions.

The main differences between 'Uva Rey' and 'Palomino Fino' cultivars grape musts were related to the physicochemical parameters sugar (°Bé), YAN (mg/L), malic acid (g/L) and TPA consistency (Nmm) and cohesiveness. The pH values obtained for both cultivars as well as for the two vintages were all similar. However, both cultivars exhibited very similar acidity in both vintages, with slightly higher values in 2017 (ANOVA *p*-adjust < 0.05). Regarding grape sugar content, it was significantly higher in 'Palomino Fino' grapes than in 'Uva Rey' from the two vintages studied (ANOVA *p*-adjust < 0.05). Again, greater sugar values (°Bé) as well as total acidity were measured in 2016 grapes from both cultivars (Table 3). Consequently, Ripening Index (RI) values obtained were significantly greater in 'Palomino Fino' than in 'Uva Rey'. However, very different content levels in both cultivars were obtained for YAN, where 'Palomino Fino' showed significantly higher concentrations of YAN than 'Uva Rey' (ANOVA *p*-adjust < 0.05), which yielded the same content level in the two vintages under study (Table 3).

Regarding organic acids content, it could be observed that tartaric acid represents over 75% of their total acidity. It can be seen that this particular acid content follows the same trend as the total acidity of the grapes. With respect to citric acid concentration, it was significantly lower in 'Palomino Fino' than in 'Uva Rey' cultivar and did not exceed 150 mg/L in either case. However, 'Uva Rey' showed a significantly higher content of malic acid than 'Palomino Fino' in both of the vintages studied (ANOVA *p*-adjust < 0.05).

With respect to the results obtained from the TPA, 'Uva Rey' obtained higher values for consistency, WoP and cohesiveness than 'Palomino Fino' in both vintages (ANOVA *p*-adjust < 0.05). However, no differences were observed between cultivars or vintages with regards to grape berry hardness.

4. Discussion

To identify grapevine cultivars, nuclear microsatellite markers are the most widely used tool, as was demonstrated by the European projects GENRES 081 and GrapeGen06. Regardless of the high degree of heterozygosity existing in the grapevine, the genotype with six microsatellite loci (VVMD5, VVMD7, VVMD27, VVS2, VrZAG62 and VrZAG79) is enough to establish the identity of a variety [6], with the exception of the peculiar case of closely related varieties [35] which requires analysis of more loci. For this reason, as a result of the GrapeGen06 project, an international consensus was established to increase the number of microsatellite loci to 20, located in different binding groups for correct identification. In this study, the analysis was extended to 22 microsatellite loci. It is very important to use the same microsatellite loci in different studies in order to be able to compare genotypes later. The identification of 'Uva Rey' genotype allowed us to confirm the synonyms of this cultivar with both 'De Rey' and 'Mantúo de Pilas', which have already been registered in the *Vitis* International Variety Catalogue (VIVC) at seven loci SSR [34]. The genetic profile for 15 additional loci is presented in this study and the synonymy between 'De Rey' and 'Uva Rey' is confirmed for the first time with the analysis at 22 microsatellite loci. Along with the cultivar genetic identification and, according to the recommendation for the adequate characterisation of *Vitis* genetic material, an ampelographic description was carried out [45]. Such morphological description has been the method previously used by different countries to have a particular cultivar included in the official lists [45]. The phenotype obtained for the cultivar 'Uva Rey' showed some differences with 'Mantuo de Pilas' as described by García de Luján et al. [46]. Some differences were found in OIV 007, OIV 008, OIV 051, OIV 053, OIV 070, OIV 074, OIV 075, OIV 087, OIV 202 and OIV 221 descriptors. It is worth mentioning, the differences in erect hairs density on main veins on lower side of blade in mature leaves (OIV 087). 'Uva Rey' showed a very high density unlike 'Mantuo de Pilas' with a very low one. Similar phenotypic differences have been found between other cultivars such as 'Garnacha' and 'Garnacha Peluda' [47], both considered as somatic variants.

Due to the high temperatures associated to the current global warming, the period during which the minimal temperatures required for the physiological activities of vines is reached is longer than it used to be, and hence, there is an increment in metabolic rates that have an impact on metabolite accumulation [48,49]. In the last 10–30 years, some major changes have been observed in grape development and ripening patterns, such as premature budbreak, flowering and fruit maturity due to agroclimatic changes [50]

The differences between the two cultivars with regards to pH and total acidity can be attributed to climate variations between the two years studied, as such differences can be found in both cultivars (Figures S1 and S2). RI values confirm the above-mentioned differences between cultivars (ANOVA *p*-adjust < 0.05), with significant differences between both cultivars regardless of the vintage analysed. The variations of these parameters associated to grape ripening processes may be related with each cultivar's phenological stages. 'Uva Rey' is, unlike 'Palomino Fino' a long cycle cultivar [51]. For this reason, grape ripening stages are not reached at the same time.

Organic acids content in each cultivar could be due to their phenological cycle differences [51]. With regard to tartaric acid content, the values remained similar except for 'Palomino Fino' cultivar in the 2016 year. During the grape ripening process, the production of malic acid decreases [52] since this carboxylic acid is also used by the plant at this stage for energy production [53]. In this way, the different malic acid content levels in each cultivar could be explained by their aforementioned asynchronous phenological cycles. Such difference in malic acid content levels could be relevant to prospective winemaking process, where malolactic fermentation (MLF) could result in wines with a greater microbiological stability and sensory complexity [54]. Some authors argue that higher weather temperatures due to global warming may lead to grape musts with a higher pH, which in turn may promote oxidation reactions [50,55]. In this sense, grapevine cultivars with similar characteristics to those presented by 'Mantúo de Pilas' could lead to the production of wines through oxidative ageing.

The YAN values that have been observed in 'Palomino Fino' musts were higher than those observed in 'Uva Rey' for both vintages. Such differences between the two cultivars may be related to the variations observed in their ripening processes, since YAN content increases in grape berries when ripening [56]. In any case, YAN values remained at a sufficient level for a proper alcoholic fermentation (AF) [57]. Yeast assimilable nitrogen (YAN) is a fundamental element for the correct AF of grape musts; since nitrogen is essential for the completion of some yeasts, its presence is compulsory for yeasts to develop in normal conditions during this biological process [58].

According to the TPA, the two vintages of 'Uva Rey' in the study had a higher consistency, hardness, WoP and cohesiveness. It should be noted that cohesiveness depends on the strength of the pulp internal bonds of the grape berries. This parameter is highly related to the OIV 235 descriptor [40], which is employed for the sensory evaluation of grapes during their ripening process. The results obtained from the TPA could be explained by the lack of synchrony between both cultivars phenological cycles. 'Uva Rey' berries, with a longer cycle, were less ripe and therefore presented a greater turgidity at the time of analysis. Such superior berry turgidity plus its higher consistency and WoP could contribute to protect grape berries from dehydration under Andalusian warm weather conditions (SW Spain). When these results are compared to those obtained by Giacosa et al. [59], it can be observed that 'Palomino Fino' presents similar cohesiveness to 'Perle von Csaba' cultivar (Hungarian white vinification grape). Nonetheless, 'Uva Rey' showed a higher degree of similarity with the cultivar 'Sultanina' (a Turkish white table grape). In view of its grape berry TPA, 'Uva Rey' could be considered as a cultivar with a greater resistance than 'Palomino Fino', mainly because of its greater pulp cohesiveness and consistency. These results might be influenced by the phenological cycle differences observed between the two cultivars studied, where the higher values correspond to less ripe berries. In this sense, these phenotypical traits could increase the cultivar's resistance to drought and to high temperatures, which would make it a more appropriate cultivar for warm dry areas and for global warming conditions.

5. Conclusions

Microsatellite analysis confirmed that 'Uva Rey' is a synonym of 'De Rey' cultivar and a somatic variant of 'Mantuo de Pilas'. With respect to the physicochemical grape must characterization, major differences were found in YAN and malic acid concentration. The TPA showed that 'Uva Rey' grape berries are more cohesive and consistent than 'Palomino Fino' ones. In this sense, 'Uva Rey' can be stated as an autochthonous grapevine cultivar with a long phenological cycle. This study recognizes Uva Rey as a somatic variant of 'Mantuo de Pilas' and as such, supports any actions towards its recovery. According to the results obtained from the different analysis that have been completed on 'Uva Rey' grape berries and musts from two consecutive vintages, this autochthonous cultivar should be further studied and included in the Spanish official register to allow is cultivation.

Supplementary Materials: The following are available online at http://www.mdpi.com/2073-4395/9/9/563/s1, Figure S1. (a) Temperature (°C) (Tª_max, Tª_min, Tª_avg), (b) humidity (%) (H_max, H_min, H_avg) and (c) radiation (W/m^2) and rainfall (L/m^2) between July and September 2016. Figure S2. (a) Temperature (°C) (Tª_max, Tª_min, Tª_avg), (b) humidity (%) (H_max, H_min, H_avg) and (c) radiation (W/m^2) and rainfall (L/m^2) between July and September 2017.

Author Contributions: P.S.-G., A.A.-A., V.P. and A.J.-C. conceived and designed the experiments. P.S.-G. and A.J.-C. performed the experiments. All authors analysed the data and wrote the paper.

Funding: This research received no external funding.

Acknowledgments: The authors want to thank the private winery from Chiclana de la Frontera (Cádiz) for grape support.

Conflicts of Interest: The authors declare no conflict of interest.

References

1. Dzhambazova, T.; Tsvetkov, I.; Atanassov, I.; Rusanos, K.; Martínez Zapater, J.M.; Atanassov, A.; Hvarleva, T. Genetic diversity in native Bulgarian grapevine germplasm (*Vitis vinifera* L.) based on nuclear and chloroplast microsatellite polymorphisms. *Vitis* **2009**, *48*, 115–121.

2. Marsal, G.; Mateo-Sanz, J.M.; Canals, J.M.; Zamora, F.; Fort, F. SSR analysis of 338 Accessions Planted in Penedes (Spain) Reveals 28 Unreported Molecular Profiles of *Vitis vinifera* L. *Am. J. Enol. Vitic.* **2016**, *67*, 466–470. [CrossRef]

3. Santana, J.C.; Heuertz, M.; Arranz, C.; Rubio, J.A.; Martínez-Zapater, J.M.; Hidalgo, E. Genetic structure, origins, and relationships of grapevine cultivars from the Castilian Plateau of Spain. *Am. J. Enol. Vitic.* **2010**, *61*, 214–224.

4. Jiménez-Cantizano, A.; Amores-Arrocha, A.; Gutiérrez-Escobar, R.; Palacios, V. Identification and relationship of the autochthonous 'Romé' and 'Rome Tinto' grapevine cultivars. *Span. J. Agric. Res.* **2018**, *16*, e07SC02. [CrossRef]

5. International Organisation of Vine and Wine (OIV). State of the Vitiviniculture World Market: State of the Sector in 2018. Available online: http://www.oiv.int/public/medias/6679/en-oiv-state-of-the-vitiviniculture-world-market-2019.pdf (accessed on 9 July 2019).

6. This, P.; Lacombe, T.; Thomas, M.R. Historical origins and genetic diversity of wine grapes. *Trends Genet.* **2006**, *22*, 511–519. [CrossRef] [PubMed]

7. De los Salmones, G. *La Invasión Filoxérica en España y las Cepas Americanas*, 1st ed.; Tipolitografía de Luis Tasso: Barcelona, Spain, 1893.

8. Buhner-Zaharieva, T.; Moussaoui, S.; Lorente, M.; Andreu, J.; Núñez, R.; Ortiz, J.M.; Gorgocena, Y. Preservation and molecular characterization of ancient varieties in Spanish grapevine germplasm collections. *Am. J. Enol. Vitic.* **2010**, *61*, 557–562. [CrossRef]

9. Balda, P.; Ibáñez, J.; Sancha, C.; Martínez de Toda, F. Characterization and identification of minority red grape varieties recovered in Rioja, Spain. *Am. J. Enol. Vitic.* **2014**, *65*, 148–152. [CrossRef]

10. García de Luján, A.; Lara, M. *La Colección de Vides del Rancho de la Merced*, 1st ed.; Consejería de Agricultura y Pesca, Junta de Andalucía: Seville, Spain, 1997; pp. 7–13.

11. Fernández-González, M.; Mena, A.; Izquierdo, P.; Martínez, J. Genetic characterization of grapevine (*Vitis vinifera* L.) cultivars from Castilla la Mancha (Spain) using microsatellite markers. *Vitis* **2007**, *46*, 126–130.

12. González-Andrés, F.; Martín, J.P.; Yuste, J.; Rubio, J.A.; Arranz, C.; Ortiz, J.M. Identification and molecular biodiversity of autochthonous grapevine cultivars in the "Comarca del Bierzo", León, Spain. *Vitis* **2007**, *46*, 71–76.

13. Iorizzo, M.; Vicenzo, M.; Testa, B.; Lombardi, S.J.; De Leonardis, A. Physicochemical and sensory characteristics of red wines from the rediscovered autochthonous Tintilia grapevine grown in the Molise region (Italy). *Eur. Food Res. Technol.* **2014**, *238*, 1037–1048. [CrossRef]

14. Jaillon, O.; Aury, J.M.; Noel, B.; Policriti, A.; Clepet, C.; Casagrande, A.; Choisne, N.; Aubourg, S.; Vitulo, N.; Jubin, C.; et al. The grapevine genome sequence suggests ancestral hexaploidization in major angiosperm phyla. *Nature* **2007**, *449*, 463–468. [PubMed]

15. Martínez-Zapater, J.M.; Carmona, M.J.; Díaz-Riquelme, J.; Fernández, L.; Lijavetzky, D. Grapevine genetics after the genome sequence: Challenges and limitations. *Aust. J. Grape Wine R.* **2010**, *16*, 33–46. [CrossRef]

16. Wolkovich, E.M.; García de Cortázar-Atauri, I.; Morales-Castilla, I.; Nicholas, K.A.; Lacombe, T. From Pinot to Xinomavro in the world's future wine-growing regions. *Nat. Clim. Chang.* **2018**, *8*, 29–37. [CrossRef]

17. Real Decreto 1338/2018 Boletín Oficial del Estado (BOE). Available online: https://www.boe.es/eli/es/rd/2018/10/29/1338 (accessed on 4 July 2019).

18. Intrigliolo, D.S.; Llacer, A.; Revert, J.; Esteve, M.D.; Climent, M.D.; Palau, D.; Gómez, I. Early defoliation reduces cluster compactness and improves grape composition in Mandó, an autochthonous cultivar of *Vitis vinifera* from southeastern Spain. *Sci. Hortic.* **2014**, *167*, 71–75. [CrossRef]

19. Fraga, H.; Santos, J.A.; Malheiro, A.C.; Oliveira, A.A.; Moutinho-Pereira, J.; Jones, G.V. Climatic suitability of Portuguese grapevine varieties and climate change adaptation. *Int. J. Climatol.* **2016**, *36*, 1–12. [CrossRef]

20. Alonso de Herrera, G. *Agricultura General*, 1st ed.; Editorial Real Sociedad Económica Matritense: Madrid, Spain, 1819.

21. Roxas Clemente, S. *Ensayo Sobre las Variedades de vid que Vegetan en Andalucía*, 1st ed.; Imprenta de Villapando: Madrid, Spain, 1807; pp. 111–113.

22. Abela y Sainz de Andino, E.J. *El Libro del Viticultor*, 1st ed.; Editorial Maxtor: Valladolid, Spain, 1805; pp. 256–265.

23. Jiménez-Povedano, M.V.; Cantos-Villar, E.; Jiménez-Hierro, M.J.; Casas, J.A.; Trillo, L.M.; Guimera, S.; Puertas, B. Influence of Soil on the Characteristics of Musts, Wines and Distillates from cv L Palomino fino. In Proceedings of the XII Terroir International Congress, Zaragoza, Spain, 18–22 June 2018.

24. Santesteban, L.G.; Miranda, C.; Royo, J.B. Vegetative Growth, Reproductive Development and Vineyard Balance. In *Methodologies and Results in Grapevine Research*, 1st ed.; Delrot, S., Medrano, H., Or, E., Bavaresco, L., Grando, S., Eds.; Springer: New York, NY, USA, 2010; pp. 45–56.

25. Di Gaspero, G.; Peterlunger, E.; Testolin, R.; Edwards, K.J.; Cipriani, G. Conservation of microsatellite loci within the genus *Vitis. Theor. Appl. Genet.* **2000**, *101*, 301–308. [CrossRef]

26. Bowers, J.E.; Dangl, G.S.; Vignani, R.; Meredith, C.P. Isolation and characterization of new polymorphic simple sequence repeat loci in grape (*Vitis vinifera* L.). *Genome* **1996**, *39*, 628–633. [CrossRef]

27. Bowers, J.E.; Dangl, G.S.; Meredith, C.P. Development and characterization of additional microsatellite markers for grape. *Am. J. Enol. Vitic.* **1999**, *50*, 243–246.

28. Thomas, M.R.; Scott, N.S. Microsatellite repeats in grapevine reveal DNA polymorphisms when analyzed as sequence-tagged sites (STSs). *Theor. Appl. Genet.* **1993**, *86*, 985–990. [CrossRef]

29. Merdinoglu, S.; Coste, P.; Dumas, V.; Haetty, S.; Butterlin, G.; Greif, C.; Adam-Blondon, A.-F.; Bouquet, A.; Pauquet, J. Genetic analysis of downy mildew resistance derived from Muscadinia rotundifolia. *Acta Hortic.* **2003**, *603*, 451–456. [CrossRef]

30. Vargas, A.M.; Velez, M.D.; de Andrés, M.T.; Laucou, V.; Lacombe, T.; Bourisquot, J.M.; Borrego, J.; Ibáñez, J. Corinto bianco: A seedless mutant of Pedro Ximenes. *Am. J. Enol. Vitic.* **2007**, *58*, 540–543.

31. Sefc, K.M.; Regner, F.; Tureschek, E.; Glössl, J.; Steinkellner, H. Identification of microsatellite sequences in *Vitis riparia* and their application for genotyping of different Vitis species. *Genome* **1999**, *42*, 367–373. [CrossRef]

32. Jiménez-Cantizano, A.; García de Lujan, A.; Arroyo-García, R. Molecular characterization of table grape varieties preserved in the Rancho de la Merced Grapevine Germplasm Bank. *Vitis* **2018**, *57*, 93–101.

33. Lacombe, T.; Audeguin, L.; Boselli, M.; Bucchetti, B.; Cabello, F.; Chatelet, P.; Crespan, M.; D'onofrio, C.; Eiras-Dias, J.; Ercisli, S.; et al. Grapevine European Catalogue: Towards a comprehensive list. *Vitis* **2011**, *50*, 65–68.

34. Vitis International Variety Catalogue. Available online: www.vivc.de (accessed on 13 June 2019).

35. Jiménez-Cantizano, A. Caracterización Molecular del Banco de Germoplasma de vid del Rancho de la Merced. Ph.D. Thesis, Universidad de Cadiz, Cadiz, Spain, 2014.

36. Ibáñez, J.; Vargas, M.A.; Palancar, M.; Borrego, J.; de Andrés, M.T. Genetic relationships among table-grape varieties. *Am. J. Enol. Vitic.* **2009**, *60*, 35–47.

37. De Andrés, M.T.; Benito, A.; Pérez-Rivera, G.; Ocete, R.; López, M.A.; Gaforio, L.; Muñoz, G.; Cabello, F.; Martínez-Zapater, J.M.; Arroyo-García, R. Genetic diversity of wild grapevine populations in Spain and its genetic relationships with cultivated grapevines. *Mol. Ecol.* **2012**, *21*, 800–816. [CrossRef] [PubMed]

38. Park, S.D.E. Trypanotolerance in West African Cattle and the Population Genetic Effects of Selection. Ph.D. Thesis, University of Dublin, Dublin, Ireland, 2001.

39. Benito, A.; Muñoz-Organero, G.; de Andrés, M.T.; Ocete, R.; García-Muñoz, S.; López, M.A.; Arroyo-García, R.; Cabello, F. Ex situ ampelographical characterisation of wild *Vitis vinifera* from fifty-one Spanish populations. *Aust. J. Grape Wine Res.* **2017**, *23*, 143–152. [CrossRef]

40. Organisation Internationale de la Vigne et du Vin (OIV). *OIV Descriptor List for Grape Varieties and Vitis Species*, 2nd ed.; OIV: Paris, France, 2009.

41. *Recuéil des Méthodes Internationales D'analyse des vins et des Moûts*; OIV Office International de la Vigne et du Vin: Paris, France, 2014.

42. Hidalgo, J. Vendimia. Recepción de Uva en la Bodega. Índices de maduración químicos. In *Tratado de Enología*, 3rd ed.; Hernández-Úbeda, I., Ed.; Editorial Mundi-Prensa: Madrid, Spain, 2018; Volume I, pp. 238–240.

43. Aerny, J. Composés azotés des moûts et des vins. *Rev. Suisse Vitic. Arboric. Hortic.* **1997**, *28*, 161–168.

44. Sancho-Galán, P.; Amores-Arrocha, A.; Jiménez-Cantizano, A.; Palacios, V. Use of Multiflora Bee Pollen as a Flor Velum Yeast Growth Activator in Biological Aging Wines. *Molecules* **2019**, *24*, 1763. [CrossRef]

45. García-Muñoz, S.; Muñoz-Organero, G.; De Andrés, M.T.; Cabello, F. Ampelography: An old technique with future uses, the case of minor varieties of *Vitis vinifera* L. from the Balearic Islands. *J. Int. Sci. Vigne Vin* **2011**, *45*, 125–137. [CrossRef]

46. García de Luján, A.; Puertas, B.; Lara, M. *Variedades de vid en Andalucía*, 1st ed.; Consejería de Agricultura y Pesca, Junta de Andalucía: Seville, Spain, 1990; pp. 69–76.

47. Carbonell-Bejerano, P.; Royo, C.; Mauri, N.; Ibáñez, J.; Martínez-Zapater, J.M. Somatic Variation and Cultivar Innovaion in Grapevine. In *Advances in Grape and Wine Biotechnology*, 1st ed.; Morata, A., Loira, I., Eds.; Intechopen: London, UK, 2016; p. 8.

48. Coombe, B. Influence of temperature on composition and quality of grapes. In Proceedings of the International Symposium on Grapevine Canopy and Vigor Management, Davis, CA, USA, 11 August 1986.

49. Winkler, A.J.; Cook, J.A.; Kliewer, W.M.; Lider, L.A. *General Viticulture*; University California Press: Berkeley, CA, USA, 1974.

50. Mira de Orduña, R. Climate change associate effects on grape and wine quality and production. *Food Res. Intl.* **2010**, *43*, 1844–1855. [CrossRef]

51. Cabello, F.; Ortiz, J.M.; Muñoz, G.; Rodríguez, I.; Benito, A.; Rubio, C.; García, S.; Saiz, R. *Variedades de vid en España*, 1st ed.; Editorial Agrícola Española, S.A: Madrid, Spain, 2011.

52. Lakso, A.N.; Kliewer, W.M. The influence on malic acid metabolism in grape berries. Temperature responses of net dark CO_2 fixation and malic acids pools. *Am. J. Enol. Vitic.* **1978**, *29*, 145–148.

53. Mullins, M.G.; Bouquet, A.; Williams, L.E. *The Biology of the Grapevine*, 1st ed.; Cambridge University Press: Cambridge, UK, 2008; pp. 80–146.

54. Palacios, V.; Roldan, A.; Jiménez-Cantizano, A.; Amores-Arrocha, A. Physicochemical and microbiological characterization of the sensory deviation responsable for the origin of the special sherry wines "palo cortado" type. *PLoS ONE* **2018**, *13*, e0208330. [CrossRef] [PubMed]

55. Boulton, R.B.; Singleton, V.L.; Bisson, L.F.; Kunkee, R.E. *Principles and Practices of Winemaking*; Chapman and Hall: New York, NY, USA, 1996.

56. Belle, S.J. The Effect of Nitrogen Fertilisation on the Growth Yield and Juice Composition of *Vitis vinifera* cv. Cabernet Sauvignon Grapevines. Ph.D. Thesis, University of Western Australia, Perth, Australia, 1994.

57. Ribéreau-Gayon, P.; Dubourdieu, D.; Donéche, B.; Donéche, B.; Lonvaud, A.; Glories, Y.; Maugean, A. *Traité D'oenologie: Microbiologie du vin. Vinifications*, 3rd ed.; Dunod Editions: Paris, France, 2017.

58. Tesnière, C.; Brice, C.; Blondin, B. Responses of Saccharomyces cerevisiae to nitrogen starvation in wine alcoholic fermentation. *Appl. Microbiol. Biotechnol.* **2015**, *99*, 7025–7034. [CrossRef] [PubMed]

59. Giacosa, S.; Torchio, F.; Rio-Segade, S.; Giust, M.; Tomasi, D.; Gerbi, V.; Rolle, L. Selection of a Mechanical Property for Flesh Firmness of Table Grapes in Accordance with an OIV Ampelographic Descriptor. *Am. J. Enol. Vitic.* **2014**, *65*, 206–214. [CrossRef]

agronomy

MDPI

Article

Postharvest Preservation of the New Hybrid Seedless Grape, 'BRS Isis', Grown Under the Double-Cropping a Year System in a Subtropical Area

Saeed Ahmed [1], Sergio Ruffo Roberto [1,*], Khamis Youssef [2], Ronan Carlos Colombo [1], Muhammad Shahab [1], Osmar José Chaves Junior [1], Ciro Hideki Sumida [1] and Reginaldo Teodoro de Souza [3]

[1] Agricultural Research Center, Londrina State University, Celso Garcia Cid Road, km 380, P.O. Box 10.011, 86057-970 Londrina, Brazil
[2] Agricultural Research Center, Plant Pathology Research Institute, 9 Gamaa St., Giza 12619, Egypt
[3] Embrapa Grape and Wine, 515 Livramento Drive, 95700-000 Bento Gonçalves, Brazil
*   Correspondence: sroberto@uel.br; Tel.: +554-333-714-774

Received: 29 July 2019; Accepted: 16 September 2019; Published: 1 October 2019

Abstract: 'BRS Isis' is a new hybrid seedless table grape tolerant to downy mildew with a good adaptation to the tropical and subtropical climates. Gray mold, caused by *Botrytis cinerea* Pers. ex Fr. is known as the most important postharvest mold in table grapes, causing extensive losses worldwide. As the postharvest behavior of 'BRS Isis' is still unknown, the objective of this work was to evaluate the postharvest preservation and *B. cinerea* mold control of this new grape cultivar, grown under the double-cropping a year system. Grape bunches were purchased from a field of 'BRS Isis' seedless table grapes trained on overhead trellises located at Marialva, state of Parana (South Brazil). Grapes were subjected to the following treatments in a cold room at 1 ± 1 °C: (i) Control; (ii) SO_2-generating pad; (iii) control with bunches inoculated with the pathogen suspension; (iv) SO_2-generating pad with bunches inoculated with the pathogen suspension. The completely randomized experimental design was used with four treatments, each including five replicates. The incidence of gray mold and other physicochemical variables, including bunch mass loss, shattered berries, skin color index, soluble solids (SS), titratable acidity (TA), and SS/TA ratio of grapes, were evaluated at 50 days after the beginning of cold storage and at seven days at room temperature (22 ± 2 °C). The 'BRS Isis' seedless grape, packaged with SO_2-generating pads and plastic liners, has a high potential to be preserved for long periods under cold storage, at least for 50 days, keeping very low natural incidence of gray mold, mass loss, and shattered berries.

Keywords: *Vitis vinifera* (L.); SO_2 pads; *B. cinerea* mold; grape quality

1. Introduction

'BRS Isis' is a new hybrid seedless table grape obtained by the crossing of CNPUV 681–29 (Arkansas 1976 × CNPUV 147-3 ('Niagara White' × 'Venus')) × 'BRS Linda'. This cultivar was released in 2013 and is tolerant to downy mildew *Plasmopara viticola* (Berk. & M.A. Curtis) Berl. & De Toni, the main vine disease in subtropical humid areas. It presents high bud fertility with 2–3 great inflorescences per shoot, with a natural weight of 375 g, and without the use of growth regulators, making it a high yielding grape. The bunch is medium-sized and predominantly cylindrical-winged, while the berry is medium-sized, reddish, elliptical, and firm and has colorless flesh and neutral flavor with traces of large, fleshy rudimentary seeds [1]. This new seedless and early season cultivar has the ability to gain the attention of consumers from domestic and international markets as there has been a significant demand of table grape supply for extended periods throughout the year worldwide.

Grapes are non-climacteric fruits with a relatively low physiological activity and are subject to serious postharvest problems during cold storage, such as mold, mass loss stem browning, shattered berries, wilting, and shriveling of berries. Thus, these factors are the main barriers for long-term storage of table grapes [2–4].

Botrytis cinerea Pers. Fr., is known to be the most important postharvest pathogen causing gray mold of table grapes [5,6]. Infection caused by this fungus remains inactive in the field unless it gets favorable environmental conditions, i.e., fruit injuries that assist pathogen development [7,8]. Even a small infection on a single berry can damage the whole lot of grapes, and if it is not noticed at pre-harvest stage, during packaging, or during shipment, it may progress and spread the infection in postharvest or during the cold storage period of table grapes, even at low temperatures [9–11].

Cold storage, where only temperature and relative humidity are controlled in the chamber, is one of the main methods to maintain the fruit quality. Thus, the reduction of temperature, up to a certain limit, increases the quality preservation and extends the period of fruit supply to the consumer market [12]. After harvesting, bunches are pre-cooled as soon as possible to remove field heat and reduce water loss [11,13]. For extended export and shipment purposes, the cold storage temperature must be kept optimum and constant because any disturbance can initiate the growth of fungi, mainly *B. cinerea* [14].

The postharvest control of this pathogen is difficult, as most countries no longer allow the application of synthetic fungicides on bunches. Combined with cold storage, different pre- and postharvest techniques can be used to control gray mold, such as the use of sulfur dioxide (SO_2) generating pads, which is the most common method worldwide [15–17]. The slow release SO_2-generating pads contain sodium metabisulfite ($Na_2S_2O_5$) as am active ingredient enclosed in a sheet of plastic and paper, which used in packing materials by releasing a low and continual dose of SO_2 with contact to humidity to eliminate/reduce *B. cinerea* spores.

The SO_2-generating pads are highly effective in controlling and killing the spores of *B. cinerea*, but also can result in unwanted situations, such as bleaching and shattered berries. Other studies have also shown that grape hairline splits, commonly associated with significant water loss, are also induced by excessive SO_2 doses. However, high levels of SO_2 can also result in fruit damage, unpleasant aftertaste, and allergies. Based on these findings, it is recommended to use a minimal dose of SO_2 that allows adequate protection from mold without reducing the berry quality in order to avoid these situations [18,19].

As there is a lack of information regarding the cold storage of the 'BRS Isis' seedless grape, it is very important to know the behavior of this new hybrid cultivar grown under the double-cropping a year system, especially for long-distance and international markets. Under this system, two crops per year are achieved (summer and off-season crops). Summer crops start from the end of grapevine dormancy in late winter and harvest is obtained in summer, while, for off-season crops, vines are pruned after summer crops and forced to sprout once more using budburst stimulators, and harvest occurs through autumn. The core difference between both crops is that in the summer crop, the rate of some fungal infection is quite low, while on the other hand, in an off-season crop, the incidence of fungus diseases is high because of favorable environmental conditions that promote the infection and can restrain long-distance transportation of table grapes [20,21].

The objective of this work was to evaluate the postharvest preservation and control gray mold of the 'BRS Isis' seedless grape grown under the double-cropping a year system in subtropical conditions.

2. Materials and Methods

2.1. Experimental Location

Table grapes were purchased from a field of 'BRS Isis' seedless grapes, grafted on 'IAC 766' rootstock from 2-year-old vines, trained on overhead trellises located at Marialva, state of Parana (PR) (South Brazil) (23°29 S, 51°47 W, elevation 570 m), with a history of gray mold. The vines were grown under the double-cropping a year system, and the fruit samples were collected from two consecutive crop seasons.

2.2. Treatments and Storage

Grapes were harvested at full ripeness when the content of the berry soluble solids reached around 14 °Brix [21,22]. Bunches were selected free from any disorders and standardized according to bunch shape, size, and mass, and subjected to the following treatments into a cold chamber at $1 \pm 1°C$: (i) Control; (ii) SO_2-generating pad; (iii) control with bunches inoculated with *B. cinerea* suspension; (iv) SO_2-generating pad with bunches inoculated with *B. cinerea* suspension. The slow release SO_2-generating pad used in treatments (ii) and (iv) (Osku Hellas®, Grapeguard, Santiago, Chile) contain 73.5% of the active ingredient ($Na_2S_2O_5$), with 26 cm × 36 cm of dimensions.

A fungal suspension was prepared, according to the standard protocol, using a *B. cinerea* isolate (BCUEL-1), isolated from infected grapes with representative symptoms of the disease, according to Youssef and Roberto [23]. The suspensions were diluted with sterilized distilled water to get a final concentration of 10^6 conidia mL^{-1} using a hemocytometer with 1/10 mm deep (Neubauer Boeco, Hamburg, Germany). As the incidence of gray mold can be low, depending on the season, the grapes from treatments (iii) and (iv) were inoculated with a pathogen suspension, according to Youssef and Roberto [23]. A volume of 200 mL of inoculums was sprayed on each 50 kg of grapes until dripping, using a plastic sprayer. The control consists of bunches treated only with distilled water. All bunches were air dried at room temperature (RT) before packaging.

The grapes of all treatments were packaged as follows: A micro-perforated plastic liner (1% of the ventilated area, Suragra S.A., San Bernardo, Chile) was placed inside carton boxes; grapes were placed inside the box; an SO_2-releasing pad was placed on top only for treatments (ii) and (iv); and the liner was sealed. The SO_2-releasing pad fully covered the grapes.

The boxes were placed in a cold room storage at 1 ± 1 °C and at high relative humidity (>95%). As 'BRS Isis' is a new cultivar and there is no information available regarding its cold storage performance, after 30 days of cold storage, the boxes were opened for inspection, and as the bunches of all treatments were intact, with fresh and green stems, free of any mold or injuries, it was decided to keep the boxes in the chamber for an extended period, i.e., 50 days, followed by 7 days of shelf-life at RT (22 ± 2 °C). The completely randomized experimental design was used as a statistical model with four treatments and five replications, and each plot consisted of one carton box (each measuring 23 cm × 16 cm × 9 cm (4 kg capacity)).

2.3. Evaluation of Gray Mold Incidence

The incidence of gray mold on grapes was evaluated at 50 days after the beginning of cold storage and at 7 days at 22 ± 2 °C after the end of cold storage. The disease incidence was then calculated: Disease incidence (% of diseased berries) = (number of infected berries/total number of berries) × 100 [23].

2.4. Physicochemical Analysis

The grape physicochemical analysis was evaluated twice: (i) 50 days after the beginning of cold storage; (ii) at seven days at 22 ± 2 °C following the cold storage period, using 10 berries for each box (replication). The bunch mass loss as a percentage was calculated as follows: Mass loss (%) = ((mi−ms)/mi) × 100, where mi is the initial mass and ms is the mass at the examined time [24]. Shattered berries were evaluated by calculating the separated grape berries from the bunch stem and were expressed as a percentage of the total number of berries: Shattered berries (% of diseased berries) = (number of shattered berries/total number of berries) × 100.

The berry color was investigated using a colorimeter CR-10 (Konica Minolta®, Tokyo, Japan) to get the following variables from the equatorial portion of grape berries ($n = 2$ per berry): $L*$ (lightness), $C*$ (chroma) and $h°$ (hue angle). The color index for red grapes (CIRG) was then calculated using the formula CIRG = $(180 − h°)/(L* + C*)$ [25]. Ten berries were collected from each replicate to be investigated. Lightness rates range from 0 (black) to 100 (white). Chroma indicates the purity or intensity of color and the distance from gray (achromatic) toward a pure chromatic color and is measured from the a^* and b^* values of the CIELab scale system, starting from zero for a completely neutral color, and does not have an arbitrary end, but the intensity increases with magnitude. Hue refers to the color wheel and is calculated in angles; green, yellow, and red correspond to 180, 90, and 0°, respectively [26–28].

For the chemical analysis, 10 berries were collected from each replicate. To determine soluble solid (SS) content and titratable acidity (TA), samples were crushed, and the juice was used. For SS, some juice drops were analyzed using a digital refractometer (Krüss DR301-95; A. Krüss Optronic, Hamburg, Germany) with automatic temperature compensation at 20 ± 1 °C, and the results were presented as °Brix. TA was determined using a dropwise titration with 0.1 N NaOH using 10 mL of grape juice diluted in 40 mL of distilled H_2O, and pH = 8.2 was considered as the endpoint. The results were presented as tartaric acid (%) [29]. The SS/TA ratio was used to express the maturation index of grape berries.

2.5. Statistical Analysis

All data were subjected to an analysis of variance (ANOVA) by using Sisvar® software (UFLA, Lavras, Brazil). The mean values of treatments were compared by using Fisher's protected least significant difference (LSD) test and judged at $p \leq 0.05$ levels. Percentage data were arcsine transformed to normalized variance. Data in the tables or charts are the untransformed percentage of rotted grape berries.

3. Results

3.1. Incidence of Gray Mold

The disease incidence found at 50 days of cold storage was considered low in both seasons, and no significant differences were observed when grapes were subjected to control and SO$_2$-generating pad treatments only (Figures 1 and 2). On the other hand, when grapes were inoculated with *B. cinerea* suspension, the SO$_2$-generating pads significantly decreased the incidence of gray mold of grapes harvested in the summer crop season, as compared to the control with bunches inoculated with *Botrytis*. In the case of the off-season crop, although the incidence of gray mold was higher in grapes of the control and SO$_2$-generating pads, both inoculated with *B. cinerea* suspension, no significant differences were observed between them.

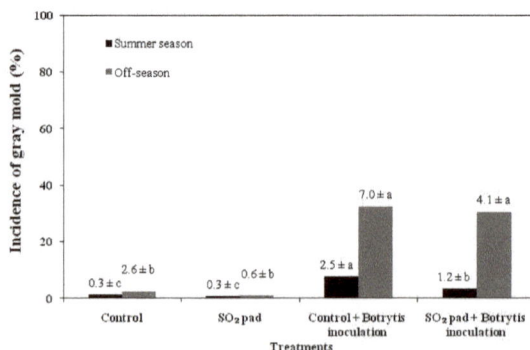

Figure 1. Incidence of gray mold (% of diseased berries) at 50 days of cold storage of 'BRS Isis' seedless table grapes during summer and off-season crops. Columns followed by different letters, in relation to the treatments within each individual crop, are statistically different, according to Fisher's protected LSD test ($p \leq 0.05$).

Figure 2. Bunches of 'BRS Isis' seedless table grapes at 50 days under cold storage. (**A**) Control; (**B**) SO$_2$-generating pads.

It was observed that the incidence of gray mold was higher in the off-season crop (~30%) when grapes were inoculated with *B. cinerea* suspension. This situation was also found after the 7 day period at 22 ± 2 °C, where the gray mold incidence (~50%) was higher as compared to the 50 day period of cold storage (Figure 3).

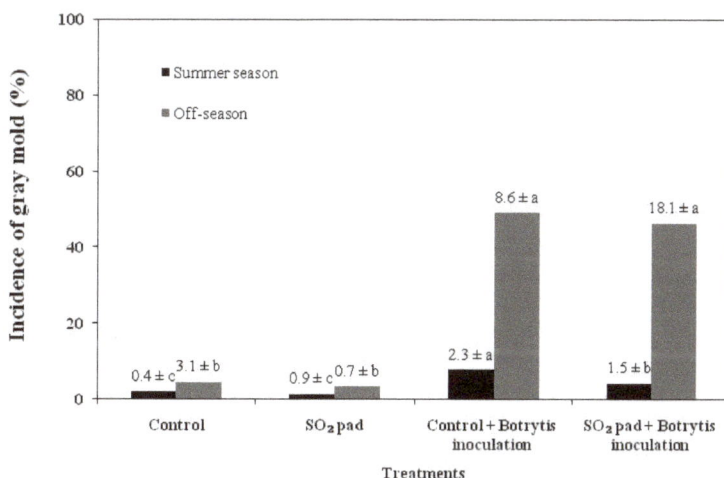

Figure 3. Incidence of gray mold (% of diseased berries) at seven days at 22 ± 2 °C of 'BRS Isis' seedless table grapes during summer and off-season crops. Columns followed by different letters, in relation to the treatments within each individual crop, are statistically different, according to Fisher's protected LSD test ($p \leq 0.05$).

After seven days of 22 ± 2 °C, no significant differences were found among treatments with SO$_2$-generating pads in comparison to the control (with no *B. cinerea* inoculation) in both crop seasons. However, when grapes were inoculated with *B. cinerea* suspension, significant differences were observed in the summer season crop, where the SO$_2$-generating pads resulted in lower gray mold incidence (4.2%) in comparison to the control with inoculated grapes (7.8%), while in the case of the off-season crop, no differences were found (Figures 3 and 4).

Figure 4. Bunches of 'BRS Isis' seedless table grapes at seven days at 22 ± 2 °C; (**A**) control with *B. cinerea* inoculation; (**B**) SO$_2$-generating pads with *B. cinerea* inoculation.

3.2. Physical Characteristics of Grapes

There were no significant differences among treatments for mass loss at 50 days of cold storage in the summer crop, and means varied from 5.4 to 7.0%, while in the case of the off-season crop, significant differences were noted, where both *B. cinerea* inoculated treatments (the control and SO$_2$-generating pad) showed higher mass loss as compared to non-inoculated treatments (Table 1).

Table 1. Mass loss (%), shattered berries (%), and color index for red grapes (CIRG) of the 'BRS Isis' seedless table grape at 50 days of cold storage and at seven days at 22 ± 2 °C during summer and off-season crops.

Treatments	Mass loss (%)		Shattered berries (%)		CIRG	
	At 50 days of cold storage					
	Summer season	Off-season	Summer season	Off-season	Summer season	Off-season
Control	6.3 ± 2.4	3.1 ± 0.4 b	0.0 ± 0.0	1.2 ± 2.5	3.8 ± 0.4	5.1 ± 0.3
SO$_2$ pad	5.6 ± 1.4	3.0 ± 0.2 b	0.3 ± 0.3	0.7 ± 0.5	4.4 ± 0.2	5.4 ± 0.3
Control + Bot	5.4 ± 0.9	3.7 ± 0.3 a	0.7 ± 0.6	3.1 ± 1.0	4.5 ± 0.5	5.4 ± 0.4
SO$_2$ pad + Bot	7.0 ± 2.6	3.6 ± 0.3 a	0.2 ± 0.4	2.1 ± 1.1	4.6 ± 0.5	5.5 ± 0.2
F value	0.7 ^{NS}	5.4 *	2.5 ^{NS}	2.6 ^{NS}	3.4 ^{NS}	1.5 ^{NS}
	At seven days at 22 ± 2 °C					
	Summer season	Off-season	Summer season	Off-season	Summer season	Off-season
Control	5.8 ± 0.3c	1.2 ± 0.1c	0.1 ± 0.3b	1.8 ± 2.2b	4.0 ± 0.6	4.8 ± 0.4
SO$_2$ pad	7.2 ± 0.8ab	1.6 ± 0.4ab	0.7 ± 0.6b	1.5 ± 1.3b	3.8 ± 0.3	4.6 ± 0.3
Control + Bot	7.6 ± 0.5a	1.3 ± 0.2bc	1.5 ± 0.5a	12.1 ± 9.3a	4.4 ± 0.2	4.8 ± 0.3
SO$_2$ pad + Bot	6.8 ± 0.5b	1.8 ± 0.2a	0.6 ± 0.6b	11.9 ± 4.3a	4.2 ± 0.5	4.8 ± 0.2
F value	10.1 *	3.8 *	6.3 *	6.4 *	2.0 ^{NS}	0.3 ^{NS}

Means within columns followed by the same letters are not statistically different by Fisher's protected LSD test ($p \leq 0.05$). Non-significant (^{NS}), *: significant at 5% level of significance.

At seven days at 22 ± 2 °C, significant differences were observed among treatments and the control treatment showed the lowest mass loss in both seasons (5.8% and 1.2% for summer and off-season crops, respectively). In the summer season, a higher mass loss (7.6%) was observed in control with grapes inoculated with *B. cinerea* and in the case of the off-season, a higher mass loss (1.8%) was recorded in SO$_2$-generating pads with grapes inoculated with *B. cinerea* suspension (Table 1).

No significant differences were found in terms of shattered berries for both seasons, and at 50 days of cold storage, the means ranged from 0.0 to 0.7% and from 0.7 to 3.1% for the summer season and the off-season, respectively (Table 1). However, statistically differences were found in both seasons at seven days at 22 ± 2 °C, where the control with grapes inoculated with *B. cinerea* suspension showed higher shattered berries (1.5%). On the other hand, in the off-season crop, when grapes were inoculated with *B. cinerea* suspension, combined or not with SO_2-generating pads, a higher percentage of shattered berries was found (~12%). However, the percentage of shattered berries was high at seven days at 22 ± 2 °C, especially for the off-season crop.

Additionally, there was no change in the berry color index among treatments in both evaluated seasons (Table 1). In the summer season, the berry color ranged from 3.8 to 4.6 (red), while in the off-season crop, the means ranged from 5.1 to 5.5 (red-violet) [30]. During the off-season crop, the anthocyanin accumulation develops under a higher diurnal temperature variation in subtropics, which intensifies berry color and explains these variations. Nevertheless, the original color of the 'BRS Isis' seedless grape was well-preserved in both storage periods.

3.3. Chemical Characteristics of Grapes

Regarding berry SS content, even though differences among treatments have only been observed at 50 days of cold storage for the summer season, the observed means are in an acceptable range (~14 °Brix), which is a standard for local and international markets of some table grape cultivars [31]. There was no difference in terms of TA and SS/TA of berries among treatments in both evaluated seasons (Table 2).

Table 2. Soluble solids (SS), titratable acidity (TA), and SS/TA ratio of 'BRS Isis' seedless table grapes at 50 days of cold storage and at seven days at 22 ± 2 °C during summer and off-season crops.

Treatments	SS (°Brix)		TA (%)		SS/TA	
	At 50 days of cold storage					
	Summer season	Off-season	Summer season	Off-season	Summer season	Off-season
Control	14.5 ± 0.4a	14.3 ± 0.2	0.6 ± 0.1	0.9 ± 0.1	23.1 ± 3.4	16.8 ± 1.2
SO$_2$ pad	13.9 ± 0.5b	14.2 ± 0.5	0.6 ± 0.03	0.8 ± 0.1	23.2 ± 1.4	17.7 ± 1.0
Control + Bot	13.6 ± 0.4b	14.1 ± 0.2	0.6 ± 0.1	0.9 ± 0.1	23.5 ± 2.0	16.6 ± 1.7
SO$_2$ pad + Bot	14.0 ± 0.3ab	14.1 ± 0.4	0.6 ± 0.03	0.8 ± 0.03	23.3 ± 0.9	18.3 ± 0.6
F value	3.8 *	0.3 NS	0.8 NS	1.8 NS	0.04 NS	2.1 NS
	At seven days at 22 ± 2 °C					
	Summer season	Off-season	Summer season	Off-season	Summer season	Off-season
Control	15.0 ± 0.4	13.9 ± 0.2	0.7 ± 0.1	0.9 ± 0.1	22.9 ± 1.9	15.9 ± 1.9
SO$_2$ pad	14.4 ± 0.6	14.1 ± 0.5	0.7 ± 0.01	0.7 ± 0.03	22.1 ± 1.2	19.1 ± 0.6
Control + Bot	14.3 ± 0.5	13.8 ± 0.4	0.6 ± 0.03	0.7 ± 0.03	22.2 ± 1.3	19.0 ± 1.1
SO$_2$ pad + Bot	14.6 ± 0.4	14.0 ± 1.1	0.7 ± 0.03	0.8 ± 0.1	21.8 ± 0.6	18.8 ± 3.4
F value	2.1 NS	0.1 NS	1.1 NS	2.6 NS	0.7 NS	2.8 NS

Means within columns followed by the same letters are not statistically different by Fisher's protected LSD test ($p \leq 0.05$). Non-significant (NS), *: significant at 5% level of significance.

4. Discussion

Botrytis cinerea Pers. ex Fr., is the most crucial postharvest pathogen attacking table grapes worldwide, causing severe losses of the crop after harvest. It was also observed earlier that the incidence of gray mold was higher, especially in the off-season crop (~30%) when grapes were inoculated with *B. cinerea* suspension. This could be explained by the occurrence of some invisible minor cracks or spots on berry skin caused by powdery mildew *Uncinula necator* (Schwein). Burrill, which is more prevalent in this season [32]. In the case of the seven day period at 22 ± 2 °C, the incidence was also found high (~50%) in comparison to the 50 day period of cold storage. These results are related with the fact that when grapes are subjected to RT, the disease incidence increases because of the more favorable environmental conditions for fungi development, especially due to the higher air temperature.

Our findings confirm that SO_2-generating pads gave better results in controlling *B. cinerea* mold of 'BRS Isis' seedless grapes at 50 days of cold storage and at seven days at 22 ± 2 °C. This new hybrid seedless cultivar showed to be non-sensitive to the amount of SO_2 gas released by the evaluated pads, since, at high concentrations, this compound can cause bleaching or premature stem browning and may also damage the fruits, resulting in unwanted conditions. Considering these aspects, combined with cold storage, the SO_2-generating pads can be used as a tool to control the gray mold of 'BRS Isis' seedless grapes, at least for a period of 50 days. A similar performance has also been observed in rthe maximum reduction of the disease incidence of 'BRS Vitoria' and 'Italia' table grapes at 50 days of cold storage and at seven days at 22 ± 2 °C, respectively [17,22].

The mass loss is also concerned as it is one of the key factors that determine excellence and quality of table grapes; the more water lost from the produce, the more it develops quality deterioration problems. During the experiment, it was observed that, in the off-season crops, both *B. cinerea* inoculated treatments (the control and SO_2-generating pads) showed higher mass loss as compared to non-inoculated treatments. This behavior can be related to the fact that the incidence of gray mold was higher in the off-season crop, which may have caused higher mass loss in inoculated treatments.

Even though during the cold storage period, temperature and relative humidity are controlled to reduce mass loss and extend the shelf-life of table grapes, sometimes mass loss can vary depending on different aspects, i.e., the grape cultivar, harvesting conditions, storage period, and packing materials used. Among other factors, fungal infection, absence, delay in pre-cooling, or high temperatures with low humidity are also the main causes of mass loss [33,34]. For most fresh produce, mass loss percentage should be low to not affect quality attributes (wilting or wrinkling), and the same behavior was observed in the current study of 'BRS Isis' table grapes. The SO_2-generating pads were also found to reduce mass loss in both evaluated situations, i.e., grapes inoculated or not with *B. cinerea* suspension. When grapes are subjected to RT, water loss and percentage of shattered berries increases because of the favorable environmental conditions, especially the higher air temperature that reduces the fruit quality, resulting in negatively affecting the grape bunch quality [35].

Regarding chemical characteristics as grape ripening develops under different weather conditions in summer and off-season crops, a slight change between them usually occurs in terms of the main berry chemical properties (Table 2) but does not decrease the grape quality. During cold storage, the recommended temperature for grapes is around 0 °C because most of the variables, such as SS, TA, and SS/TA, remain stable in different grape cultivars at this temperature with controlled atmosphere [36,37].

Table grapes intended for long periods of storage are kept in cold chambers, but each cultivar has a different behavior, i.e., each one has a different storage performance that may comprise from a few days to few weeks, which is determined by its susceptibility to quality defects under low temperatures. Regarding the behavior of the new hybrid seedless grape, 'BRS Isis', this cultivar showed to have a high potential to be stored for long periods under cold chambers, since, after 50 days under these conditions, the bunches packaged with SO_2-generating pads and liners were virtually intact. Additionally, the natural incidence of gray mold was found to be very low, which indicates that the natural incidence of *B. cinerea* in this hybrid grape, unlike in some *Vitis vinifera* table grape cultivars, is not a major concern. Even with the use of SO_2-generation pads, unwanted situations like bleaching, hairline cracking, and berry softening were not found on the surface of the berries, as these were the main barriers for grape post-harvest quality and maintenance. Shattered berries were also noticed in low levels, which contributes to a better storability and marketability of this grape cultivar in markets.

The period from harvest to marketing of the table grape has a significant importance regarding the maintenance of fruit quality. The results obtained herein showed that the 'BRS Isis' seedless grape has a large potential for domestic, long distance, and international markets, because a high quality of bunches can be achieved under cold storage for at least 50 days. For domestic markets, including when a long distance transportation is required, the 'BRS Isis' grapes, after being properly packaged, can be transported in refrigerated trucks and easily kept in cold chambers of the market chains and gradually exposed to the consumers with a minimum loss quality. The same could also be applied when the intention is to export this grape overseas, to the European Community or even to North American countries. As long as the cold chain is retained, and considering that it takes up to three weeks to transport a refrigerated container by ship from South America to these regions, 'BRS Isis' seems to fit well for this type of international trade due its high capacity of storage in cold chambers during long periods of up to 50 days or longer. However, since a large proportion of table grapes can be traded overseas, more attention is required for better shipment and quality management.

Finally, for long-term storage and transportation, packaging materials, such as SO_2-generating pads and proper liners, also play a crucial role to preserve 'BRS Isis' grapes under cold storage, reducing some unwanted situations and providing a higher efficiency of the SO_2 gas for controlling gray mold.

5. Conclusions

The new hybrid 'BRS Isis' seedless grape, packaged with SO_2-generating pads and plastic liners, has a high potential to be preserved for long periods under cold storage at $1 \pm 1\ °C$, at least for 50 days, keeping a very low natural incidence of gray mold, mass loss, and shattered berries.

Author Contributions: S.R.R. conceived the research idea. S.A., K.Y., R.C.C., M.S., O.J.C.J., C.H.S., and R.T.d.S. helped to collect the data. S.A. and K.Y. also analyzed the data and wrote the paper.

Funding: The authors are grateful for the financial support provided by the Brazilian National Council for Scientific and Technological Development (CNPq) and The World Academy of Sciences (TWAS) for financial support (Grant #190015/2014-4).

Acknowledgments: The author gives heartfelt thanks to Antonio Peres for providing grapes and other materials.

Conflicts of Interest: The authors declare no conflict of interest.

References

1. Ritschel, P.S.; Maia, J.D.G.; Camargo, U.A.; de Souza, R.T.; Fajardo, T.V.M.; de Naves, R.L.; Girardi, C.L. *BRS Isis Nova Cultivar de Uva de Mesa Vermelha, Sem Sementes e Tolerante ao Míldio*; Comunicado Técnico 143; Embrapa: Brasilia, Brazil, 2013; pp. 6–11.
2. Daudt, C.E.; Fogaca, A.O. Phenolic compounds in Merlot wines from two wine regions of Rio Grande do Sul, Brazil. *Food Sci. Technol.* **2013**, *33*, 355–361. [CrossRef]
3. Silva-Sanzanaa, C.; Balica, I.; Sepúlvedaa, P.; Olmedoa, P.; Leóna, G.; Defilippi, B.G.; Blanco-Herreraa, F.; Campos-Vargasa, R. Effect of modified atmosphere packaging (MAP) on rachis quality of 'Red Globe' table grape variety. *Postharvest Biol. Technol.* **2016**, *119*, 33–40. [CrossRef]
4. Sen, F.; Oksar, R.; Kesgin, M. Effects of shading and covering on 'Sultana Seedless' grape quality and storability. *J. Agric. Sci. Technol.* **2016**, *18*, 245–254.
5. Hashim, A.F.; Youssef, K.; Abd-Elsalam, K.A. Ecofriendly nanomaterials for controlling gray mold of table grapes and maintaining postharvest quality. *Eur. J. Plant Pathol.* **2019**, *154*, 377–388. [CrossRef]
6. Youssef, K.; de Oliveira, A.G.; Tischer, C.A.; Hussain, I.; Roberto, S.R. Synergistic effect of a novel chitosan/silica nanocomposites-based formulation against gray mold of table grapes and its possible mode of action. *Int. J. Biol. Macromol.* **2019**, *141*, 247–258. [CrossRef] [PubMed]
7. Carisse, O.; Van Der Heyden, H. Relationship of airborne Botrytis cinerea conidium concentration to tomato flower and stem infections: A threshold for de-leafing operations. *Plant Dis.* **2015**, *99*, 137–142. [CrossRef] [PubMed]
8. Feliziani, E.; Romanazzi, G. Postharvest decay of strawberry fruit: Etiology, epidemiology, and disease management. *J. Berry Res.* **2016**, *6*, 47–63. [CrossRef]

9.   Crisosto, C.H.; Garner, D.; Crisosto, G. Carbon dioxide-enriched atmosphere during cold storage limit losses from Botrytis but accelerate Rachis browning of 'Red globe' table grapes. *Postharvest Biol. Technol.* **2002**, *26*, 181–189. [CrossRef]
10.   Celik, M.; Kalpulov, T.; Zutahy, Y.; Ish-Shalom, B.; Lurie, S.; Litcher, A. Quantitative and qualitative analysis of Botrytis inoculated on table grapes by qPCR and antibodies. *Postharvest Biol. Technol.* **2009**, *52*, 235–239. [CrossRef]
11.   Romanazzi, G.; Joseph, L.S.; Erica, F.; Droby, S. Integrated management of postharvest gray mold on fruit crops. *Postharvest Biol. Technol.* **2016**, *113*, 69–76. [CrossRef]
12.   Sen, F.; Altun, A.; Kesgin, M.; Inan, M.S. Effect of different shading practices used in the pre-harvest stage on quality and storage life of sultana seedless grapes. *J. Agric. Sci. Technol.* **2012**, *2*, 1234–1240.
13.   Brackmann, A.; Ceretta, M.; Pinto, J.A.V.; Venturini, T.L.; Lucio, A.D.L. Tolerância de maçãs 'Gala' a baixas temperaturas durante o armazenamento. *Ciência Rural* **2010**, *40*, 1909–1915. [CrossRef]
14.   Liguori, G.; Sortino, G.; De Pasquale, C.; Inglese, P. Effects of modified atmosphere packaging on quality parameters of minimally processed table grapes during cold storage. *Adv. Hortic. Sci.* **2015**, *29*, 152–154.
15.   Lichter, A.; Zutahy, Y.; Kaplunov, T.; Lurie, S. Evaluation of table grape storage in boxes with sulfur dioxide releasing pads with either an internal plastic liner or external wrap. *HortTechnology* **2008**, *18*, 206–214. [CrossRef]
16.   Melgarejo-Flores, B.G.; Ortega-Ramírez, L.A.; Silva-Espinoza, B.A.; González-Aguilar, G.A.; Miranda, M.R.A.; Ayala-Zavala, J.F. Antifungal protection and antioxidant enhancement of table grapes treated with emulsions, vapors, and coatings of cinnamon leaf oil. *Postharvest Biol. Technol.* **2013**, *86*, 321–328. [CrossRef]
17.   Domingues, A.R.; Roberto, S.R.; Ahmed, S.; Shahab, M.; Junior, O.J.C.; Sumida, C.H.; Souza, R.T. Postharvest Techniques to Prevent the Incidence of Botrytis Mold of 'BRS Vitoria' Seedless Grape under Cold Storage. *Horticulturae* **2018**, *4*, 17. [CrossRef]
18.   Lurie, S.; Pesis, E.; Gadiyeva, O.; Feygenberg, O.; Ben-Arie, R.; Kaplunov, T.; Zutachi, Y.; Lichter, A. Modified ethanol atmosphere to control decay of table grapes during storage. *Postharvest Biol. Pathol.* **2006**, *42*, 222–227. [CrossRef]
19.   Zutahy, Y.; Lichter, A.; Kaplunov, T.; Lurie, S. Extended storage of 'Red Globe' grapes in modified SO_2 generating pads. *Postharvest Biol. Technol.* **2008**, *50*, 12–17. [CrossRef]
20.   Ricce, W.S.; Caramori, P.H.; Roberto, S.R. Potencial climático para a produção de uvas em sistema de dupla poda anual no estado do Paraná. *Bragantia* **2013**, *72*, 408–415. [CrossRef]
21.   Youssef, K.; Roberto, S.R.; Chiarotti, F.; Koyama, R.; Hussain, I.; De Souza, R.T. Control of Botrytis mold of the new seedless grape 'BRS Vitoria' during cold storage. *Sci. Hortic.* **2015**, *193*, 316–321. [CrossRef]
22.   Ahmed, S.; Roberto, S.R.; Domingues, A.R.; Shahab, M.; Junior, O.J.C.; Sumida, C.H.; Souza, R.T. Effects of Different Sulfur Dioxide Pads on Botrytis Mold in 'Italia' Table Grapes under Cold Storage. *Horticulturae* **2018**, *4*, 29. [CrossRef]
23.   Youssef, K.; Roberto, S.R. Applications of salt solutions before and after harvest affect the quality and incidence of postharvest gray mold of 'Italia' table grapes. *Postharvest Biol. Technol.* **2014**, *87*, 95–102. [CrossRef]
24.   Mattiuz, B.; Miguel, A.C.A.; Galati, V.C.; Nachtigal, J.C. Efeito da temperatura no armazenamento de uvas apirênicas minimamente processadas. *Rev. Bras. Frutic.* **2009**, *31*, 44–52. [CrossRef]
25.   Carreño, J.; Martínez, A.; Almela, L.; Fernández-López, J.A. Proposal of an index for the objective evaluation of the color of red table grapes. *Food Res. Int.* **1995**, *28*, 373–377. [CrossRef]
26.   Mcguire, R.G. Reporting of objective color measurements. *HortScience* **1992**, *27*, 1254–1255. [CrossRef]
27.   Lancaster, J.E.; Lister, C.; Reay, P.F.; Triggs, C.M. Influence of pigment composition on skin color in a wide range of fruits and vegetables. *Am. Soc. Hortic. Sci.* **1997**, *122*, 594–598. [CrossRef]
28.   Peppi, M.C.; Fidelibus, M.W.; Dokoozlian, N. Abscisic acid application timing and concentration affect firmness, pigmentation and color of 'Flame Seedless' grapes. *HortScience* **2006**, *41*, 1440–1445. [CrossRef]
29.   Youssef, K.; Roberto, S.R. Salt strategies to control Botrytis mold of 'Benitaka' table grapes and to maintain fruit quality during storage. *Postharvest Biol. Technol.* **2014**, *95*, 95–102. [CrossRef]
30.   Carreño, J.; Martinez, A.; Almela, L.; Fernández-López, J.A. Measuring the color of table grapes. *Color Res. Appl.* **1998**, *21*, 50–54. [CrossRef]

31. UNECE—United Nations Economic Commission for Europe. *Standard FFV-19 Concerning the Marketing and Commercial Quality Control of Table Grapes*; United Nations: New York, NY, USA; Geneva, Switzerland, 2010; p. 8.

32. Tessmann, D.J.; Vida, J.B.; Genta, W.; Roberto, S.R.; Kishino, A.Y. Doenças e Seu Manejo. In *Viticultura Tropical: O sistema de produção de uvas de mesa do Paraná*; Kishino, A.Y., Carvalho, S.L.C., Roberto, S.R., Eds.; IAPAR: Londrina, Brazil, 2018; pp. 453–548.

33. Valverde, J.M.; Guillén, F.; Martínez-Romero, D.; Castillo, S.; Serrano, M.; Valero, D. Improvement of table grapes quality and safety by the combination of modified atmosphere packaging (MAP) and eugenol, menthol, or thymol. *J. Agric. Food Chem.* **2005**, *53*, 7458–7464. [CrossRef]

34. Nelson, K.E. Retarding deterioration of table grapes with in–package sulfur dioxide generators with and without refrigeration. *Acta Hortic.* **1983**, *138*, 121–130. [CrossRef]

35. Wright, H.; Delong, J.; Lada, R.; Prange, R. The relationship between water status and chlorophyll a fluorescence in grapes (*Vitis* spp.). *Postharvest Biol. Technol.* **2009**, *51*, 193–199. [CrossRef]

36. Artés-HernándeZ, F.; Tomàs-Barberán, F.A.; Artés, F. Modified atmosphere packaging preserves quality of SO_2-free 'Superior seedless' table grapes. *Postharvest Biol. Technol.* **2006**, *39*, 146–154. [CrossRef]

37. Rosales, R.; Romero, I.; Fernandez-Caballero, C.; Escribano, M.I.; Merodio, C.; Sanchez-Ballesta, M.T. Low Temperature and Short-Term High CO_2 treatment in postharvest storage of table grapes at two maturity stages: Effects on Transcriptome Profiling. *Front. Plant Sci.* **2016**, *7*, 1020. [CrossRef] [PubMed]

MDPI

Article

Anthocyanin Accumulation and Color Development of 'Benitaka' Table Grape Subjected to Exogenous Abscisic Acid Application at Different Timings of Ripening

Muhammad Shahab [1], Sergio Ruffo Roberto [1,*], Saeed Ahmed [1], Ronan Carlos Colombo [1], João Pedro Silvestre [1], Renata Koyama [1] and Reginaldo Teodoro de Souza [2]

[1] Agricultural Research Center, Department of Agronomy, Londrina State University, Londrina, PR 86057-970, Brazil; mshahab78@gmail.com (M.S.); saeeddikhan@gmail.com (S.A.); ronancolombo@yahoo.com.br (R.C.C.); jp.silvestre@gmail.com (J.P.S.); emykoyama@hotmail.com (R.K.)
[2] Embrapa Grape and Wine, Tropical Station Unit., Bento Gonçalves, RS 95700-000, Brazil; reginaldo.souza@embrapa.br
* Correspondence: sroberto@uel.br; Tel.: +55-43-3371-4774

Received: 14 January 2019; Accepted: 9 February 2019; Published: 28 March 2019

Abstract: In colored table grapes, the anthocyanin contents are inhibited by the high temperature during ripening and berries suffer a lack of skin color, thus affecting their market value. In order to overcome this issue, a research study was planned to evaluate the influence of (*S*)-*cis*-abscisic acid (*S*-ABA) on rates of anthocyanin accumulation in table grapes when applied at different timings of ripening, and to quantify the gradual increase of berry color. The study was conducted in a commercial vineyard of 'Benitaka' table grapes (*Vitis vinifera* L.), grown under double annual cropping system in a subtropical area. The trials were carried out during two consecutive seasons (i.e., summer season of 2015 and off-season of 2016). The treatments used for the experiments contained 400 mg L^{-1} *S*-ABA applied at different timings of veraison (the onset of ripening), as follows: control (with no application); at pre-veraison (PRV); at veraison (V); and at post-veraison (POV). For all *S*-ABA treatments, a second application was performed 10 days after the first application. Berries were analyzed for weekly and daily anthocyanin accumulations, weekly and daily color index development (CIRG), total soluble solids (TSS) content, titratable acidity (TA), and maturation index (TSS/TA). Grapes subjected to exogenous application of *S*-ABA at any time of veraison, especially at PRV or at V, significantly increased the anthocyanin accumulation as well as berry color index development. Other chemical properties of grapes (i.e., TSS, TA, and TSS/TA evolution) were not affected by the use of *S*-ABA and followed a predictable pattern in relation to days of berries ripening.

Keywords: *Vitis vinifera* L.; production system; *S*-ABA; rate of anthocyanin accumulation; CIRG; bioactive compounds

1. Introduction

Table grapes are a rich source of phenolic compounds with antioxidant and anti-inflammatory properties, which are helpful in preventing several human diseases [1–3]. These secondary metabolites are present in different parts of berries, where skin is enriched with anthocyanins, pigments responsible for the red, pink, or black color [4,5], and in some cultivars these pigments can also be found in the flesh [6,7].

However, when colored table grapes are grown in subtropical areas, high temperatures during ripening may inhibit anthocyanin accumulation and prevent color development, thus negatively affecting the market value of the table grapes, since the skin color is a very important economic feature [7–9]. In addition, in some subtropical regions, due to the mild winter and the use of bud burst stimulators, a double annual cropping of grapes can be achieved. Therefore, besides the summer season crop, an off-season crop is obtained when there is no or less supply of fresh grapes in the market [10].

'Benitaka' (*Vitis vinifera* L.) is one of the most important colored table grapes developed from the bud sport of 'Italia' grape [11]. Interest in growing this cultivar has been increasing due to its dark pink color and uniform, large, and crunchy berries, and in some regions, such as the Brazilian subtropics, this grape represents more than 50% of the area cultivated by table grapes. However, lack of skin color is an issue while growing this cultivar in this kind of subtropical warm climates [9,10].

In grapes, the anthocyanin starts accumulating at the time when abscisic acid (ABA) also starts to increase in berries, and is reportedly responsible for the anthocyanin biosynthesis [6,7]. In recent years, it has been demonstrated that the application of the enantiomer (*S*)-*cis*-abscisic acid (*S*-ABA) can increase the anthocyanin contents of grapes and improve their color [6,7,9,12–16].

Exogenous application of *S*-ABA is effective around the time of veraison (onset of ripening), a time when physiological changes start to appear in grapes, such as the increase of soluble solids, berry softening, and coloring [15,17]. However, in most of the cases, application of *S*-ABA at the time of veraison is a difficult task, especially because the onset of these changes does not occur simultaneously and may widely vary among cultivars [18]. Additionally, large growing areas and unfavorable climatic conditions, such as prolonged rainfall periods, make it difficult to apply this plant growth regulator at veraison in a short period of time over the whole area, since it is a time-consuming operation and only the bunches are subjected to the application.

Considering these aspects, an evaluation of the effect of *S*-ABA application over a longer period of time (i.e., from pre- to post-veraison) on color development has not been explored yet, especially regarding the weekly and daily rates of anthocyanin development, which could provide information leading to a better understanding of the responses of berries towards such treatments in different circumstances. In order to overcome this issue, a research study was planned to evaluate the influence of *S*-ABA on anthocyanin accumulation when applied at different timings of veraison, and to quantify the gradual increase of berry color in 'Benitaka' table grapes grown under double annual cropping in a subtropical area.

2. Materials and Methods

2.1. Experimental Area and Pre-Conditions

The study was conducted in an 11-year-old commercial vineyard of 'Benitaka' table grapes (*Vitis vinifera* L.), grafted on 'IAC 766 Campinas' rootstock located at Marialva city, in the state of Parana, Brazil (23°29'52.8" S, 51°47'58" W, elevation 570 m), under double annual cropping system. The climate of this area is classified as Cfa by Köppen (i.e., subtropical humid) with winter mean temperature below 18 °C, summer mean temperature above 22 °C and 1596 mm of rainfall, which occurs mostly during summer [19]. The trials were carried out during two consecutive seasons (i.e., summer season of 2015 and off-season of 2016). The total precipitation and average temperature during the grape ripening period of the 2015 summer season and 2016 off-season were 462 mm and 23.0 °C and 290 mm and 17.5 °C, respectively. The vines were spaced at a distance of 3.0 × 6.0 m and cane-pruned with eight buds per cane. For uniform bud burst, 2.5% of hydrogen cyanamide was applied on the two terminal buds. Other practices like fertilizer application, weed control, pest and disease management were carried out according to the local practices used [9].

2.2. Treatments and Statistical Design

The isomer (*S*)-*cis*-abscisic acid (*S*-ABA) was provided by Valent BioSciences® Co. (Illinois, Libertyville, IL, USA), containing 100 g L^{-1} of active ingredient. The experiments were conducted in a randomized block design, where the treatments were replicated five times, and each plot consisted of one single vine. Ten representative bunches per each plot were marked for further analysis.

The treatments contained S-ABA 400 mg L^{-1} [9,15] applied at different timings of veraison, as follows: control (with no application); at pre-veraison (PRV); at veraison (V); and at post-veraison (POV). For all S-ABA treatments, a second application was performed 10 days after the first application in order to potentialize anthocyanin accumulation, according to previous works with this and other grape cultivars [9,15]. Application timings were identified considering the total soluble solids (TSS) of berries. The first treatment (at PRV) was applied when the berry TSS contents suddenly jumped from 4.0 to 5.7 Brix. Similarly, for the second treatment (at V), TSS was 7.3 Brix and at least 50% of the berries showed change in color, performed 7 days after PRV. The last treatment (at POV) was also applied 7 days after V, where the TSS was 9.5 Brix.

For S-ABA application, only the bunches of the 'Benitaka' table grapes were sprayed using a knapsack sprayer at a pressure of 568.93 psi (39.22 bar) with hollow cone nozzle tips, model JA-1 (1 mm of diameter) (Jacto Group, Pompeia, Brazil), at a volume of 800 L ha^{-1} in a complete and uniform way. In addition, 0.3 mL L^{-1} of BreakThru®(Evonik Industries, Essen, Germany), a non-ionic surfactant, was added to all treatments for uniformity of the treatment.

2.3. Sampling and Analyses

During both evaluated seasons, anthocyanin contents, color index development, TSS, titratable acidity (TA), and maturation index (TSS/TA) were evaluated on weekly basis starting right from the application of first treatment (at PRV). For this purpose, 30 berries were randomly selected from each plot (i.e., three from each marked bunch with one from the top, one from the middle, and one from the bottom of each bunch). These samples were then split into three subsamples ($n = 10$) for further evaluation. For both seasons, bunches were harvested when the TSS of the berries stabilized at around 14.0 Brix.

2.4. Anthocyanin Evaluations

To determine the anthocyanin content in berries from the first treatment application (at PRV; i.e., 7 days before veraison), samples of 3 g of berry skin were used from each plot, which were gently separated from the flesh using a sterile blade and washed with distilled and de-ionized water. The skins were than dried with a sterilized tissue, and added to 30 mL of acidified methanol (HCl 1% + methanol 99%) and left in the dark for 48 hrs. Spectrophotometer Genesys™ 10S UV-VIS® (Thermo Scientific, Waltham, MA, USA) at 520 nm was used for evaluating the samples, whereas results were expressed in milligrams of total anthocyanins as malvidin-3-glucoside per gram of berry skin (mg g^{-1}) [8]. For evaluating the weekly rate of anthocyanin accumulation (from one week after the treatments application), readings from earlier samples were subtracted from the later ones and divided by the total number of days (i.e., 7 days), and the results were expressed as milligrams of malvidin-3-glucoside per gram of berry skin per day (mg g^{-1} of skin).

2.5. Skin Color Evaluations

A colorimeter CR-10 (Minolta®, Tokyo, Japan) was used for skin color evaluation. For each plot, 10 berries were analyzed for color development by recording their L^* (lightness), C^* (chroma), and $h°$ (hue angle). The values of light may range from 0 (black) to 100 (white). Chroma is calculated from the a^* and b^* of the CIELab scale system. Chroma signifies color purity or color intensity from achromatic (grey) towards chromatic color that starts from zero without any possible end point, but the intensity increases with magnitude. Hue angle refers to the color wheel (i.e., green, yellow, and red in regard to the values of 180, 90, and 0, respectively) [8,12,13,16]. For color index of red grapes (CIRG) from the first treatment application (at PRV), the formula CIRG = $(180 - h°)/(L^* + C^*)$ was used [20]. The weekly rate of color index of red grape (CIRG), from one week after the application of the treatments, was calculated by subtracting the final CIRG values from the initial readings, and then dividing them by the total number of days (i.e., 7 days).

2.6. Total Soluble Solids (TSS), Titratable Acidity (TA), and Maturation Index (TSS/TA)

A digital refractometer DR301-95 (Krüss Optronic, Hamburg, Germany) was used for the TSS evaluation. For this purpose, juice was extracted from 10 berries of each plot, and the results were expressed as Brix. For titratable acidity (TA) determination, a semi-automatic titrator was used, where juice extracted from the berries was titrated with 0.1 N NaOH. The results are presented as percentage of tartaric acid [21]. The maturation index was calculated from the ratio of TSS and TA.

2.7. Statistical Analysis

The collected data was analyzed using analysis of variance (ANOVA) and Tukey's HSD (Honest significant difference) test was used to calculate mean significant differences at 5% probability level [22] for all the variables, including rates of anthocyanin and CIRG development. Furthermore, regression analyses were carried out for TSS, TA, and TSS/TA. These procedures were carried out using statistical software SISVAR® version 5.8 build 80 (Lavras Federal University, Lavras, Brazil) and MS Excel (Microsoft, Washington, WA, USA).

3. Results

3.1. Total and Weekly Rate of Anthocyanin Accumulation

Total anthocyanin accumulation was significantly affected in both seasons by the use of exogenous *S*-ABA applied at different timings of veraison. After two weeks of the application at PRV, the grape berries started to show increases in the anthocyanin concentration of the skin (Figure 1). During both seasons, all treatments, regardless of their application timing (i.e., at PRV, at V, and at POV), presented a significant increase throughout the berry ripening until harvest. This increase was superior among *S*-ABA-treated berries in comparison with control treatments from the start of veraison until harvest. It was observed that during both seasons, although at PRV and at V presented higher means as compared with at POV during the process of the berries ripening, in both cases the final means of these treatments were statistically similar to each other and significantly higher than the control treatment.

Moreover, regardless of the different timings of *S*-ABA application, it was clear the importance of the second application (i.e., 10 days after first one) to keep the accumulation of anthocyanin over time, and this behavior was observed during both growing seasons.

Although the final means of total anthocyanin were similar during both seasons, the development pattern of anthocyanin accumulation was slightly different. During the 2015 summer season, the anthocyanin accumulation was fast after the application of *S*-ABA, but at the end of the cycle, the increase seemed to stabilize. On the other hand, during 2016 off-season, this behavior varied, where the anthocyanin buildup was initially slow but acquired momentum, and even at harvest, berries showed a tendency towards producing more anthocyanin content, unlike the summer season of 2015, where anthocyanin accumulation stabilized at the time of harvest.

This phenomenon can be more clearly observed from the weekly rate of anthocyanin accumulation (Figure 1). It can be observed that during the summer season of 2015, the weekly rate of anthocyanin accumulation during early ripening stages (at 14 and 21 days after veraison - DAV) was faster as compared to later stages (28 and 35 DAV), where this development stabilized. On the other hand, during the off-season of 2016, the rate of weekly anthocyanin accumulation was not significantly high during the early stages (14 DAV) but gradually increased and was still increasing at harvest. The rate of weekly anthocyanin accumulation during the off-season of 2016 was more efficient compared to the summer season of 2015, whereas during both seasons berries treated with *S*-ABA at PRV showed a significantly higher accumulation rate followed by at V application.

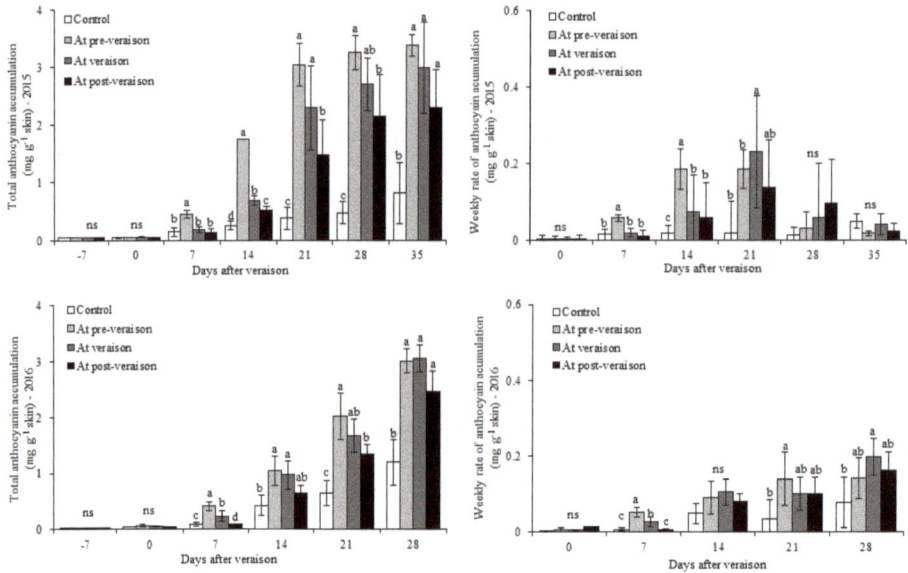

Figure 1. Total anthocyanin accumulation (mg g^{-1} of skin) and weekly rate of anthocyanin accumulation (mg g^{-1} of skin) of 'Benitaka' berries (*Vitis vinifera* L.) subjected to (*S*)-*cis*-abscisic acid (*S*-ABA 400 mg L^{-1}) at different timings of ripening. A second application of *S*-ABA 400 mg L^{-1} was performed for all treatments 10 days after the first application, except for the control. Means within columns for the same letter followed by different letters differ significantly by Tukey's test (*p* < 0.05). ns: non-significant.

3.2. Berries Color and Weekly Rate of Color Development

Color development followed the same pattern observed for total anthocyanin accumulation, with a very slight variation during the 2016 off-season. Like anthocyanin color development, it started to increase more quickly in *S*-ABA-treated berries as compared with non-treated berries (Figure 2). During both seasons, right after the application of *S*-ABA, the treated berries showed significantly darker color in comparison to control with application at PRV and at V being the superior treatments in terms of color development (Figure 3). Multiple *S*-ABA applications showed similar effect on berries color development as anthocyanin concentration, as previously discussed.

During the 2015 summer season, early *S*-ABA applications (at PRV and at V) were recorded with higher anthocyanin accumulation throughout the ripening of berries, where at harvest, all treatments were significantly at par with each other, including at POV, except control. On the other hand, for the 2016 off-season, the same behavior was observed for the treatments, but with a little difference, where at harvest, *S*-ABA applied at POV treatment showed significantly similar results to that of at V application, but lower to at PRV. Overall, all the treatments during off-season were higher than control, but the treatments applied at PRV and at V were recorded with higher means.

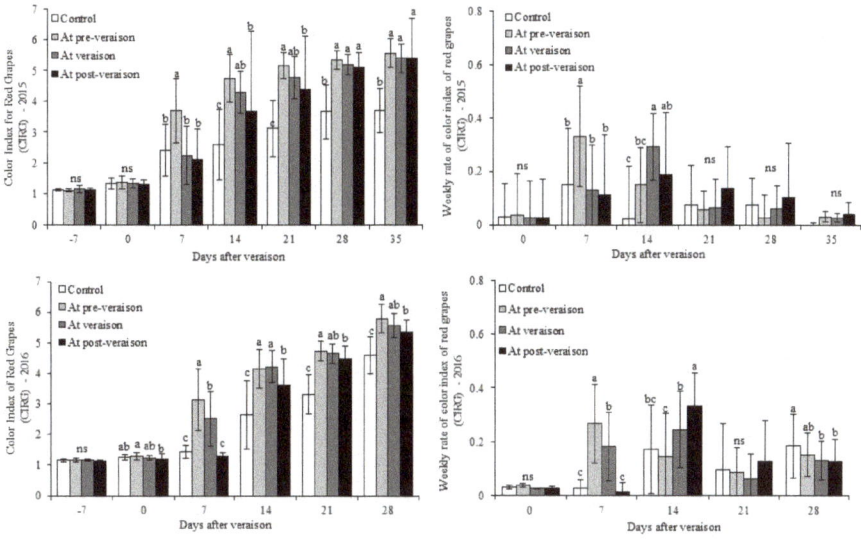

Figure 2. Color index of red grapes (CIRG) and weekly rate of color index for red grape (CIRG) of 'Benitaka' berries (*Vitis vinifera* L.) subjected to (*S*)-*cis*-abscisic acid (*S*-ABA 400 mg L^{-1}) at different timings of ripening. A second application of *S*-ABA 400 mg L^{-1} was performed for all treatments 10 days after the first application, except for the control. Means within columns for the same letter followed by different letters differ significantly by Tukey's test ($p < 0.05$). ns: non-significant.

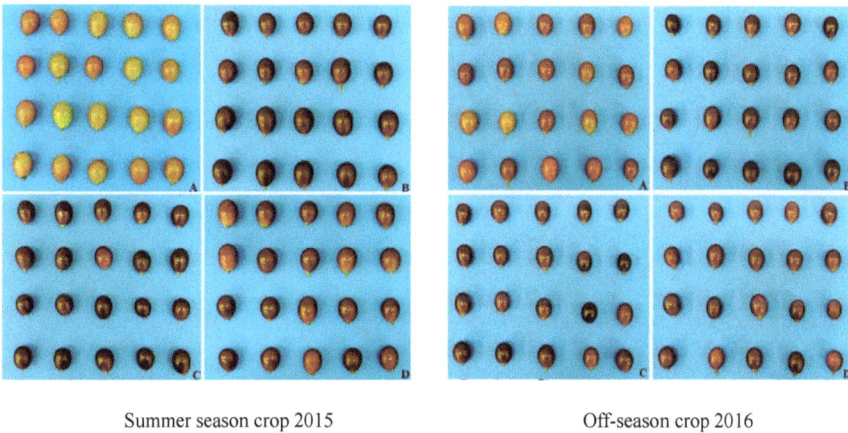

Summer season crop 2015   Off-season crop 2016

Figure 3. Berries of 'Benitaka' table grapes subjected to different treatments with (*S*)-*cis*-abscisic acid (*S*-ABA 400 mg L^{-1}). **A**: control (no application); **B**: at pre-veraison; **C**: at veraison; **D**: at post-veraison. A second application of *S*-ABA 400 mg L^{-1} was performed for all treatments 10 days after the first application, except for the control.

3.3. Total Soluble Solids (TSS), Titratable Acidity (TA), and Maturation Index (TSS/TA)

Regression analysis showed that TSS contents of 'Benitaka' berries (Figure 4) developed with no effect from the *S*-ABA application. TSS development through the course of ripening showed a linear regression with days of berries ripening (Figure 4). All of the samples reached maximum TSS around 14 Brix during both seasons with no influence from the use of the plant growth regulator.

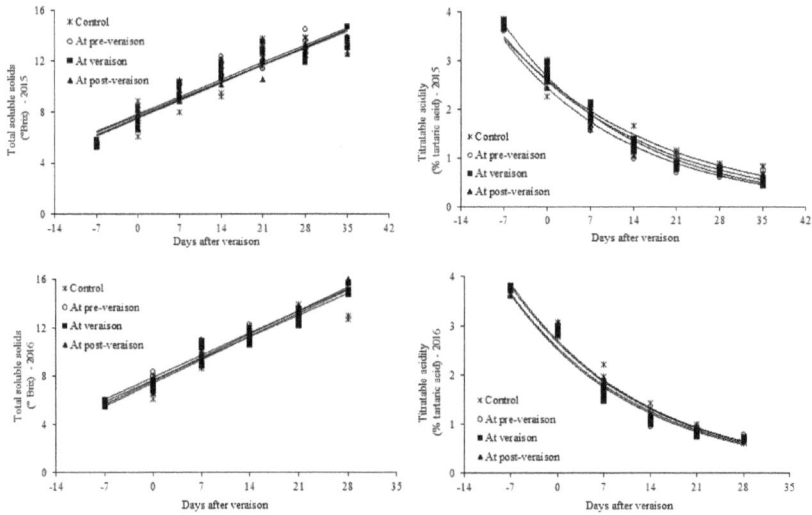

Figure 4. Evolution of total soluble solids (TSS; Brix) and titratable acidity (% tartaric acid) of 'Benitaka' berries (*Vitis vinifera* L.) subjected to *(S)-cis*-abscisic acid (*S*-ABA 400 mg L^{-1}) at different timings of veraison. A second application of *S*-ABA 400 mg L^{-1} was performed for all treatments 10 days after the first application, except for the control. Data were originated from polynomial regression.

Regression analysis of TA showed a negative polynomial behavior through the course of berries ripening in both seasons (Figure 4). Like TSS, TA was also not influenced by the use of exogenous *S*-ABA application to the berries. The same can be observed for the maturation index of 'Benitaka' berries, where the regression analysis showed a positive polynomial regression as the berries matured (Figure 5, Supplementary tables).

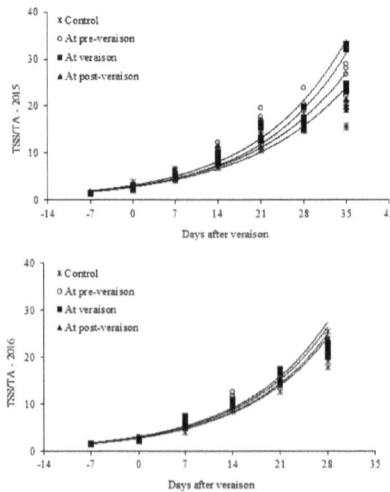

Figure 5. Evolution of maturation index (TSS/TA) of 'Benitaka' berries (*Vitis vinifera* L.) subjected to *(S)-cis*-abscisic acid (*S*-ABA 400 mg L^{-1}) at different timings of veraison. A second application of *S*-ABA 400 mg L^{-1} was performed for all treatments 10 days after the first application, except for the control. Data were originated from polynomial regression.

4. Discussion

It has been demonstrated that exogenous applications of *S*-ABA around the time of veraison increase the anthocyanin contents in the berry skin of several grape cultivars [8,12,13,15–17]. This can be attributed to the effects of *S*-ABA on the expression of genes related to anthocyanin biosynthesis and accumulation of metabolites in grape berries [7,18,23]. The use of ABA enhances the accumulation of the myb-related transcription factor *VvMYBA1*, regulator of *UFGT* (flavonoid 3-O-glucosyltransferase) gene expression [24–27], whereas the *UFGT* acts specifically for the production of anthocyanins [28,29].

Some table grape cultivars may respond well to a single application of *S*-ABA, such as 'Crimson Seedless' [13], whereas others, like 'Rubi' and 'Niagara Rosada', may need multiple applications to get such benefits [30,31].

Rapid anthocyanin accumulation was observed among *S*-ABA-treated 'Cabernet Sauvignon' berries during early stages of ripening. Also, the anthocyanin synthesis differed among the two growing seasons of the trial, which can be attributed to the climatological differences among the two seasons [32]. Several environmental factors influence the anthocyanin accumulation in grape skin, including temperature, solar radiation, and the interaction between temperature and solar radiation [33]. Thus, the different pattern of anthocyanin accumulation during the two seasons may have occurred due to differences in the climatic conditions of the two seasons, especially the temperature and its diurnal temperature variation.

The early increase in anthocyanin contents may have occurred due to the fact that less mature berries respond well to *S*-ABA as compared to more ripe berries, in terms of anthocyanin accumulation and color development [16]. However, some cultivars respond well to late *S*-ABA application as well [12,13]. Meanwhile, in the current study, the timing of applying *S*-ABA had a longer range, where starting from PRV to POV all the treatments showed significant improvement in the anthocyanin contents of the berries, thus allowing a longer period of time for applying the plant growth regulator [16].

Regarding weekly rate of color development (Figure 2), treated berries tend to develop color faster. It can also be observed that berries that produce high color during early stages tend to produce lower daily color development during later stages of berry ripening. Decreases in berry color at later stages may be attributed to the degradation of anthocyanins by glycosidases and peroxidases [32]. During both seasons, the same pattern was observed, where the earlier *S*-ABA application (at PRV) showed a fast increase during early stages, but decreased over time, and similarly after each application the rate of color development increased in such manner. This behavior is related to the development of anthocyanins in the early stages of ripening. Since berries produced a rapid accumulation of anthocyanins at early stages of berry ripening due to early veraison applications (at PRV and at V) [16,31], this is the reason why the color development was also fast during early ripening stages.

The berries showed a predictable pattern of TSS improvement during both seasons. Similarly, no effect of *S*-ABA was observed over the TSS contents of 'Crimson Seedless' grapes, which ranged from 14 to 15 Brix [34]. The TSS of grapes is not usually influenced by the use of *S*-ABA [12,34], but rather on the environmental conditions and cultural practices [35]. As sugar is the dominant component (90%) of TSS [35], its accumulation is mostly dependent on photosynthesizing leaves and woody storage parts [36] rather than use of *S*-ABA, and that is why a linear behavior was observed in the TSS development of 'Benitaka' berries. Application of *S*-ABA at different concentrations around veraison did not alter the chemical characteristics of 'Sovereign Coronation' grape berries [37]. The TA contents of 'Crimson Seedless' decreased during the course of berry ripening but with no influence from the *S*-ABA treatments [34]. Unlike TA, the maturation index increased in the same gradual manner that TA decreased, and with high maturation index at harvest of 'Monastrell' [38] as well as 'Chambourcin' grapes [39], where no effect of ABA application was found on the physicochemical properties of the berries. Several factors can influence the development of these variables and the results can vary according to the cultivar and the environmental conditions in the bunch ripening [40]. In berries of 'Flame Seedless', the application of exogenous *S*-ABA reduced the TA of berries [41].

This work was focused on investigating the anthocyanin accumulation, as well as the color development, during the course of berry ripening in response to exogenous S-ABA treatments at different timings of veraison. The study also focused on new aspects of anthocyanin and color development, such as the weekly rate of anthocyanin accumulation and weekly rate of CIRG, which have never been explored before. It was observed that S-ABA significantly improved the color development and anthocyanin accumulation in 'Benitaka' table grapes. Early application of the regulator produced faster anthocyanin accumulation during early weeks of veraison, which stabilized at the time of harvest.

During the 2016 off-season, this behavior slightly differed, but the role of S-ABA in the accumulation of anthocyanin and color development was the same in both cases. Daily rates of anthocyanin accumulation and CIRG followed a similar pattern. These analyses demonstrate that after the exogenous application of S-ABA the rate of daily anthocyanin peaked in response to the treatments, whereas the control treatment did not show any such response. A second application of S-ABA resulted in anthocyanin accumulation over time, as well as the color development of the 'Benitaka' grapes. This result supports the previous findings regarding 'Benitaka' and 'Rubi' table grapes grown in subtropical area [9,15], since the levels of ABA decreased after one week of the exogenous S-ABA application. Thus, the second application of this plant growth regulator keeps the gene expression related to anthocyanin accumulation in higher levels, resulting in a better color coverage [29]. However, depending on the cultivar, a single application can be sufficient for color improvement, although when grown in warm areas, a second application is often necessary. The response of the berries towards S-ABA clearly showed that multiple applications of S-ABA are necessary to get such effects. The weekly rate of color and anthocyanin development provides a better idea regarding the amount of anthocyanin that accumulates in the berry skin on a daily basis and helps to establish a better understanding of the behavior of anthocyanin accumulation and its effect on the color development of berry skin. Similar to total anthocyanin accumulation and CIRG, the weekly rates of both variables also varied slightly between the two seasons, but this difference can be more attributed to the climate rather than the use of regulator itself, where a slight change in weather and climate can cause a significant difference in observations [35]. The regression analysis of physicochemical properties (i.e., TSS, TA, and maturation index) reveals that these variables depend on the natural phenology of the vine, as well as other factors including the environment, genotype, cultural practices, etc. However, he use of S-ABA had no observable impact on these variables, which followed a predictable pattern that is usually observed for this cultivar under similar environmental conditions.

5. Conclusions

For both the summer and off-season crops, treating 'Benitaka' grape berries pre- or at veraison with S-ABA significantly increased the anthocyanin accumulation as well as the color development of the berries. Other chemical properties of grapes (i.e., TSS, TA, and TSS/AT), were not affected by the use of S-ABA and followed a predictable pattern in relation to days of berry ripening.

Supplementary Materials: The following are available online at http://www.mdpi.com/2073-4395/9/4/164/s1, Figure; Table S1. Regression equations for total soluble solids (TSS), titratable acity (TA) and TSS/AT evolution of 'Benitaka' berries (*Vitis vinifera* L.) subjected to (S)-cis-abscisic acid (S-ABA 400 mg L^{-1}) at different timings of veraison. Summer season crop 2015, Table S2. Regression equations for total soluble solids (TSS), titratable acity (TA) and TSS/AT evolution of 'Benitaka' berries (*Vitis vinifera* L.) subjected to (S)-cis-abscisic acid (S-ABA 400 mg L^{-1}) at different timings of veraison. Off-season crop 2016.

Author Contributions: S.R.R. conceived the research idea. M.S., S.A., R.C.C., J.P.S., R.K., and R.T.d.S. helped during collection of the data. M.S. also analyzed the data and wrote the paper.

Funding: The authors are grateful for the financial support provided by Brazilian National Council for Scientific and Technological Development (CNPq) and The World Academy of Sciences (TWAS) (Grant #190194/2013-8).

Acknowledgments: The authors give their heartfelt thanks to the Londrina State University (UEL), Agronomy Department, Claudemir Zucarelli, Ines Cristina de Batista Fonseca, Fabio Yamashita, Antonio Martinez Peres, Josi Bonifacio, and Weda Westin for their help.

Conflicts of Interest: The authors declare no conflict of interest.

References

1. Pezzuto, J.M. Grapes and human health: A perspective. *J. Agr. Food Chem.* **2018**, *56*, 6777–6784. [CrossRef] [PubMed]
2. Katalinić, V.; Možina, S.; Skroza, D.; Generalić, I.; Abramoviĉ, H.; Miloš, M.; Boban, M. Polyphenolic profile, antioxidant properties and antimicrobial activity of grape skin extracts of 14 *Vitis vinifera* varieties grown in Dalmatia (Croatia). *Food Chem.* **2010**, *119*, 715–723. [CrossRef]
3. Nixdorf, S.L.; Hermosín-Gutiérrez, I. Brazilian red wines made from the hybrid grape cultivar Isabel: Phenolic composition and antioxidant capacity. *Anal. Chim. Acta.* **2010**, *659*, 208–215. [CrossRef] [PubMed]
4. Lacampagne, S.; Gagné, S.; Gény, L. Involvement of Abscisic Acid in Controlling the Proanthocyanidin Biosynthesis Pathway in Grape Skin: New Elements Regarding the Regulation of Tannin Composition and Leucoanthocyanidin Reductase (LAR) and Anthocyanidin Reductase (ANR) Activities and Expres. *J. Plant Growth Regul.* **2010**, *29*, 81–90. [CrossRef]
5. Flamini, R.; Mattivi, F.; De Rosso, M.; Arapitsas, P.; Bavaresco, L. Advanced knowledge of three important classes of grape phenolics: anthocyanins, stilbenes and flavonols. *Int. J. Mol. Sci.* **2013**, *14*, 19651–19669. [CrossRef]
6. Owen, S.J.; Lafond, M.D.; Bowen, P.; Bogdanoff, C.; Usher, K.; Abrams, S.R. Profiles of abscisic acid and its catabolites in developing Merlot grape (*Vitis vinifera*) berries. *Am. J. Enol. Viticult.* **2009**, *60*, 277–284.
7. Koyama, K.; Sadamatsu, K.; Goto-Yamamoto, N. Abscisic acid stimulated ripening and gene expression in berry skins of the Cabernet Sauvignon grape. *Funct. Integ. Genomics* **2010**, *10*, 367–381. [CrossRef] [PubMed]
8. Peppi, M.C.; Fidelibus, M.W.; Dokoozlian, N. Abscisic acid application timing and concentration affect firmness, pigmentation and color of 'Flame Seedless' grapes. *HortScience* **2006**, *41*, 1440–1445. [CrossRef]
9. Roberto, S.R.; Assis, A.M.; Yamamoto, L.Y.; Miotto, L.C.V.; Sato, A.J.; Koyama, R.; Genta, W. Application timing and concentration of abscisic acid improve color of 'Benitaka' table grape. *Sci. Hortic.* **2012**, *142*, 44–48. [CrossRef]
10. Kishino, A.A.; Marur, C.J.; Roberto, S.R. Caracteristicas da planta. In *Viticultura Tropical: o sistema de produção de uvas de mesa do Paraná*; Kishino, A.Y., Carvalho, S.L.C., Roberto, S.R., Eds.; IAPAR Publishing: Londrina, Brazil, 2018; pp. 155–249.
11. Leão, P.C.S.; Soares, J.M.; Rodrigues, B.L. Principais cultivares. In *A Vitivinicultura no Semiárido Brasileiro*; Soares, J.M., Leão, P.C.S., Eds.; Embrapa Informação Tecnológica: Brasília, Brazil, 2009; pp. 151–214.
12. Peppi, M.C.; Fidelibus, M.W.; Dokoozlian, N. Application timing and concentration of abscisic acid affect the quality of 'Redglobe' grapes. *J. Hortic. Sci. Biotechnol.* **2007**, *82*, 304–310. [CrossRef]
13. Peppi, M.C.; Fidelibus, M.W.; Dokoozlian, N. Timing and concentration of abscisic acid applications affect the quality of 'Crimson Seedless' grapes. *Int.l J. Fruit Sci.* **2007**, *7*, 71–83. [CrossRef]
14. Deis, L.; Canagnaro, B.; Rubens, B.; Wuilloud, R.; Silva, M.F. Water deficit and exogenous ABA significantly affect grape and wine phenolic composition under in field and in vitro conditions. *Plant Growth Reg.* **2011**, *65*, 11–21. [CrossRef]
15. Roberto, S.R.; De Assis, A.M.; Yamamoto, L.Y.; Miotto, L.C.V.; Koyama, R.; Sato, A.J.; Borges, R.S. Ethephon use and application timing of abscisic acid for improving color of 'Rubi' table grape. *Pesqui. Agropecu. Bras.* **2013**, *48*, 797–800. [CrossRef]
16. Yamamoto, L.Y.; Koyama, R.; De Assis, A.M.; Borges, W.F.S.; De Oliveira, I.R.; Roberto, S.R. Color of berry and juice of 'Isabel' grape treated with abscisic acid in different ripening stages. *Pesqui. Agropecu. Bras.* **2015**, *50*, 1160–1167. [CrossRef]
17. Koyama, R.; Assis, A.M.; Yamamoto, L.Y.; Borges, W.F.; Borges, R.S.; Prudêncio, S.H.; Roberto, S.R. Exogenous abscisic acid increases the anthocyanin concentration of berry and juice from 'Isabel' grapes (*Vitis labrusca* L.). *HortScience* **2014**, *49*, 460–464. [CrossRef]
18. Giribaldi, M.; Hartung, W.; Schubert, A.S. The effects of abscisic acid on grape berry ripening are affected by the timing of treatment. *J. Int. Sci. Vigne. Vin.* **2010**, *44*, 9–15.
19. Caviglione, J.H.; Kiihl, L.R.B.; Caramori, P.H.; Oliveira, D. *Cartas climáticas do Paraná*; IAPAR Publishing: Londrina, Brazil, 2000.

20. Carreño, J.; Martínez, A.; Almela, L.; Fernández-López, J.Á. Proposal of an index for the objective evaluation of the color of red table grapes. *Food Res. Int.* **1995**, *28*, 373–377. [CrossRef]
21. Instituto Adolfo Lutz. *Normas analíticas do Instituto Adolfo Lutz: métodos químicos e físicos para análise dos alimentos*, 3rd ed.; Instituto Adolfo Lutz: São Paulo, Brazil, 2008.
22. Tukey, J.W. Comparing individual means in the analysis of variance. *Biometrics* **1949**, *5*, 99–114. [CrossRef]
23. Villalobos-González, L.; Peña-Neira, A.; Ibáñez, F.; Pastenes, C. Long term effects of abscisic acid (ABA) on the grape berry phenylpropanoid pathway: Gene expression and metabolite content. *Plant Physiol. Bioch.* **2016**, *105*, 213–223. [CrossRef]
24. Boss, P.K.; Davies, C.; Robinson, S.P. Expression of anthocyanin biosynthesis pathway genes in red and white grapes. *Plant Mol. Biol.* **1996**, *32*, 565–569. [CrossRef]
25. Kobayashi, S.; Ishimaru, M.; Hiraoka, C.; Honda, C. MYB-related genes of the Kyoho grape (*Vitis labruscana*) regulate anthocyanin biosynthesis. *Plant Sci.* **2002**, *215*, 924–933.
26. Jeong, S.T.; Goto-Yamamoto, N.; Kobayashi, S.; Esaka, M. Effects of plant hormones and shading on the accumulation of anthocyanins and the expression of anthocyanin biosynthetic genes in grape berry skin. *Plant Sci.* **2004**, *167*, 247–252. [CrossRef]
27. Azuma, A.; Kobayashi, S.; Yakushiji, H.; Yamada, M.; Mitani, N.; Sato, A. *VvMYBA1* genotype determines grape skin color. *Vitis* **2007**, *46*, 154–155.
28. Roubelakis-Angelakis, K.A. *Grapevine Molecular Physiology and Biotechnology*, 2nd ed.; Springer: Dordrecht, The Netherlands, 2009.
29. Koyama, R.; Roberto, S.R.; de Souza, R.T.; Borges, W.F.S.; Anderson, M.; Waterhouse, A.L.; Cantu, D.; Fidelibus, M.W.; Blanco-Ulate, B. Exogenous Abscisic Acid Promotes Anthocyanin Biosynthesis and Increased Expression of Flavonoid Synthesis Genes in *Vitis vinifera* × *Vitis labrusca* Table Grapes in a Subtropical Region. *Front. Plant Sci.* **2018**, *9*, 323. [CrossRef] [PubMed]
30. Domingues-Neto, F.J.; Junior, S.P.; Borges, C.V.; Cunha, S.R.; Callili, D.; Lima, G.P.P.; Roberto, S.R.; Leonel, S.; Tecchio, M.T. The exogenous application of abscisic acid induces accumulation of anthocyanins and phenolic compounds of the 'Rubi' grape. *Am. J. Plant Sci.* **2017**, *8*, 2422–2432. [CrossRef]
31. Tecchio, M.A.; Neto, F.J.D.; Junior, A.P.; Da Silva, M.J.R.; Roberto, S.R.; Smarsi, R.C. Improvement of color and increase in anthocyanin content of 'Niagara Rosada' grapes with application of abscisic acid. *Afr. J. Biotechnol.* **2017**, *16*, 1400–1403.
32. Gagné, K.; Esteve, K.; Deytieux, C.; Saucier, C.; Geny, L. Influence of abscisic acid in triggering « véraison » in grape berry skins of *Vitis vinifera* L. cv. Cabernet Sauvignon. *J. Int. Sci. Vigne Vin* **2006**, *38*, 7–14. [CrossRef]
33. Tarara, J.; Lee, J.; Spayd, S.; Scagel, S. Berry temperature and solar radiation alter acylation, proportion, and concentration of anthocyanin in Merlot grapes. *Am. J. Enol. Vitic.* **2008**, *59*, 235–247.
34. Ferrara, G.; Mazzeo, A.; Matarrese, M.A.S.; Pacucci, C.; Punzi, R.; Faccia, M.; Trani, A.; Gambacorta, G. Application of abscisic acid (*S*-ABA) and sucrose to improve color, anthocyanin content and antioxidant activity of cv. crimson seedless grape berries. *Aust. J. Grape Wine Res.* **2015**, *21*, 18–29. [CrossRef]
35. Keller, M. *The Science of Grapevines: Anatomy and Physiology*, 2nd ed.; Elsevier Academic Press: London, UK, 2015.
36. Rebucci, B.; Poni, S.; Intrieri, C.; Magnanini, E.; Lakso, N.A. Effect of manipulated grape berry transpiration and post-veraison sugar accumulation. *Aust. J. Grape Wine Res.* **1997**, *3*, 57–65. [CrossRef]
37. Reynolds, A.; Robbins, N.; Lee, H.S.; Kotsaki, E. Impacts and Interactions of Abscisic Acid and Gibberellic Acid on Sovereign Coronation and Skookum Seedless Table Grapes. *Am. J. Enol. Vitic.* **2016**, *67*, 327–338. [CrossRef]
38. Ruiz-García, Y.; Gil-Muñoz, R.; López-Roca, J.M.; Martínez-Cutillas, A.; Romero-Cascales, I.; Gómez-Plaza, E. Increasing the Phenolic Compound Content of Grapes by Preharvest Application of Abscisic Acid and a Combination of Methyl Jasmonate and Benzothiadiazole. *J. Agr. Food Chem.* **2013**, *61*, 3978–3983.
39. Zhang, Y.; Dami, I. Improving freezing tolerance of 'Chambourcin' grapevines with exogenous abscisic acid. *HortScience* **2012**, *47*, 1750–1757. [CrossRef]

40. Jackson, R.S. *Wine Science: Principles and Applications*, 3rd ed.; Elsevier: Amsterdam, The Netherlands, 2008.
41. Peppi, M.C.; Walker, M.A.; Fidelibus, M.W. Application of abscisic acid rapidly upregulated *UFGT* gene expression and improved color of grape berries. *Vitis* **2008**, *47*, 11–14.

agronomy

MDPI

Article

Elevated CO$_2$ Levels Impact Fitness Traits of Vine Mealybug *Planococcus ficus* Signoret, but Not Its Parasitoid *Leptomastix dactylopii* Howard

Maria Schulze-Sylvester [1,*] **and Annette Reineke** [2]

[1] Instituto para el Estudio de la Biodiversidad de Invertebrados (IEBI), Facultad de Ciencias Naturales, Universidad Nacional de Salta, Av. Bolivia 5150, Salta CP 4400, Argentina
[2] Geisenheim University, Department of Crop Protection, Von-Lade-Str. 1, D-65366 Geisenheim, Germany; annette.reineke@hs-gm.de
* Correspondence: schulzesylvester@gmail.com; Tel.: +54-387-425-5472

Received: 29 April 2019; Accepted: 11 June 2019; Published: 20 June 2019

Abstract: Carbon dioxide (CO$_2$) is one of the primary factors driving climate change impacts on plants, pests, and natural enemies. The present study reports the effects of different atmospheric CO$_2$ concentrations on the vine mealybug *Planococcus ficus* (Signoret) and its parasitoid wasp *Leptomastix dactylopii* (Howard). We investigated the life-history parameters of both species on grapevine *Vitis vinifera* (L.) plants grown under elevated (eCO$_2$) and ambient (aCO$_2$) CO$_2$ levels in a greenhouse and in a vineyard free-air carbon dioxide enrichment (FACE) facility. The greenhouse experiments with an eCO$_2$ level of around 800 ppm showed a significant increase in survival rates, a strong trend towards declining body size, and an increasing fecundity of female mealybugs, while fertility and development time did not change. However, none of these parameters were altered by different CO$_2$ concentrations in the VineyardFACE facility (eCO$_2$ level around 450 ppm). On the other hand, the parasitism success, development time and sex ratio of *L. dactylopii*, reared on *P. ficus* under eCO$_2$ or aCO$_2$, varied neither in the greenhouse nor in the FACE facility. These results suggest that future CO$_2$ levels might cause small-scale changes in vine mealybug fitness; however, this is not necessarily reflected by parasitoid performance.

Keywords: climate change; elevated CO$_2$; grapevine pest; mealybug; parasitoid; FACE

1. Introduction

Atmospheric CO$_2$ levels are on the rise, with the latest reports published by the Intergovernmental Panel on Climate change (IPCC) reporting an increase of 20 ppm per decade, resulting in an increase of over 35% since pre-industrial times. Current atmospheric CO$_2$ levels are close to 400 ppm, and this value is expected to double by the end of this century [1]. Plants react to elevated atmospheric CO$_2$ levels with a whole range of morphological and physiological adaptations. Most C3 plants increase their photosynthesis rates and primary production [2,3]. This also applies to grapevine plants (*Vitis vinifera* L.). For example, Bindi et al. [4] noted that atmospheric CO$_2$ enrichment stimulated grapevine growth and enhanced fruit and total biomass. Stimulated growth and yield, as well as enhanced stomatal conductance and transpiration, under elevated CO$_2$, were also reported from a vineyard free-air carbon dioxide enrichment (FACE) facility in Geisenheim, Germany [5].

Alterations in temperature, precipitation, and other climatic factors are expected to have substantial direct impacts on grape insect pests and pathogens, as well as on the suitability of a grapevine as a host plant for a range of organisms [6]. The increased availability of carbon (C) vs. nitrogen (N) leads to an accumulation of carbohydrates in the foliage and consequently to a higher C:N ratio and a lower nutritional value for herbivores [2,3,7]. This has negative effects, especially on leaf-chewing herbivores, while phloem-feeders seem to be less affected. Several studies report an improved aphid fitness under elevated CO_2 (eCO_2) [3,8], although the mechanisms behind this are not yet fully understood. While aphids are a primary pest in many agricultural and horticultural systems, mealybugs, such as the vine mealybug *Planococcus ficus* Signoret (Hemiptera: Pseudococcidae), are a much bigger concern in grapevine production.

Planococcus ficus is an invasive phloem-feeding insect from the Mediterranean area, which has become a serious invasive pest in many grape-growing regions worldwide [9,10]. Mealybugs affect grapevines both directly and indirectly. By feeding on the phloem sap of all plant organs, mealybugs weaken the plants' vigour. Furthermore, the excreted honeydew promotes the growth of sooty mould on leaves and fruits, reducing photosynthesis, grape marketability and wine quality [9,11,12]. *Planococcus ficus* is also known to transmit grapevine leafroll-associated virus (GLRaV) and other diseases, which reduce the crop yield and wine quality [13,14]. Mealybug control was based on repeated applications of broad-spectrum insecticides, although with limited success, as mealybugs feed not only on the canopy, but also under the bark and in the root area [15], where they are inaccessible to contact-active pesticides. Additionally, the reiterated use of broad-spectrum insecticides is associated with negative effects on non-target organisms, including biological control agents, and the risk of future pesticide resistance [10,16,17]. Therefore, alternative methods for mealybug control include employing new pesticides, pheromone-based mating disruption, and biological control [18–22]. Models predict that under future climate change scenarios, *P. ficus* will thrive and continue to expand its range [23,24]. The only available study on the effects of eCO_2 on scale insects concludes that temperature, rather than CO_2, alters the performance of the Madeira mealybug, *Phenacoccus madeirensis* (Green) [25].

The biological control of mealybugs with natural enemies forms an important part of sustainable pest management programs. *Planococcus ficus* can be parasitised by several encyrtid species, such as *Anagyrus pseudococci* (Girault), *Leptomastix dactylopii* (Howard), *Leptomastidea abnormis* (Girault), *Coccidoxenoides perminutus* Girault, and *Coccidoxenoides peregrinus* (Timberlake) [10]. *Leptomastix dactylopii* (Hymenoptera: Encyrtidae) is a solitary, arrhenotokous, koinobiont endoparasitoid, probably native to eastern Africa [26], which has been introduced into Europe, the United States, Pakistan, India, and Australia [27]. *Leptomastix dactylopii* has been used in biological control programs against the citrus mealybug, *Planococcus citri* (Risso) [28,29], but it has also been found in vineyards infested with *P. ficus* and other vineyard mealybugs in California (US), Iran, South Africa and Tunisia [9,30–32]. Laboratory trials showed that *P. ficus* is a suitable host for *L. dactylopii* and wasps might even have an innate preference for this host [26,33]. *Leptomastix dactylopii* is susceptible to low temperatures, and its geographical range might expand due to global warming, similar to other encyrtid wasps [24]. There are no studies available on the effects of eCO_2 on *L. dactylopii* or other Encyrtidae. Considering other related pest-natural enemy complexes, some studies investigated the knock-on effects of eCO_2 on aphid parasitoids, but results vary between neutral, beneficial, and detrimental effects [34,35]. If host quality decreases under eCO_2, parasitoids will be adversely affected. If eCO_2 increases host quality, parasitoid performance may be maintained or increased. However, the number of available studies is too low to derive general patterns.

Besides host quality, parasitism success depends on the parasitoid's ability to locate possible hosts. When attacked by herbivores, plants release attack-specific volatiles which help natural enemies to locate these herbivores [36]. While chewing insects stimulate the jasmonic acid-signalling pathway, phloem-feeding insects trigger the salicylic acid signalling pathway [34]. Considering the reactions of grapevine plants to attacks by *P. ficus*, the transcriptional response of vine plants is rather weak [37]. However, elevated CO_2 stimulates the production of salicylic acid, which might favour parasitoid host location of phloem-feeders and improve parasitism success [38].

Systematic investigations on the consequences of changing temperature, precipitation or CO_2 concentration on grapevine diseases and pests are few in number [6,39] and multi-trophic interactions of grapevine pests and their natural enemies have never been studied under elevated CO_2 levels. The present study aims to investigate the effects of elevated CO_2 on the performance of *L. dactylopii* and its host, the vine mealybug, reared on grapevine plants in an FACE system and in the greenhouse.

We hypothesised that (I) vine mealybugs reared under eCO$_2$ would react like other phloem-feeders, as aphids, predicting that (II) life history parameters would not be altered or even improve from higher atmospheric CO$_2$ levels, and (III) the parasitoid *L. dactylopii* would either be unaffected or benefit from eCO$_2$, depending on the mealybug performance.

2. Materials and Methods

Insects were obtained from the Department of Crop Protection at Geisenheim University, Germany. *Planococcus ficus* was reared on sprouted potatoes at 26 ± 1 °C and 60–70% relative humidity (RH) in darkness in an incubator. *Leptomastix dactylopii* cultures were reared on *P. ficus* on sprouted potatoes in transparent plastic containers (37 × 22 × 25 cm) at 26 °C, 60–70% RH, and a photoperiod of 16:8 (L:D) in an environmental chamber. Containers were also supplied with honey agar as an additional food for wasps.

In order to answer the question of whether life-history parameters of *P. ficus* vary under different CO$_2$ concentrations, we conducted greenhouse experiments as well as field studies in the VineyardFACE facility (Geisenheim University, Geisenheim, Germany). The greenhouse experiment was performed with potted Riesling grapevine plants exposed to ambient and elevated CO$_2$ concentrations of 400 CO$_2$ and 800 ppm CO$_2$, respectively (Table S1). At the onset of the experiment, plants were 12 weeks old and approximately 40 cm high. Grapevine plants used in the greenhouse experiments in this study were grown under ambient CO$_2$ conditions for 12 weeks before exposure to eCO$_2$ started. Each plant was infested with ca. 100 *P. ficus* nymphs using leaf disc transport [40]. Briefly, two to three ovisacs harvested from *P. ficus* females were transferred to 2 × 2 cm paper towel squares and placed on 5 cm vine leaf discs. Leaf disks were placed on water-soaked cotton wool and 5 cm filter papers to avoid leaf edges entangling in the cotton wool. After 48 h, paper towel squares were removed and the number of crawlers on the leaf disks was counted. Then, two to three leaf discs were placed on each experimental plant, applying a total number of approx. 100 crawlers per plant. Plants were covered with a fabric gardening bag (60 × 40 cm, 19 g/m^2, Classic80, HECO Textilverlag, Memmingen, Germany), ensuring oxygen and light supply while preventing mealybugs from escaping and being attacked by natural enemies. A total of 40 plants were placed in a greenhouse chamber with elevated CO$_2$ concentrations, while another 40 plants served as the control in an ambient CO$_2$ greenhouse chamber. Half of the 40 plants in each cabin were used to study life history parameters of *P. ficus*, while the other 20 plants were used to raise adult *P. ficus* females for the parasitism experiments with *L. dactylopii*. During the course of the experiments (mid-July to late August 2016), an average temperature of 22 °C was reached in the greenhouse (average minimum temperature 16.0 °C, average maximum temperature 30 °C) (Table S2), and plants were watered twice per week with rainwater.

The VineyardFACE facility enhances open-air CO$_2$ concentrations using ring-like structures with a diameter of 12 m, which are placed over the rows of an actual vineyard. Three rings are sparged with CO$_2$ reaching eCO$_2$ levels of approx. 450 ppm, while the other three rings use air and serve as the aCO$_2$ (ca. 400 ppm) control. For a detailed description of the VineyardFACE design, see [5,41] and Supplementary Figure S1. Field experiments were conducted between mid-July and mid-September 2016. During the experiments described here, CO$_2$ concentrations were measured by using two LI-8100 analyser control units (Li-CorLI8100/8150 Multiplexer, Li-Cor Biosciences, Lincoln, NE, USA) installed at two heights (1.7 m and 0.75 m) in the grapevine canopy. Within aCO$_2$ rings, an average level of 401 ± 1.3 ppm was reached during the experimental periods, while in eCO$_2$ rings, air was enriched during daylight hours to approximately 12% to 18% above the ambient CO$_2$ (456 ± 16.1 ppm), which is the concentration predicted for the mid-21st century. Supplementary Figure S2 illustrates CO$_2$ concentrations in aCO$_2$ and eCO$_2$ rings during the first period of the experiments (mid-July to beginning of August 2016) described here. Data on weather conditions during the experimental periods are provided in the Supplementary Table S3. During the course of the experiments, an average temperature of 20.5 °C was reached in the field (average minimum temperature 14.0 °C, average maximum temperature 27 °C). Experiments were conducted on 10 five-year-old Riesling grapevine

274

plants per ring, resulting in three replicates, with 10 sub-replicates each. Similar to the greenhouse experiment, two canes per plant were inoculated with approximately 100 1st instar mealybug nymphs and covered with gardening fabric. Of the two infested canes per plant, one was used for the study on mealybug life-history parameters, while the other one served to provide adult mealybugs reared under eCO_2 as hosts for *L. dactylopii*.

In both the greenhouse and FACE experiments, the development time, body size, fecundity, fertility, and survival of *P. ficus* females were recorded. The developmental stages of mealybugs were determined according to Walton & Pringle [9]. Three weeks after infesting vines with mealybug nymphs, greenhouse and FACE plants were assessed for the presence of females reaching the oviposition stage. Plants were checked every other day and the first ten (greenhouse) or five (VineyardFACE) ovipositing females from every plant were collected and stored in ethanol until further analysis. The infestation and collection dates for each female were used to determine the development time. Body length (from head to anal lobes) of the collected *P. ficus* females was measured under a stereomicroscope. To determine the fecundity and fertility, one additional female at the onset of oviposition was sampled from each treated and untreated plant (Greenhouse: $n = 20$; VineyardFACE: $n = 3$). These females were transferred individually to a small piece of paper towel in a sealed Petri dish and incubated in the respective greenhouse chambers (eCO_2 and aCO_2) for two weeks. Then, Petri dishes were frozen, and the number of crawlers and unhatched eggs was counted under a microscope. The sum of unhatched eggs and hatched crawlers accounted for fecundity, while fertility was determined by the percentage of hatched crawlers. Greenhouse experiments finished after six weeks, while the duration of the VineyardFACE experiments extended over 9 weeks, due to the slower mealybug development under field conditions. At the end of the experiments, the number of surviving adult females was recorded on each plant. In accordance with Ross et al. [42] and Cocco et al. [43], we assumed a baseline survival rate of 60% of females in our *P. ficus* culture. Hence, the survival of females was calculated as

$$\text{(Number of surviving female adults post-experiment + number of females collected for body size + 1 female collected for the fertility)/Number of applied} \quad (1)$$
$$P.\ ficus \text{ nymphs} \times 0.6$$

A second experiment aimed to answer the question of whether the parasitism success, development time, or sex ratio of *L. dactylopii* vary under different CO_2 concentrations in the greenhouse or in the VineyardFACE facility. In the greenhouse, 20 potted cv. Riesling grapevine plants were inoculated with approximately 100 1st instar nymphs of *P. ficus* and placed in respective greenhouse chambers with ambient or elevated CO_2. Likewise, in the VineyardFACE facility, one cane of each of the ten vine plants per FACE ring was infested with 100 1st instar *P. ficus* nymphs. When female mealybugs reached the 3rd instar to preoviposition stage, one leaf of each plant was placed in a small, water filled tube; glued to the ground of a round, transparent experimental arena (diameter 15 cm, height 15 cm); and covered with a detachable lid. Subsequently, five mealybugs, reared from the same plant from which the leaf was collected, were transferred to the leaf using a sable brush. Mealybugs were allowed to settle overnight; then, a male and a female individual of *L. dactylopii* were introduced into the arena for 24 h (greenhouse) or 48 h (VineyardFACE) and then removed. Experimental arenas were placed in their respective treatment greenhouse chambers or FACE ring (eCO_2 or aCO_2). Water was refilled in the tubes, when necessary. After two weeks, arenas from the VineyardFACE facility were moved to the greenhouse and placed in the respective aCO_2 and eCO_2 treatments, since temperatures in the field decreased to unfavourable levels for *L. dactylopii* development. Subsequently, arenas were visually checked for parasitised mealybugs in order to determine parasitism success. In the following weeks, arenas were checked for *L. dactylopii* adult emergence every 24 h and wasps were removed. The date of emergence was recorded to determine the development time. Emerged individuals were transferred to glass vials and sexed. The experiment finished after 5 weeks when wasp emergence ended.

To detect the possible effects of eCO_2 on *P. ficus* and *L. dactylopii*, data of each measured parameter were tested for normality, followed by an unpaired t-test. Only *P. ficus* fertility data from the FACE experiment did not follow a normal distribution and was analysed with a Mann-Whitney test for non-parametric data sets. Contingency tables and Fisher's exact test were used to detect possible differences in categorical variables, i.e., the sex ratio of newly emerged *L. dactylopii*. Pearson's correlation analysis was conducted to investigate the relation of female body size and development time (onset of oviposition) of female *P. ficus* mealybugs reared under elevated or ambient CO_2 conditions. All analyses were performed with Graphpad Prism version 7.00 for Windows (GraphPad Software, La Jolla, CA, USA).

3. Results

3.1. Life History Parameters of P. ficus Under Different CO2 Levels

3.1.1. Greenhouse Experiments

The mean body size of mealybugs reared on grapevine plants under aCO_2 was larger compared to those reared under eCO_2. However, this difference was not significant, even though there was a tendency ($p = 0.05$) towards smaller sized mealybugs under eCO_2 (Figure 1A; Tables S4 and S5). There was a significant negative correlation between the time of exposure and body size of ovipositing females in the eCO_2 treatment ($R^2 = 0.63$, $p = 0.01$), while such a correlation was not detected with aCO_2 (Figure 2).

The survival of adult females was significantly greater for mealybugs reared on grapevine plants under eCO_2 than for mealybugs under aCO_2 conditions (53.2% and 40.1% survival, respectively) ($p = 0.04$) (Figure 1B; Tables S4 and S6). There was no difference between aCO_2 and eCO_2 in terms of the development time from the 1st instar to oviposition, neither for the mean development time (Figure 1C, Tables S4 and S5) nor considering the first ovipositing female of each plant. Fecundity, on the other hand, resulted in 230.9 (aCO_2) and 290.1 (eCO_2) mealybug eggs. The improved fecundity under eCO_2 was not significant, but might indicate a strong trend ($p = 0.05$); while fertility, calculated as the percentage of 1st instar nymphs hatched from these eggs, was not different between the CO_2 treatments (Figure 1D; Tables S4 and S7).

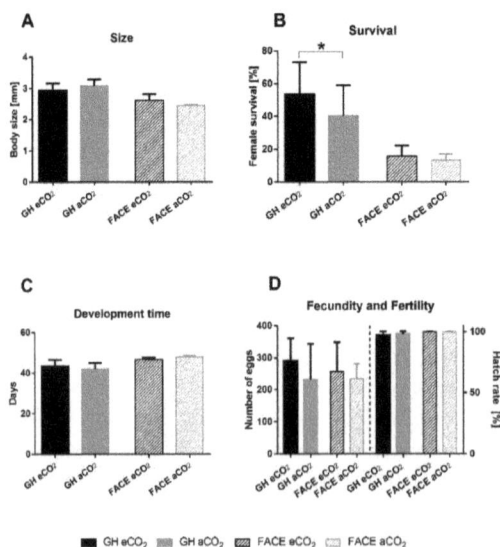

Figure 1. Life history parameters of mealybug females reared under eCO_2 and aCO_2 in the greenhouse and in the VineyardFACE facility. Data displayed as means + SD. Statistically significant differences between CO_2 treatments are marked with *. Solid columns display greenhouse data (GH), and hatched columns display data from the VineyardFACE experiment. (**A**) Body size of females at the onset of oviposition. (**B**) Survival of females. (**C**) Development time of females from 1st instar to oviposition. (**D**) Fecundity, shown as the total number of eggs laid per female, and fertility, shown as the percentage of hatched crawlers from these eggs.

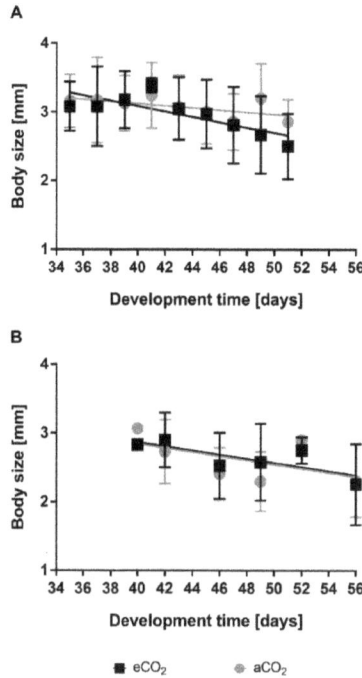

Figure 2. Pearson's correlation analysis of female body size and sampling date of female *P. ficus* mealybugs at the onset of oviposition reared under elevated (eCO_2, black squares) or ambient (aCO_2 grey circles) CO_2 conditions. Data displayed as means + SD. (**A**) Greenhouse data: eCO_2 R^2 = 0.63, p = 0.01; aCO_2 R^2 = 0.37, p = 0.08. (**B**) VineyardFACE data: eCO_2 R^2 = 0.60, p = 0.07; aCO_2 R^2 = 0.31, p = 0.25.

3.1.2. VineyardFACE Experiments

Field data on mealybug survival, body size, development time from 1st instar to oviposition stage, fecundity, and fertility were compared between aCO_2 and eCO_2 VineyardFACE rings (n = 3), but none of these parameters showed significant differences (Figure 1; Tables S8–S11). Generally, the field results followed similar trends to the results obtained in greenhouse experiments, but did not reach statistical significance. The relationship between body size and the onset of oviposition was not significant in the FACE experiment. (Figure 2B).

3.2. Parasitism by L. dactylopii

The parasitism experiment with *L. dactylopii* did not reveal any differences between parasitism success, development time, or sex ratio under eCO_2 and aCO_2 treatments, neither in the greenhouse nor in the VineyardFACE facility (Table 1). In the greenhouse, the aCO_2 treatment resulted in a total of 15 parasitized *P. ficus* females, from which 14 wasps emerged: 10 females (69%) and 3 males, while 1 escaped before being sexed. Under eCO_2 greenhouse, 16 mealybug females were parasitized, all of which emerged: 12 females (80%) and 3 males, while 1 escaped before being sexed (Table 1, Table S12). In the field and under FACE eCO_2, 40 mealybugs were parasitised and 36 wasps hatched (58% females) (Table 1, Table S13). In the FACE aCO_2 treatment, a total of 52 mealybugs were parasitized and 44 wasps hatched, 61% of which were female. The development time of *L. dactylopii* from egg to adult was approximately 24 days in both eCO_2 and aCO_2 greenhouse treatments, while wasps from the VineyardFACE facility needed 25 days (eCO_2) and 26 days (aCO_2) to complete their development.

Parasitism success was 15% in the aCO_2 treatment and 16% under eCO_2 conditions in the greenhouse, while in the field, more mealybugs were parasitised (35% under aCO_2, and 29% under eCO_2).

Table 1. Results from the parasitism experiment with *L. dactylopii* and *P. ficus* in a greenhouse and field (VineyardFACE) under ambient (aCO_2) and elevated (eCO_2) conditions. Values are expressed as means ± SD.

Site	Treatment	Parasitism Success (%)	Development Time (days)	Sex Ratio (% Females; % Males)
Greenhouse	aCO_2	15 ± 17, $n = 20$	23.86 ± 1.35, $n = 14$	69.23; 30.77, $n = 13$
	eCO_2	16 ± 22.1, $n = 20$	23.69 ± 0.79, $n = 16$	80; 20, $n = 15$
	p	0.87	0.67	0.67
VineyardFACE	aCO_2	34.67 ± 34.01, $n = 30$	25.67 ± 3.34, $n = 44$	61.36; 38.63, $n = 44$
	eCO_2	29.33 ± 27.66, $n = 30$	25.13 ± 2.54, $n = 36$	58.33; 41.67, $n = 36$
	p	0.51	0.41	0.82

4. Discussion

Here, we report for the first time on the possible effects of eCO_2 in a grapevine-based pest-parasitoid system using the vine mealybug *P. ficus* and its parasitoid *L. dactylopii* as model organisms. Our greenhouse experiments with eCO_2 showed a significant increase in survival rates, and strong trends for a decreased body size and increased fecundity of *P. ficus* females, while fertility did not change under eCO_2 concentrations. Body size was negatively correlated with sampling date, as females that started oviposition later were of a smaller size. Body size is usually positively related to fecundity in mealybugs [42,43], but we did not find evidence for this in the present study. It is possible that the number of eggs suffered a decline over time, similar to mealybug size, but this was not evaluated. On the contrary, fecundity did increase under eCO_2, although this trend was not significant.

The field experiment at the VineyardFACE site did not detect any differences between aCO_2 and eCO_2 treatment groups. There are several possible explanations for this. Firstly, field experiments take place in a more complex environment, including a whole series of uncontrolled variables, e.g., humidity, wind, and putatively other natural enemies. While this more realistic scenario is important to put laboratory results into context, it also makes it difficult to detect small-scale effects, as they might occur in the case of *P. ficus* under eCO_2. Secondly, VineyardFACE eCO_2 concentrations were much lower (ca. 450 ppm) than in the greenhouse (ca. 800 ppm), due to the open-air nature of the facility. The differences detected under greenhouse eCO_2 were rather small; hence it is not surprising that the lower VineyardFACE eCO_2 levels did not achieve comparable results. Thirdly, despite its large scale and technical sophistication, the Geisenheim VineyardFACE facility only allows a limited number of replicates. There are only three independent test rings for each CO_2 treatment, which complicates the detection of subtle differences *per se*. To gain statistical power, we also analysed the ten biological replicates (subreplicates) per ring, see Supplement Tables S9–S11, S13). These subreplicate-based results are limited in their implications and no general conclusions should be drawn from them. Even so, none of the analysed life history parameters of *P. ficus* showed significant differences between aCO_2 and eCO_2. This combined evidence suggests that VineyardFACE eCO_2 levels did not affect *P. ficus* and it appears unlikely that a higher number of independent treatments (i.e., FACE rings) would change this result.

Since the experimental conditions were substantially different, no statistical testing was done to compare field and greenhouse data. There are some differences which might be attributed to other environmental stress factors, such as temperature and precipitation amounts. Our field experiment yielded low survival rates, but they are comparable to those measured in a screenhouse experiment with *P. ficus* [43]. Female mealybugs in the field needed more time to reach the oviposition state and were slightly smaller than those from the greenhouse, especially in the aCO_2 treatment. Mealybug fecundity and fertility were similar at both sites. Interestingly, a negative correlation between size and development time was detected under eCO_2 at both experimental set-ups, although this relation was not significant ($p = 0.07$) in the VineyardFACE facility. It would be interesting to study these findings in more detail. The overall difference between the obtained field (FACE) and greenhouse data might also be attributed to an overall plus of an average temperature of 2–3 °C in the greenhouse, which might have been more favourable for mealybug development.

Higher atmospheric levels of CO_2 have been shown to result in a higher biomass, increased yield, and lower nutritional values in grapevine and other plants [2,5,44]. However, plant tissues and fluids vary in their response to elevated CO_2 [2,3,45] and influence different feeding guilds of herbivores in different ways [3,7]. Leaf-chewing herbivores generally perform worse and phloem-feeders have been shown to be less affected by rising CO_2 concentrations [3,46]. Most literature on the effects of CO_2 and climate change on Hemiptera focuses on aphids. In the only study on mealybugs, Chong et al. [25] found that temperature rather than elevated CO_2 influenced the survival, development time and fecundity of *P. madeirensis*. Aphids can benefit from elevated CO_2, showing increases in fecundity and survival, and decreases in development time [3,46]. However, several studies showed that the direction and size of the effect of eCO_2 depend on the specific combination of host plant and insect species [47–50]. Hughes & Bazzaz [47] tested five aphid species and their host plants under eCO_2; one species was negatively affected, another positively affected, and no significant effects were found for the remaining three.

Cocco et al. [43] investigated the performance of female mealybugs reared on a grapevine with increasing levels of nitrogen (N) fertilisation. There are parallels between the results of the study by Cocco et al. [43] on *P. ficus* under elevated N and the results of our study on *P. ficus* under eCO_2. In both studies, survival rates increased, and fecundity showed an increasing tendency, although this tendency was not significant in the present study. Fertility was not affected in both studies; body size augmented with increasing N levels, but decreased under eCO_2. Cocco et al. [43] attributed their findings to a higher nutritional value of the phloem sap caused by fertilisation. It is known that the effects of eCO_2 can be mitigated by elevated temperatures, drought [25,46] or fertilisation [3]. Clearly, these three factors and eCO_2 interact in future agriculture, especially in perennial systems such as vineyards. A study by Sudderth et al. [8] investigated the effects of the interaction of fertilisation and eCO_2 on aphids by manipulating soil N and atmospheric CO_2 levels. At ambient CO_2 levels, high soil N increased the aphid population size, similar to the mealybugs in the study by Cocco et al. [43]. However, the aphid population size also increased in response to elevated CO_2 on plants grown under low soil N [8], which might relate to the higher survival rates of mealybugs in the present study. Apparently, elevated CO_2 and increased N fertilisation affect phloem-feeders in a similar manner, but this does not correspond to changes in the C:N ratios of the leaf tissue [8]. Studies on the nutritional quality of phloem sap suggest that nitrogen composition (predominantly free amino acids in phloem) is probably the most important determinant of aphid performance [45,51].

While the present study observed a decreasing mealybug body size with the duration of the eCO_2 treatment, we also found that elevated CO_2 levels were positively related to survival, but development time, fertility and fecundity were not significantly affected. Hence, we detected indicators for reduced fitness (body size), as well as factors that support the assumption that fitness was unaffected (development time, fertility and fecundity) or even improved (survival). These mixed results might also be caused by the experimental design. Grapevine plants used in the greenhouse experiments in this study were grown under ambient CO_2 conditions for 12 weeks before exposure to eCO_2 started. Elevated CO_2 levels affect sap-feeding insects mainly through changes in the host plant [3]. These changes do not happen instantly, hence more clear-cut results on their influence on life history parameters of *P. ficus* might take longer than the duration of the present experiment. It would be interesting to study the long-term effects of eCO_2 on *P. ficus* and *L. dactylopii* on grapevines grown under eCO_2 for a longer period than in the present study. Long-term studies could make changes of life-history of both species become more evident.

Host size is known to influence the sex ratio of parasitoid wasps, with females being more likely to emerge from bigger hosts according to de Jong & Alphen [28]. Despite the subsequent size decrease of *P. ficus* females at the onset of oviposition in the greenhouse experiments in our study, no significant differences in the sex allocation of wasps were observed between eCO_2 and aCO_2 treatments. This might also be attributed to the fact that host size differences were rather small. In fact, none of the evaluated variables of *L. dactylopii* (parasitism success, development time, sex ratio) were altered by CO_2 levels, neither in the greenhouse nor in the VineyardFACE. Parasitism success was greater in the field than in the greenhouse, because wasps remained for 48 h in the VineyardFACE host arenas, compared to 24 h in the greenhouse. To the authors' best knowledge, there have been no studies investigating the performance of mealybug parasitoids under elevated CO_2 conditions. Studies on natural enemies of aphids showed mixed results [34,35]. Hymenopteran aphid parasitoids display the whole range of possible reactions: *Myzus persicae* (Sulzer) and *Brevicoryne brassicae* (L.) showed unaltered and improved life history parameters without changing the performance of the parasitoid wasp *Diaeretiella rapae* (MacInstosh). In a study by Klaiber et al. [52], however, the same parasitoid suffered a decrease in longevity and rates of parasitism on *B. brassicae,* which were lower quality hosts when reared under eCO_2. On the other hand, the biocontrol efficiency of *Aphidius picipes* (Nees) against *Sitobion avenae* (Fabricius) was enhanced under elevated CO_2, although elevated CO_2 had adverse effects on the growth and development of *A. picipes* [53].

Elevated atmospheric CO_2 levels affect parasitoids mainly through plant-mediated changes in host quality which cascade upwards [35]. Our results indicate a tendency towards a non-altered or even improved mealybug fitness under eCO_2, which is, as hypothesised, in accordance with reports on other phloem-feeders reared in similar conditions under eCO_2. The results of the present study also support the hypothesis that parasitoid performance seemed to be related to mealybug performance.

Climate change influences grapevine plants, their pests, and natural enemies today and in the future [6]. Rising temperatures affect grapevine pests as *P. ficus* directly, speeding up development and voltinism [10]. Gutierrez et al. [24] modelled the future distributions of *P. ficus* and its parasitoids in California based on weather data. Without taking into account elevated CO_2 concentrations, the model explains how mealybugs will boom at elevated temperatures. *Planococcus ficus* will seek cooler sites under the bark and in the root zone, to compensate for its rather narrow optimal temperature range. These refuges also function as a shelter from natural enemies and certain pesticides. Elevated CO_2 levels possibly benefit *P. ficus* and also increase the biomass in vine plants, which might offer more refuge sites and complicate host findings for parasitoids. It has been suggested that, in general, measured responses of manipulated systems to global change decrease over greater spatial and temporal experimental scales, as well as with the number of climate change drivers studied [35].

5. Conclusions

Rising carbon dioxide levels affect agricultural systems worldwide. Crop growers are especially interested in the possible effects of climate change on pest species and their respective antagonists used in biological control programs. This study is the first to test the effects of elevated CO_2 concentrations on a grapevine pest under field conditions and in the greenhouse. It is also the only study to investigate a mealybug-based pest-parasitoid complex under eCO_2 in a multitrophic approach. Our results suggest a trend towards a non-altered or improved fitness of *P. ficus* in the greenhouse with eCO_2 levels around 800 ppm, but not in the field, where eCO_2 concentrations reached approximately 450 ppm. Meanwhile, its parasitoid *L. dactylopii* did not seem to be affected in either of the scenarios tested. Further research should include prolonged eCO_2 exposure of greenhouse plants and investigate the response of grapevine plants to mealybug attack under different CO_2 levels. Additionally, climate change is characterised by a combination of multiple abiotic factors, including rising temperatures and CO_2 levels, as well as varying precipitation patterns. None of these factors act by themselves upon plants, pests and natural enemies and future studies should take this complexity into account. Future research should aim to integrate several trophic levels and environmental stress factors.

Supplementary Materials: The following can be found at http://www.mdpi.com/2073-4395/9/6/326/s1. Figure S1: (A) Aerial view of the FACE facility; (B) Schematic illustration of one FACE ring; Figure S2: FACE CO_2; Table S1: Greenhouse CO_2; Table S2: Greenhouse Temperature; Table S3: FACE climate data; Table S4: GreenhouseResults (analysed); Table S5: Greenhouse size and develop; Table S6: Greenhouse Survival; Table S7: Greenhouse fertility fecund; Table S8 FACE results (analysed); Table S9: FACE size and development; Table S10: FACE Survival; Table S11: FACE Fertility Fecundity; Table S12: Greenhouse Leptomastix; Table S13: FACE Leptomastix.

Author Contributions: Conceptualisation, M.S.-S. and A.R.; Data curation, M.S.-S.; Formal analysis, M.S.-S.; Investigation, M.S.-S.; Methodology, M.S.-S. and A.R.; Project administration, M.S.-S. and A.R.; Resources, A.R.; Supervision, A.R.; Validation, M.S.-S. and A.R; Visualization, M.S.-S.; Writing—original draft, M.S.-S.; Writing—review & editing, M.S.-S. and A.R.

Acknowledgments: We acknowledge Mirjam Hauck and Olivia Herczynski for technical assistance. The authors would like to thank the anonymous reviewers for their comments and suggestions.

Conflicts of Interest: The authors declare no conflict of interest.

References

1. Pachauri, R.K.; Allen, M.R.; Barros, V.R.; Broome, J.; Cramer, W.; Christ, R.; Church, J.A.; Clarke, L.; Dahe, Q.; Dasgupta, P.; et al. *IPCC Climate Change 2014: Synthesis Report. Contribution of Working Groups I, II and III to the Fifth Assessment Report of the Intergovernmental Panel on Climate Change*; IPCC: Geneva, Switzerland, 2014; Available online: https://epic.awi.de/id/eprint/37530/ (accessed on 31 January 2019).

2. Long, S.P.; Ainsworth, E.A.; Rogers, A.; Ort, D.R. Rising atmospheric carbon dioxide: Plants FACE the future. *Annu. Rev. Plant Biol.* **2004**, *55*, 591–628. [CrossRef] [PubMed]

3. Robinson, E.A.; Ryan, G.D.; Newman, J.A. A meta-analytical review of the effects of elevated CO_2 on plant-arthropod interactions highlights the importance of interacting environmental and biological variables. *New Phytol.* **2012**, *194*, 321–336. [CrossRef] [PubMed]

4. Bindi, M.; Fibbi, L.; Miglietta, F. Free air CO_2 enrichment (FACE) of grapevine (*Vitis vinifera* L.): II. Growth and quality of grape and wine in response to elevated CO_2 concentrations. *Eur. J. Agron.* **2001**, *14*, 145–155. [CrossRef]

5. Wohlfahrt, Y.; Smith, J.P.; Tittmann, S.; Honermeier, B.; Stoll, M. Primary productivity and physiological responses of *Vitis vinifera* L. cvs. under Free Air Carbon dioxide Enrichment (FACE). *Eur. J. Agron.* **2018**, *101*, 149–162. [CrossRef]

6. Reineke, A.; Thiéry, D. Grapevine insect pests and their natural enemies in the age of global warming. *J. Pest Sci. (2004)* **2016**, *89*, 313–328. [CrossRef]

7. Bezemer, T.M.; Jones, T.H. Plant-insect herbivore interactions in elevated atmospheric CO_2: Quantitative analyses and guild effects. *Oikos* **1998**, *82*, 212–222. [CrossRef]

8. Sudderth, E.A.; Stinson, K.A.; Bazzaz, F.A. Host-specific aphid population responses to elevated CO_2 and increased N availability. *Glob. Chang. Biol.* **2005**, *11*, 1997–2008. [CrossRef]

9. Walton, V.M.; Pringle, K.L. Vine mealybug, *Planococcus ficus* (Signoret) (Hemiptera: Pseudococcidae), a key pest in South African vineyards: A Review. *S. Afr. J. Enol. Vitic.* **2004**, *25*, 54–62. [CrossRef]

10. Daane, K.M.; Almeida, R.P.P.; Bell, V.A.; Walker, J.T.S.; Botton, M.; Fallahzadeh, M.; Mani, M.; Miano, J.L.; Sforza, R.; Walton, V.M. Biology and management of mealybugs in vineyards. In *Arthropod Management in Vineyards*; Bostanian, N.J., Vincent, C., Isaacs, R., Eds.; Springer: New York, NY, USA, 2012; pp. 271–307.

11. Chiotta, M.L.; Ponsone, M.L.; Torres, A.M.; Combina, M.; Chulze, S.N. Influence of *Planococcus ficus* on *Aspergillus* section *Nigri* and ochratoxin A incidence in vineyards from Argentina. *Lett. Appl. Microbiol.* **2010**, *51*, 212–218. [CrossRef]

12. Bordeu, E.; Troncoso, D.O.; Zaviezo, T. Influence of mealybug (*Pseudococcus* spp.)-infested bunches on wine quality in Carmenere and Chardonnay grapes. *Int. J. Food Sci. Technol.* **2012**, *47*, 232–239. [CrossRef]

13. Engelbrecht, D.J.; Kasdorf, G.G.F. Field spreas of corky bark, fleck, leafroll and Shiraz decline diseases and associated viruses in South African grapevines. *Phytophylactica* **1990**, *22*, 347–354.

14. Almeida, R.P.P.; Daane, K.M.; Bell, V.A.; Blaisdell, G.K.; Cooper, M.L.; Herrbach, E.; Pietersen, G. Ecology and management of grapevine leafroll disease. *Front. Microbiol.* **2013**, *4*, 94. [CrossRef] [PubMed]

15. Becerra, V.; Gonzalez, M.; Herrera, M.E.; Miano, J.L. Population dynamics of vine mealybug *Planococcus ficus* Sign. in vineyards. Mendoza. *Rev. Fac. Ciencias Agrar. Univ. Nac. Cuyo* **2006**, *1*, 1–6.

16. Prabhaker, N.; Gispert, C.; Castle, S.J. Baseline susceptibility of *Planococcus ficus* (Hemiptera: Pseudococcidae) from California to select insecticides. *J. Econ. Entomol.* **2012**, *105*, 1392–1400. [CrossRef] [PubMed]

17. Venkatesan, T.; Jalali, S.K.; Ramya, S.L.; Prathibha, M. Insecticide resistance and its management in mealybugs. In *Mealybugs and Their Management in Agricultural and Horticultural crops*; Springer: New Delhi, India, 2016; pp. 223–229.

18. Muscas, E.; Cocco, A.; Mercenaro, L.; Cabras, M.; Lentini, A.; Porqueddu, C.; Nieddu, G. Effects of vineyard floor cover crops on grapevine vigor, yield, and fruit quality, and the development of the vine mealybug under a Mediterranean climate. *Agric. Ecosyst. Environ.* **2017**, *237*, 203–212. [CrossRef]

19. Cocco, A.; Muscas, E.; Mura, A.; Iodice, A.; Savino, F.; Lentini, A. Influence of mating disruption on the reproductive biology of the vine mealybug, *Planococcus ficus* (Hemiptera: Pseudococcidae), under field conditions. *Pest Manag. Sci.* **2018**, *74*, 2806–2816. [CrossRef] [PubMed]

20. Walton, V.M.; Daane, K.M.; Bentley, W.J.; Millar, J.G.; Larsen, T.E.; Malakar-Kuenen, R. Pheromone-based mating disruption of *Planococcus ficus* (Hemiptera: Pseudococcidae) in California vineyards. *J. Econ. Entomol.* **2006**, *99*, 1280–1290. [CrossRef]

21. Mansour, R.; Grissa Lebdi, K.; Rezgui, S. Assessment of the performance of some new insecticides for the control of the vine mealybug *Planococcus ficus* in a Tunisian vineyard. *Entomol. Hell.* **2017**, *19*, 21–33. [CrossRef]

22. Daane, K.M.; Bentley, W.J.; Walton, V.M.; Malakar-Kuenen, R.; Millar, J.G.; Ingels, C.A.; Weber, E.A.; Gispert, C. New controls investigated for vine mealybug. *Calif. Agric.* **2006**, *60*, 31–38. [CrossRef]

23. Gutierrez, A.P.; Ponti, L.; D'Oultremont, T.; Ellis, C.K. Climate change effects on poikilotherm tritrophic interactions. *Clim. Chang.* **2008**, *87*, 167–192. [CrossRef]

24. Paul Gutierrez, A.; Daane, K.M.; Ponti, L.; Walton, V.M.; Ellis, C.K. Prospective evaluation of the biological control of vine mealybug: Refuge effects and climate. *J. Appl. Ecol.* **2008**, *45*, 524–536. [CrossRef]

25. Chong, J.-H.; van Iersel, M.W.; Oetting, R.D. Effects of elevated carbon dioxide levels and temperature on the life history of the Madeira mealybug (Hemiptera: Pseudococcidae). *J. Entomol. Sci.* **2004**, *39*, 387–397. [CrossRef]

26. Kol-Maimon, H.; Ghanim, M.; Franco, J.C.; Mendel, Z. Evidence for gene flow between two sympatric mealybug species (Insecta; Coccoidea; Pseudococcidae). *PLoS ONE* **2014**, *9*, 1–13. [CrossRef] [PubMed]

27. Noyes, J.S.; Hayat, M. *Oriental mealybug parasitoids of the Anagyrini (Hymenoptera: Encyrtidae)*; CAB International: Wallingford, UK, 1994.

28. Jong, P.W.; Alphen, J.J.M. Host size selection and sex allocation in *Leptomastix dactylopii*, a parasitoid of *Planococcus citri*. *Entomol. Exp. Appl.* **1989**, *50*, 161–169. [CrossRef]

29. Krishnamoorthy, A.; Singh, S.P. Biological control of citrus mealybug, *Planococcus citri* with an introduced parasite, *Leptomastix dactylopii* in India. *Entomophaga* **1987**, *32*, 143–148. [CrossRef]

30. Fallahzadeh, M.; Japoshvili, G.; Saghaei, N.; Daane, K.M. Natural enemies of *Planococcus ficus* (Hemiptera: Pseudococcidae) in Fars Province vineyards, Iran. *Biocontrol Sci. Technol.* **2011**, *21*, 427–433. [CrossRef]

31. Mahfoudhi, N.; Dhouibi, M.H. Survey of Mealybugs (Hemiptera: Pseudococcidae) and their Natural Enemies in Tunisian Vineyards. *Afr. Entomol.* **2009**, *17*, 154–160. [CrossRef]

32. Daane, K.M.; Cooper, M.L.; Triapitsyn, S.V.; Andrews, J.W.; Ripa, R. Parasitoids of obscure mealybug, *Pseudococcus viburni* (Hem.: Pseudococcidae) in California: Establishment of *Pseudaphycus flavidulus* (Hym.: Encyrtidae) and discussion of related parasitoid species. *Biocontrol Sci. Technol.* **2008**, *18*, 43–57. [CrossRef]

33. Marras, P.M.; Cocco, A.; Muscas, E.; Lentini, A. Laboratory evaluation of the suitability of vine mealybug, *Planococcus ficus*, as a host for *Leptomastix dactylopii*. *Biol. Control* **2016**, *95*, 57–65. [CrossRef]

34. Hentley, W.T.; Wade, R.N. Global change, herbivores and their natural enemies. In *Global Climate Change and Terrestrial Invertebrates*; John Wiley & Sons, Ltd.: Chichester, UK, 2016; pp. 177–200.

35. Facey, S.L.; Ellsworth, D.S.; Staley, J.T.; Wright, D.J.; Johnson, S.N. Upsetting the order: How climate and atmospheric change affects herbivore–enemy interactions. *Curr. Opin. Insect Sci.* **2014**, *5*, 66–74. [CrossRef]

36. Mumm, R.; Dicke, M. Variation in natural plant products and the attraction of bodyguards involved in indirect plant defense. *Can. J. Zool.* **2010**, *88*, 628–667. [CrossRef]

37. Timm, A.E.; Reineke, A. First insights into grapevine transcriptional responses as a result of vine mealybug *Planococcus ficus* feeding. *Arthropod. Plant. Interact.* **2014**, *8*, 495–505. [CrossRef]

38. DeLucia, E.H.; Nabity, P.D.; Zavala, J.A.; Berenbaum, M.R. Climate change: Resetting plant-insect interactions. *Plant Physiol.* **2012**, *160*, 1677–1685. [CrossRef] [PubMed]

39. Boudon-Padieu, É.; Maixner, M. Potential effects of climate change on distribution and activity of insect vectors of grapevine pathogens. In Proceedings of the International and multi-disciplinary Colloquim-Global warming, which potential impacts on the vineyards? Beaune, France, 28–30 March 2007; p. 23. [CrossRef]

40. Hogendorp, B.K.; Cloyd, R.A.; Swiader, J.M. Effect of nitrogen fertility on reproduction and development of Citrus mealybug, *Planococcus citri* Risso (Homoptera: Pseudococcidae), feeding on two colors of coleus, *Solenostemon scutellarioides* L. Codd. *Environ. Entomol.* **2006**, *35*, 201–211. [CrossRef]

41. Reineke, A.; Selim, M. Elevated atmospheric CO_2 concentrations alter grapevine (*Vitis vinifera*) systemic transcriptional response to European grapevine moth (*Lobesia botrana*) herbivory. *Sci. Rep.* **2019**, *9*, 2995. [CrossRef] [PubMed]

42. Ross, L.; Dealey, E.J.; Beukeboom, L.W.; Shuker, D.M. Temperature, age of mating and starvation determine the role of maternal effects on sex allocation in the mealybug *Planococcus citri*. *Behav. Ecol. Sociobiol.* **2011**, *65*, 909–919. [CrossRef] [PubMed]

43. Cocco, A.; Marras, P.M.; Muscas, E.; Mura, A.; Lentini, A. Variation of life-history parameters of *Planococcus ficus* (Hemiptera: Pseudococcidae) in response to grapevine nitrogen fertilization. *J. Appl. Entomol.* **2015**, *139*, 519–528. [CrossRef]

44. Bindi, M.; Fibbi, L.; Lanini, M.; Miglietta, F. Free air CO_2 enrichment (FACE) of grapevine (*Vitis vinifera* L.): I. Development and testing of the system for CO_2 enrichment. *Eur. J. Agron.* **2001**, *14*, 135–143. [CrossRef]

45. Ryan, G.D.; Shukla, K.; Rasmussen, S.; Shelp, B.J.; Newman, J.A. Phloem phytochemistry and aphid responses to elevated CO_2, nitrogen fertilization and endophyte infection. *Agric. For. Entomol.* **2014**, *16*, 273–283. [CrossRef]

46. Zvereva, E.L.; Kozlov, M.V. Consequences of simultaneous elevation of carbon dioxide and temperature for plant-herbivore interactions: A metaanalysis. *Glob. Chang. Biol.* **2006**, *12*, 27–41. [CrossRef]

47. Hughes, L.; Bazzaz, F.A. Effects of elevated CO_2 on five plant-aphid interactions. *Entomol. Exp. Appl.* **2001**, *99*, 87–96. [CrossRef]

48. Awmack, C.; Harrington, R.; Leather, S. Host plant effects on the performance of the aphid *Aulacorthum solani* (Kalt.) (Homoptera: Aphididae) at ambient and elevated CO_2. *Glob. Chang. Biol.* **1997**, *3*, 545–549. [CrossRef]

49. Poorter, H. Interspecific variation in the growth response of plants to an elevated ambient CO_2 concentration. In *CO_2 and Biosphere*; Springer: Dordrecht, The Netherlands, 1993; pp. 77–98.

50. Bezemer, T.M.; Knight, K.J.; Newington, J.E.; Jones, T.H. How general are aphid responses to elevated atmospheric CO_2? *Ann. Entomol. Soc. Am.* **1999**, *92*, 724–730. [CrossRef]

51. Ryan, G.D.; Sylvester, E.V.A.; Shelp, B.J.; Newman, J.A. Towards an understanding of how phloem amino acid composition shapes elevated CO_2-induced changes in aphid population dynamics. *Ecol. Entomol.* **2015**, *40*, 247–257. [CrossRef]

52. Klaiber, J.; Najar-Rodriguez, A.J.; Dialer, E.; Dorn, S. Elevated carbon dioxide impairs the performance of a specialized parasitoid of an aphid host feeding on *Brassica* plants. *Biol. Control* **2013**, *66*, 49–55. [CrossRef]

53. Jun Chen, F.; Wu, G.; Parajulee, M.N.; Ge, F. Impact of elevated CO_2 on the third trophic level: A predator *Harmonia axyridis* and a parasitoid *Aphidius picipes*. *Biocontrol Sci. Technol.* **2007**, *17*, 313–324. [CrossRef]

MDPI

St. Alban-Anlage 66

4052 Basel

Switzerland

Tel. +41 61 683 77 34

Fax +41 61 302 89 18

www.mdpi.com

Agronomy Editorial Office

E-mail: agronomy@mdpi.com

www.mdpi.com/journal/agronomy

www.ingramcontent.com/pod-product-compliance
Lightning Source LLC
Chambersburg PA
CBHW051719210326
41597CB00032B/5534